Chaos in Hydrology

Bellie Sivakumar

Chaos in Hydrology

Bridging Determinism and Stochasticity

Bellie Sivakumar
School of Civil and Environmental
Engineering
The University of New South Wales
Sydney, NSW
Australia

and

Department of Land, Air and Water
Resources
University of California
Davis, CA
USA

ISBN 978-94-024-1314-4 ISBN 978-90-481-2552-4 (eBook)
DOI 10.1007/978-90-481-2552-4

Softcover reprint of the hardcover 1st edition 2016

Printed on acid-free paper

This Springer imprint is published by Springer Nature
The registered company is Springer Science+Business Media B.V. Dordrecht

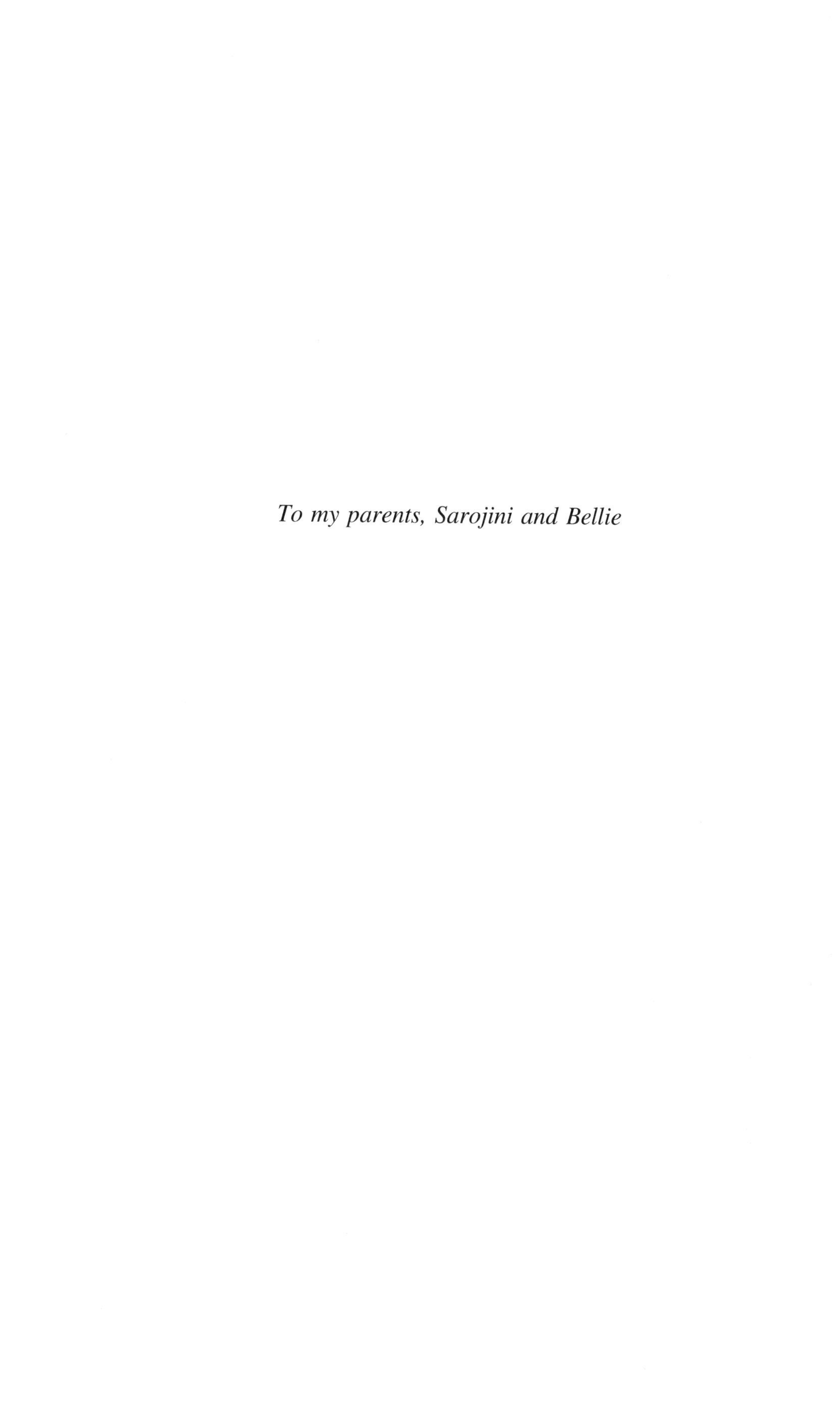

To my parents, Sarojini and Bellie

Preface

It is possible that you have this book in your hands because of its intriguing name (Chaos) or simply by accident, but I hope that you will continue to read it for its contents and then also recommend it to others.

In common parlance, the word 'chaos,' derived from the Ancient Greek word Χάος, typically means a state lacking order or predictability; in other words, chaos is synonymous to 'randomness.' In modern dynamic systems science literature, however, the term 'chaos' is used to refer to situations where complex and 'random-looking' behaviors arise from simple deterministic systems with sensitive dependence on initial conditions; therefore, chaos and randomness are quite different. This latter definition has important implications for system modeling and prediction: randomness is irreproducible and unpredictable, while chaos is reproducible and predictable in the short term (due to determinism) but irreproducible and unpredictable only in the long term (due to sensitivity to initial conditions).

The three fundamental properties inherent in the above definition of chaos, namely (a) nonlinear interdependence; (b) hidden order and determinism; and (c) sensitivity to initial conditions, are highly relevant in almost all real systems. In hydrology, for instance: (a) nonlinear interactions are dominant among the components and mechanisms in the hydrologic cycle; (b) determinism and order are prevalent in daily temperature and annual river flow; and (c) contaminant transport in surface and sub-surface waters is highly sensitive to the time (e.g., rainy or dry season) at which the contaminants were released. The first property represents the 'general' nature of hydrologic phenomena, whereas the second and third represent their 'deterministic' and 'stochastic' natures, respectively. Further, despite their complexity and random-looking behavior, hydrologic phenomena may be governed only by a few degrees of freedom, another basic idea of chaos theory; for instance, runoff in a well-developed urban catchment depends essentially on rainfall.

This book is intended to address a fundamental question researchers in hydrology commonly grapple with: is the complex, irregular, and random-looking behavior of hydrologic phenomena simply the outcome of random (or stochastic) system dynamics, or is there some kind of order and determinism hidden behind? In other words, since simple deterministic systems can produce complex and random-looking outputs, as has been shown through numerous synthetic examples, is it reasonable then to ask if hydrologic systems can also belong to this category? A reliable answer to this question is important for proper identification of the type and complexity of hydrologic models to be developed, evaluation of data and computer requirements, determination of maximum predictability horizon for hydrologic processes, and assessment, planning, and management of water resources.

I approach the above question in a very systematic manner, by first discussing the general and specific characteristics of hydrologic systems, next reviewing the tools available at our disposal to study such systems, and then presenting the applications of such tools to various hydrologic systems, processes, and problems. In the end, I argue that chaos theory offers a balanced and middle-ground approach between the deterministic and stochastic extreme paradigms that are prevalent in hydrology (and in almost every other field) and, thus, serves as a bridge connecting the two paradigms.

The book is divided into four major parts, focusing on specific topics that I deem necessary to meet the intended goal. Part A (Hydrologic Systems and Modeling) covers the introduction to hydrology (Chap. 1), characteristics of hydrologic systems (Chap. 2), stochastic time series methods (Chap. 3), and modern nonlinear time series methods (Chap. 4). Part B (Nonlinear Dynamics and Chaos) details the fundamentals of chaos theory (Chap. 5), chaos identification and prediction (Chap. 6), and issues associated with chaos methods (Chap. 7), especially in their applications to real data. Part C (Applications of Chaos Theory in Hydrology) details the applications of chaos theory in hydrology, first with an overview of hydrologic applications (Chap. 8), followed by applications to rainfall (Chap. 9), river flow (Chap. 10), and other hydrologic data (Chap. 11), and then with studies on hydrologic data-related issues (Chap. 12). Part D (A Look Ahead) summarizes the current status (Chap. 13), offers future directions (Chap. 14), and includes a broader discussion of philosophical and pragmatic views of chaos theory in hydrology (Chap. 15).

I must emphasize that this book is about hydrology (and *not* about chaos theory), with focus on the applications of nonlinear dynamic and chaos concepts in hydrologic systems. Consequently, a significant portion of the presentation is devoted to hydrologic system characteristics, time series modeling in hydrology, relevance of nonlinear dynamic and chaos concepts in hydrology, and their applications and advances in hydrology, especially from an engineering perspective. The presentation about the fundamentals of chaos theory, methods for identification and prediction, and relevant issues in their applications is by no means exhaustive, and is deliberately kept to a minimum level that is needed to meet the above goal. However, the amount of literature cited on the theoretical aspects of chaos theory and methodological developments is extensive, which should guide the interested reader to further details. For the benefit of the reader, and especially for someone new to the field, I also attempt to be descriptive in reviewing the theoretical concepts, detailing the applications, and interpreting the outcomes. All this, I believe, makes this book suitable for both experienced researchers and new ones in hydrology and water resources engineering, and beyond.

Sydney, Australia and Davis, USA Bellie Sivakumar

Acknowledgments

This book is a result of my research in the area of chaos theory in hydrology over the last two decades, starting from my doctoral degree at the National University of Singapore. During this time, I have benefited from numerous colleagues and friends, funding agencies, research fellowships, and other invited research visits. The list is too long to mention here. Therefore, I will limit the list mostly to those that have directly contributed to the preparation of this book and to a few others that have been a great encouragement and support throughout.

The idea for writing a book on chaos theory in hydrology arose many years ago. However, the actual planning for this book occurred during one of my visits to Inha University, Korea, a few years ago. My sincere and special thanks to Hung Soo Kim (Inha University) and Ronny Berndtsson (Lund University, Sweden) for their support and contributions in planning for this book. They provided useful inputs in identifying the areas and topics to focus in this book and in outlining and organizing the contents. Apart from this book, both Ronny and Hung Soo have and continue to play important roles in advancing my research and career through our research collaborations.

Several colleagues and students provided help in the preparation of the book. Fitsum Woldemeskel offered significant help with the adoption and modification of a number of figures from existing publications, including my own. I am grateful to Fitsum for his time, effort, and generosity, at a critical time in the preparation of the book. Hong-Bo Xie provided Figs. 5.4, 5.6, and 5.7. Jun Niu provided Fig. 4.3. Carlos E. Puente provided Fig. 5.2. Seokhyeon Kim and R. Vignesh helped in the preparation of a few figures, especially in Chap. 12. Peter Young and Jun Niu offered useful inputs in the preparation of Sects. 4.3 and 4.6, respectively. Jun Niu and V. Jothiprakash also offered reviews for the manuscript. My sincere thanks to all of them for their time, effort, and generosity.

Many colleagues and friends have and continue to offer great encouragement and support to my academic, research, and professional activities. Their encouragement and support, especially during the earlier years, played an important role in writing this book. Among these are (in alphabetical order) Ronny Berndtsson,

Ji Chen, Eric Gaume, Thomas Harter, A.W. Jayawardena, Kenji Jinno, V. Jothiprakash, Akira Kawamura, Hung Soo Kim, Upmanu Lall, Chih-Young Liaw, Shie-Yui Liong, Rajeshwar Mehrotra, Jun Niu, Jonas Olsson, Kok-Kwang Phoon, Carlos E. Puente, Ashish Sharma, and Wesley W. Wallender. I am grateful to all of them, for their encouragement and support. Ji Chen and Carlos Puente deserve particular mention, especially for the regular contacts, interactions, and discussions, despite the geographic distances. Raj Mehrotra also deserves particular mention for his constant encouragement. My numerous conversations with Raj, especially over coffee(!) almost on a daily basis, have greatly helped me to keep my focus on the book.

Writing a book like this consumes an enormous amount of time. The material for this book has been gathered over many years and constantly updated over time. However, a significant part of the writing has been undertaken only during the past 2–3 years. I would like to thank the Australian Research Council (ARC) for the financial support through the Future Fellowship grant (FT110100328). This support has allowed me the time and flexibility to focus on the book more than it would have been otherwise possible.

My sincere thanks to the entire Springer team for their encouragement and support throughout the preparation of this book. Petra van Steenbergen and Hermine Vloemans deserve special mention, not only for their encouragement and support, but also for their patience and understanding.

Finally, I am grateful to my family, for their constant love, encouragement, and support, without which writing a book would have remained only a dream.

Sydney, Australia and Davis, USA Bellie Sivakumar

Contents

Part IV A Look Ahead

About the Author

Bellie Sivakumar received his Bachelor degree in Civil Engineering from Bharathiar University (India) in 1992, Master degree in Hydrology and Water Resources Engineering from Anna University (India) in 1994, and Ph.D. degree in Civil Engineering from the National University of Singapore in 1999. After a one-year postdoctoral research at the University of Arizona, Tucson, USA, he joined University of California, Davis (UCDavis). At UCDavis, he held the positions of postgraduate researcher and Associate Project Scientist, and now holds an Associate position. He joined the University of New South Wales, Sydney, Australia in 2010, where he is currently an Associate Professor.

Bellie Sivakumar's research interests are in the field of hydrology and water resources, with particular emphasis on nonlinear dynamics, chaos, scaling, and complex networks. He has authored one book and more than 130 peer-reviewed journal papers. He has been an associate editor for several journals, including Hydrological Sciences Journal, Journal of Hydrology, Journal of Hydrologic Engineering, and Stochastic Environmental Research and Risk Assessment. He has received a number of fellowships throughout his career, including the ICSC World Laboratory Fellowship, Japan Society for the Promotion of Science Fellowship, Korea Science and Technology Societies' Brainpool Fellowship, and Australian Research Council Future Fellowship.

List of Figures

List of Tables

Part I
Hydrologic Systems and Modeling

Chapter 1
Introduction

Abstract In simple terms, hydrology is the study of the waters of the Earth, including their occurrence, distribution, and movement. The constant circulation of water and its change in physical state is called the *hydrologic cycle*. The study of water started at least a few thousands years ago, but the modern scientific approach to the hydrologic cycle started in the seventeenth century. Since then, hydrology has witnessed a tremendous growth, especially over the last century, with significant advances in computational power and hydrologic data measurements. This chapter presents a general and introductory account of hydrology. First, the concept of the hydrologic cycle is described. Next, a brief history of the scientific development of hydrology is presented. Then, the concept of hydrologic system is explained, followed by a description of the hydrologic system model and model classification. Finally, the role of hydrologic data and time series modeling as well as the physical basis of time series modeling are highlighted.

1.1 Definition of Hydrology

The name 'hydrology' was derived from the Greek words 'hydro' (water) and 'logos' (study), and roughly translates into 'study of water.' Different textbooks may offer different definitions, but all of them generally reflect the following working definition:

> Hydrology is the science that treats the waters of the Earth, their occurrence, circulation and distribution, their chemical and physical properties, and their interactions with their environments, including their relations to living things.

Within hydrology, various sub-fields exist. In keeping with the essential ingredients of the above definition, these sub-fields may depend on the region (e.g. over the land surface, below the land surface, mountains, urban areas) or property (e.g. physical, chemical, isotope) or interactions (e.g. atmosphere, environment, ecosystem) or other aspects (e.g. tools used for studies) of water. There may also be significant overlaps between two or more sub-fields, and even inter-change of terminologies depending on the emphasis for water in studies of the

B. Sivakumar, *Chaos in Hydrology*, DOI 10.1007/978-90-481-2552-4_1

Earth-ocean-atmospheric system. Some of the popular sub-fields within hydrology are:

- Surface hydrology—study of hydrologic processes that operate at or near the Earth's surface
- Sub-surface hydrology (or Groundwater hydrology or Hydrogeology)—study of the presence and movement of water below the Earth's surface
- Vadose zone hydrology—study of the movement of water between the top of the Earth's surface and the groundwater table
- Hydrometeorology—study of the transfer of water and energy between land and water body surfaces and the lower atmosphere
- Hydroclimatology—study of the interactions between climate processes and hydrologic processes
- Paleohydrology—study of the movement of water and sediment as they existed during previous periods of the Earth's history
- Snow hydrology—study of the formation, movement, and effects of snow
- Urban hydrology—study of the hydrologic processes in urban areas
- Physical hydrology—study of the physical mechanisms of hydrologic processes
- Chemical hydrology—study of the chemical characteristics of water
- Isotope hydrology—study of the isotopic signatures of water
- Ecohydrology (or Hydroecology)—study of the interactions between hydrologic processes and organisms
- Hydroinformatics—the adaptation of information technology to hydrology and water resources applications.

1.2 Hydrologic Cycle

The constant movement of water and its change in physical state on the Earth (in ocean, land, and atmosphere) is called the *hydrologic cycle* or, quite simply, *water cycle*. The hydrologic cycle is the central focus of hydrology. A schematic representation of the hydrologic cycle is shown in Fig. 1.1. A description of the hydrologic cycle can begin at any point and return to that same point, with a number of processes continuously occurring during the cycle; however, oceans are usually considered as the origin. In addition, depending upon the scope or focus of the study, certain processes (or components) of the hydrologic cycle may assume far more importance over the others and, hence, such may be described in far more detail. In what follows, the hydrologic cycle is described with oceans as the origin and processes on and above/below the land surface assuming more importance. For further details, including other descriptions of the hydrologic cycle, the reader is referred to Freeze and Cherry (1979), Driscoll (1986), Chahine (1992), Maidment (1993), and Horden (1998), among others.

Water in the ocean evaporates and becomes atmospheric water vapor (i.e. moisture). Some of this water vapor is transported and lifted in the atmosphere until

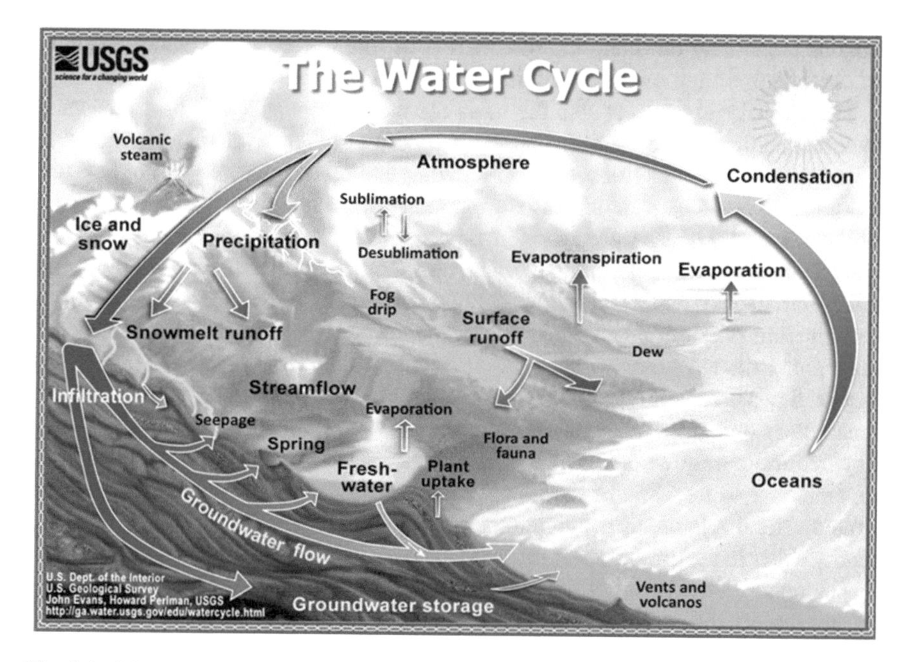

Fig. 1.1 Schematic representation of hydrologic cycle (*source* US Geological Survey, http://water.usgs.gov/edu/watercycle.html; accessed May 5, 2015)

it condenses and falls as precipitation, which sometimes evaporates or gets intercepted by vegetation before it can reach the land surface. Of the water that reaches the land surface by precipitation, some may evaporate where it falls, some may infiltrate the soil, and some may run off overland to evaporate or infiltrate elsewhere or to enter streams. The water that infiltrates the ground may evaporate, be absorbed by plant roots, and then transpired by the plants, or percolate downward to groundwater reservoirs (also called *aquifers*). Water that enters groundwater reservoirs may either move laterally until it is close enough to the surface to be subject to evaporation or transpiration, reach the land surface and form springs, seeps or lakes, or flow directly into streams or into the ocean. Stream water can accumulate in lakes and surface reservoirs, evaporate or be transpired by riparian vegetation, seep downward into groundwater reservoirs or flow back into the ocean, where the cycle begins again.

Although the concept of the hydrologic cycle is simple, the phenomenon is enormously complex and intricate. It is not just one large cycle but rather composed of many inter-related cycles of continental, regional, and local extent. Each phase of the hydrologic cycle also provides opportunities for temporary accumulation and storage of water, such as snow and ice on the land surface, moisture in the soil and groundwater reservoirs, water in ponds, lakes, and surface reservoirs, and vapor in the atmosphere. Although the total volume of water in the global hydrologic cycle remains essentially constant, the distribution of this water is continually changing on continents, in regions, and within local drainage basins.

The hydrologic cycle is also the basis for the concept of water as a renewable resource. Hydrology recognizes that the natural hydrologic cycle can be altered by human and natural activities, as the following examples indicate. Geologic forces that raise mountains can increase orographic precipitation on one side of the mountains and decrease precipitation on the other side with all of the attendant changes in streamflow, flooding, etc. The development and use of water modify the natural circulatory pattern of the hydrologic cycle, such as the use of surface water for irrigation possibly resulting in downward seepage from reservoirs, canals, ditches, and irrigated fields, adding to the groundwater. Diversions of streamflows impact downstream flows which, if transferred to other watersheds, impact the streamflows and groundwater systems in the other watersheds. Pumping from wells may reduce the flow of water from springs or seeps, increase the downward movement of water from the land surface and streams, reduce the amount of natural groundwater discharge by evaporation and transpiration, induce the inflow of poorer quality water to the groundwater reservoir, or have a combination of all these effects.

The natural circulation of the hydrologic cycle may also be changed by actions not related to direct water use. Among these actions are weather modification activities (e.g. cloud seeding), drainage of swamps and lakes, water-proofing of the land surface by buildings and pavements, and major changes in vegetative cover (e.g. removal of forests).

1.3 Scientific Development of Hydrology

Humans have been concerned with managing water at least since the first civilizations developed along river banks over 8000 years ago. Hydraulic engineers built functioning canals, levees, dams, water conduits, and wells along the Indus in Pakistan, the Tigris and Euphrates in Mesopotamia, the Hwang Ho in China, and the Nile in Egypt as early as 5000–6000 B.C. Flow monitoring was started by the Egyptians around 3800 B.C., and the first rainfall measurements were made by Kautilya of India around 2400 B.C. (Eagleson et al. 1991).

The concept of global hydrologic cycle started perhaps around 3000 B.C. (Nace 1974), when King Solomon wrote in Ecclesiastes 1:7 that

> All the rivers run into the sea; yet the sea is not full; unto the place from whence the rivers come, thither they return again.

Early Greek philosophers, such as Thales, Anaxagoras, Herodotus, Hippocrates, Plato, and Aristotle also embraced the basic idea of the hydrologic cycle. However, while some of them had reasonable understandings of certain hydrologic processes, they postulated various fanciful underground mechanisms by which water returned from sea to land and entered rivers. The Romans had considerable practical knowledge of hydrology (and especially hydraulics) and constructed extensive aqueduct systems, but their scientific ideas were largely based on those of the Greeks.

Independent thinking occurred in ancient Asian civilizations (UNESCO 1974). The Chinese recorded observations of rain, sleet, snow, and wind on Anyang oracle bones as early as 1200 B.C. They probably used raingages around 1000 B.C., and established systematic raingaging around 200 B.C. In India, the first quantitative measurements of rainfall date back to the fourth century B.C. The concept of a dynamic hydrologic cycle may have arisen in China by 900 B.C, in India by 400 B. C., and in Persia by the tenth century, but these ideas had little impact on Western thought (Chow et al. 1988).

In the meantime, the theories of the Greek philosophers continued to dominate Western thought until much of the Renaissance, which spanned roughly from the 12th to the 17th century A.D. Then, Leonardo da Vinci (about 1500 A.D.) in Italy and Bernard Palissy (about 1550 A.D.) in France asserted, based on field observations, that the water in rivers comes from precipitation (Adams 1938; Biswas 1970). With this initiation, the modern scientific approach to the hydrologic cycle was taken up in the seventeenth century by the Frenchmen Pierre Perrault and Edmé Marriotte, who published, in the 1670s and 1680s, measurements and calculations that quantitatively verified the rainfall origin of streamflow. Shortly after that (around 1700), Edmund Halley, an English scientist, extended the quantification of the hydrologic cycle through estimation of the amounts of water involved in the ocean-atmosphere-rivers-ocean cycle of the Mediterranean Sea and surrounding islands.

The eighteenth century witnessed significant advances in the applications of mathematics to fluid mechanics and hydraulics, notably by Henri Pitot, Daniel Bernoulli, Leonhard Euler, Antoine de Chézy, and other Europeans. In fact, use of the term 'hydrology' in approximately its current meaning also began during this time (around 1750). By about 1800, the nature of evaporation and the present concepts of the global hydrologic cycle were firmly established by the English physicist and chemist John Dalton (Dalton 1802), and Charles Lyell, James Hutton, and John Playfair published scientific works on the fluvial erosion of valleys. Routine network measurements of precipitation were begun before 1800 in Europe and the United States, and established there and in India by 1820s.

Until mid-nineteenth century, one of the barriers to understanding the hydrologic cycle was the ignorance of the groundwater flow process. This changed in 1856, when the French engineer Henry Darcy established the basic phenomenological law of flow through porous media (Darcy 1856). The nineteenth century also saw further advances in fluid mechanics, hydraulics, and sediment transport by Jean-Louis Poiseuille, Jules DuPuit, Paul DuBoys, George Stokes, Robert Manning, William Reynolds, and others, whose names have become associated with particular laws or principles. Details of these efforts can be seen in Manning (1891), among others.

Treaties on various aspects of hydrology, beginning with Nathaniel Beardmore's *Manual of Hydrology* in 1851 (Beardmore 1851), appeared with increasing frequency in the second half of the nineteenth century. Many of these works examined relations between rainfall amounts and streamflow rates, because of the need to estimate flood flows for the design of bridges and other hydraulic structures. This was also the beginning of a close association between hydrology and civil engineering; as a matter of fact, the first English-language texts in hydrology by Daniel Mead in 1904 and Adolf Meyer in 1919 were written for civil engineers (Eagleson et al. 1991).

The first half of the twentieth century saw great progress in many aspects of hydrology. With the formation of the Section of Scientific Hydrology in the International Union of Geodesy and Geophysics (in 1922) and the Hydrology Section of the American Geophysical Union (in 1930), hydrology received formal scientific recognition for the first time. During this time, there were many notable contributions to advances in specific areas: Allen Hazen, Emil Gumbel, Harold Hurst, and Walter Langbein in the application of statistics to hydrologic data; Oscar Meinzer, Charles Theis, Charles Slichter, and Marion King Hubbert in the development of the theoretical and practical aspects of groundwater hydraulics, and especially Lorenzo Richards in the development of governing equation for unsaturated flow; Ludwig Prandtl, Theodor von Kármán, Hunter Rouse, Ven Te Chow, Grove Karl Gilbert, and Hans Einstein in stream hydraulics and sediment transport; Robert Horton and Luna Leopold in understanding runoff processes and quantitative geomorphology; Charles Warren Thornthwaite and Howard Penman in understanding climatic aspects of hydrology and modeling evapotranspiration; and Abel Wolman and Robert Garrels in the understanding and modeling of water quality. Details of these developments can be seen in Richards (1931), Horton (1933, 1945), Gumbel (1941), and Hurst (1951), among others.

The 1960s witnessed the beginning of stochastic concepts in hydrology, notably applications of *linear* stochastic methods to hydrologic data (Thomas and Fiering 1962; Yevjevich 1963; Fiering 1967). Assisted by the discovery of self-similarity concept during this decade (Mandelbrot 1967), the ideas of scale in hydrology also gained more recognition (Mandelbrot and Wallis 1968, 1969). With advances in stochastic time series methods and fractal concepts in the 1970s (Box and Jenkins 1970; Mandelbrot 1975), the linear stochastic and scaling concepts in hydrology started to proliferate and are now prevalent in hydrology (e.g. Yevjevich 1972; Mandelbrot 1977; Gupta et al. 1986; MacNeill and Umphrey 1987; Gelhar 1993; Salas et al. 1995; Kalma and Sivapalan 1996; Rodriguez-Iturbe and Rinaldo 1997; Govindaraju 2002).

The revolutionary advances since the 1970s in computer and measurement technologies (e.g. supercomputers, remote sensors, and geographic information systems) have facilitated the emergence of various nonlinear concepts and the development of a host of nonlinear time series methods as well as others. In addition to the nonlinear stochastic ones, these methods include: data-based mechanistic models, artificial neural networks, wavelets, entropy theory, support vector machines, genetic programming, fuzzy logic, and nonlinear dynamics and chaos. Applications of these concepts in hydrology roughly began in the late 1980s and early 1990s and have tremendously amplified since then (e.g. Foufoula-Georgiou and Kumar 1994; Young and Beven 1994; Bardossy and Duckstein 1995; Babovic

1996; Singh 1997; Govindaraju and Rao 2000; Sivakumar 2000; Dibike et al. 2001; see also Sivakumar and Berndtsson 2010), largely under the umbrella of 'Hydroinformatics.' These advances have also led to the development of numerous lumped, semi-distributed, and distributed hydrologic models, such as the TANK model, SWMM (Storm Water Management Model), TOPMODEL (Topographic model), HEC-HMS (Hydrologic Engineering Center Hydrologic Modeling System) and HEC-RAS (River Assessment System), SHE (Système Hydrologique Européen) and MIKE-SHE, SLURP (Semi-distributed Land Use Runoff Process), and SWAT (Soil and Water Assessment Tool) models. Extensive details about these models are already available in the literature (e.g. Metcalf and Eddy 1971; HEC 1995, 1998; Singh 1995; Abbott and Refsgaard 1996; Beven 1997; Neitsch et al. 2005). Since the complex semi-distributed and distributed models incorporate more and more processes and, thus, require calibration of more and more parameters, the problems of parameter estimation and the associated uncertainties have also become significant, and have been important areas of hydrologic studies since the 1980s (Sorooshian and Gupta 1983; Beven 1993; Duan et al. 2003).

Advances in the above areas, and still in many others, are continuing at a much faster rate than at any other time in the history of hydrology. Despite these, however, our understanding of hydrologic systems and the associated processes and problems is still far from complete. In fact, it is fair to say that the advances we have made thus far have brought in more questions than answers (e.g. Klemeš 1986; Sivakumar 2008c). There are some major challenges in several areas, and so are great opportunities. These include: simplification in our modeling practice, uncertainty estimation in hydrologic models, formulation of a hydrologic classification framework, scale issues, predictions in ungaged basins, assessment of the impacts of global climate change on our future water resources, connections between hydrologic data and system physics, translations and interpretations of our mathematical models and methods for better understanding of hydrologic systems and processes (e.g. Beven 2002, 2006; Sivapalan et al. 2003; McDonnell and Woods 2004; Kirchner 2006; Gupta et al. 2007; Sivakumar 2008a, b, c; IPCC 2014). There is no doubt that studying these issues will be an important part of hydrologic theory and practice in the coming decades and centuries.

1.4 Concept of Hydrologic System

Hydrologic phenomena are enormously complex, and are not fully understood. In the absence of perfect knowledge, a simplified way to represent them may be through the concept of *system*. There are many different definitions of a system, but perhaps the simplest may be: 'a system is a set of connected parts that form a whole.' Chow (1964) defined a system as an aggregate or assemblage of parts, being either objects or concepts, united by some form of regular interaction or inter-dependence. Dooge (1967a), however, defined a system as: "any structure, device, scheme, or procedure, real or abstract, that inter-relates in a given time reference, an input, cause, or stimulus, of matter, energy, or information and an output, effect, or response of information, energy, or matter." This definition by Dooge is much more comprehensive and instructive and it brings out, among others, the following important characteristics of the system: (1) a system can consist of more than one component; (2) these components are separate, and they may be inter-dependent; (3) these components are put together following some sort of scheme, i.e. a system is an ordered arrangement; (4) a system inter-relates input and output, cause and effect, or stimulus and response; (5) a system does not require that input and output be alike or have the same nature; and (6) a system can be composed of a number of sub-systems, each of which can have a distinct input-output linkage.

With this system concept, the entire hydrologic cycle may be regarded as a hydrologic system, whose components might include precipitation, interception, evaporation, transpiration, infiltration, detention storage or retention storage, surface runoff, interflow, and groundwater flow, and perhaps other phases of the hydrologic cycle. Each component may be treated as a sub-system of the overall cycle, if it satisfies the characteristics of a system set out in its definition. Thus, the various components of the hydrologic system can be regarded as hydrologic sub-systems. To analyze the total system, the simpler sub-systems can be treated separately and the results combined according to the interactions between the sub-systems (especially with the assumption of linearity). Whether a particular component is to be treated as a system or sub-system depends on the objective of the inquiry (Singh 1988; see also Sivakumar and Singh 2012).

Considering in our aim a fair balance between reduction in system complexity and incorporation of necessary system details, the global hydrologic cycle system may roughly be divided into three sub-systems (Chow et al. 1988).

- The *atmospheric water system* containing the processes of precipitation, interception, evaporation, and transpiration. This sub-system is studied under the sub-field of *hydrometeorology*;
- The *surface water system* containing the processes of overland flow, surface runoff, sub-surface and groundwater flow, and runoff to streams and the ocean. This sub-system is studied under the sub-field of *surface hydrology*; and
- The *sub-surface water system* containing the processes of infiltration, groundwater recharge, sub-surface flow and groundwater flow. This sub-system is studied under the sub-field of *sub-surface hydrology*. Sub-surface flow takes place in the soil near the land surface, while groundwater flow occurs deeper in the soil or rock strata.

In fact, for most practical problems, only a few processes of the hydrologic cycle are considered at a time, and then only considering a small portion of the Earth's surface. For such treatment, a more restricted system definition than the global hydrologic system may be appropriate, with the concept of the *control volume*, as is the case in the field of fluid mechanics. In this context, a hydrologic system can be defined as a structure or volume in space, surrounded by a boundary, that accepts water and other inputs, operates on them internally, and produces them as outputs. The structure (for surface or sub-surface flow) or volume in space (for atmospheric moisture flow) is the totality of the flow paths through which the water may pass as *throughput* from the point it enters the system to the point it leaves. The boundary is a continuous surface defined in three dimensions enclosing the volume or structure. A *working medium* enters the system as input, interacts with the structure and other media, and leaves as output. Physical, chemical, and biological processes operate on the working media within the system.

The procedure for developing working equations and models of hydrologic phenomena is similar to that in fluid mechanics, where mass, momentum, and energy principles serve as bases. In hydrology, however, there is generally a greater degree of approximation in applying physical laws because the systems are larger and more complex, and may involve several working media, whose properties may change tremendously in time and/or space. It must also be noted that many hydrologic systems are normally treated as random because their major input is precipitation, which is a highly variable and often unpredictable phenomenon, although there are non-random ways of treating precipitation behavior (e.g. Rodriguez-Iturbe et al. 1989; Sivakumar et al. 2001; see also Chap. 9 for further details). Consequently, statistical analysis plays a large role in hydrologic analysis. Because of these complications, and many others, it is not possible to describe some hydrologic processes with exact physical laws. The system concept helps in the construction of a model that relates inputs and outputs, rather than the extremely difficult task of exact representation of the system details, and thus has significant practical advantage.

1.5 Hydrologic System Model

As mentioned just now, the goal of the system concept is to establish an input-output relationship that can be used for reconstructing past events or prediction of future events. In this systems approach, we are concerned with the system operation, not the nature of the system itself (its components, their connection with one another, and so on) or the physical laws governing its operation. A system model is an approximation of the actual system (i.e. prototype); its inputs and outputs are measurable variables and its structure is an equation (or a set of equations) linking the inputs and outputs. Central to the model structure is the concept of a *system transformation*.

Let us assume that precipitation (P) over a river basin produces some flow (Q) at the outlet of the basin. It can then be said that the river basin system performs a transformation of precipitation (P) into flow (Q), which can be represented by:

$$Q = f(P) \tag{1.1}$$

where f is the transformation function, or simply *transfer function.* Generally speaking, f is a transfer function between the input (cause) and the output (effect). The cause and effect can be either internal to the system or external to the system or a combination, depending on the 'boundaries' of the system. In Eq. (1.1), Q is the 'dependent' variable, and P is the 'independent' variable.

Equation (1.1) is the simplest form of the equation for the flow process in a river basin, with precipitation serving as the only input variable. Since a host of other variables also influences the flow process (depending upon the river basin characteristics), this relationship is oftentimes a gross approximation at best. If, for example, infiltration (I) and evaporation (E) also act as influencing variables, then Eq. (1.1) must be changed to:

$$Q = f(P, I, E) \tag{1.2}$$

Assuming that data are available (and in good quality), Eq. (1.2) is certainly a more accurate representation of the flow process than Eq. (1.1). At the same time, however, Eq. (1.2) is also more complex than Eq. (1.1) and, thus, is more difficult to solve.

It must be clear now that inclusion of any additional influencing variable(s) will result in an even more complex equation, which will be even more difficult to solve. It is important, therefore, to be mindful of the complexity of the equation (i.e. model), not only for its solution but also for its data requirements. The problems associated with the development of more and more complex hydrologic models have been extensively discussed in the literature (e.g. Sorooshian and Gupta 1983; Konikow and Bredehoeft 1992; Beven 1993, 2002; Young et al. 1996; Duan et al. 2003), especially under the topics of parameter estimation and uncertainty.

It must also be noted that the variables in Eq. (1.1) represent the total values (i.e. total amount of precipitation and total volume of flow) over the basin over a period of time. However, since precipitation changes (and, hence, the flow) with time (t), Eq. (1.1) must be modified to incorporate this as:

$$Q_t = f(P_t) \tag{1.3}$$

where P_t is the rainfall intensity and Q_t is the flow rate. This 'time' or 'dynamic' factor (e.g. precipitation intensity) brings in further complexity to the precipitation-flow relationship, and so does the distribution of the variables in 'space.' This then requires a spatio-temporal perspective to the overall precipitation-flow relationship. Consideration of both space and time factors in all the variables influencing the flow process [such as an extension of Eq. (1.2)] will result in a highly complex spatio-temporal relationship. This is why modeling and prediction of hydrologic processes is often a tremendously difficult task.

The transfer function f in the above equations may be linear or nonlinear, depending on the properties of the input variables and on the characteristics of the river basin. In simple terms, 'linear' means output is proportional to the input (e.g. double the amount of rainfall producing double the amount of flow), and 'nonlinear' means output is not proportional to the input. Looking at the general non-proportionality between hydrologic inputs and outputs, it is fair to say that most, if not all, hydrologic processes are nonlinear in nature. The nonlinear nature of hydrologic processes had indeed been recognized as early as in the 1960s (e.g. Minshall 1960; Jacoby 1966; Amorocho 1967; Dooge 1967b; Amorocho and Brandstetter 1971). However, much of early hydrologic analysis (during the 1960s–1980s), especially based on time series methods (see Sect. 1.7), assumed the transfer functions as linear, perhaps due to the lack of data and computational power. This situation, however, changed dramatically in the 1980s, with the development of nonlinear time series methods facilitated by the availability of more data and computational power. At the current time, both linear and nonlinear transfer functions are assumed in hydrology, depending upon whether linear or nonlinear time series analysis method is employed. Further details about linearity and nonlinearity (and several other characteristics of hydrologic systems and processes) will be discussed in Chap. 2, and some of the popular linear and nonlinear time series methods applied in hydrology will be discussed in Chaps. 3 and 4, respectively.

1.6 Hydrologic Model Classification

As of now, there is no universally accepted hydrologic model classification. As different people perceive, conceptualize, and understand hydrologic systems in different ways, hydrologic models may be classified in different ways too; see Snyder and Stall (1965), Dawdy (1969), Dawdy and Kalinin (1969), DeCoursey (1971), Snyder (1971), Woolhiser (1971, 1973, 1975), Miller and Woolhiser (1975) for some early studies. Consequently, any classification or grouping of hydrologic models can be only rather arbitrary. For purposes of simplicity and convenience, however, hydrologic models may generally be grouped under two broad categories: *physical* models and *abstract* models.

1.6.1 Physical Models

A physical model is a representation of the system in a reasonably physically realistic manner. Physical models include scale models and analog models.

A scale model is a model that represents the system in a different size (enlarged or reduced) than the prototype. In hydrology, scale models are normally on a reduced size. A hydraulic model of a dam spillway is one example of a scale model, and an open channel hydrologic laboratory model of a river is another.

An analog model is a model that uses another physical system having properties similar to those of the system under study but is much easier to work with. It does not physically resemble the actual system but depends on the correspondence between the symbolic models describing the prototype and the analog system. The Hele-Shaw model (Hele-Shaw 1898) is one example of an analog model, as it uses the movement of a viscous fluid between two closely-spaced parallel plates to model seepage in an aquifer or embankment. An electrical analog model for watershed response is another example.

1.6.2 Abstract Models

An abstract model is a representation of the actual system in mathematical form, and thus it is also called as a mathematical model. In this model, the system operation is described by a set of equations linking the input and the output variables. Abstract or mathematical models may be divided into three groups (Dooge 1977): (1) empirical models; (2) theoretical models; and (3) conceptual models.

An empirical model is merely a representation of the facts based on the available data; if the conditions change, it has no predictive capability. Therefore, all empirical models have some chance of being fortuitous and, in principle, should not be used outside the range of data from which they were derived. One example of empirical models is the rational method (Kuichling 1889)

$$Q = CiA \tag{1.4}$$

where Q is flow rate (cubic feet per second), i is rainfall intensity (inches per hour), A is area (acre), and C is a constant (runoff coefficient) that can range from 0 to 1. Other examples include the unit hydrograph models based on harmonic analysis (O'Donnell 1960), the least squares method (Snyder 1955), and the Laguerre polynomials (Dooge 1965).

A theoretical model is presumably a consequence of the most important laws governing the phenomena. It has a logical structure similar to the real-world system and may be helpful under changed circumstances. Examples of theoretical models may include infiltration models based on two-phase flow theory of porous media (Morel-Seytoux 1978), evaporation models based on theories of turbulence and diffusion (Brutsaert and Mawdsley 1976), and groundwater models based on fundamental transport equations (Freeze 1971).

A conceptual model is an intermediate between an empirical and a theoretical model, although it can be used broadly to embrace both of these types of models. Generally, conceptual models consider physical laws but in highly simplified form. Examples of conceptual models may include rainfall-runoff models based on the spatially lumped form of the continuity equation and the storage-discharge relationship (Nash 1958; Dooge 1959), and models derived from linear diffusion analogy and linearized versions of St. Venant equations (Harley 1967; O'Meara 1968).

1.6.3 Remarks

Further sub-grouping of hydrologic models may be made on the basis of their nature or complexity or other aspects. These include: linear and nonlinear, time-invariant and time-variant, lumped and distributed, and deterministic and stochastic. Some of these will be discussed in Chaps. 2–4, as part of the characteristics of hydrologic systems and processes and popular models that have been in existence. An excellent essay on empirical and physical models in hydrology was presented by Klemeš (1982), which provides in detail a philosophical perspective on: (1) hydrologic modeling in general; (2) merits and demerits of empirical modeling; (3) why hydrologic models work; and (4) drawbacks, dangers, and potential benefits of causal modeling. Singh (1988) also presented an extensive account of the various types of models in existence in hydrology.

It must be emphasized that models are only approximate representations of actual systems. Their development depends not only on our limited understanding of the actual systems but also on limited observations and computational/structural powers. Therefore, it is fair to say that all models are wrong, or will be proven to be wrong. The only perfect model of a physical system is the system itself. There is no such thing as a perfect model because an abstract quantity cannot perfectly represent a physical entity (Singh 1988).

1.7 Hydrologic Data and Time Series Modeling

Whether physical or abstract hydrologic models, observations or data on relevant hydrologic variables (e.g. rainfall, evaporation, streamflow) as inputs/outputs play a key role in model formulation and model validation. A variable can be observed at a particular location over time or at a particular time at different locations or at different locations over time. A set of observations of a hydrologic variable made at a particular location over time is called a hydrologic time series (Fig. 1.2). For instance, if precipitation P is the variable of interest at a given location and is observed over time N, then the time series of precipitation may be denoted as P_i, $i \leq N$. Observations over time at a particular location are helpful for studying the temporal dynamics; those made at a particular time at different locations are helpful for studying the spatial patterns; and observations made at different locations over time are helpful for studying the spatio-temporal dynamics.

The timescale of a hydrologic time series may be either discrete or continuous. A discrete timescale would result from observations at specific times with the times of the observations separated by i or from observations that are some function of the values that actually occurred during i. Most hydrologic time series fall in this latter category, such as the one shown in Fig. 1.2. Examples would be the annual peak discharge (i = 1 year), monthly precipitation (i = 1 month), and average daily flow in a stream (i = 1 day). A continuous timescale results when data is recorded

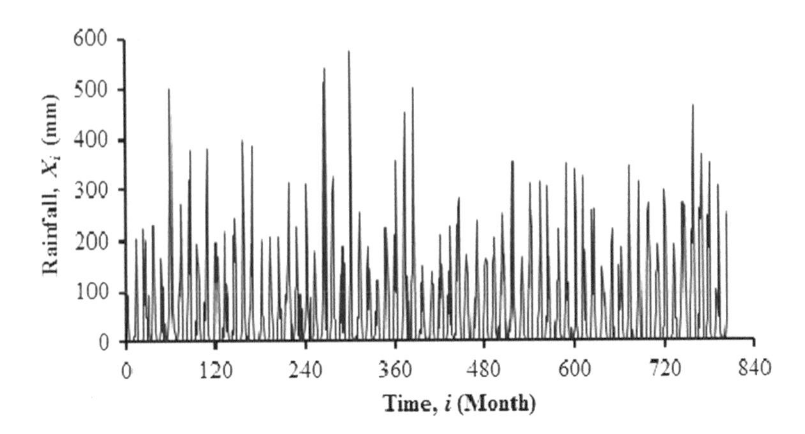

Fig. 1.2 An example of a hydrologic time series: monthly rainfall data

continuously with time, such as the stage at a streamgaging location. Even when a continuous timescale is used for collecting the data, the analysis is usually done by selecting values at specific time intervals. For example, raingage charts are usually analyzed by reading the data at selected times (i.e. every 5 min) or at 'break points' (here i is not a constant).

Depending upon the hydrologic variable and the period of observation, a hydrologic time series may be composed of only deterministic events (e.g. no precipitation over a certain period) or only stochastic events (e.g. precipitation with significant variability over a period of time) or a combination of the two (e.g. precipitation over a period of time with a certain amount of variability as well as trend). Generally speaking, and especially considering a reasonably long period of observations, a hydrologic time series is of the third category, oftentimes with stochastic components superimposed on deterministic components; for example, the series composed of average monthly streamflows at some location would contain the deterministic component of seasonal cycle plus the stochastic component arising from random deviations from the seasonal values.

Despite this knowledge, hydrologic modeling efforts have mostly adopted either only a deterministic approach or only a stochastic approach but not their combination (e.g. Yevjevich 1968; Quimpo 1971; Vogel 1999; Sivakumar 2008a, c). In the deterministic approach, the hydrologic system is described and represented by theoretical and/or empircal physical relationships; that is, there is always a unique correspondence between the input, say precipitation, and the output, say streamflow. On the other hand, in the stochastic approach, a type of model is assumed aiming to represent the most relevant statistical characteristics of the historic series. Further details on the deterministic and stochastic components in hydrologic time series and the available models can be seen in Haan (1994), Salas et al. (1995), and Beven (2001), among others.

Over the past half a century, analysis of hydrologic time series has become a fascinating endeavor and an important part of hydrologic studies. Two factors have contributed mainly to this: developments in measurement technology (e.g. improved precipitation and streamgages, remote sensors, geographic information systems) and advances in data analysis concepts and tools (e.g. novel mathematical and statistical methods, computers, softwares). Numerous time series analysis methods have found their applications in hydrology, such as the linear and nonlinear stochastic techniques, scaling and fractal methods, artificial neural networks, data-based mechanistic methods, wavelets, entropy-based methods, support vector machines, genetic programming, fuzzy logic, and nonlinear dynamics and chaos. All of these techniques may be put under the broad umbrella of 'data-based' or 'data-driven' approaches.

A plethora of literature is already available on these time series analysis techniques and their applications in hydrology. Examples include: Box and Jenkins (1970), Yevjevich (1972), and Salas et al. (1995) for stochastic methods; Mandelbrot and Wallis (1969), Mandelbrot (1983), and Rodriguez-Iturbe and Rinaldo (1997) for scaling methods; Haykin (1999), ASCE Task Committee (2000a, b), Govindaraju and Rao (2000) for artificial neural networks; Young (1984), and Young and Beven (1994), Young et al. (1996) for data-based mechanistic models; Chui (1992), Foufoula-Georgiou and Kumar (1994), and Labat (2005) for wavelets; Singh (1997, 1998, 2013) for entropy concepts; Vapnik (1995), Cristianini and Shawe-Taylor (2000), and Dibike et al. (2001) for support vector machines; Holland (1975), Koza (1992), and Babovic (1996) for evolutionary algorithms and genetic programming; Zadeh (1965), Bardossy and Duckstein (1995), and Zadeh et al. (1996) for fuzzy logic; Tsonis (1992), Sivakumar (2000), and Kantz and Schreiber (2004) for nonlinear dynamics and chaos. A comprehensive review of many of these data-based methods and their applications in hydrology is presented in Sivakumar and Berndtsson (2010).

Among these methods, the linear stochastic methods are much more popular and established, partly due to their developments earlier than the others and partly due to our assumption that hydrologic processes are stochastic in nature. The stochastic assumption of hydrologic processes, brought about mainly due to their 'complex' and 'random' behaviors, was a significant deviation from the deterministic paradigm that was dominant during the first half of the twentieth century. However, fast developments in data measurement and computer technologies have made the other methods equally attractive as well, such as the ones based on nonlinear dynamic and chaos theories, which are the main focus of this book.

1.8 Physical Basis of Time Series Modeling

Although time series models have become immensely popular and found wide applications in hydrology, they are oftentimes treated only as 'black-box' models. This is because, time series models, despite their ability to represent even highly complex and nonlinear input-output relationships based on data alone, often lack the physical connections between the model structure/function and the catchment details/physics. The difficulty comes from the fact that the actual mechanisms in a catchment occur at various (and all) scales but the data are usually measured only at the catchment scale. Numerous attempts have been made to establish connections between time series models and catchment physics, and a short list is presented below.

From the viewpoint of stochastic time series models in hydrology, Yevjevich (1963) and Fiering (1967) tried to set the physical basis of stochastic modeling, at least for the case of autoregressive models. Moss and Bryson (1974) tried to establish the physical basis of seasonal stochastic models. O'Connor (1976) attempted to relate the unit hydrograph and flood routing models to autoregressive and moving average models. Pegram (1977) and Selvalingam (1977) provided the physical justification of continuous stochastic streamflow models. Other studies offering physical explanation for stochastic time series models include Klemeš (1978), Salas and Smith (1981), Parlange et al. (1992).

From a scaling perspective, Gupta et al. (1996) offered a theory to establish connections between physics of floods and power laws. Their goal was first to understand how spatial peak-discharge power laws are connected to physical processes during rainfall-runoff events and then to extend this understanding to longer timescales. These results have then been generalized in many directions (e.g. Gupta and Waymire 1998; Menabde and Sivapalan 2001; Menabde et al. 2001; Morrison and Smith 2001; Veitzer and Gupta 2001; Ogden and Dawdy 2003; Furey and Gupta 2005, 2007). Gupta et al. (2007) provide an overview of progress that had been made during the previous 20 years in understanding the physical origins of spatial power laws that are observed, on average, in floods.

A few studies have attempted to provide a physical basis for the use of artificial neural networks in hydrology. In their attempt to model daily river flow using neural networks, Jain et al. (2004) suggested that two of the hidden units used in their network were clearly capturing the baseflow component, as they were strongly correlated with past river discharge, baseflow, and soil moisture. Sudheer and Jain (2004) explained the internal structure and behavior of neural networks similar to the function of the flow-duration curve. See et al. (2008) performed a similar investigation, and results from different methods of analysis applied to both internal and external outputs depicted comparable organization of hidden units into baseflow, surface flow, and quick flow components. For futher details, the reader is also referred to Abrahart et al. (2010).

Many studies employing time series methods in hydrology have adopted the concept of 'thresholds' in many different forms, such as 'critical states' in the search for self-organization in landscapes and river networks (e.g. Rodriguez-Iturbe and Rinaldo 1997), 'characteristic patterns' of rainfall behavior for prediction of streamflow using self-organizing maps (e.g. Hsu et al. 2002), and 'regimes' in the prediction of streamflow dynamics using nonlinear dynamic and chaos methods (e.g. Sivakumar 2003). Sivakumar (2005) tried to offer a physical explanation on the role of 'thresholds' in catchments through an analogy between catchment behavior and human behavior.

Despite these attempts and advances, establishing connections between data and catchment physics continues to be a tremendously challenging task, and a coherent approach to deal with this issue remains by and large elusive (e.g. Kirchner 2006; Sivakumar 2008a). There are sufficient grounds to believe that a general 'disconnection' that has been present between researchers employing 'physics-based' approaches and those employing 'data-based approaches' has partially contributed to this problem (e.g. Sivakumar 2008a). The difficulties in 'communication' (even among those employing time series methods), largely because of the use of different 'jargons' in the literature, has made the situation only worse (e.g. Sivakumar 2005).

Current efforts that attempt to reconcile the upward (process-based) approaches and the downward (data-based) approaches are encouraging towards establishing connections between data and catchment physics. As scaling properties are essentially related to the physics of the basins, scaling theories could also lead to a better understanding of hydrologic systems and processes (e.g. Gupta 2004; Dawdy 2007). Further, since different time series methods possess different advantages, another possibility to establish relations between data and catchment physics may be to integrate two or more methods, each of which is suitable for the purpose at hand but not as effective as their combination. The study by Young and Ratto (2009), introducing some extensions to the data-based mechanistic approach (e.g. Young and Beven 1994) and coupling the hypothetico-deductive approach of simulation modeling with the inductive approach of data-based modeling, is a good example to this idea of integration. Such integration of methods may also lead to simplification in hydrologic modeling (e.g. Sivakumar 2004b). However, research in these directions is still in a state of infancy, and there is certainly some distance to go.

1.9 Scope and Organization of the Book

The central focus of this book is the role and applications of chaos theory and related ideas in the field of hydrology. In the nonlinear science literature, the term 'chaos' refers to situations where complex and random-looking behaviors arise from simple nonlinear deterministic systems with sensitive dependence on initial conditions (Lorenz 1963), and the converse also applies. The three fundamental properties inherent in this definition: (1) nonlinear inter-dependence; (2) hidden determinism and order; and (3) sensitivity to initial conditions are highly relevant in hydrologic systems and processes. For example: (1) components and mechanisms involved in the hydrologic cycle act in a nonlinear manner and are also inter-dependent; (2) daily cycle in temperature and annual cycle in river flow possess determinism and order; and (3) contaminant transport phenomena in surface and sub-surface waters largely depend upon the time (e.g. rainy or dry season) at which the contaminants are released at the source, which themselves may not be known. The first property represents the 'general' nature of hydrologic phenomena, whereas the second and third represent their 'deterministic' and 'stochastic' natures, respectively. Further, despite their complexity and random-looking behavior, hydrologic phenomena may also be governed by a very few degrees of freedom (e.g. runoff in a well-developed urban catchment depends essentially on rainfall), another fundamental idea of chaos theory (e.g. Sivakumar 2004a).

In view of these, chaos theory has found a growing number of applications in hydrology during the last two decades or so. Sivakumar (2000, 2004a, 2009) present extensive reviews of these applications and also address the current issues and future challenges [see also Sivakumar and Berndtsson (2010)]. Much of the advances thus far on chaos theory in hydrology has come in terms of time series analysis. Therefore, a good understanding of the salient characteristics of hydrologic systems and processes and of the fundamentals of time series analysis is needed in the first place to appreciate the role of chaos theory in hydrology and the advances made (this is one reason for highlighting, in this chapter, the time series analysis methods and their physical basis). With this in mind, the rest of this book is organized as follows.

Chapter 2 discusses the salient characteristics of hydrologic systems and processes, including complexity, correlation, trend, periodicity, cyclicity, and seasonality, intermittency, stationarity and nonstationarity, linearity and nonlinearity, determinism and randomness, scale and scale-invariance, self-organization and self-organized criticality, threshold, emergence, feedback, sensitivity to initial conditions, and the class of nonlinear determinism and chaos. Chapter 3 reviews some popular conventional linear stochastic time series analysis methods (both parametric and nonparametric; e.g. autoregressive, autoregressive moving average, Markov, and *k*-nearest neighbor) and their applications in hydrology. In Chap. 4, a brief account of some popular modern nonlinear time series methods (e.g. nonlinear stochastic methods, data-based mechanistic methods, artificial neural networks, support vector machines, wavelets, evolutionary computing, fuzzy logic, entropy methods, nonlinear dynamic and chaos methods) and their hydrologic applications is presented.

Part B is devoted to the theoretical aspects of nonlinear dynamic and chaos concepts, with Chap. 5 on the fundamental ideas (e.g. dynamic systems, attractors, bifurcations and chaos, phase space, dimension), Chap. 6 on the methods for identification and prediction of chaos (e.g. phase space reconstruction, correlation dimension method, false nearest neighbor algorithm, close returns plot, nonlinear local approximation method), and Chap. 7 on the important issues associated with the application of chaos techniques to finite (especially small) and noisy time series, which are often the type of time series observed in hydrology.

Part C provides an extensive review of chaos theory applications in hydrology. After an overview in Chap. 8, applications to rainfall and river flow are presented in Chaps. 9 and 10, respectively. Applications to other processes (e.g. rainfall-runoff, lake volume and level, sediment transport, ground water) are reviewed in Chap. 11, while Chap. 12 presents details of studies on methodological and data issues in the applications of chaos theory in hydrology (e.g. delay time, data size, data noise, presence of zeros).

Part D looks to the future of chaos theory in hydrology. After re-visiting the current status (e.g. successes, failures, limitations, concerns) in Chap. 13, some potential directions for further advances (e.g. parameter identification, multi-variable analysis, model simplification and integration, reconstruction of system equations, linking data and physics), especially in light of future challenges (including the impacts of climate change), are highlighted in Chap. 14. Finally, Chap. 15 offers some thoughts on the philosophy and pragmatism in studying hydrology and argues in favor of chaos theory as a balanced and middle-ground approach to our dominant extreme views of determinism and stochasticity.

References

Abbott MB, Refsgaard JC (ed) (1996) Distributed hydrological modelling. Kluwer Academic Publishers, Dordrecht, The Netherlands

Abrahart RJ, See LM, Dawson CW, Shamseldin AY, Wilby RL (2010) Nearly two decades of neural network hydrologic modeling. In: Sivakumar B, Berndtsson R (eds) Advances in data-based approaches for hydrologic modeling and forecasting. World Scientific Publishing Company, Singapore, pp 267–346

Adams FD (1938) The birth and development of the geological sciences. Williams and Wilkins, Baltimore, MD

Amorocho J (1967) The nonlinear prediction problems in the study of the runoff cycle. Water Resour Res 3(3):861–880

Amorocho J, Brandstetter A (1971) Determination of nonlinear functional response functions in rainfall-runoff processes. Water Resour Res 7(5):1087–1101

ASCE Task Committee (2000a) Artificial neural networks in hydrology. I: Preliminary concepts. ASCE. J Hydrol Eng 5(2):115–123

ASCE Task Committee (2000b) Artificial neural networks in hydrology. II: Hydrologic applications. ASCE. J Hydrol Eng 5(2):124–137

Babovic V (1996) Emergence, evolution, intelligence; hydroinformatics. Balkema Publishers, Rotterdam

Bardossy A, Duckstein L (1995) Fuzzy rule-based modeling with applications to geophysical, biological and engineering systems. CRC Press, Boca Raton, FL, USA, p 256

Beardmore N (1851) Manual of hydrology. Waterlow & Sons, London

Beven KJ (1993) Prophecy, reality and uncertainty in distributed modeling. Adv Water Resour 16:41–51

Beven KJ (1997) Distributed hydrological modelling: applications of the TOPMODEL concept. John Wiley, Chichester, UK

Beven KJ (2001) Rainfall-runoff modelling: the primer. Wiley, Chichester, UK

Beven KJ (2002) Uncertainty and the detection of structural change in models of environmental systems. In: Beck MB (ed) Environmental foresight and models: a manifesto. Elsevier, The Netherlands, pp 227–250

Beven KJ (2006) On undermining the science? Hydrol Process 20:3141–3146

Biswas AK (1970) History of hydrology. North-Holland Publishing, Amsterdam

Box GEP, Jenkins G (1970) Time series analysis, forecasting and control. Holden-Day, San Francisco

Brutsaert W, Mawdsley JA (1976) The applicability of planetary boundary layer theory to calculate regional evapotranspiration. Water Resour Res 12:852–858

Chahine MT (1992) The hydrological cycle and its influence on climate. Nature 359:373–380
Chow VT (ed) (1964) Handbook of applied hydrology. McGraw-Hill, New York
Chow VT, Maidment DR, Mays LW (1988) Applied hydrology. McGraw-Hill, Singapore
Chui CK (1992) An introduction to wavelets. Academic Press, Boston
Cristianini N, Shawe-Taylor J (2000) An introduction to support vector machines. Cambridge University Press, Cambridge, UK
Dalton J (1802) Experimental essays on the constitution of mixed gases; on the force of steam or vapor from waters and other liquids, both in a Torricellian vacuum and in air; on evaporation; and on the expansion of gases by heat. Mem Proc Manch Lit Phil Soc 5:535–602
Darcy H (1856) Les fontaines publiques de la ville de Dijon. V. Dalmont, Paris
Dawdy DR (1969) Considerations involved in evaluating mathematical modeling of urban hydrologic systems. U. S. Geological Survey Water Supply Paper 1591-D, U. S. Department of Interior, Washington, D. C., D1-D18
Dawdy DR (2007) Prediction versus understanding (The 2007 Ven Te Chow Lecture). ASCE J Hydrol Eng 12:1–3
Dawdy DR, Kalinin GP (1969) Mathematical modeling in hydrology. International Association of Scientific Hydrology Report, Mid-Decade Conference of the International Hydrological Decade, held in August in Surány, Hungary
DeCoursey DG (1971) The stochastic approach to watershed modeling. Nordic Hydrol 11:186–216
Dibike YB, Velickov S, Solomatine DP, Abbott M (2001) Model induction with support vector machines: introduction and applications. ASCE J Comput Civil Eng 15(2):208–216
Dooge JCI (1959) A general theory of the unit hydrograph. J Geophys Res 64(2):241–256
Dooge JCI (1965) Analysis of linear systems by means of Laguerre functions. Soc Indust Appl Math (SIAM) J Cont 2(3):396–408
Dooge JCI (1967a) The hydrologic cycle as a closed system. Int Assoc Sci Hydrol Bull 13(1):58–68
Dooge JCI (1967b) A new approach to nonlinear problems in surface water hydrology: hydrologic systems with uniform nonlinearity. Int Assoc Sci Hydrol Publ 76:409–413
Dooge JCI (1977) Problems and methods of rainfall-runoff modeling. In: Ciriani TA, Maione U, Wallis JR (eds) Mathematical models for surface water hydrology. John Wiley, New York, pp 71–108
Driscoll FG (ed) (1986) Groundwater and wells. Johnson Division, St. Paul, Minnesota
Duan Q, Gupta HV, Sorooshian S, Rousseau AN, Turcotte R (2003) Calibration of watershed models. Water Science and Application Series. American Geophysical Union, Washington vol 6, pp 1–346
Eagleson PS, Brutsaert WH, Colbeck SC, Cummins KW, Dozier J, Dunne T, Edmond JM, Gupta VK, Jacoby JC, Manabe S, Nicholson SE, Nielsen DR, Rodríguez-Iturbe I, Rubin J, Sposito G, Swank WT, Zipser EJ, Burges S (1991) Opportunities in the hydrologic sciences. National Academy Press, Washington, DC, p 348
Fiering MB (1967) Streamflow synthesis. Harvard University Press, Cambridge, Massachusetts
Foufoula-Georgiou E, Kumar P (eds) (1994) Wavelets in geophysics. Academic Press, San Diego, California, p 373
Freeze RA (1971) Three-dimensional, transient, saturated-unsaturated flow in a groundwater basin. Water Resour Res 7(2):347–366
Freeze RA, Cherry JA (1979) Groundwater. Prentice-Hall, Englewood Cliffs, New Jersey
Furey PR, Gupta VK (2005) Effects of excess-rainfall on the temporal variability of observed peak discharge power laws. Adv Water Resour 28:1240–1253
Furey PR, Gupta VK (2007) Diagnosing peak-discharge power laws observed in rainfall–runoff events in Goodwin Creek experimental watershed. Adv Water Resour 30:2387–2399
Gelhar LW (1993) Stochastic subsurface hydrology. Prentice-Hall, Englewood Cliffs, New Jersey

Govindaraju RS (ed) (2002) Stochastic methods in subsurface contaminant hydrology. ASCE, New York

Govindaraju RS, Rao AR (2000) Artificial neural networks in hydrology. Kluwer Acadmic Publishers, Amsterdam

Gumbel EJ (1941) The return period of flood flows. Ann Math Stat 12(2):163–190

Gupta VK (2004) Emergence of statistical scaling in floods on channel networks from complex runoff dynamics. Chaos Soliton Fract 19:357–365

Gupta VK, Waymire E (1998) Spatial variability and scale invariance in hydrologic regionalization. In: Sposito G (ed) Scale dependence and scale invariance in hydrology. Cambridge University Press, Cambridge, pp 88–135

Gupta VK, Rodríguez-Iturbe I, Wood EF (eds) (1986) Scale problems in hydrology: runoff generation and basin response. Reidel Publishing Company, FD, p 244

Gupta VK, Castro SL, Over TM (1996) On scaling exponents of spatial peak flows from rainfall and river network geometry. J Hydrol 187:81–104

Gupta VK, Troutman BM, Dawdy DR (2007) Towards a nonlinear geophysical theory of floods in river networks: an overview of 20 years of progress. In: Tsonis AA, Elsner JB (eds) Twenty years of nonlinear dynamics in geosciences. Springer Verlag

Haan CT (1994) Statistical methods in hydrology. Iowa University Press, Iowa

Harley BM (1967) Linear routing in uniform flow channels. M. Eng. Sc. Thesis, National University of Ireland, University College, Cork, Ireland

Haykin S (1999) Neural networks: a comprehensive foundation. Prentice Hall International Inc, New Jersey

Hele-Shaw HS (1898) The flow of water. Nature 58:34–36

Holland JH (1975) Adaptation in natural and artificial systems. University of Michigan Press, Ann Arbor, MI

Horden RH (1998) The hydrologic cycle. In: Herschy RW, Fairbridge RW (eds) Encyclopedia of hydrology and water resources. Kluwer Academic Publishers, Dordrecht, The Netherlands, pp 400–404

Horton RE (1933) The role of infiltration in the hydrologic cycle. Trans Am Geophys Union 14:446–460

Horton RE (1945) Erosional development of streams and their drainage basins: Hydrophysical approach to quantitative morphology. Bull Geol Soc Am 56:275–370

Hsu KL, Gupta HV, Gao X, Sorooshian S, Imam B (2002) Self-organizing linear output map (SOLO): an artificial neural network suitable for hydrologic modeling and analysis. Water Resour Res 38(12). doi:10.1029/2001WR000795

Hurst HE (1951) Long-term storage capacity of reservoirs. Trans Am Soc Civ Eng 116:770–799

Hydrologic Engineering Center (1995) HEC-HMS: Hydrologic modeling system user's manual. U. S, Army Corps of Engineers, Davis, CA

Hydrologic Engineering Cener (1998) HEC-RAS: river analysis system user's manual. U.S. Army Corps of Engineers, Davis, CA

IPCC (2014) Climate change 2014—impacts, adaptation and vulnerability. Contribution of Working Group II to the Fifth Assessment Report of the Intergovernmental Panel on Climate Change [Field CB, Barros VR, Dokken DJ, Mach KJ, Mastrandrea MD, Bilir TE, Chatterjee M, Ebi KL, Estrada YO, Genova RC, Girma B, Kissel ES, Levy AN, MacCracken S, Mastrandrea PR, White LL (eds.)], Cambridge University Press, Cambridge

Jacoby SLS (1966) A mathematical model for nonlinear hydrologic systems. J Geophys Res 71 (20):4811–4824

Jain A, Sudheer KP, Srinivasulu S (2004) Identification of physical processes inherent in artificial neural network rainfall runoff models. Hydrol Process 18(3):571–581

Kalma JD, Sivapalan M (1996) Scale issues in hydrological modeling. Wiley, London

Kantz H, Schreiber T (2004) Nonlinear time series analysis. Cambridge University Press, Cambridge

Kirchner JW (2006) Getting the right answers for the right reasons: linking measurements, analyses, and models to advance the science of hydrology. Water Resour Res 42:W03S04. doi:10.1029/2005WR004362
Klemeš V (1978) Physically based stochastic hydrologic analysis. Adv Hydrosci 11:285–352
Klemeš V (1982) Empirical and causal models in hydrology. Scientific basis of water-resources management. National Academic Press, Washington, DC, pp 95–104
Klemeš V (1986) Dilettantism in hydrology: Transition or destiny? Water Resour Res 22(9):177S–188S
Konikow LF, Bredehoeft JD (1992) Ground-water models cannot be validated. Adv Water Resour 15:75–83
Koza JR (1992) Genetic programing: on the programming of computers by natural selection. MIT Press, Cambridge, MA
Kuichling E (1889) The relation between the rainfall and the discharge of sewers in populous districts. Trans Am Soc Civ Eng 20:1–56
Labat D (2005) Recent advances in wavelet analyses: Part 1. A review of concepts. J Hydrol 314:275–288
Lorenz EN (1963) Deterministic nonperiodic flow. J Atmos Sci 20:130–141
MacNeill IB, Umphrey GJ (1987) Stochastic hydrology. Kluwer Academic Publishers, Boston
Maidment DR (ed) (1993) Handbook of hydrology. McGraw-Hill, New York
Mandelbrot BB (1967) How long is the coast of Britain? Statistical self-similarity and fractional dimension. Science 156:636–638
Mandelbrot BB (1975) On the geometry of homogeneous turbulence, with stress on the fractal dimension of the isosurfaces of scalars. J Fluid Mech 72:401–416
Mandelbrot BB (1977) Fractals: form, chance and dimension. W. H, Freeman and Co, New York
Mandelbrot BB (1983) The fractal geometry of nature. Freeman, New York
Mandelbrot BB, Wallis JR (1968) Noah, Joseph and operational hydrology. Water Resour Res 4 (5):909–918
Mandelbrot BB, Wallis JR (1969) Some long run properties of geophysical records. Water Resour Res 5(2):321–340
Manning R (1891) On the flow of water in open channels and pipes. Trans Inst Civ Eng Ireland 20:161–207
McDonnell JJ, Woods RA (2004) On the need for catchment classification. J Hydrol 299:2–3
Menabde M, Sivapalan M (2001) Linking space-time variability of river runoff and rainfall fields: a dynamic approach. Adv Wat Resour 24:1001–1014
Menabde M, Veitzer S, Gupta VK, Sivapalan M (2001) Tests of peak flow scaling in simulated self-similar river networks. Adv Wat Resour 24:991–999
Metcalf and Eddy, Inc., University of Florida, and Water Resources Engineers, Inc., (1971) Storm water management model, vol. 1. Final Report, 11024DOC07/71 (NTIS PB-203289), U.S. EPA, Washington, DC
Miller WA, Woolhiser DA (1975) Choice of models. In: Yevjevich VM (ed) Unsteady flow in open channels. Water Resources Publications, Littleton, Colorado
Minshall NE (1960) Predicting storm runoff on small experimental watersheds. J Hydraul Div Am Soc Eng 86(HYB):17–38
Morel-Seytoux HJ (1978) Derivation of equations for variable rainfall infiltration. Water Resour Res 14(4):561–568
Morrison JE, Smith JA (2001) Scaling properties of flood peaks. Extremes 4(1):5–22
Moss ME, Bryson MC (1974) Autocorrelation structure of monthly streamflows. Water Resour Res 10(4):737–744
Nace RL (1974) General evolution of the concept of the hydrological cycle. Three centuries of scientific hydrology. UNESCO-World Meteorological Organization-International Association of Hydrological Sciences, Paris, pp 40–48
Nash JE (1958) Determining runoff from rainfall. Proc Instit Civil Eng Ireland 10:163–184

Neitsch SL, Arnold JG, Kiniry JR, Williams JR (2005) Soil and water assessment tool theoretical documentation, Version 2005. USDA.ARS Grassland, Soil and Water Research Laboratory, Temple, TX

O'Connor KM (1976) A discrete linear cascade model for hydrology. J Hydrol 29:203–242

O'Donnell T (1960) Instantaneous unit hydrograph derivation by harmonic analysis. Int Assoc Sci Hydrol Publ 51:546–557

O'Meara WA (1968) Linear routing of lateral inflow in uniform open channels. M. E. Sc. Thesis, The National University of Ireland, University College, Cork, Ireland

Ogden FL, Dawdy DR (2003) Peak discharge scaling in a small hortonian watershed. J Hydrol Eng 8(2):64–73

Parlange MB, Katul GG, Cuenca RH, Kavvas ML, Nielsen DR, Mata M (1992) Physical basis for a time series model of soil water content. Water Resour Res 28(9):2437–2446

Pegram GGS (1977) Physical justification of continuous streamflow model. In: Morel-Seytoux HJ, Salas JD, Sanders TG, Smith RE (eds) Modeling hydrologic processes. Proceedings of Fort Collins III international hydrol Symposium, pp 270–280

Quimpo R (1971) Structural relation between parametric and stochastic hydrology models. In: Mathematical models in hydrology, Warsaw Symposium (IAHS Publication 100) 1:151–157

Richards LA (1931) Capillary conduction of liquids through porous mediums. Physics A 1:318–333

Rodriguez-Iturbe I, Rinaldo A (1997) Fractal river basins: chance and self-organization. Cambridge University Press, Cambridge

Rodriguez-Iturbe I, De Power FB, Sharifi MB, Georgakakos KP (1989) Chaos in rainfall. Water Resour Res 25(7):1667–1675

Salas JD, Smith RA (1981) Physical basis of stochastic models of annual flows. Water Resour Res 17(2):428–430

Salas JD, Delleur JW, Yevjevich V, Lane WL (1995) Applied modeling of hydrologic time series. Water Resources Publications, Littleton, Colorado

See LM, Jain A, Dawson CW, Abrahart RJ (2008) Visualisation of hidden neuron behaviour in a neural network rainfall-runoff model. In: Abrahart RJ, See LM, Solomatine DP (eds) Practical hydroinformatics: computational intelligence and technological developments in water applications. Springer-Verlag, Berlin, Germany, pp 87–99

Selvalingam S (1977) ARMA and linear tank models. In: Morel-Seytoux HJ, Salas JD, Sanders TG, Smith RE (eds) Modeling hydrologic processes. Proceedings of Fort Collins III international hydrol symposium, pp 297–313

Singh VP (1988) Hydrologic systems: vol 1. Rainfall-runoff modeling, Prentice Hall, New Jersey

Singh VP (1995) Computer models of watershed hydrology. Water Resources Publications, Highlands Ranch CO

Singh VP (1997) The use of entropy in hydrology and water resources. Hydrol Process 11:587–626

Singh VP (1998) Entropy-based parameter estimation in hydrology. Kluwer Academic, Dordrecht, The Netherlands

Singh VP (2013) Entropy theory and its application in environmental and water engineering. John Wiley and Sons, Oxford, UK

Sivakumar B (2000) Chaos theory in hydrology: important issues and interpretations. J Hydrol 227 (1–4):1–20

Sivakumar B (2003) Forecasting monthly streamflow dynamics in the western United States: a nonlinear dynamical approach. Environ Model Softw 18(8–9):721–728

Sivakumar B (2004a) Chaos theory in geophysics: past, present and future. Chaos Soliton Fract 19 (2):441–462

Sivakumar B (2004b) Dominant processes concept in hydrology: moving forward. Hydrol Process 18(12):2349–2353

Sivakumar B (2005) Hydrologic modeling and forecasting: role ofthresholds. Environ Model Softw 20(5):515–519

Sivakumar B (2008a) Dominant processes concept, model simplification and classification framework in catchment hydrology. Stoch Env Res Risk Assess 22:737–748

Sivakumar B (2008b) Undermining the science or undermining Nature? Hydrol Process 22 (6):893–897

Sivakumar B (2008c) The more things change, the more they stay the same: the state of hydrologic modeling. Hydrol Process 22:4333–4337

Sivakumar B (2009) Nonlinear dynamics and chaos in hydrologic systems:latest developments and a look forward. Stoch Environ Res Risk Assess 23:1027–1036

Sivakumar B, Berndtsson R (2010) Advances in data-based approaches for hydrologic modeling and forecasting. World Scientific Publishing Company, Singapore

Sivakumar B, Singh VP (2012) Hydrologic system complexity and nonlinear dynamic concepts for a catchment classification framework. Hydrol Earth Syst Sci 16:4119–4131

Sivakumar B, Sorooshian S, Gupta HV, Gao X (2001) A chaotic approach to rainfall disaggregation. Water Resour Res 37(1):61–72

Sivapalan M, Takeuchi K, Franks SW, Gupta VK, Karambiri H, Lakshmi V, Liang X, McDonnell JJ, Mendiondo EM, O'Connell PE, Oki T, Pomeroy JW, Schertzer D, Uhlenbrook S, Zehe E (2003) IAHS decade on predictions in ungauged basins (PUB), 2003–2012: shaping an exciting future for the hydrological sciences. Hydrol Sci J 48(6): 857–880

Snyder WM (1955) Hydrograph analysis by the method of least squares. J Hyd Div, Proc Ame Soc Civ Eng 81(793), 25 pp

Snyder WM (1971) The parametric approach to watershed modeling. Nordic Hydrol 11:167–185

Snyder WM, Stall JB (1965) Men, models, methods, and machines in hydrologic analysis. J Hyd Div, Proc Ame Soc Civ Eng 91(HY2):85–99

Sorooshian S, Gupta VK (1983) Automatic calibration of conceptual rainfall-runoff models: the question of parameter observability and uniqueness. Water Resour Res 19(1):251–259

Sudheer KP, Jain A (2004) Explaining the internal behavior of artificial neural network river flow models. Hydrol Process 18(4):833–844

Thomas HA, Fiering MB (1962) Mathematical synthesis of streamflow sequences for the analysis of river basins by simulation. In: Mass A et al (eds) Design of water resource systems. Harvard University Press, Cambridge, Massachusetts, pp 459–493

Tsonis AA (1992) Chaos: from theory to applications. Plenum Press, New York

UNESCO (1974) Contributions to the development of the concept of the hydrological cycle. Sc. 74/Conf.804/Col. 1, Paris

Veitzer SA, Gupta VK (2001) Statistical self-similarity of width function maxima with implications to floods. Adv Wat Resour 24:955–965

Vapnik V (1995) The nature of statistical learning theory. Springer-Verlag, Germany

Vogel RM (1999) Stochastic and deterministic world views. J Water Resour Plan Manage 125 (6):311–313

Woolhiser DA (1971) Deterministic approach to watershed modeling. Nordic Hydrol 11:146–166

Woolhiser DA (1973) Hydrologic and watershed modeling: state of the art. Trans Ame Soc Agr Eng 16(3):553–559

Woolhiser DA (1975) The watershed approach to understanding our environment. J Environ Qual 4(1):17–21

Yevjevich VM (1963) Fluctuations of wet and dry years. Part 1. Research data assembly and mathematical models. Hydrology Paper 1, Colorado State University, Fort Collins, Colorado, pp 1–55

Yevjevich VM (1968) Misconceptions in hydrology and their consequences. Water Resour Res 4 (2):225–232

Yevjevich VM (1972) Stochastic processes in hydrology. Water Resour Publ, Fort Collins, Colorado
Young PC (1984) Recursive estimation and time-series analysis. Springer-Verlag, Berlin
Young PC, Beven KJ (1994) Database mechanistic modeling and rainfall-flow non-linearity. Environmetrics 5(3):335–363
Young PC, Ratto M (2009) A unified approach to environmental systems modeling. Stoch Environ Res Risk Assess 23:1037–1057
Young PC, Parkinson SD, Lees M (1996) Simplicity out of complexity in environmental systems: Occam's Razor revisited. J Appl Stat 23:165–210
Zadeh LA (1965) Fuzzy sets. Inform. Control 8(3):338–353
Zadeh LA, Klir GJ, Yuan B (1996) Fuzzy sets, fuzzy logic and fuzzy systems. World Scientific Publishers, Singapore

Chapter 2
Characteristics of Hydrologic Systems

Abstract The dynamics of hydrologic systems are governed by the interactions between climate inputs and the landscape. Due to the spatial and temporal variability in climate inputs and the heterogeneity in the landscape, hydrologic systems exhibit a wide range of characteristics. While some characteristics may be specific to certain systems and situations, most hydrologic systems often exhibit a combination of these characteristics. This chapter discusses many of the salient characteristics of hydrologic systems, including complexity, correlation, trend, periodicity, cyclicity, seasonality, intermittency, stationarity, nonstationarity, linearity, nonlinearity, determinism, randomness, scale and scale-invariance, self-organization and self-organized criticality, threshold, emergence, feedback, and sensitivity to initial conditions. The presentation focuses on the occurrence, form, and role of each of these characteristics in hydrologic system dynamics and the methods for their identification. At the end, a particularly interesting property of hydrologic systems, wherein simple nonlinear deterministic systems with sensitive dependence on initial conditions can give rise to complex and 'random-looking' dynamic behavior, and popularly known as 'chaos,' is also highlighted.

2.1 Introduction

Hydrologic systems are often complex heterogeneous systems and function in synchronization with other Earth systems. Hydrologic phenomena arise as a result of interactions between climate inputs and landscape characteristics that occur over a wide range of space and time scales. Considering the spatial scale, our interest in hydrologic studies may be the entire hydrologic cycle or a continental-scale river basin or a medium-size catchment or a small creek in a forest or a 10 m × 10 m plot on a farm or some area even finer. Similarly, considering the temporal scale, we may be interested in studying processes at decadal or annual or monthly or daily or hourly or even finer intervals. The appropriate spatial and temporal scales for hydrologic studies are often dictated by the purpose at hand. For example, for medium- to long-term water resources planning and management for a region,

B. Sivakumar, *Chaos in Hydrology*, DOI 10.1007/978-90-481-2552-4_2

monthly or annual or decadal scale in time and river-basin scale in space are normally more appropriate than other scales; on the other hand, for design of drainage structures in a city, hourly or even finer-resolution temporal scale and a spatial scale in the order of a kilometer or even less are generally more appropriate than others.

Models that can adequately mimic real hydrologic systems are vital for reliable assessment of the overall landscape changes, understanding of the specific processes, and forecasting of the future events. However, development of such models crucially depends on our ability to properly identify the level of complexity of hydrologic systems and understand the associated processes in the first place. This has always been an extremely difficult task, and will probably become even more challenging, especially with the continuing explosion in human population and changes to our landscapes and rivers, not to mention the influence of various external factors, including those related to climate and other Earth systems with which hydrologic systems interact.

This chapter presents information about some of the inherent and salient characteristics of hydrologic systems, which could offer important clues as to the development of appropriate models. The term 'system' herein is defined as a combination of hydrologic process, scale, and purpose, as appropriate, in addition to 'catchment' in its general sense. These features may include, among others: complexity, correlation, trend, seasonality, cyclicity, stationarity, nonstationarity, linearity, nonlinearity, periodicity, quasi-periodicity, non-periodicity, intermittency, determinism, randomness, scaling or fractals, self-organized criticality, thresholds, emergence, and sensitivity to initial conditions and chaos. Depending upon the system under consideration, any or all of these characteristics may come into the picture. The discussion herein on these characteristics is with particular emphasis on hydrologic time series, consistent with the focus of this book. Extensive details on these characteristics with particular reference to hydrologic systems can be found in many books, including those by Yevjevich (1972), Chow et al. (1988), Isaaks and Srivastava (1989), Maidment (1993), Haan (1994), Salas et al. (1995), Rodriguez-Iturbe and Rinaldo (1997), and McCuen (2003).

2.2 Complexity

Although the words 'complex' and 'complexity' are widely used both in scientific theory and in common practice, there is no general consensus on the definition. The difficulty in arriving at a consensus definition comes from the fact that it is oftentimes subjective; what is 'complex' for one person may not be complex at all for another person, or even when viewed by the same person from a different perspective or at a different time. Nevertheless, one workable definition may be: "something (or some situation) with inter-connected or inter-woven parts." Such a definition is often tied to the concept of a 'system' (see Chap. 1, Sect. 1.4). For physical and dynamic systems, such as the ones encountered in hydrology, the term 'complexity' often refers to the 'degree' to which the components engage in organized structured interactions. With this definition, however, it is also important to clarify why the nature of a complex system is inherently related to its parts, since simple systems are also formed out of parts. Therefore, to explain the difference between 'simple' and 'complex' systems, the terms 'interconnected' or 'interwoven' are essential.

Qualitatively, to understand the behavior of a complex system, we must understand not only the behavior of the parts but also how they act together to form the behavior of the whole. This is because: (1) we cannot describe the whole without describing each part; and (2) each part must be described also in relation to other parts. As a result, complex systems are often difficult to understand. This is relevant to another definition of 'complex': 'not easy to understand or analyze.' These qualitative ideas about what a complex system is can be made more quantitative. Articulating them in a clear way is both essential and fruitful in pointing out the way toward progress in understanding the universal properties of these systems.

For a quantitative description, the central issue again is defining quantitatively what 'complexity' means. In the context of systems, it may perhaps be useful to ask: (1) What do we mean when we say that a system is complex? (2) What do we mean when we say that one system is more complex than another? and (3) Is there a way to identify the complexity of one system and to compare it with the complexity of another system? To develop a quantitative understanding of complexity, a variety of tools can be used. These may include: statistical (e.g. coefficient of variation), nonlinear dynamic (e.g. dimension), information theoretic (e.g. entropy) or some other measure.

In this book, the complexity of a system is essentially taken to be a quantitative measure of the variability of time series under consideration. Further, in the specific context of nonlinear dynamic methods (see Part B) and their applications in hydrology (see Part C), the variability is generally represented by reconstruction of a time series and determination of the 'dimensionality' (or related measure) and, thus, the complexity is related to the number of variables dominantly governing the system that produced the time series; in other words, the amount of information necessary to describe the system.

During the past few decades, numerous attempts have been made to define, qualify, and quantify 'complexity' and also to apply complexity-based theories for studying natural and physical systems. Extensive details on these can be found in Ferdinand (1974), Cornacchio (1977), Nicolis and Prigogine (1989), Waldrop (1992), Cilliers (1998), Buchanan (2000), Barabási (2002), McMillan (2004), Johnson (2007), and Érdi (2008), among others.

Due to the tremendous variabilities and heterogeneities in climatic inputs and landscape properties, hydrologic systems are often highly variable and complex at all scales (although simplicity is also possible). Consequently, they are not fully understood; indeed, it is even hard to arrive at an acceptable definition of a 'hydrologic system.' Sivakumar (2008a) suggests that hydrologic systems may be viewed from three different, but related, angles: process, scale, and purpose of interest. Examples of hydrologic processes are rainfall, streamflow, groundwater flow, and evaporation. Scale is generally considered in terms of space (e.g. plot scale, landscape scale, river basin scale) and time (e.g. daily, monthly, annual). Examples of purposes in hydrology are characterization, prediction, and aggregation/disaggregation or upscaling/downscaling.

Depending upon the angle at which they are viewed, hydrologic systems may be either simple or complex; for example, the rainfall occurrence in a desert (even at different spatial and temporal scales) may be treated as an extremely simple system since there may be no rainfall at all, while the runoff system in a large river basin may be highly complex due to the basin complexities and heterogeneities, in addition to rainfall variability. The complexity of the hydrologic systems has important implications for hydrologic modeling, since it is the property that essentially dictates the complexity of the model to be developed (and type and amount of data to be collected and computational power required) to obtain reliable results. Consequently, hydrologic modeling must also be viewed from the above three angles; in other words, the appropriate model to represent a given hydrologic system may also be either simple or complex. The obvious question, however, is: how simple or how complex the models should be? There is a plethora of literature on the question of complexity of hydrologic models; see, for example, Jakeman and Hornberger (1993), Young et al. (1996), Grayson and Blöschl (2000), Perrin et al. (2001), Beven (2002), Young and Parkinson (2002), Sivakumar (2004b, 2008a, b), Wainwright and Mulligan (2004), Sivakumar et al. (2007), Sivakumar and Singh (2012), and Jenerette et al. (2012), among others. This issue of complexity in hydrologic systems is extensively addressed in this book in the discussion of nonlinear dynamic and chaos methods and their applications in hydrology, especially through estimation of variability of time series ('dimensionality') and, thus, determination of the number of variables dominantly governing the underlying system dynamics.

2.3 Correlation and Connection

Generally speaking, correlation refers to relation between two or more things, say variables. However, measuring correlation between totally unrelated variables (at least for well-known situations) has no practical relevance. Therefore, correlation must be essentially viewed in the context of dependence or connection between variables; this dependence may be in only one direction (i.e. a variable either influences another *or* is influenced by another) or in both directions (i.e. a variable influences another *and* is also influenced by another). In other words, correlation must be associated with causation, *at least* in one direction (e.g. streamflow is correlated to rainfall). Therefore, in scientific and engineering studies, the purpose of correlation analysis is to measure how strongly pairs of variables (or entities, more broadly) are related or connected, if at all.

Hydrologic systems are complex systems governed by a large number of influencing variables that are also often interdependent. The nature and extent of interdependencies among hydrologic variables are often different for different systems. Indeed, such interdependencies may even be different for the same system under different conditions or when different components are considered. The hydrologic cycle is a perfect example: each and every component is connected to (i.e. influences and is influenced by) every other component, but the way the components are connected among themselves (e.g. direct or indirect) and the strength of such connections (e.g. strong or weak) vary greatly. For instance, rainfall and streamflow, being components of the hydrologic cycle, are connected, influencing and being influenced by each other. However, the influence of rainfall on streamflow is direct, far more pronounced, and often immediate (notwithstanding evaporation and infiltration), whereas the influence of streamflow on rainfall is not easily noticeable and also very slow (often having to pass through many steps in the functioning of the hydrologic cycle). Unraveling the nature and extent of connections in hydrologic systems, as well as their interactions with other systems, has always been a fundamental challenge in hydrology.

Correlation analysis plays a basic and vital role in the study of hydrologic systems, in almost every context imaginable. For instance, correlation analysis between data collected for a single hydrologic variable at successive times (i.e. temporal correlation) is useful for prediction. Correlation analysis between data collected at different locations for the same hydrologic variable (i.e. spatial correlation) is useful for interpolation and extrapolation. Similarly, correlation analysis between data collected for different hydrologic variables at the same location or at different locations is useful for very many purposes.

Correlation analysis may be performed in different ways and using different methods, depending upon the variable, data, and task at hand. Correlation analysis may involve a single variable, two variables, or multiple variables, and similarly a single location, two locations, or multiple locations. The existing methods for correlation analysis may broadly be classified into linear and nonlinear methods, and include autocorrelation function (Yule 1896), Pearson product moment correlation (Pearson 1895), Spearman's rank correlation (Spearman 1904), Kendall tau rank correlation (Kendall 1938), regression (e.g. Legendre 1805; Galton 1894), mutual information function (Shannon 1948), and distance correlation (Székely et al. 2007), among others. Extensive accounts of these methods are already available in the literature and, therefore, details are not reported herein. However, the autocorrelation function and mutual information function methods will be briefly discussed in Chap. 7, in the specific context of delay time selection for data reconstruction for chaos analysis.

Due to the significance of correlation in establishing possible connections, correlation analysis has been a key component of research on hydrologic systems. Numerous studies have applied a wide range of correlation and regression techniques to explore hydrologic connections in space, time, space-time, and between/among variables and to use such information for various applications; see, for example, Douglas et al. (2000), McMahon et al. (2007), Skøien and Blöschl (2007), and Archfield and Vogel (2010), among others.

2.4 Trend

A trend is a slow, gradual change in a system over a period of observation, and can be upward, downward or other. A trend thus generally represents a deterministic component and is a useful property to identify system changes and to make future predictions. At the same time, a trend does not repeat, or at least does not repeat within the time range of our system observations (which is normally not that long). A trend may be in a linear form or in a nonlinear form or even a combination of both. In hydrologic systems (and in many other real systems), trend is most often in a nonlinear form.

There are no proven 'automatic' methods to identify the trend component in a system. In fact, the presence of a trend cannot be readily identified, since trends and minor system fluctuations are oftentimes indistinguishable. Many methods to identify trends (in time series) have been developed, including the Mann-Kendall test, the linear and nonlinear least-squares regression methods, the moving average method, the exponential smoothing, and the spectral method. Extensive details of these are already available in the literature (e.g. Mann 1945; Box and Jenkins 1970; Kendall 1975; Haan 1994; Chatfield 1996).

In hydrologic systems, trends may be observed in various ways, and a few example situations are as follows. Natural climatic changes may result in gradual changes to the hydrologic environment over a very long period of time. Gradual changes to the landscape over a long or medium timescale, whether natural or man-made, may cause gradual changes to the hydrologic processes. Changes in basin conditions over a period of several years can result in corresponding changes in streamflow characteristics, mainly because of the basin's response to rainfall. Urbanization and deforestation on a large scale may result in gradual changes in precipitation amounts over time.

Since the presence of trend helps in identifying system evolution and prediction, examination of trend forms one of the most basic analyses in hydrology. Numerous studies have applied the above different trend analysis methods (and their variants) to identify trends in hydrologic systems, in many different contexts and for many different purposes. Some studies have also proposed modifications to the standard trend analysis methods, to be appropriate for hydrologic systems. Extensive details of these studies can be found in, for example, Hirsch et al. (1982), Lettenmaier et al. (1994), Lins and Slack (1999), Douglas et al. (2000), Burn and Hag Elnur (2002), Hamed (2008), and Şen (2014).

2.5 Periodicity, Cyclicity, and Seasonality

In addition to correlation and trend, a system may exhibit a host of properties that are generally deterministic in nature and, thus, facilitate prediction of its evolution. The exact nature and extent of determinism associated with each of such properties may vary, especially depending upon the scale of consideration. However, a commonality among almost all of these properties is the presence of some kind of 'repetition' or 'cycle.' These properties include periodicity, cyclicity, and seasonality. The absence of such properties are then called aperiodicity, acyclicity, and non-seasonality. These are defined below.

Periodicity refers to the property of being repeated at certain regular and periodic intervals, but without exact repetition. Aperiodicity refers to the property of lacking any kind of periodicity. Quasi-periodicity is a property that displays irregular periodicity. Cyclicity refers to the property of being repeated in a cyclic manner. When cyclicity is absent, the property is called acyclicity. Seasonality refers to the property of being repeated according to seasons. When seasonality is absent, then the property is called non-seasonality.

Almost all natural, physical, and social systems exhibit, in one way or another, periodic, cyclic, or seasonal properties, depending upon the temporal and spatial scales. The hydroclimate system is a good example for this, including its somewhat repetitive nature at annual, seasonal, and daily scales that allow modeling and prediction of its evolution. Indeed, these properties are intrinsic to the functioning of complex systems. Due to the often nonlinear interactions among the various system components, these properties drive and are driven by several other key properties of complex systems, including self-organization, threshold, emergence, and feedback.

There exist numerous methods and models for identifying periodic, cyclic, and seasonal properties of dynamic systems. Such methods range from very simple ones to highly sophisticated ones, including those based on autocorrelation function, power spectrum, maximum likelihood, runs test, information decomposition, Rayleigh analysis, wavelets, singular value decomposition, and empirical mode decomposition, as well as their variants and hybridizations. These methods have been extensively used to study systems in many different fields.

The properties of periodicity, cyclicity, and seasonality are commonplace in hydrologic systems at many different scales, due to the influence of the climate system that drives hydrologic processes and the nature of the hydrologic cycle in itself. For instance, rainy and flow seasons indicate one form of periodicity, as they occur and span over a certain time every year. Periodicity is reflected in the occurrence of high precipitation and high runoff during some months (e.g. summer) and low precipitation and low runoff during some other months (e.g. winter). Evaporation is another form of periodicity, as it is predominantly a day-time process. Similarly, cyclicity in hydrologic processes also exists within a give year. On the other hand, monthly streamflow series are marked by the presence of seasonality according to the geophysical year.

As these properties play crucial roles in the evolution of a system, their identification is often a fundamental step in the analysis of hydrologic systems. To this end, numerous studies have employed various methods to identify these properties in hydrologic systems. Such an identification is also particularly prominent in the application of stochastic time series methods. There exists a plethora of literature on this, and the interested reader is referred to Yevjevich (1972), Rao and Jeong (1992), Maidment (1993), and Salas et al. (1995), among others.

2.6 Intermittency

Intermittency is the irregular alternation of phases of behavior, i.e. occurrence (say, non-zero) and non-occurrence (say, zero) at irregular intervals. Intermittency is observed in numerous natural, physical, and socio-economic systems.

Due to its basic nature of irregular changes in behavior, intermittency is an extremely challenging property to understand, model, and predict. Conventional time series approaches generally adopt separation of a system (time series) into structural components (trend, periodicity, cyclicity, seasonality) and error (noise) that facilitates the use of standard modeling techniques; see, for example, Box and Jenkins (1970). While such approaches generally work well for non-intermittent systems, they are not suitable for intermittent systems. The fact that there might also be different degrees of intermittency brings additional challenges to modeling intermittent systems. Research over the past few decades have resulted in many different approaches and numerous models to study intermittency. Notable among the models are the point process model, cluster process model, Cox process model, renewal process model, random multiplicative cascades, resampling, normal quantile transform, and many others.

Intermittency is very common in hydrologic systems. For instance, rainfall that is observed in a recording raingage is an intermittent time series. Hourly, daily, and weekly rainfall in many parts of the world are typically intermittent time series. In semi-arid and arid regions, even monthly and annual rainfall and monthly and annual runoff are also often intermittent. In view of this, numerous studies have addressed the issue of intermittency in hydrologic systems. Many studies have also proposed different methods for modeling intermittent time series. Extensive details of such studies are available in Todorovic and Yevjevich (1969), Richardson (1977), Kavvas and Delleur (1981), Waymire and Gupta (1981), Smith and Carr (1983, 1985), Yevjevich (1984), Rodriguez-Iturbe et al. (1987), Delleur et al. (1989), Isham et al. (1990), Copertwait (1991, 1994), Gupta and Waymire (1993), Salas et al. (1995), Gyasi-Agyei and Willgoose (1997), Verhoest et al. (1997), Montanari (2005), Burton et al. (2010), Pui et al. (2012), and Paschalis et al. (2013), among others.

2.7 Stationarity and Nonstationarity

Stationarity is defined as a property of a system where the statistical properties of the system (e.g. mean, variance, autocorrelation) do not change with time. This means that there are no trends. The most important property of a stationary system is that the autocorrelation function depends on lag alone and does not change with the time at which the function is calculated. Weak stationarity refers to a constant mean and variance. True stationarity or strong stationarity means that all higher-order moments (including variance and mean) are constant.

Stationarity is a very common assumption in stochastic methods; see, for example, Cramer (1940), Yevjevich (1972), and Box and Jenkins (1970) for some early details. Such an assumption, however, may not be appropriate for all systems under all conditions, as the statistical properties of most systems, especially complex systems, change over time. In view of this, there has been an increasing attention, in recent years, to address the nonstationarity conditions in stochastic methods. In particular, attempts have been made to extend the classical stochastic approaches to accommodate certain types of nonstationarity. On the other hand, if the type of nonstationarity can be identified and modeled, one can remove such to arrive at a stationary time series to suit the stationarity assumption. However, since nonstationarity can take various forms, such a separation may also be tremendously challenging.

There are two broad approaches for testing stationarity/nonstationarity in a system: parametric and nonparametric. Parametric approaches are based on certain prior assumptions about the nature of the data and are usually used when working in the time domain. Nonparametric approaches do not make any prior assumptions about the nature of the data and are usually used when working in the frequency domain. In recent years, there have also been advances in the time-frequency domain analysis.

There exist many different ways and methods for testing stationarity/nonstationarity in a time series. Among the commonly used methods are the augmented Dickey-Fuller (ADF) unit root test (Dickey and Fuller 1979; Said and Dickey 1984), multi-taper method (Thomson 1982), maximum entropy method (Childers 1978), evolutionary spectral analysis (Priestley 1965), wavelet analysis (Daubechies 1992), and the KPSS test (Kwiatkowski et al. 1992; Shin and Schmidt 1992).

The assumption of stationarity is very common in hydrology, especially with the applications of stochastic methods that have been dominant over several decades now; see, for instance, Thomas and Fiering (1962), Chow (1964), Dawdy and Matalas (1964), Yevjevich (1972), Kottegoda (1980), Hipel and McLeod (1994), Salas et al. (1995), Hubert (2000), Chen and Rao (2002) for details. However, with the recognition of nonstationarity in real time series, studies on nonstationarity in hydrology have also been growing in recent decades (e.g. Potter 1976; Kottegada 1985; Rao and Hu 1986; Hamed and Rao 1998; Young 1999; Cohn and Lins 2005; Coulibaly and Baldwin 2005; Koutsoyiannis 2006; Clarke 2007; Kwon et al. 2007). With recognition of the significant changes in global climate and the anticipated impacts on hydrology and water resources, especially in the form of extreme hydroclimatic events, the need to move away from the traditional stationarity-based approaches and to develop nonstationarity-based approaches for hydrologic modeling, prediction, and design is increasingly realized at the current time (e.g. Milly et al. 2008).

2.8 Linearity and Nonlinearity

In simple terms, linearity represents a situation where changes in inputs will result in proportional changes in outputs. For instance, linearity is a situation in which if a change in any variable at some initial state produces a change in the same or some other variable at some later time, then twice as large a change at the same initial time will produce twice as large a change at the same later time. It follows that if the later values of any variable are plotted against the associated initial values of any variable on graph paper, the points will be on a straight line—hence the name. Nonlinearity, on the other hand, represents a situation where changes in inputs would not produce proportional changes in outputs.

Although the above definitions of linearity and nonlinearity are theoretically accurate, they cannot, and are not, strictly adopted in practice. This is because, according to these definitions, true linearity may exist only in (simple) artificially-created systems; it does not exist in natural systems at all. For instance, any change in the quantity of food purchased will result in a proportional change in the cost (provided that there is no discount for purchasing large quantities!). However, any change in rainfall amount would not result in a proportional change in streamflow (even in catchments that are not complex), since many other factors also influence the conversion of rainfall (input) to streamflow (output).

The fact that a strict definition of linearity does not apply to natural systems does not mean that the assumption of linearity and development of linear models are not relevant and useful for such systems. The importance of the assumption of linearity lies in a combination of two circumstances. First, many tangible phenomena behave approximately linearly over restricted periods of time or restricted areas of space or restricted ranges of variables, so that useful linear mathematical models can simulate their behavior. Second, linear equations can be handled by a wide variety of techniques that do not work with nonlinear equations. These aspects, not to mention the constraints in computational power and measurement technology, mostly led, until at least the mid-twentieth century, to the development and application of linear approaches to natural systems. In recent decades, however, nonlinear approaches have been gaining consideration attention.

Depending on the specific definition of a system (in terms of process, scale, and purpose; see Sect. 2.2), a hydrologic system may be treated as either linear or nonlinear; for example, overland flow in a desert over a few hours of time may be treated as a purely linear system (since a small change in rainfall may not result in any overland flow), while overland flow in a partially-developed or fully-developed catchment over the same period of time is almost always nonlinear (due to the influence of both rainfall and land use properties). In a holistic perspective of “inter-connected or inter-woven parts,” however, all hydrologic systems are inherently nonlinear. The nonlinear behavior of hydrologic systems is evident in various ways and at almost all spatial and temporal scales. For instance, the hydrologic cycle itself is an example of a system exhibiting nonlinear behavior, with almost all of the individual components themselves exhibiting nonlinear behavior at different temporal and spatial scales. Nevertheless, linearity is also present over certain parts, depending upon the variable and time period, among others.

For the reasons mentioned above, much of the past research in hydrologic systems, including those developing and applying time series methods, had essentially resorted to linear approaches (e.g. Thomas and Fiering 1962; Harms and Campbell 1967; Yevjevich 1972; Valencia and Schaake 1973; Klemeš 1978; Beaumont 1979; Kavvas and Delleur 1981; Salas and Smith 1981; Srikanthan and McMahon 1983; Bras and Rodriguez-Iturbe 1985; Salas et al. 1995), although the nonlinear nature of hydrologic systems had already been known for some time (e.g. Minshall 1960; Jacoby 1966; Amorocho 1967, 1973; Dooge 1967; Amorocho and Brandstetter 1971; Bidwell 1971; Singh 1979). In the last few decades, however, a number of nonlinear approaches have been developed and applied for hydrologic systems; see Young and Beven (1994), Kumar and Foufoula-Georgiou (1997), Singh (1997, 1998, 2013), ASCE Task Committee (2000a, b), Govindaraju and Rao (2000), Dibike et al. (2001), Kavvas (2003), Sivakumar (2000, 2004a, 2009), Gupta et al. (2007), Young and Ratto (2009), Şen (2009), Abrahart et al. (2010), and Sivakumar and Berndtsson (2010), among others. More details about the linear approaches and nonlinear approaches, especially in the context of time series methods, are presented in Chaps. 3 and 4, respectively. Among the nonlinear approaches, a particular class is that of nonlinear determinism and chaos, i.e. systems with sensitivity to initial conditions (e.g. Lorenz 1963; Gleick 1987). Section 2.17 presents a brief account of this class, which is also the main focus of this book, as can be seen from the extensive details presented in Part B, Part C, and Part D.

It is appropriate to mention, at this point, that there is still some confusion on the definition of 'nonlinearity' in hydrology, and perhaps in many other fields as well. This is highlighted, for example, by Sivapalan et al. (2002), who discuss two definitions of nonlinearity that appear in the hydrologic literature, especially with respect to catchment response. One is with respect to the dynamic property, such as the rainfall-runoff response of a catchment, where nonlinearity refers to a nonlinear dependence of the storm response on the magnitude of the rainfall inputs (e.g. Minshall 1960; Wang et al. 1981), which is also generally the basis in time series methods. The other definition is with respect to the dependence of a catchment statistical property, such as the mean annual flood, on the area of the catchment (e.g. Goodrich et al. 1997). The ideas presented in this book are mainly concerned with the dynamic property of hydrologic processes.

2.9 Determinism and Randomness

Determinism represents a situation where the evolution from an earlier state to a later state(s) occurs according to a fixed law. Randomness (or stochasticity), on the other hand, represents a situation where the evolution from one state to another is not according to any fixed law but is independent. In natural systems, it is almost impossible to find a completely deterministic or a completely random situation, especially when considered over a long period of time. In such systems, determinism and randomness often co-exist, even if over different times.

In light of determinism and randomness in nature, there have been two corresponding dominant approaches in modeling. In the deterministic modeling approach, deterministic mathematical equations based on well-known scientific laws are used to describe system evolution. In the stochastic approach, probability distributions based on probability concepts are used to assure that certain properties of the system are reproduced. Either approach has its own merits and limitations when applied to natural systems. For instance, the deterministic approach is particularly useful if the purpose is to make accurate prediction of the system evolution, but it also requires accurate knowledge of the governing equations and system details. The stochastic approach, on the other hand, is particularly useful when one is interested in generating possible future scenarios of system properties, but it cannot reproduce important physical processes.

Both the deterministic approach and the stochastic approach have clear merits for studying hydrologic systems. For instance, the deterministic approach has merits considering the 'permanent' nature of the Earth and the 'cyclical' nature of the associated processes. Similarly, the stochastic approach has merits considering that hydrologic systems are often governed by complex interactions among various components in varying degrees and that we have only 'limited ability to observe' the detailed variations. In view of their relevance and usefulness, both these approaches have been extensively employed to study hydrologic systems, but almost always independently (e.g. Darcy 1856; Richards 1931; Sherman 1932; Horton 1933, 1945; Nash 1957; Thomas and Fiering 1962; Yevjevich 1963, 1972; Fiering 1967; Mandelbrot and Wallis 1969; Woolhiser 1971; Srikanthan and McMahon 1983; Bras and Rodriguez-Iturbe 1985; Dooge 1986; Gelhar 1993; Salas and Smith 1981; Salas et al. 1995; Govindaraju 2002).

In spite of, and indeed because of, their differences, the deterministic approach and the stochastic approach can actually be complementary to each other to study hydrologic systems. For instance, in the context of river flow, the deterministic approach is useful to represent the significant deterministic nature present in the form of seasonality and annual cycle, while the stochastic approach is useful to represent the randomness brought by the varying degrees of nonlinear interactions among the various components involved. Therefore, the question of whether the deterministic or the stochastic approach is better for hydrologic systems is often meaningless, and is really a philosophical one. What is more meaningful is to ask whether the two approaches can be coupled to increase their advantages and limit their limitations, for practical applications to specific situations of interest (e.g. Yevjevich 1974; Sivakumar 2004a). This is where ideas of nonlinear deterministic dynamic and chaos theories can be particularly useful to bridge the gap, as they encompass nonlinear interdependence, hidden determinism and order, and sensitivity to initial conditions (e.g. Sivakumar 2004a). Additional details on this are provided in Sect. 2.17 and in Part B.

2.10 Scale, Scaling, and Scale-invariance

The term 'scale' may be defined as a characteristic dimension (or size) in either space or time or both. For instance, 1 km × 1 km (1 km^2) area is a scale in space, a day is a scale in time, and their combination is a scale in space-time. The term 'scaling' is used to represent the link (and transformation) of things between different scales. For instance, the link in a process between 1 km × 1 km (1 km^2) and 10 km × 10 km (100 km^2) is scaling in space, between daily and monthly scales is scaling in time, and their combination is scaling in space-time. The term 'scale-invariance' is used to represent a situation where such links do not change across different scales.

Scale, scaling, and scale-invariance are key concepts and properties in studying natural systems. Their relevance and significance can be explained as follows. Natural phenomena occur at a wide range of spatial and temporal scales. They are generally governed by a large number of components (e.g. variables) that often interact in complex ways. Each of these components and the interactions among themselves may or may not change across different spatial and temporal scales. Therefore, for an adequate understanding of such systems, observations at many different temporal and spatial scales are necessary. Although significant progress has been made in measurement technology and data collection, it is almost impossible to make observations at all the relevant scales, due to technological, financial, and many other constraints.

An enormous amount of effort has been made to study scale-related issues in natural systems, especially in the development and application of methods for downscaling (transferring information from a given scale to a smaller scale) and upscaling (transferring information from a given scale to a larger scale). In this regard, ideas gained from the modern concept of 'fractal' or 'self-similarity' (e.g. Mandelbrot 1977, 1983) combined with the earlier ideas from the concept of 'topology' (e.g. Cantor 1874; Poincaré 1895; Hausdorff 1919) have been extensively used. There exist several methods for identifying scale-invariant behavior and for transformation of data from one scale to another. These methods may largely be grouped under mono-fractal and multi-fractal methods, and include box counting method, power spectrum method, variogram method, empirical probability distribution function method, statistical moment scaling method, and probability distribution multiple scaling method, among others.

The scale-related concepts and issues are highly relevant for hydrologic systems, since hydrologic phenomena arise as a result of interactions between climate inputs and landscape characteristics that occur over a wide range of space and time scales. For instance, unsaturated flows occur in a 1 m soil profile, while floods in major river systems occur over millions of kilometers; similarly, flash floods occur over several minutes only, while flows in aquifers occur over hundreds of years. Hydrologic processes span about eight orders of magnitude in space and time (Klemeš 1983). At least six causes of scale problems with regard to hydrologic responses have been identified (Bugmann 1997; Harvey 1997): (1) spatial heterogeneity in surface processes; (2) nonlinearity in response; (3) processes require threshold scales to occur; (4) dominant processes change with scale; (5) evolution of properties; and (6) disturbance regimes.

In the study of hydrologic systems, three dominant types of scales are relevant: (1) Process scales—Process scales are defined as the scales at which hydrologic processes occur. These scales are not fixed, but vary with process; (2) Observational scales—Observational scales are the scales at which we choose to collect samples of observations and to study the phenomenon concerned. They are determined by logistics (e.g. access to places of observation), technology (e.g. cost of state-of-the-art instrumentation), and individuals' perception (i.e. what is perceived to be important for a study at a given point in time); and (3) Operational scales—Operational scales are the working scales at which management actions and operations focus. These are the scales at which information is available. These three scales seldom coincide with each other: we are not able to make observations at the scales hydrologic processes actually occur, and the operational scales are determined by administrative rather than by purely scientific considerations.

Yet another type is the 'modeling scales,' which are also 'working scales.' They are generally agreed upon within the scientific community and are partly related to processes and partly to the applications of hydrologic models. Typical modeling scales in space are: the local scale—1 m; the hillslope (reach) scale—100 m; the catchment scale—10 km; and the regional scale—1000 km. Typical modeling scales in time are: the event scale—1 day; the seasonal scale—1 year; and the long-term scale—100 years; see, for example, Dooge (1982, 1986). With increasing need to understand hydrologic processes at very large and very small scales, and with the availability of observations at these scales, there are also changes to our modeling scales. However, oftentimes, the modeling scale is much larger or much smaller than the observation scale. Therefore, 'scaling' is needed to bridge this gap.

During the past few decades, an extensive amount of research has been devoted for studying the scale issues in hydrologic systems and for transferring hydrologic information from one scale to another (e.g. Mandelbrot and Wallis 1968, 1969; Klemeš 1983; Gupta and Waymire 1983, 1990; Gupta et al. 1986; Stedinger and Vogel 1984; Salas et al. 1995; Kalma and Sivapalan 1996; Puente and Obregon 1996; Tessier et al. 1996; Dooge and Bruen 1997; Rodriguez-Iturbe and Rinaldo 1997; Tarboton et al. 1998; Sivakumar et al. 2001; Sposito 2008). In addition to downscaling and upscaling for transfer of hydrologic information from one scale to another, regionalization is also used to transfer information from one catchment (location) to another, including in the context of ungaged basins (e.g. Merz and Blöschl 2004; Oudin et al. 2008; He et al. 2011); see also Razavi and Coulibaly (2013) for a review. Regionalization may be satisfactory if the catchments are similar (in some sense), but error-prone if they are not (Pilgrim 1983). To this end, there has also been great interest in recent years to identify similar catchments, within the specific context of catchment classification (e.g. Olden and Poff 2003; Snelder et al. 2005; Isik and Singh 2008; Moliere et al. 2009; Kennard et al. 2010; Ali et al. 2012; Sivakumar and Singh 2012), although catchment classification had been attempted in many past studies (e.g. Budyko 1974; Gottschalk et al. 1979;

Haines et al. 1988; Nathan and McMahon 1990); see also Sivakumar et al. (2015) for a review. With increasing interest in studying the impacts of global climate change on water resources, downscaling of coarse-scale global climate model outputs to fine-scale hydrologic variables has also been gaining considerable attention; see Wilby and Wigley (1997), Prudhomme et al. (2002), Wood et al. (2004), and Fowler et al. (2007) for details. Despite our progress in addressing the scale issues in hydrologic systems, many challenges still remain. One of the factors that make scaling so difficult is the heterogeneity of catchments and the variability of hydrologic processes, not to mention the uncertainties in climate inputs and hydrologic data measurements.

2.11 Self-organization and Self-organized Criticality

Self-organization has various and often conflicting definitions. In its most general sense, however, self-organization refers to the formation of patterns arising out of the internal dynamics of a system (i.e. local interactions between the system components), independently of external controls or inputs. Because such an organizing process may offset or intensify the effects of external forcings and boundary conditions, self-organization is often a source of nonlinearity in a system.

The original principle of self-organization was formulated by Ashby (1947), which states that any deterministic dynamic system will automatically evolve towards a state of equilibrium that can be described in terms of an attractor in a basin of surrounding states; see also Ashby (1962). According to von Foerster (1960), self-organization is facilitated by random perturbations (i.e. noise) that let the system explore a variety of states in its state space, which, in turn, increases the chance that the system would arrive into the basin of a 'strong' and 'deep' attractor, from which it would then quickly enter the attractor itself. This, in other words, is 'order from noise.' A similar principle was formulated by Nicolis and Prigogine (1977) and Prigogine and Stenders (1984). The concept of self-organization was further advanced by Bak et al. (1987, 1988), through the introduction of 'self-organized criticality' (SOC) to explain the behavior of a cellular automaton (CA) model. Self-organized criticality is a property of dynamic systems that have a critical, or emergent, point as an attractor, through natural processes. Self-organized criticality is linked to fractal structure, $1/f$ noise, power law, and other signatures of dynamic systems; see also Bak (1996), Tang and Bak (1988a, b), and Vespignani and Zapperi (1998) for some additional details.

Hydrologic systems generally exhibit complex nonlinear behaviors and are also often fractal. These properties combine to make hydrologic systems organize themselves in many different ways. For instance, drainage patterns in a landscape and its properties, such as slope, topography, channel properties, soil texture, and vegetation, are the result of very long-term and nonlinear interactions between geology, soils, climate, and the biosphere and, therefore, often organize themselves. The hydroclimatic system, governed by the numerous individual components of land, ocean, atmosphere and their complex nonlinear interactions, organizes itself as well. Similar observations can be made regarding the ecologic-hydrologic interactions. Therefore, self-organization and self-organized criticality are highly relevant for hydrologic systems and their interactions with other Earth systems.

In light of these, numerous studies have investigated the existence of self-organization and self-organized critical behavior in hydrologic systems. While a significant majority of these studies have been on river networks, many other systems have also been studied, including land-atmosphere interactions, soil moisture, and rainfall; see Rinaldo et al. (1993), Rigon et al. (1994), Rodriguez-Iturbe et al. (1994), Stolum (1996), Rodriguez-Iturbe and Rinaldo (1997), Andrade et al. (1998), Rodriguez-Iturbe et al. (1998, 2006), Phillips (1999), Sapozhnikov and Foufoula-Georgiou (1996, 1997, 1999), Talling (2000), Baas (2002), Garcia-Marin et al. (2008), Caylor et al. (2009), Jenerette et al. (2012), and Bras (2015), among others.

Another concept that is also highly relevant in the context of self-organization and SOC in hydrologic systems, especially in the context of river networks, is 'optimal channel networks' (OCNs) (Rodriguez-Iturbe et al. 1992a). Drainage basins organize themselves to convey water and sediment from upstream to downstream in the most efficient way possible. Optimum channel networks are based on three principles: (1) minimum energy expenditure in any link of the network; (2) equal energy expenditure per unit area of channel anywhere in the network; and (3) minimum energy expenditure in the network as a whole. Since the study by Rodriguez-Iturbe et al. (1992a), there has been significant interest in the study of optimal channel networks in hydrology (e.g. Rinaldo et al. 1992, 2014; Rodriguez-Iturbe et al. 1992b; Rigon et al. 1993, 1998; Maritan et al. 1996; Colaiori et al. 1997; Molnar and Ramirez 1998; Banavar et al. 2001; Briggs and Krishnamoorthy 2013).

2.12 Threshold

In simple terms, a threshold is the point at which a system's behavior changes. More accurately, however, it refers to the point where the system *abruptly* changes its behavior from one state to another even when the influencing factors change only *progressively*. Thresholds in a system may be either extrinsic or intrinsic. Extrinsic thresholds are associated with, and responses to, an external influence, i.e. a progressive change in an external factor triggers abrupt changes or failure within the system. In this case, the threshold exists within the system, but it will not be crossed and change will not occur without the influence of an external factor. Intrinsic thresholds, on the other hand, are associated with the inherent structure or dynamics of the system, without any external influences whatsoever.

Threshold behavior in a system can be deemed as an extreme form of nonlinear dynamics, such as, for example, when the system dynamics are highly intermittent. Consequently, threshold behavior drastically reduces our ability to make predictions at different levels, including at the level of: (1) an individual process; (2) the response of larger units that involve interactions of many processes; and (3) the long-term functioning of the whole systems. The fact that threshold behavior can often be different at different levels of a system, its identification at each and every level of the system of interest is often a tremendously challenging problem.

There are many different ways and methods to identify thresholds. These include: (1) histogram-based methods (e.g. convex hull, peak-and-valley, shape-modeling); (2) clustering-based methods (iterative, clustering, minimum error, fuzzy clustering); (3) entropy-based methods (entropic, cross-entropic, fuzzy entropic); (4) attribute similarity-based methods (e.g. moment preserving, edge field matching, fuzzy similarity, maxium information); (5) spatial methods (co-occurrence, higher-order entropy, 2-D fuzzy partitioning); and (6) locally adapative methods (e.g. local variance, local contrast, kriging), among others. However, one of the most effective ways to study threshold behavior is through catastrophe theory (Thom 1972; Zeeman 1976). Catastrophe theory describes the discontinuities (sudden changes) in dependent variables of a dynamic system as a function of continuous changes (progressive changes) in independent variables.

Hydrologic systems exhibit abrupt changes in behavior in many ways, even when the influencing factors change only progressively. Indeed, threshold behavior in the form of intermittency is one of the most important characteristics of hydrologic systems, especially at finer scales. Therefore, the concept of thresholds is highly relevant for hydrologic systems. Thresholds in hydrologic systems may be either intrinsic or extrinsic. They are observed in many different ways and at many different levels. For instance, surface and subsurface runoff generation processes at the local, hillslope, and catchment scales exhibit threshold behavior. Similarly, particle detachment and soil erosion (influenced by rainfall intensity, shear stress due to overland flow, and soil stability) is a threshold process. Threshold behavior is also discussed in the context of the long-term development of soil structures and landforms, fluvial morphology, rill and gully erosion, and formation and growth of channel networks. Soil moisture and land-atmosphere interaction processes also exhibit threshold behavior.

The significance of threshold behavior in hydrologic systems has led to a large number of studies on its identification, nature, causes, and effects in such systems and other systems with which they interact, in the specific context of thresholds (e.g. Dunne et al. 1991; Grayson et al. 1997; Hicks et al. 2000; Toms and Lesperance 2003; Blöschl and Zehe 2005; Sivakumar 2005; Phillips 2006, 2014; Pitman and Stouffer 2006; Tromp-Van-Meerveld and McDonnell 2006a, b; Emanuel et al. 2007; Lehmann et al. 2007; McGrath et al. 2007; O'Kane and Flynn 2007; Zehe et al. 2007; Andersen et al. 2009; Zehe and Sivapalan 2009), while numerous other studies have addressed the role of thresholds in many other contexts as well. Towards the identification of thresholds, the use of the 'range of variability approach' (RVA), which considers the ecologic flow regime characteristics (i.e. magnitude, frequency, duration, timing, and rate of change of flow) (Richter et al. 1996, 1997, 1998), has been gaining considerable attention in recent years (e.g. Shiau and Wu 2008; Kim et al. 2011; Yin et al. 2011; Yang et al. 2014). The usefulness of catastrophe theory to study hydrologic systems has also been investigated (Ghorbani et al. 2010).

2.13 Emergence

In simple terms, emergence is a property of a system in which larger entities, patterns, and regularities occur through interactions among smaller ones that themselves do not exhibit such properties. Emergence is a key property of complex systems. It is impossible to understand complex systems without recognizing that simple and separate entities (e.g. atoms) in large numbers give rise to complex collective behaviors that have patterns and regularity, i.e. emergence. How and when this occurs is the simplest and yet the most profound problem in studying complex systems. There are two types of emergence: (1) local emergence, where collective behavior appears in a small part of the system; and (2) global emergence, where collective behavior pertains to the system as a whole. The significance of the concept of emergence in studying natural, physical, and social systems can be clearly recognized when considered against our traditional view of reductionism, where the part defines the whole.

Although the concept of emergence has a long history, recent developments in complex systems science, especially in the areas of nonlinear dynamics, chaos, and complex adaptive systems, have provided a renewed impetus to its studies; see, for example, Waldrop (1992), Crutchfield (1994), O'Connor (1994), Holland (1998), Kim (1999), and Goldstein (2002) for some details. Such developments have led to a number of approaches for studying emergence, especially in the context of agent-based modeling, including genetic algorithms and artificial life simulations; see Holland (1975), Goldberg (1989), and Fogel (1995), among others. Dimensionality-reduction methods, such as self-organizing maps, local linear embedding and its variants, and isomap, are among the useful tools for an initial, exploratory investigation of the dynamics, or in the subsequent visual representation and description of the dynamics.

Emergent properties are inherent in hydrologic systems, as such systems are governed by various Earth-system components and their interactions in nonlinear ways at different spatial and temporal scales. For instance, vertical vadose zone processes or macropore influences are dominant at small plot scales, whereas topography begins to dominate runoff processes at the hillslope scale, and the stream network may begin to dominate catchment organization, spatial soil moisture variations, and patterns of runoff generation at the catchment scale (e.g. Blöschl and Sivapalan 1995). A similar situation occurs through timescale changes; for instance, from the diurnal scale to the event scale to the annual scale to the decadal scale. Consequently, many studies have attempted to address the emergent properties in hydrologic systems and associated ones in different ways (e.g. Levin 1992; Lansing and Kremer 1993; Young 1998, 2003; Eder et al. 2003; Lehmann et al. 2007; Rodriguez-Iturbe et al. 2009; Phillips 2011, 2014; Yeakel et al. 2014; Moore et al. 2015).

Emergent properties are generally associated with many other properties of complex systems, such as scale, nonlinear interactions, self-organization, threshold, and feedback. As all these properties play vital roles in hydrologic system dynamics, study of emergent properties is key to advance our understanding, modeling, and prediction of such systems.

2.14 Feedback

Feedback is a mechanism by which a change in something (e.g. a variable) results in either an amplification or a dampening of that change. When the change results in an amplification, it is called 'positive feedback;' when the change results in dampening, it is called 'negative feedback.' Complex systems are influenced by countless interacting processes at many scales and levels of system organization. These interactions mean that changes rarely occur in linear and incremental ways but happen in a nonlinear way, often driven by feedbacks. Due to their nature, and especially in amplification, feedbacks are also described as a threshold concept for understanding complex systems.

A positive feedback, due to amplification of changes, generally leads to destabilization of the system and moves it into another state, such as a regime shift. For example, when the atmospheric temperature rises, evaporation increases. This causes an increase in atmospheric water vapor concentration, resulting in an additional rise in atmospheric temperature through the greenhouse effect, which causes more evaporation, and the process continues. If there were no other factors contributing to atmospheric temperature, then this rise in temperature would spiral out of control. Therefore, a positive feedback is generally not good for a system.

A negative feedback, due to suppression of changes, generally leads to a stabilizing effect on a system. For example, if increased water in the atmosphere leads to greater cloud cover, there will be an increase in the percentage of sunlight reflected away from the Earth (albedo). This leads to a fall in the atmospheric temperature and a decrease in the rate of evaporation (Schmidt et al. 2010). A negative feedback is, therefore, generally good for a system.

The significance of feedback mechanisms in hydrologic systems has been known for a long time (e.g. Dooge 1968, 1973). Indeed, the hydrologic cycle itself serves as a perfect example of the feedback mechanisms, since every component in this cycle is connected to every other component, which leads to feedback processes, both positive and negative, at different times and at different scales. With the increasing recognition of anthropogenic influences on hydrologic systems and our increasing interest in understanding the interactions between hydrologic systems and the associated Earth and socio-economic systems, which bring their own and additional feedback mechanisms, the significance of feedbacks in hydrology has been increasingly realized in recent times. Consequently, there have been numerous attempts to study the causes, nature, and impacts of feedbacks in hydrologic systems and in their interactions with others (e.g. Dooge 1986; Brubaker and Entekhabi 1996; Hu and Islam 1997; Dooge et al. 1999; Yang et al. 2001; Hall 2004; Steffen et al. 2004; Dirmeyer 2006; Maxwell and Kollet 2008; Francis et al. 2009; Kastens et al. 2009; Roe 2009; Ferguson and Maxwell 2010, 2011; Brimelow et al. 2011; Runyan et al. 2012; Van Walsum and Supit 2012; D'Odorico et al. 2013; Butts et al. 2014; Blair and Buytaert 2015; Di Baldassarre et al. 2015).

Although the term 'feedback' has not been specifically used in a large number of these studies, the study of feedback mechanisms and the reported outcomes clearly indicate the advances made in understanding feedbacks in hydrology.

2.15 Sensitivity to Initial Conditions

The concept of sensitive dependence on initial conditions has been popularized in the so-called "butterfly effect;" i.e. a butterfly flapping its wings in one location (say, New York) could change the weather in a far off location (say, Tokyo). The underlying message in this is that even as inconsequent as the simple flap of a butterfly's wings could be enough to change the initial conditions of the Earth's atmosphere and, consequently, could have profound effects on global weather patterns. Edward Lorenz discovered this effect when he observed that runs of his weather model with initial condition data that was rounded in a seemingly inconsequential manner would fail to reproduce the results of runs with the unrounded initial condition data (Lorenz 1963); see also Gleick (1987) for additional details. The main reason for this effect is the presence of a strong level of interdependence among the components of the underlying (climate) system and deterministic nonlinearity in each and every component and in their interactions, as well as the possibility for signal amplification via feedback. Sensitive dependence on initial conditions of a system may place serious limits on the predictability of its dynamic evolution.

The property of sensitive dependence on initial conditions is highly relevant for complex systems, since such systems often exhibit a strong level of interdependence, nonlinearity, and feedback mechanisms among the components. The existence of such a property has, consequently, far reaching implications for the modeling, understanding, prediction, and control of complex systems. This led to the investigation of this property in complex dynamic systems, especially in fluid turbulence (Ruelle 1978; Farmer 1985; Lai et al. 1994; Faisst and Eckhardt 2004). There exist many ways to quantify the sensitive dependence of initial conditions. One of the most popular methods is the Lyapunov exponent method (e.g. Wolf et al. 1985; Eckhardt and Yao 1993). Lyapunov exponents are the average exponential rates of divergence (expansion) or convergence (contraction) of nearby orbits in the phase space; see Chap. 6 for details.

Since hydrologic systems are made up of highly interconnected components that also exhibit nonlinearity, the property of sensitive dependence on initial conditions are certainly relevant to such systems and their modeling and predictions. Such a property may be observed in different ways in different hydrologic systems at different scales. For instance, overland flow is highly sensitive not only to small changes in rainfall but also small changes in catchment properties. Similarly, contaminant transport phenomena in surface and sub-surface waters largely depend upon the time (e.g. rainy or dry season) at which the contaminants were released at the source. In light of the significance of the property of sensitive dependence, a number of studies have investigated such a property in hydrologic systems and associated ones (e.g. Stephenson and Freeze 1974; Rabier et al. 1996; Zehe and Blöschl 2004; Zehe et al. 2007; DeChant and Moradkhani 2011; Fundel and Zappa 2011). Some studies have also addressed this property in the specific context of nonlinear dynamic and chaotic properties, especially using the Lyapunov exponent method (e.g. Rodriguez-Iturbe et al. 1989; Jayawardena and Lai 1994; Puente and Obregon 1996; Shang et al. 2009; Dhanya and Nagesh Kumar 2011).

2.16 The Class of Nonlinear Determinism and Chaos

Although nonlinearity represents a situation where changes in inputs would not produce proportional changes in outputs (see Sect. 2.8), it does not necessarily mean that there is complete absence of determinism/predictability. Indeed, nonlinearity may contain inherent determinism on one hand, but may also be sensitively dependent on initial conditions on the other. While the former allows accurate predictions in the short term, the latter eliminates the possibility of accurate predictions in the long term. This class of nonlinearity is popularly termed as 'deterministic chaos' or simply 'chaos' (e.g. Lorenz 1963). This class of nonlinearity is also particularly interesting because it, despite the inherent determinism, is essentially 'random-looking.' For instance, time series generated from such nonlinear deterministic systems are visually indistinguishable from those generated from purely random systems, and some basic (linear) tools that are widely used for identification of system behavior (e.g. autocorrelation function, power spectrum) often cannot distinguish the two time series either; see, for example, Lorenz (1963), Henon (1976), May (1976), Rössler (1976), Tsonis (1992), and Kantz and Schreiber (2004) for some details.

The intriguing nature of chaos and its possible existence in various natural, physical, and socio-economic systems led to the development of many different methods for its identification, since the 1980s. These include correlation dimension method (e.g. Grassberger and Procaccia 1983a), Kolmogorov entropy method (Grassberger and Procaccia 1983b), Lyapunov exponent method (Wolf et al. 1985), nonlinear prediction method (e.g. Farmer and Sidorowich 1987; Casdagli 1989), false nearest neighbor method (e.g. Kennel et al. 1992), and close returns plot (e.g. Gilmore 1993), among others. These methods have been extensively applied in numerous fields; see, for example, Tsonis (1992), Strogatz (1994), Kaplan and Glass (1995), Kiel and Elliott (1996), and Kantz and Schreiber (2004), among others.

The fundamental properties inherent in the definition of 'chaos:' (1) nonlinear interdependence; (2) hidden determinism and order; and (3) sensitivity to initial conditions, are highly relevant for hydrologic systems, as is also clear from the observations made in the preceding sections. For example: (1) components and mechanisms involved in the hydrologic cycle act in a nonlinear manner and are also interdependent; (2) daily cycle in temperature and annual cycle in river flow possess determinism and order; and (3) contaminant transport phenomena in surface and sub-surface waters largely depend upon the time (i.e. rainy or dry season) at which the contaminants were released at the source, which themselves may not be known (Sivakumar 2004a). The first property represents the 'general' nature of hydrologic systems, whereas the second and third represent their 'deterministic' and 'stochastic' natures, respectively. Further, despite their complexity and random-looking behavior, hydrologic systems may also be governed by a very few degrees of freedom (e.g. runoff in a well-developed urban catchment depends essentially on rainfall), another fundamental idea of chaos theory.

In view of these, numerous studies have applied the concepts and methods of chaos theory in hydrology. Such studies have analyzed various hydrologic time series (e.g. rainfall, streamflow, lake volume, sediment), addressed different hydrologic problems (e.g. system identification, prediction, scaling and disaggregation, catchment classification), and examined a host of data-related issues in chaos studies in hydrology (e.g. data size, data noise, presence of zeros); see Sivakumar (2000, 2004a, 2009) for comprehensive reviews. Chaos theory and its applications in hydrology are the focus of this book.

2.17 Summary

Hydrologic phenomena arise as a result of interactions between climate inputs and the landscape. The significant spatial and temporal variability in climate inputs and the complex heterogeneous nature of the landscape often give rise to a wide range of characteristics in the resulting phenomena. This chapter has discussed many of these characteristics. Some of these characteristics are easy to identify, but some others are far more difficult. Over the past century, numerous methods have been developed to identify, model, and predict these characteristics, especially based on data (i.e. time series). Early methods were mostly based on the assumption of linearity. In recent decades, however, advances in computational power and data measurements have allowed development of a number of nonlinear methods. Both types of methods have been found to be very useful and are now extensively applied in hydrology. The next two chapters discuss many of these methods, with Chap. 3 focusing on the linear methods (and also methods that make no prior assumption regarding linearity/nonlinearity) and Chap. 4 presenting the nonlinear methods.

References

Abrahart RJ, See LM, Dawson CW, Shamseldin AY, Wilby RL (2010) Nearly two decades of neural network hydrologic modeling. In: Sivakumar B, Berndtsson R (eds) Advances in data-based approaches for hydrologic modeling and forecasting. World Scientific Publishing Company, Singapore, pp 267–346

Ali G, Tetzlaff D, Soulsby C, McDonnell JJ, Capell R (2012) A comparison of similarity indices for catchment classification using a cross-regional dataset. Adv Water Resour 40:11–22
Amorocho J (1967) The nonlinear prediction problems in the study of the runoff cycle. Water Resour Res 3(3):861–880
Amorocho J (1973) Nonlinear hydrologic analysis. Adv Hydrosci 9:203–251
Amorocho J, Brandstetter A (1971) Determination of nonlinear functional response functions in rainfall-runoff processes. Water Resour Res 7(5):1087–1101
Andersen T, Carstensen J, Hernández-García E, Duarte CM (2009) Ecological thresholds and regime shifts: approaches to identification. Trends Ecol Evol 24(1):49–57
Andrade RFS, Schellnhuber HJ, Claussen M (1998) Analysis of rainfall records: possible relation to self-organized criticality. Phys A 254:557–568
Archfield SA, Vogel RM (2010) Map correlation method: selection of a reference streamgage to estimate daily streamflow at ungaged catchments. Water Resour Res 46:W10513. doi:10.1029/2009WR008481
ASCE Task Committee (2000a) Artificial neural networks in hydrology. I: preliminary concepts. ASCE J Hydrol Eng 5(2):115–123
ASCE Task Committee (2000b) Artificial neural networks in hydrology. II: hydrologic applications. ASCE J Hydrol Eng 5(2):124–137
Ashby WR (1947) Principles of the self-organizing dynamic system. J General Psychology 37 (2):125–128
Ashby WR (1962) Principles of the self-organizing system. In: von Foerster H, Zopf GW (eds) Principles of self-organization. Office of Naval Research, U.S, pp 255–278
Baas ACW (2002) Chaos, fractals, and self-organization in coastal geomorphology: simulating dune landscapes in vegetated environments. Geomorph 48:309–328
Bak P (1996) How nature works: the science of self-organized criticality. Springer-Verlag, New York 212 pp
Bak P, Tang C, Wiesenfeld K (1987) Self-organized criticality: an explanation of the 1/f noise. Phys Rev Lett 59(4):381–384
Bak P, Tang C, Wiesenfeld K (1988) Self-organized criticality. Phys Rev A 38(1):364–374
Banavar JR, Colaiori F, Flammini A, Maritan A, Rinaldo A (2001) Scaling, optimality, and landscape evolution. J Stat Phys 104:1–48
Barabási A-L (2002) Linked: the new science of networks. Pegasus, Cambridge, MA, USA
Beaumont C (1979) Stochastic models in hydrology. Prog Phys Geogr 3:363–391
Beven KJ (2002) Uncertainty and the detection of structural change in models of environmental systems. In: Beck MB (ed) Environmental foresight and models: a manifesto. Elsevier, The Netherland, pp 227–250
Bidwell VJ (1971) Regression analysis of nonlinear catchment systems. Water Resour Res 7:1118–1126
Blair P, Buytaert W (2015) Modelling socio-hydrological systems: a review of concepts, approaches and applications. Hydrol Earth Syst Sci Discuss 12:8761–8851
Blöschl G, Sivapalan M (1995) Scale issues in hydrological modeling—a review. Hydrol Process 9:251–290
Blöschl G, Zehe E (2005) On hydrological predictability. Hydrol Process 19(19):3923–3929
Box GEP, Jenkins G (1970) Time series analysis, forecasting and control. Holden-Day, San Francisco
Bras RL (2015) Complexity and organization in hydrology: a personal view. Water Resour Res 51 (8):6532–6548
Bras RL, Rodriguez-Iturbe I (1985) Random functions and hydrology. Addison-Wesley, Reading, Massachusetts
Brimelow JC, Hanesiak JM, Burrows WR (2011) Impacts of land–atmosphere feedbacks on deep, moist convection on the Canadian Prairies. Earth Interactions 15(31):1–29
Briggs LA, Krishnamoorthy M (2013) Exploring network scaling through variations on optimal channel networks. PNAS 110(48):19295–19300

Brubaker KL, Entekhabi D (1996) Analysis of feedback mechanisms in land-atmosphere interaction. Water Resour Res 32:1343–1357
Buchanan M (2000) Ubiquity: the science of history … or why the world is simpler than we think. Weidenfeld & Nicolson, New York, USA
Budyko MI (1974) Climate and Life. Academic Press, New York
Bugmann H (1997) Scaling issues in forest succession modelling. In: Hassol H, Katzenberger J (eds) Elements of change 1997—session one: scaling from site-specific observations to global model grids. Aspen Global Change Institute, Aspen, Colorado, USA, pp 47–57
Burn DH, Hag Elnur MA (2002) Detection of hydrologic trends and variability. J Hydrol 255:107–122
Burton A, Fowler HJ, Kilsby CG, O'Connell PE (2010) A stochastic model for the spatial-temporal simulation of nonhomogeneous rainfall occurrence and amounts. Water Resour Res 46:W11501. doi:10.1029/2009WR008884
Butts M, Morten Drews M, Larsen AD, Lerer S, Rasmussen SH, Grooss J, Overgaard J, Refsgaard JC, Christensen OB, Christensen JH (2014) Embedding complex hydrology in the regional climate system—Dynamic coupling across different modelling domains. Adv Water Resour 74:166–184
Cantor G (1874) Über eine eigenschaft des inbegriffes aller reellen algebraischen Zahlen. J Reine Angew Math 77:258–262
Casdagli M (1989) Nonlinear prediction of chaotic time series. Physica D 35:335–356
Caylor KK, Scanlon TM, Rodriguez-Iturbe I (2009) Ecohydrological optimization of pattern and processes in water-limited ecosystems: a trade-off-based hypothesis. Water Resour Res 45. doi:10.1029/2008wr007230
Chatfield C (1996) The analysis of time series. Chapman & Hall, New York, USA
Chen H-L, Rao AR (2002) Testing hydrologic time series for stationarity. J Hydrol Eng 7(2):129–136
Childers DG (1978) Modem Spectrum Analysis. IEEE Press
Chow VT (ed) (1964) Handbook of applied hydrology. McGraw-Hill, New York
Chow VT, Maidment DR, Mays LW (1988) Applied hydrology. McGraw-Hill, Singapore
Cilliers P (1998) Complexity and postmodernism: understanding complex systems. Routledge, London, UK
Clarke RT (2007) Hydrological prediction in a non-stationary world. Hydrol Earth Syst Sci 11 (1):408–414
Cohn TA, Lins HF (2005) Nature's style: naturally trendy. Geophys Res Lett 32(23):L23402
Colaiori F, Flammini A, Maritan A, Banavar JR (1997) Analytical and numerical study of optimal channel networks. Phys Rev E: Stat, Nonlin, Soft Matter Phys 55:1298–1302
Cowpertwait PSP (1991) Further developments of the Neyman-Scott clustered point process for modeling rainfall. Water Resour Res 27:1431–1438
Cowpertwait PSP (1994) A generalized point process model for rainfall. Proc R Soc London Ser A 447:23–37
Cornacchio JV (1977) Maximum entropy complexity measures. Int J General Syst 3:217–225
Coulibaly P, Baldwin CK (2005) Nonstationary hydrological time series forecasting using nonlinear dynamic methods. J Hydrol 307(1–4):164–174
Cramer H (1940) On the theory of stationary random processes. Ann Math 41:215–230
Crutchfield J (1994) The calculi of emergence: computation, dynamics, and induction. Physica D 75:11–54
Darcy H (1856) Les fontaines publiques de la ville de Dijon. V. Dalmont, Paris
Daubechies I (1992) Ten lectures on wavelets. SIAM Publications, CSBM-NSF Series Appli Math 357 pp
Dawdy DR, Matalas NC (1964) Analysis of variance covariance and time series. In: Chow VT (ed) Handbook of applied hydrology. Section 8-IIII. McGraw-Hill, New York, pp 8.68–8.90
DeChant CM, Moradkhani H (2011) Improving the characterization of initial condition for ensemble streamflow prediction using data assimilation. Hydrol Earth Syst Sci 15(11):3399–3410

Delleur JW, Chang TJ, Kavvas ML (1989) Simulation models of sequences of dry and wet days. J Irri Drain Eng 115(3):344–357
Dhanya CT, Nagesh Kumar D (2011) Predictive uncertainty of chaotic daily streamflow using ensemble wavelet networks approach. Water Resour Res 47:W06507. doi:10.1029/2010WR010173
Dibike YB, Velickov S, Slomatine D, Abbott MB (2001) Model induction with support vector machines: introduction and applications. J Comp Civil Eng 15(3):208–216
Di Baldassarre G, Viglione A, Carr G, Kuil L, Yan K, Brandimarte L, Blöschl G (2015) Debates—perspectives on socio-hydrology: capturing feedbacks between physical and social processes. Water Resour Res 51:4770–4781
Dirmeyer PA (2006) The hydrologic feedback pathway for land-climate coupling. J Hydrometeorol 7:857–867
D'Odorico P, Bhattachan A, Davis KF, Ravi S, Runyan CW (2013) Global desertification: drivers and feedbacks. Adv Water Resour 51:326–344
Dickey DA, Fuller WA (1979) Distribution of the estimators for autoregressive time series with a unit root. J Am Stat Assoc 74:423–431
Dooge JCI (1967) A new approach to nonlinear problems in surface water hydrology: hydrologic systems with uniform nonlinearity. Int Assoc Sci HydrolPubl 76:409–413
Dooge JCI (1968) The hydrologic cycle as a closed system. Int Assoc Sci Hydrol Bull 13(1):58–68
Dooge JCI (1973) Linear theory of hydrologic systems. Tech Bull 1468, U.S. department of agriculture. Washington, DC
Dooge JCI (1982) Parameterization of hydrologic processes. In: Eagleson PS (ed) Land surface processes in atmospheric general circulation models. Cambridge University Press, Cambridge, MA, pp 243–288
Dooge JCI (1986) Looking for hydrologic laws. Water Resour Res 22(9):46S–58S
Dooge JCI, Bruen M (1997) Scaling effects on moisture fluxes on unvegetated land surfaces. Water Resour Res 33(12):2923–2927
Dooge JCI, Bruen M, Parmentier B (1999) A simple model for estimating the sensitivity of runoff to long-term changes in precipitation without a change in vegetation. Adv Water Resour 23 (2):153–163
Douglas EM, Vogel RM, Kroll CN (2000) Trends in floods and low flows in the United States: impact of spatial correlation. J Hydrol 240:90–105
Dunne T, Zhang W, Aubry BF (1991) Effects of rainfall, vegetation and microtopography on infiltration and runoff. Water Resour Res 27:2271–2285
Eckhardt B, Yao D (1993) Local lyapunov exponents. Physica D 65:100–108
Eder G, Sivapalan M, Nachtnebel HP (2003) Modelling water balances in an Alpine catchment through exploitation of emergent properties over changing time scales. Hydrol Processes 17 (11):2125–2149
Emanuel RE, D'Odorico P, Epstein HE (2007) A dynamic soil water threshold for vegetation water stress derived from stomatal conductance models. Water Resour Res 43(3):W03431. doi:10.1029/2005wr004831
Érdi P (2008) Complexity explained. Springer, Berlin, Germany
Faisst H, Eckhardt B (2004) Sensitive dependence on initial conditions in transition to turbulence in pipe flow. J Fluid Mech 504:343–352
Farmer JD (1985) Sensitive dependence on parameters in nonlinear dynamics. Phys Rev Lett 55 (4):351–354
Farmer DJ, Sidorowich JJ (1987) Predicting chaotic time series. Phys Rev Lett 59:845–848
Ferdinand AE (1974) A theory of system complexity. Int J General Syst 1:19–33
Ferguson IM, Maxwell RM (2010) Role of groundwater in watershed response and land surface feedbacks under climate change. Water Resour Res 46:W00F02
Ferguson IM, Maxwell RM (2011) Hydrologic and land-energy feedbacks of agricultural water management practices. Environ Res Lett 6:014006
Fiering MB (1967) Streamflow synthesis. Harvard University Press, Cambridge, Massachusetts

Fogel DB (1995) Evolutionary computation: toward a new philosophy of machine intelligence. IEEE Press, New York

Fowler HJ, Blenkinsop S, Tebaldi C (2007) Linking climate change modeling to impacts studies: recent advances in downscaling techniques for hydrological modeling. Int J Climatol 27 (12):1547–1578

Francis JA, White DM, Cassano JJ, Gutowski WJ Jr, Hinzman LD, Holland MM, Steele MA, Vörösmarty CJ (2009) An arctic hydrologic system in transition: Feedbacks and impacts on terrestrial, marine, and human life. J Geophys Res 114:G04019. doi:10.1029/2008JG000902

Fundel F, Zappa M (2011) Hydrological ensemble forecasting in mesoscale catchments: Sensitivity to initial conditions and value of reforecasts. Water Resour Res 47(9):W09520. doi:10.1029/2010WR009996

Galton F (1894) Natural Inheritance. Macmillan and Company, New York, USA

García-Marín AP, Jiménez-Hornero FJ, Ayuso JL (2008) Applying multifractality and the self-organized criticality theory to describe the temporal rainfall regimes in Andalusia (southern Spain). Hydrol Process 22:295–308

Gelhar LW (1993) Stochastic subsurface hydrology. Prentice-Hall, Englewood Cliffs, New Jersey

Ghorbani MA, Khatibi R, Sivakumar B, Cobb L (2010) Study of discontinuities in hydrological data using catastrophe theory. Hydrol Sci J 55(7):1137–1151

Gilmore CG (1993) A new test for chaos. J Econ Behav Organ 22:209–237

Gleick J (1987) Chaos: making a new science. Penguin Books, New York

Goldberg DE (1989) Genetic algorithms in search, optimization, and machine learning. Addison-Wesley, Reading, MA

Goldstein J (2002) The singular nature of emergent levels: suggestions for a theory of emergence. Nonlinear Dynamics, Psychology, and Life Sciences 6(4):293–309

Goodrich DC, Lane LJ, Shillito RM, Miller SN, Syed KH, Woolhiser DA (1997) Linearity of basin response as a function of scale in a semiarid watershed. Water Resour Res 33(12):2951–2965

Gottschalk L, Jensen JL, Lundquist D, Solantie R, Tollan A (1979) Hydrologic regions in the Nordic countries. Hydrol Res 10(5):273–286

Govindaraju RS (2002) Stochastic methods in subsurface contaminant hydrology. ASCE, New York

Govindaraju RS, Rao AR (2000) Artificial neural networks in hydrology. Kluwer Acadmic Publishers, Amsterdam

Grassberger P, Procaccia I (1983a) Measuring the strangeness of strange attractors. Physica D 9:189–208

Grassberger P, Procaccia I (1983b) Estimation of the Kolmogorov entropy from a chaotic signal. Phys Rev A 28:2591–2593

Grayson RB, Blöschl G (2000) Spatial patterns in catchment hydrology: observations and modeling. Cambridge University Press, Cambridge, UK

Grayson RB, Western AW, Chiew FHS, Blöschl G (1997) Preferred states in spatial soil moisture patterns: local and nonlocal controls. Water Resour Res 33(12):2897–2908

Gupta VK, Waymire E (1983) On the formulation of an analytical approach to hydrologic response and similarity at the basin scale. J Hydrol 65:95–123

Gupta VK, Waymire E (1990) Multiscaling properties of spatial rainfall and river flow distributions. J Geophys Res 95(D3):1999–2009

Gupta VK, Waymire EC (1993) A statistical-analysis of mesoscale rainfall as a random cascade. J Appl Meteorol 32:251–267

Gupta VK, Rodríguez-Iturbe I, Wood EF (eds) (1986) Scale problems in hydrology: runoff generation and basin response. Reidel Publishing Company, FD, p 244

Gupta VK, Troutman BM, Dawdy DR (2007) Towards a nonlinear geophysical theory of floods in river networks: an overview of 20 years of progress. In: Tsonis AA, Elsner JB (eds) Twenty years of nonlinear dynamics in geosciences. Springer Verlag

Gyasi-Agyei Y, Willgoose GR (1997) A hybrid model for point rainfall modelling. Water Resour Res 33(7):1699–1706

Haan CT (1994) Statistical methods in hydrology. Iowa University Press, Iowa
Haines AT, Finlayson BL, McMahon TA (1988) A global classification of river regimes. Appl Geogr 8(4):255–272
Hall A (2004) The role of surface albedo feedback in climate. J Clim 17:1550–1568
Hamed KH (2008) Trend detection in hydrologic data: the Mann-Kendall trend test under the scaling hypothesis. J Hydrol 349(3–4):350–363
Hamed KH, Rao AR (1998) A modified Mann-Kendall trend test for autocorrelated data. J Hydrol 204(1):182–196
Harms AA, Campbell TH (1967) An extension to the Thomas-Fiering model for the sequential generation of streamflow. Water Resour Res 3(3):653–661
Harvey LDD (1997) Upscaling in global change research. In: Hassol H, Katzenberger J (eds) Elements of change 1997 – session one: scaling from site-specific observations to global model grids. Aspen Global Change Institute, Aspen, Colorado, USA, pp 14–33
Hausdorff F (1919) Dimension und äußeres Maß. Math Ann 79:157–179
He Y, Bárdossy A, Zehe E (2011) A review of regionalization for continuous streamflow simulation. Hydrol Earth Syst Sci 15:3539–3553
Henon M (1976) A two-dimensional mapping with a strange attractor. Commun Math Phys 50:69–77
Hicks DM, Gomez B, Trustrum NA (2000) Erosion thresholds and suspended sediment yields, Waipaoa River Basin. New Zealand. Water Resour Res 36(4):1129–1142
Hipel KW, McLeod AI (1994) Time series modelling of water resources and environmental systems. Elsevier, Amsterdam, pp 463–465
Hirsch RM, Slack JR, Smith RA (1982) Techniques of trend analysis for monthly water quality data. Water Resour Res 18:107–121
Holland JH (1975) Adaptation in Natural and Artificial Systems. MIT Press, Cambridge, MA
Holland JH (1998) Emergence. Helix Books, Reading, Massachusetts, From Chaos to Order
Horton RE (1933) The role of infiltration in the hydrologic cycle. Trans Am Geophys Union 14:446–460
Horton RE (1945) Erosional development of streams and their drainage basins: Hydrophysical approach to quantitative morphology. Bull Geol Soc Am 56:275–370
Hu Z, Islam S (1997) Evaluation of sensitivity of land surface hydrology representations with and without land–atmosphere feedback. Hydrol Process 11:1557–1572
Hubert P (2000) The segmentation procedure as a tool for discrete modeling of hydrometeorological regimes. Stoch Environ Res Risk Assess 14:297–304
Isaaks EH, Srivastava RM (1989) An introduction to applied geostatistics. Oxford University Press, New York, USA
Isik S, Singh VP (2008) Hydrologic regionalization of watersheds in Turkey. J Hydrol Eng 13:824–834
Isham S, Entekhabi D, Bras RL (1990) Parameter estimation and sensitivity analysis for the modified Bartlett-Lewis rectangular pulses model of rainfall. J Geophys Res 95(D3):2093–2100
Jacoby SLS (1966) A mathematical model for nonlinear hydrologic systems. J Geophys Res 71 (20):4811–4824
Jakeman AJ, Hornberger GM (1993) How much complexity is warranted in a rainfall-runoff model? Water Resour Res 29(8):2637–2650
Jayawardena AW, Lai F (1994) Analysis and prediction of chaos in rainfall and stream flow time series. J Hydrol 153:23–52
Jenerette GD, Barron-Gafford GA, Guswa AJ, McDonnell JJ, Villegas JC (2012) Organization of complexity in water limited ecohydrology. Ecohydrology 5(2):184–199
Johnson NF (2007) Two's company, three is complexity: a simple guide to the science of all sciences. Oneworld, Oxford, UK
Kalma JD, Sivapalan M (1996) Scale issues in hydrological modeling. Wiley, London
Kantz H, Schreiber T (2004) Nonlinear time series analysis. Cambridge University Press, Cambridge

Kaplan DT, Glass L (1995) Understanding nonlinear dynamics. Springer, New York

Kastens KA, Manduca CA, Cervato C, Frodeman R, Goodwin C, Liben LS, Mogk DW, Spangler TC, Stillings NA, Titus S (2009) How geoscientists think and learn. EOS Trans AGU 90(31):265–266

Kavvas ML (2003) Nonlinear hydrologic processes: conservation equations for determining their means and probability distributions. ASCE J Hydrol Eng 8(2):44–53

Kavvas ML, Delleur JW (1981) A stochastic cluster model of daily rainfall sequences. Water Resour Res 17:1151–1160

Kendall M (1938) A new measure of rank correlation. Biometrika 30(1–2):81–89

Kendall MG (1975) Rand correlation methods. Charles Griffin, London

Kennard MJ, Pusey BJ, Olden JD, Mackay SJ, Stein JL, Marsh N (2010) Classification of natural flow regimes in Australia to support environmental flow management. Freshwater Biol 55 (1):171–193

Kennel MB, Brown R, Abarbanel HDI (1992) Determining embedding dimension for phase space reconstruction using a geometric method. Phys Rev A 45:3403–3411

Kiel LD, Elliott E (1996) Chaos theory in the social sciences: foundations and applications. The University of Michigan Press, Ann Arbor, USA, 349 pp

Kim BS, Kim BK, Kwon HH (2011) Assessment of the impact of climate change on the flow regime of the Han River basin using indicators of hydrologic alteration. Hydrol Processes 25:691–704

Kim J (1999) Making sense of emergence. Philos Stud 95:3–36

Klemeš V (1978) Physically based stochastic hydrologic analysis. Adv Hydrosci 11:285–352

Klemeš V (1983) Conceptualization and scale in hydrology. J Hydrol 65:1–23

Kottegoda NT (1980) Stochastic water resources technology. Macmillan Press, London

Kottegoda NT (1985) Assessment of non-stationarity in annual series through evolutionary spectra. J Hydrol 76:391–402

Koutsoyiannis D (2006) Nonstationarity versus scaling in hydrology. J Hydrol 324:239–254

Kumar P, Foufoula-Georgiou E (1997) Wavelet analysis for geophysical applications. Rev Geophys 35(4):385–412

Kwiatkowski D, Phillips PCB, Schmidt P, Shin Y (1992) Testing the null of stationarity against the alternative of a unit root: How sure are we that economic time series have a unit root? J Econometrics 54:159–178

Kwon H-H, Lall U, Khalil AF (2007) Stochastic simulation model for nonstationary time series using an autoregressive wavelet decomposition: Applications to rainfall and temperature. Water Resour Res 43:W05407. doi:10.1029/2006WR005258

Lai Y-C, Celso G, Kostelich EJ (1994) Extreme final state sensitivity in inhomogeneous spatiotemporal chaotic systems. Phys Lett A 196:206–212

Lansing JS, Kremer JN (1993) Emergent properties of Balinese water temple networks: coadaptation on a rugged fitness landscape. American Anthropologist, New Series 95(1):97–114

Legendre AM (1805) Nouvelles méthodes pour la détermination des orbites des comètes. Firmin Didot, Paris

Lehmann P, Hinz C, McGrath G, Tromp-Van Meerveld HJ, McDonnell JJ (2007) Rainfall threshold for hillslope outflow: an emergent property of flow pathway connectivity. Hydrol Earth Syst Sci 11(2):1047–1063

Lettenmaier DP, Wood EF, Wallis JR (1994) Hydro-climatological trends in the continental United States 1948–1988. J Clim 7:586–607

Levin SA (1992) The problem of pattern and scale in ecology: the Robert H. MacArthur award lecture. Ecology 73:1943–1967

Lins HF, Slack JR (1999) Streamflow trends in the United States. Geophys Res Lett 26(2):227–230

Lorenz EN (1963) Deterministic nonperiodic flow. J Atmos Sci 20:130–141

Maidment DR (1993) Hydrology. In: Maidment DR (ed) Handbook of hydrology chapter 1. McGraw-Hill, New York

Mandelbrot BB (1977) Fractals: form, chance and dimension. W.H.Freeman and Co, New York
Mandelbrot BB (1983) The fractal geometry of nature. Freeman, New York
Mandelbrot BB, Wallis JR (1968) Noah, Joseph and operational hydrology. Water Resour Res 4 (5):909–918
Mandelbrot BB, Wallis JR (1969) Some long run properties of geophysical records. Water Resour Res 5(2):321–340
Mann HB (1945) Nonparametric tests again trend. Econometrica 13:245–259
Maritan A, Colaiori F, Flammini A, Cieplak M, Banavar JR (1996) Universality classes of optimal channel networks. Science 272(5264):984–986
Maxwell RM, Kollet SJ (2008) Interdependence of groundwater dynamics and land–energy feedbacks under climate change. Nat Geosci 1:665–669
May RM (1976) Simple mathematical models with very complicated dynamics. Nature 261:459–467
McCuen RH (2003) Modeling hydrologic change: statistical methods. Lewis Publishers, CRC Press, Boca Raton
McGrath GS, Hinz C, Sivapalan M (2007) Temporal dynamics of hydrological threshold events. Hydrol Earth Syst Sci 11(2):923–938
McMahon TA, Vogel RM, Peel MC, Pegram GGS (2007) Global streamflows—Part 1: characteristics of annual streamflows. J Hydrol 347:243–259
McMillan E (2004) Complexity. Organizations and Change, Routledge, London, UK
Merz R, Blöschl G (2004) Regionalization of catchment model parameters. J Hydrol 287:95–123
Milly PCD, Betancourt J, Falkenmark M, Hirsch RM, Kundzewicz ZW, Lettenmaier DP, Stouffer RJ (2008) Stationarity is dead: whither water management? Science 319(5863):573–574
Minshall NE (1960) Predicting storm runoff on small experimental watersheds. J Hydraul Div Am Soc Eng 86(HYB):17–38
Moliere DR, Lowry JBC, Humphrey CL (2009) Classifying the flow regime of data-limited streams in the wet-dry tropical region of Australia. J Hydrol 367(1–2):1–13
Molnar P, Ramirez JA (1998) Energy dissipation theories and optimal channel characteristics of river networks. Water Resour Res 34(7):1809–1818
Montanari A (2005) Deseasonalisation of hydrological time series through the normal quantile transform. J Hydrol 313:274–282
Moore JW, Beakes MP, Nesbitt HK, Yeakel JD, Patterson DA, Thompson LA, Phillis CC, Braun DC, Favaro C, Scott D, Carr-Harris C, Atlas WI (2015) Emergent stability in a large, free-flowing watershed. Ecology 96(2):340–347
Nash JE (1957) The form of the instantaneous unit hydrograph. Int Assoc Sci Hydrol Pub 45 (3):114–121
Nathan RJ, McMahon TA (1990) Identification of homogeneous regions for the purpose of regionalization. J Hydrol 121(1–4):217–238
Nicolis G, Prigogine I (1977) Self-organization in nonequilibrium systems: From dissipative structures to order through fluctuations. Wiley, New York
Nicolis G, Prigogine I (1989) Exploring Complexity: An Introduction. W. H. Freeman & Company, New York, USA
O'Connor T (1994) Emergent properties. American Philosophical Quarterly 31(2):91–104
O'Kane JP, Flynn D (2007) Thresholds, switches and hysteresis in hydrology form the pedon to the catchment scale: a non-linear systems theory. Hydrol Earth Syst Sci 11(1):443–459
Olden JD, Poff NL (2003) Redundancy and the choice of hydrologic indices for characterizing streamflow regimes. River Res Appl 19(2):101–121
Oudin L, Andréassian V, Perrin C, Michel C, Le Moine N (2008) Spatial proximity, physical similarity, regression and ungaged catchments: a comparison of regionalization approaches based on 913 French catchments. Water Resour Res 44:W03413. doi:10.1029/2007WR006240
Paschalis A, Molnar P, Fatichi S, Burlando P (2013) A stochastic model for high-resolution space-time precipitation simulation. Water Resour Res 49(12):8400–8417

Pearson K (1895) Notes on regression and inheritance in the case of two parents. Proc Royal Soc London 58:240–242
Perrin C, Michel C, Andréassian V (2001) Does a large number of parameters enhance model performance? Comparative assessment of common catchment model structures on 429 catchments. J Hydrol 242(3–4):275–301
Phillips JD (1999) Earth surface systems, complexity, order, and scale. Basil Blackwell, Oxford, U.K
Phillips JD (2006) Evolutionary geomorphology: thresholds and nonlinearity in landform response to environmental change. Hydrol Earth Syst Sci 10:731–742
Phillips JD (2011) Emergence and pseudo-equilibrium in geomorphology. Geomorphology 132:319–326
Phillips JD (2014) Thresholds, mode switching, and emergent equilibrium in geomorphic systems. Earth Surf Process Landforms 39:71–79
Pilgrim DH (1983) Some problems in transferring hydrological relationships between small and large drainage basins and between regions. J Hydrol 65:49–72
Pitman AJ, Stouffer RJ (2006) Abrupt change in climate and climate models. Hydrol Earth Syst Sci 10(6):903–912
Poincaré H (1895) Analysis situs. J de l'École Polytechnique 2(1):1–123
Potter KW (1976) Evidence for nonstationarity as physical explanation of the Hurst phenomenon. Water Resour Res 12(5):1049–1052
Priestley MB (1965) Evolutionary spectra and non-stationary processes. J Roy Stat Soc B 27:204–237
Prigogine I, Stengers I (1984) Order out of chaos: man's new dialogue with nature. Bantam Books
Prudhomme C, Reynard N, Crooks S (2002) Downscaling of global climate models for flood frequency analysis: where are we now? Hydrol Processes 16:1137–1150
Puente CE, Obregon N (1996) A deterministic geometric representation of temporal rainfall. Results for a storm in Boston. Water Resour Res 32(9):2825–2839
Pui A, Sharma A, Mehrotra R, Sivakumar B, Jeremiah E (2012) A comparison of alternatives for daily to sub-daily rainfall disaggregation. J Hydrol 470–471:138–157
Rabier F, Klinker E, Courtier P, Hollingsworth A (1996) Sensitivity of forecast errors to initial conditions. Quart J Royal Meteorol Soc 122(529):121–150
Rao AR, Hu GH (1986) Detection and nonstationarity in hydrologic time series. Manage Sci 32 (9):1206–1217
Rao AR, Jeong GD (1992) Estimation of periodicities in hydrologic data. Stoch Hydrol Hydaul 6:270–288
Razavi T, Coulibaly P (2013) Streamflow prediction in ungauged basins: review of regionalization methods. J Hydrol Eng 18:958–975
Richards LA (1931) Capillary conduction of liquids through porous mediums. Physics A 1:318–333
Richardson CW (1977) A model of stochastic structure of daily precipitation over an area. Hydrology Paper 91, Colorado State University, Fort Collins, Colorado
Richter BD, Baumgartner JV, Powell J, Braun DP (1996) A method for assessing hydrologic alteration within ecosystems. Conserv Biol 10:1163–1174
Richter BD, Baumgartner JV, Wigington R, Braun DP (1997) How much water does a river need? Freshw Biol 37:231–249
Richter BD, Baumgartner JV, Braun DP, Powell J (1998) A spatial assessment of hydrologic alteration within a river network. Regulated Rivers: Res Manage 14:329–340
Rigon R, Rinaldo A, Rodriguez-Iturbe I, Bras RL, Ijjasz-Vasquez E (1993) Optimal channel networks: a framework for the study of river basin morphology. Water Resour Res 29(6):1635–1646
Rigon R, Rinaldo A, Rodriguez-Iturbe I (1994) On landscape selforganization. J Geophys Res 99 (B6):11971–11993
Rigon R, Rodriguez-Iturbe I, Rinaldo A (1998) Feasible optimality implies Hack's law. Water Resour Res 34(11):3181–3188

Rinaldo A, Rodriguez-Iturbe I, Rigon R, Bras RL, Ijjasz-Vasquez E, Marani A (1992) Minimum energy and fractal structures of drainage networks. Water Resour Res 28(9):2183–2195
Rinaldo A, Rodriguez-Iturbe I, Rigon R, Ijjasz-Vasquez E, Bras RL (1993) Self-organized fractal river networks. Phys Rev Lett 70(6):822–825
Rinaldo A, Rigon R, Banavar JR, Maritan A, Rodriguez-Iturbe I (2014) Evolution and selection of river networks: Statics, dynamics, and complexity. PNAS 111(7):2417–2424
Rodriguez-Iturbe I, Rinaldo A (1997) Fractal river basins: chance and self-organization. Cambridge University Press, Cambridge
Rodriguez-Iturbe I, Cox D, Isham V (1987) Some models for rainfall based on stochastic point processes. Proc R Soc Lond A 410:269–288
Rodriguez-Iturbe I, De Power FB, Sharifi MB, Georgakakos KP (1989) Chaos in rainfall. Water Resour Res 25(7):1667–1675
Rodriguez-Iturbe I, Rinaldo A, Rigon R, Bras RL, Ijjasz-Vasquez E (1992a) Energy dissipation, runoff production and the three dimensional structure of channel networks. Water Resour Res 28(4):1095–1103
Rodriguez-Iturbe I, Rinaldo A, Rigon R, Bras RL, Ijjasz-Vasquez E (1992b) Fractal structures as least energy patterns: The case of river networks. Geophys Res Lett 19(9):889–892
Rodriguez-Iturbe I, Marani M, Rigon R, Rinaldo A (1994) Self-organized river basin landscapes—fractal and multifractal characteristics. Water Resour Res 30:3531–3539
Rodriguez-Iturbe I, D'Odorico P, Rinaldo A (1998) Possible selforganizing dynamics for land-atmosphere interaction. J Geophys Res-Atmos 103:23071–23077
Rodriguez-Iturbe I, Isham V, Cox DR, Manfreda S, Porporato A (2006) Space-time modeling of soil moisture: stochastic rainfall forcing with heterogeneous vegetation. Water Resour Res 42: W06D05. doi:10.1029/2005WR004497
Rodriguez-Iturbe I, Muneepeerakul R, Bertuzzo E, Levin SA, Rinaldo A (2009) River networks as ecological corridors: A complex systems perspective for integrating hydrologic, geomorphologic, and ecologic dynamics. Water Resour Res 45:1–22
Roe (2009) Feedbacks, Timescales, and Seeing Red. Ann Rev Earth and Planetary Sci 37:93–115
Rössler OE (1976) An equation for continuous chaos. Phys Lett A 57:397–398
Ruelle D (1978) Sensitive dependence on initial condition and turbulent behavior of dynamical systems. Ann NY Acad Sci 316:408–416
Runyan CW, D'Odorico P, Lawrence D (2012) Physical and biological feedbacks of deforestation. Rev Geophys 50:1–32
Said SE, Dickey D (1984) Testing for unit roots in autoregressive moving-average models with unknown order. Biometrika 71:599–607
Salas JD, Smith RA (1981) Physical basis of stochastic models of annual flows. Water Resour Res 17(2):428–430
Salas JD, Delleur JW, Yevjevich V, Lane WL (1995) Applied modeling of hydrologic time series. Water Resources Publications, Littleton, Colorado
Sapozhnikov V, Foufoula-Georgiou E (1996) Do the current landscape evolution model show self-organized criticality? Water Resour Res 32(4):1109–1112
Sapozhnikov V, Foufoula-Georgiou E (1997) Experimental evidence of dynamic scaling and self-organized criticality in braided rivers. Water Resour Res 33(8):1983–1991
Sapozhnikov V, Foufoula-Georgiou E (1999) Horizontal and vertical self-organization of braided rivers towards a critical state. Water Resour Res 35(3):843–851
Schmidt GA, Ruedy R, Miller RL, Lacis AA (2010) The attribution of the present-day total greenhouse effect. J Geophys Res 115: D20106. doi:10.1029/2010JD014287
Şen Z (2009) Fuzzy Logic and Hydrologic Modeling. CRC Press, Boca Raton, FL
Şen Z (2014) Trend identification simulation and application. J Hydrol Eng 19(3):635–642
Shang P, Na X, Kamae S (2009) Chaotic analysis of time series in the sediment transport phenomenon. Chaos Soliton Fract 41(1):368–379
Shannon CE (1948) A mathematical theory of communication. Bell Syst Tech J 27:379–423
Sherman LK (1932) Streamflow from rainfall by the unit graph method. Eng News Rec 108:501–505

Shiau JT, Wu FC (2008) A histogram matching approach for assessment of flow regime alteration: application to environmental flow optimization. River Research and Applications 24:914–928
Shin Y, Schmidt P (1992) The KPSS stationarity test as a unit root test. Econ Lett 38:387–392
Singh VP (1979) A uniformly nonlinear hydrologic cascade model. Irrigation Power 36(3):301–317
Singh VP (1997) The use of entropy in hydrology and water resources. Hydrol Process 11:587–626
Singh VP (1998) Entropy-based parameter estimation in hydrology. Kluwer Academic, Dordrecht, The Netherlands
Singh VP (2013) Entropy theory and its application in environmental and water engineering. John Wiley and Sons, Oxford, UK
Sivakumar B (2000) Chaos theory in hydrology: important issues and interpretations. J Hydrol 227 (1–4):1–20
Sivakumar B (2004a) Chaos theory in geophysics: past, present andfuture. Chaos Soliton Fract 19 (2):441–462
Sivakumar B (2004b) Dominant processes concept in hydrology: moving forward. Hydrol Process 18(12):2349–2353
Sivakumar B (2005) Hydrologic modeling and forecasting: role of thresholds. Environ Model Softw 20(5):515–519
Sivakumar B (2008a) Dominant processes concept, model simplification and classification framework in catchment hydrology. Stoch Env Res Risk Assess 22(6):737–748
Sivakumar B (2008b) The more things change, the more they stay the same: the state of hydrologic modelling. Hydrol Processes 22:4333–4337
Sivakumar B (2009) Nonlinear dynamics and chaos in hydrologic systems: latest developments and a look forward. Stoch Environ Res Risk Assess 23(7):1027–1036
Sivakumar B, Berndtsson R (2010) Advances in data-based approaches for hydrologic modeling and forecasting. World Scientific Publishing Company, Singapore
Sivakumar B, Singh VP (2012) Hydrologic system complexity and nonlinear dynamic concepts for a catchment classification framework. Hydrol Earth Syst Sci 16:4119–4131
Sivakumar B, Sorooshian S, Gupta HV, Gao X (2001) A chaotic approach to rainfall disaggregation. Water Resour Res 37(1):61–72
Sivakumar B, Jayawardena AW, Li WK (2007) Hydrologic complexity and classification: a simple data reconstruction approach. Hydrol Process 21(20):2713–2728
Sivakumar B, Singh V, Berndtsson R, Khan S (2015) Catchment classification framework in Hydrology: challenges and directions. J Hydrol Eng 20:A4014002
Sivapalan M, Jothityangkoon C, Menabde M (2002) Linearity and nonlinearity of basin response as a function of scale: discussion of alternative definitions. Water Resour Res 38(2):1012. doi:10.1029/2001WR000482
Skøien JO, Blöschl G (2007) Spatiotemporal topological kriging of runoff time series. Water Resour Res 43:W09419. doi:10.1029/2006WR005760
Smith JA, Karr AF (1983) A point process of summer reason rainfall occurrences. Water Resour Res 19:95–103
Smith JA, Karr AF (1985) Statistical inference for point process models of rainfall. Water Resour Res 21(1):73–79
Snelder TH, Biggs BJF, Woods RA (2005) Improved eco-hydrological classification of rivers. River Res Applic 21:609–628
Spearman C (1904) The proof and measurement of association between two things. Amer J Psychol 15:72–101
Sposito G (2008) Scale dependence and scale invariance in hydrology. Cambridge University Press, Cambridge, UK
Srikanthan R, McMahon TA (1983) Stochastic simulation of daily rainfall for Australian stations. Trans ASAE 26(3):754–759
Stedinger JR, Vogel RM (1984) Disaggregation procedures for generating serially correlated flow vectors. Water Resour Res 20(1):47–56

Steffen W, Sanderson RA, Tyson PD, Jäger J, Matson PA, Moore B III, Oldfield F, Richardson K, Schellnbuber H-J, Turner BL, Wasson RJ (2004) Global change and the earth system: a planet under pressure. The IGBP global change series. Springer-Verlag, Berlin

Stephenson GR, Freeze RA (1974) Mathematical simulation of subsurface flow contributions to snowmelt runoff, Reynolds Creek Watershed. Idaho. Water Resour Res 10(2):284–294

Stolum H-H (1996) River meandering as a self-organization process. Science 271(5256):1710–1713

Strogatz SH (1994) Nonlinear dynamics and chaos: with applications to physics, biology, chemistry, and engineering. Perseus Books, Cambridge

Székely GJ, Rizzo ML, Bakirov NK (2007) Measuring and testing independence by correlation of distances. Ann Stat 35(6):2769–2794

Talling PJ (2000) Self-organization of river networks to threshold states. Water Resour Res 36 (4):1119–1128

Tang C, Bak P (1988a) Critical exponents and scaling relations for self-organized critical phenomena. Phys Rev Lett 60(23):2347–2350

Tang C, Bak P (1988b) Mean field theory of self-organized critical phenomena. J Stat Phys 51(5–6):797–802

Tarboton DG, Sharma A, Lall U (1998) Disaggregation procedures for stochastic hydrology based on nonparametric density estimation. Water Resour Res 34(1):107–119

Tessier Y, Lovejoy S, Hubert P, Schertzer D, Pecknold S (1996) Multifractal analysis and modeling of rainfall and river flows and scaling, causal transfer functions. J Geophys Res 101 (D21):26427–26440

Thom R (1972) Stabilité Structurelle et Morphogénèse. Benjamin, New York

Thomas HA, Fiering MB (1962) Mathematical synthesis of streamflow sequences for the analysis of river basins by simulation. In: Mass A et al (eds) Design of water resource systems. Harvard University Press, Cambridge, Massachusetts, pp 459–493

Thomson DJ (1982) Spectrum estimation and harmonic analysis. Proc IEEE 70:1055–1096

Todorovic P, Yevjevich V (1969) Stochastic process of precipitation. Hydrol Pap 35, Colorado State University, Fort Collins

Toms JD, Lesperance ML (2003) Piecewise regression: a tool for identifying ecological thresholds. Ecology 84(8):2034–2041

Tromp-Van-Meerveld HJ, Mcdonnell JJ (2006a) Threshold relations in subsurface stormflow 1. A storm analysis of the Panola hillslope. Water Resour Res 42:W02410. doi:10.1029/2004WR003778

Tromp-Van-Meerveld HJ, Mcdonnell JJ (2006b) Threshold relations in subsurface stormflow: 2. The fill and spill hypothesis, Water Resour Res 42(2):W02411. doi:10.1029/2004wr003800

Tsonis AA (1992) Chaos: from theory to applications. Plenum Press, New York

Valencia D, Schaake JC (1973) Disaggregation processes in stochastic hydrology. Water Resour Res 9(3):211–219

Van Walsum PEV, Supit I (2012) Influence of ecohydrologic feedbacks from simulated crop growth on integrated regional hydrologic simulations under climate scenarios. Hydrol Earth Syst Sci 16:1577–1593

Verhoest N, Troch P, De Troch FP (1997) On the applicability of Bartlett-Lewis rectangular pulses models in the modeling of design storms at a point. J Hydrol 202:108–120

Vespignani A, Zapperi S (1998) How self-organized criticality works: a unified mean-field picture. Phys Rev E 57:6345

von Foerster H (1960) On self-organizing systems and their environments. In: Yovits MC, Cameron S (eds) self-organizing systems. Pergamon Press, London, pp 31–50

Wainwright W, Mulligan M (2004) Environmental modeling: finding simplicity in complexity. Wiley, London

Waldrop MM (1992) Complexity: the emerging science at the edge of order and chaos. Simon & Schuster, New York, USA

Wang CT, Gupta VK, Waymire E (1981) A geomorphologic synthesis of nonlinearity in surface runoff. Water Resour Res 17(3):545–554

Waymire E, Gupta VK (1981) The mathematical structure of rainfall representations, 2, a review of the theory of point processes. Water Resour Res 17(5):1273–1286
Wilby RL, Wigley TML (1997) Downscaling general circulation model output: a review of methods and limitations. Prog Phys Geogr 21:530–548
Wolf A, Swift JB, Swinney HL, Vastano A (1985) Determining Lyapunov exponents from a time series. Physica D 16:285–317
Wood AW, Leung LR, Sridhar V, Lettenmaier DP (2004) Hydrologic implications of dynamical and statistical approaches to downscaling climate model outputs. Clim Change 62:189–216
Woolhiser DA (1971) Deterministic approach to watershed modeling. Nordic Hydrol 11:146–166
Yang F, Kumar A, Wang W, Juang H-MH, Kanamitsu M (2001) Snow-albedo feedback and seasonal climate variability over North America. J. Climate 14:4245–4248
Yang P, Yin X-A, Yang Z-F, Tang J (2014) A revised range of variability approach considering periodicity of hydrological indicator. Hydrol Processes 28:6222–6235
Yeakel JD, Moore JW, Guimarães PR, de Aguiar MAM (2014) Synchronisation and stability in river metapopulation networks. Ecol Lett 17:273–283
Yevjevich VM (1963) Fluctuations of wet and dry years. Part 1. Research data assembly and mathematical models. Hydrology Paper 1, Colorado State University, Fort Collins, Colorado, pp 1–55
Yevjevich VM (1972) Stochastic processes in hydrology. Water Resour Publ, Fort Collins, Colorado
Yevjevich VM (1974) Determinism and stochasticity in hydrology. J Hydrol 22:225–258
Yevjevich VM (1984) Structure of daily hydrologic time series. Water Resources Publications, Littleton, USA
Yin XA, Yang ZF, Petts GE (2011) Reservoir operating rules to sustain environmental flows in regulated rivers. Water Resour Res 47:W08509. doi:10.1029/2010wr009991
Young PC (1998) Data-based mechanistic modeling of environmental, ecological, economic and engineering systems. Environ Modell Softw 13:105–122
Young PC (1999) Nonstationary time series analysis and forecasting. Progress Environ Sci 1:3–48
Young PC (2003) Top-down and data-based mechanistic modelling of rainfall-flow dynamics at the catchment scale. Hydrol Processes 17:2195–2217
Young PC, Beven KJ (1994) Data-based mechanistic modeling and rainfall-flow non-linearity. Environmetrics 5(3):335–363
Young PC, Parkinson SD (2002) Simplicity out of complexity. In: Beck MB (ed) Environmental foresight and models: a manifesto. Elsevier Science, The Netherlands, pp 251–294
Young PC, Ratto M (2009) A unified approach to environmental systems modeling. Stoch Environ Res Risk Assess 23:1037–1057
Young PC, Parkinson SD, Lees M (1996) Simplicity out of complexity in environmental systems: Occam's Razor revisited. Journal of Applied Statistics 23:165–210
Yule GU (1896) On the significance of Bravais' formulae for regression in the case of skew correlation. Proc Royal Soc London 60:477–489
Zeeman EC (1976) Catastrophe theory. Sci Am 234:65–83
Zehe E, Blöschl G (2004) Predictability of hydrologic response at the plot and catchment scales: Role of initial conditions. Water Resour Res 40(10):1–21
Zehe E, Sivapalan M (2009) Threshold behavior in hydrological systems as (human) geo-ecosystems: manifestations, controls, implications. Hydrol Earth Syst Sci 13:1273–1297
Zehe E, Elsenbeer H, Lindenmaier F, Schulz K, Blöschl G (2007) Patterns of predictability in hydrological threshold systems. Water Resour Res 43(7):1–12

Chapter 3
Stochastic Time Series Methods

Abstract Hydrology was mainly dominated by deterministic approaches until the mid-twentieth century. However, the deterministic approaches suffered from our lack of knowledge on the exact nature of hydrologic system dynamics and, hence, the exact governing equations required for models. This led to the development and application of stochastic methods in hydrology, which are based on the concepts of probability and statistics. Since the 1950s–1960s, hydrology has witnessed the development of a large number of stochastic time series methods and their applications. The existing stochastic methods can be broadly grouped into two categories: parametric and nonparametric. In the parametric methods, the structure of the models is defined a priori and the number and nature of the parameters are generally fixed in advance. On the other hand, the nonparametric methods make no prior assumptions on the model structure, and it is essentially determined from the data themselves. This chapter presents an overview of stochastic time series methods in hydrology. First, a brief account of the history of development of stochastic methods is presented. Next, the concept of time series and relevant statistical characteristics and estimators are described. Finally, several popular parametric and nonparametric methods and their hydrologic applications are discussed.

3.1 Introduction

The complex, irregular, and random-looking nature of hydrologic time series, combined with our lack of knowledge of the exact governing equations required in deterministic models, motivated the development and applications of methods based on probability and statistics for modeling and prediction of such time series. In this regard, stochastic methods are an appropriate means, since they aim at prediction (or estimation) of data in a probabilistic manner, with particular emphasis on the statistical characteristics of the data (e.g. mean, standard deviation, variance) and with proper consideration to uncertainty in such predictions.

Stochastic time series methods in hydrology has a long and rich history. The emergence of stochastic methods during the first half of the 20th century (e.g. Lévy

B. Sivakumar, *Chaos in Hydrology*, DOI 10.1007/978-90-481-2552-4_3

1925, 1948; Doob 1934, 1938, 1940, 1945; Berndstein 1938; Itô 1944, 1946; Bochner 1949) led to their introduction in hydrology in the 1950s (Hurst 1951, 1956; Hannan 1955; Le Cam 1961), especially for studying storage in reservoirs and rainfall modeling. Further advances in the development of many stochastic time series methods around the middle of the 20th century (e.g. Wiener 1949; Feller 1950; Bartlett 1955; Box and Jenkins 1970; Brillinger 1975) led to a real impetus in applying stochastic methods in hydrology during 1960s–1970s (e.g. Thomas and Fiering 1962; Matalas 1963a, b, 1967; Yevjevich 1963, 1972; Roesner and Yevjevich 1966; Fiering 1967; Harms and Campbell 1967; Quimpo 1967; Mandelbrot and Wallis 1968, 1969; Carlson et al. 1970; Valencia and Schaake 1973; McKerchar and Delleur 1974; Haan et al. 1976; Hipel et al. 1977; Hipel and McLeod 1978a, b; Lawrence and Kottegoda 1977; Lettenmaier and Burges 1977; Delleur and Kavvas 1978; Klemeš 1978; Hirsch 1979); see also Salas et al. (1995) for a comprehensive account.

Since then, stochastic methods in hydrology have witnessed tremendous advances, both in terms of theoretical development and in terms of applications in many different areas and problems in hydrology. Studies during the 1980s and early 1990s mainly focused on the development and applications of parametric methods in hydrology (e.g. Gupta and Waymire 1981; Kavvas and Delleur 1981; Salas and Smith 1981; Salas and Obeysekera 1982; Srikanthan and McMahon 1983; Rao and Rao 1984; Rodriguez-Iturbe et al. 1987; Koutsoyiannis and Xanthopoulos 1990); see also Bras and Rodriguez-Iturbe (1985), Gelhar (1993), and Salas (1993) for some comprehensive accounts. However, the development of nonparametric approaches (e.g. Efron 1979; Silverman 1986; Eubank 1988; Härdle and Bowman 1988; Efron and Tibishirani 1993) to overcome the difficulties associated with the estimation of parameters in parametric methods led to their applications in hydrology. Such applications have been gaining significant momentum during the past two decades or so (e.g. Kendall and Dracup 1991; Lall 1995; Lall and Sharma 1996; Vogel and Shallcross 1996; Tarboton et al. 1998; Buishand and Brandsma 2001; Sharma and O'Neill 2002; Prairie et al. 2006; Mehrotra and Sharma 2010; Salas and Lee 2010; Wilks 2010; Li and Singh 2014).

In more recent years, stochastic time series methods have been finding increasing applications in the study of hydrologic extremes, parameter estimation in hydrologic models, and downscaling of global climate model outputs, among others (e.g. Fowler et al. 2007; Vrugt et al. 2008; Maraun et al. 2010; Mehrotra and Sharma 2012; Pui et al. 2012; Grillakis et al. 2013; Bordoy and Burlando 2014; D'Onofrio et al. 2014; Sikorska et al. 2015; Wasko et al. 2015; Langousis et al. 2016). Although many modern nonlinear approaches have found important places in hydrology in recent times (see Chap. 4 for details), stochastic time series methods continue to dominate, for various reasons, including for their great flexibility. Extensive details of the applications of stochastic methods in hydrology are available in Kottegoda (1980), Bras and Rodriguez-Iturbe (1985), Gelhar (1993), Clarke (1994), Haan (1994), Hipel and McLeod (1994), Salas et al. (1995), Govindaraju (2002), and McCuen (2003), among others.

This chapter presents an overview of stochastic time series methods in hydrology. First, a brief account of the history of development and fundamentals of stochastic methods is presented. Next, the definition and classification of hydrologic time series are provided, followed by a description of their relevant statistical characteristics and estimators. Then, several parametric methods and nonparametric methods that have found widespread applications in hydrology are described. The chapter ends with a brief summary of stochastic methods in hydrology.

Due to the domination of stochastic methods for over half a century, the literature on stochastic methods in hydrology is enormous. Therefore, it is impossible to cover all the available literature here. For additional details, the reader is directed to the many studies cited in this chapter and the references therein. Furthermore, the stochastic methods described in this chapter are those that either assume linearity in the process or make no prior assumptions regarding linearity/nonlinearity. There are indeed many stochastic methods that explicitly assume nonlinearity of the process. A brief account of such nonlinear stochastic methods is presented in Chap. 4, along with many other modern nonlinear time series methods.

3.2 Brief History of Development of Stochastic Methods

The term 'stochastic' was derived from the Greek word 'Στόχος' (stochos), meaning 'target.' However, in the context of modern science, 'stochastic' generally means 'random' or refers to the presence of randomness. Stochastic methods, in essence, aim at predicting (or estimating) the value of some variable at non-observed times or at non-observed locations in a probabilistic manner, while also stating how uncertain the predictions are. The methods place emphasis on the statistical characteristics (e.g. mean, standard deviation, variance) of relevant processes.

Probabilistic approaches with emphasis on statistical characteristics of data observed in nature have a very long history. However, stochastic models, as they are seen in their current form, originated around the mid-twentieth century; see Wiener (1949) and Feller (1950) for some early accounts. Such developments were based on many earlier developments, including stochastic differential equations, stochastic integrals, Markov chains, Brownian motion, and diffusion equations (e.g. Berndstein 1938; Doob 1938, 1940, 1945; Lévy 1937, 1948; Itô 1944, 1946; Mann 1945; Bochner 1949). The 1950s–1960s witnessed some important developments, including advances in diffusion processes, harmonic analysis, Markov processes, branching processes, and random walk theory; see Doob (1953), Bartlett (1955), Bochner (1955), Loève (1955), Dynkin (1959), Spitzer (1964), and McKean (1969) for details.

A major shift in stochastic methods and their applications occurred around the 1970s. This could mainly be attributed to the work of Box and Jenkins (1970), who presented an exhaustive account of stochastic time series methods until then, and several others (e.g. Jazwinski 1970; Brillinger 1975; Karlin and Taylor 1975). The methods presented in Box and Jenkins (1970) have and continue to be widely followed in various scientific fields, including hydrology. Some of the models that have found widespread applications are the autoregressive (AR) models, moving average (MA) models, autoregressive moving average (ARMA) models, Markov chain models, point process models, and their variants. Parallel theoretical developments in the concepts of fractional Brownian motion (fBm), fractional Gaussian noise (fGn), Lévy processes, and broken line processes (e.g. Mandelbrot and Van Ness 1968; Mandlbrot and Wallis 1969; Mandelbrot 1972) also resulted in many different types of models and subsequently found numerous applications in different fields. Indeed, this period also witnessed a significant shift in the study of scale-invariance or self-similarity of natural processes, especially with the works of Mandelbrot and with the coining of the term 'fractal' (e.g. Mandelbrot 1967, 1977); see also Mandelbrot (1983) for an exhaustive coverage.

These advances further propelled the development of stochastic methods and their applications in different fields; see Chatfield (1989), Brockwell and Davis (1991), Ross (1996), and Gardiner (2009) for details. However, due to the mainly parametric nature of most of these methods, the models generally suffered from certain important limitations, such as: (1) since the structure of the models is defined a priori and the number and nature of the parameters are generally fixed in advance, there is little flexibility; (2) the parameter estimation procedure is often complicated, especially when the number of parameters is large; and (3) the models are often not able to capture several important properties of the time series. This led to the proposal of many nonparametric approaches since the 1980s (Efron 1979; Silverman 1986; Eubank 1988; Härdle and Bowman 1988) and resulted in another major shift in the development and applications of stochastic time series methods. The nonparametric approaches make no prior assumptions on the model structure. Instead, the model structure is determined from the data themselves. Although nonparametric methods may also involve parameters, the number and nature of the parameters are not fixed in advance and, thus, are flexible. Since their emergence, many nonparametric methods have been developed, including those based on kernel density estimate, block bootstrap, *k*-nearest neighbor bootstrap, *k*-nearest neighbor with local polynomial, hybrid models, and others. Comprehensive accounts of the concepts and applications of nonparametric methods can be found in Efron and Tibishirani (1993), Higgins (2003), Sprent and Smeeton (2007), and Hollander et al. (2013) among others.

3.3 Hydrologic Time Series and Classification

A hydrologic time series is a set of observations of a hydrologic variable (e.g. rainfall, streamflow) made at a particular location over time (i). A hydrologic time series can be either continuous or discrete. For instance, a streamflow monitoring device at a location in a river provides a continuous record of river stage and discharge $X(i)$ through time. A plot of the flow hydrograph $X(i)$ versus time i is a *continuous time series*. Sampling this continuous time series at discrete points in time or integrating it over successive time intervals results in a *discrete time series*, denoted by X_i. Hydrologic time series are generally measured and studied at discrete timescales (e.g. hourly, daily, monthly, annual). Such a discrete time series may be denoted as X_i, $i = 1, 2, \ldots, N$, where N is the total length of the observed time and, thus, often represents the total number of data (or points) in the time series. As the study of discrete time series is the widely used practice in hydrology, the presentation in this chapter (and in the rest of this book) focuses on such a time series only.

In addition to continuous and discrete time series, hydrologic time series may be further classified into different categories depending upon the nature of the observations and the general/specific properties. A *single or scalar or univariate time series* is a time series of a hydrologic variable at a given location. A *multiple or multivariate time series* may represent either time series measured at more than one location or time series of multiple variables at one location (or more). In a single time series, if the value of a variable at time i, i.e. X_i, depends on the value of the variable at other (earlier) times i–1, i–2, ..., then the time series is called *autocorrelated, serially correlated or correlated in time*; otherwise, it is *uncorrelated*. A similar explanation also goes in a spatial sense. For instance, considering a multiple time series (e.g. rainfall, streamflow), if the value of a variable (e.g. streamflow) at time i at one location, X_i, depends on the value of another variable (e.g. rainfall) at the same location (Y_i) (or on the value of the same variable at another location) at times i, i–1, i–2,..., then the two time series are *cross-correlated*. Autocorrelation and cross-correlation in hydrologic time series can be simple or complex.

A *seasonal time series* is a series that corresponds to a time interval that is a fraction of a year, such as muliples of a month. Since seasonality is an intrinsic property of hydrologic systems, it is often useful, and even necessary, to study the seasonal aspects of hydrologic systems. An *intermittent time series* is one when the variable of interest takes on non-zero and zero values throughout the length of the time series. Rainfall measured at finer timescales (e.g. hourly or daily) is often an excellent example of an intermittent time series. A *stationary time series* is a time series that is free of trends, shifts, or periodicity, implying that the statistical parameters of the series (e.g. mean, variance) remain constant through time. If such parameters do not remain constant through time, then the time series is called a *nonstationary time series*. If the measurements are available at a regularly spaced interval, then the time series is called a *regularly spaced or regular time series*. If the data are at irregular intervals, then the time series is called an *irregularly spaced or irregular time series*.

Depending upon the properties or components, a time series can be partitioned or decomposed into its component series. Some of the generic components of hydrologic time series are trend, shift, and seasonality (in mean and variance as well as in correlation). These components are often removed before employing stochastic time series methods.

3.4 Relevant Statistical Characteristics and Estimators

Hydrologic systems and, hence, the observed hydrologic data exhibit a wide range of characteristics. Chapter 2 presented an overview of many of the salient characteristics, including complexity, correlation, trend, periodicity, cyclicity, seasonality, intermittency, stationarity, nonstationarity, linearity, nonlinearity, determinism, randomness, scale and scale-invariance, self-organization and self-organized criticality, threshold, emergence, feedback, and sensitivity to initial conditions. Some of these characteristics have found particular significance in the development and application of stochastic time series methods in hydrology. This section briefly describes some of these characteristics and estimators, especially those that have generally formed the basis for model development and for model evaluation.

The most commonly used statistical properties for analyzing stationary or nonstationary hydrologic time series are the sample mean, variance, coefficient of variation, skewness coefficient, lag-τ autocorrelation function, and the power spectrum. These are briefly described here.

3.4.1 *Mean*

The mean is the fundamental property considered in time series methods. The mean is the first moment measured about the origin. It is also the average of all observations on a random variable. For a discrete time series X_i, $i = 1, 2, \ldots, N$, the sample mean $\bar{X}$ (or population mean μ) is calculated as:

$$\bar{X} = \frac{1}{N}\sum_{i=1}^{N} X_i \tag{3.1}$$

Although the mean conveys certain information, it does not completely characterize a time series.

3.4.2 *Variance*

The variance is another basic property considered in time series methods. It is the second moment about the mean, and is an indicator of the closeness of the values of a time series to its mean. For a discrete time series X_i, $i = 1, 2, \ldots, N$, the sample variance s^2 (or population variance σ^2) is given as:

$$s^2 = \frac{1}{N-1}\sum_{i=1}^{N} (X_i - \bar{X})^2 \tag{3.2}$$

Although variance is used in all aspects of statistical analysis, its use as a descriptor is limited because of its units. Specifically, the units of the variance are not the same as those of the time series. The square root of variance is the standard deviation, s.

3.4.3 Coefficient of Variation

The coefficient of variation (CV) or the relative standard deviation (RSD) of a time series is defined as the ratio of the standard deviation to the mean of the time series. It represents the extent of variability in relation to the mean of the time series. It is written as:

$$CV = \frac{s}{\bar{X}} \tag{3.3}$$

The coefficient of variation is also particularly useful when comparing the variability of different time series. For instance, considering two time series, the time series with a higher level of variability will have a greater CV value than the other. This is a significant advantage in using CV, since the mean and the variance cannot be used to actually compare the variability of two time series.

3.4.4 Skewness Coefficient

The skew is the third moment measured about the mean. For a discrete time series X_i, $i = 1, 2, \ldots, N$, the sample skewness coefficient g (or population skewness coefficient γ) is given as:

$$g = \frac{N \sum_{i=1}^{N} (X_i - \bar{X})^3}{(N-1)(N-2)s^3} \tag{3.4}$$

The skew is a measure of symmetry. For a symmetric distribution, the skew will be zero. For a nonsymmetric distribution, the skew will be positive or negative depending on the location of the tail of the distribution. If the more extreme tail of the distribution is to the right of the mean, the skew is positive. If the more extreme tail of the distribution is to the left of the mean, the skew is negative.

3.4.5 Autocorrelation Function

The autocorrelation in a time series represents correlations between the values. It is determined by the autocorrelation function, which is a normalized measure of the linear correlation among successive values in a time series. Autocorrelation function is useful to determine the degree of dependence present in the values. For a discrete time series X_i, $i = 1, 2, \ldots, N$, and for a given lag time τ, the sample autocorrelation function r_τ (or the population autocorrelation function ρ_τ) is given by:

$$r_\tau = \frac{\sum_{i=1}^{N-\tau} X_i X_{i+\tau} - \frac{1}{N-\tau}\sum_{i=1}^{N-\tau} X_{i+\tau} \sum_{i=1}^{N-\tau} X_i}{\left[\sum_{i=1}^{N-\tau} X_i^2 - \frac{1}{N-\tau}\left(\sum_{i=1}^{N-\tau} X_i\right)^2\right]^{1/2} \left[\sum_{i=1}^{N-\tau} X_{i+\tau}^2 - \frac{1}{N-\tau}\left(\sum_{i=1}^{N-\tau} X_{i+\tau}\right)^2\right]^{1/2}} \quad (3.5)$$

In general, for a periodic process, the autocorrelation function is also periodic, indicating the strong relation between values that repeat over and over again. For a purely random process, the autocorrelation function fluctuates randomly about zero, indicating that the process at any certain instance has no 'memory' of the past at all. Other stochastic processes generally have decaying autocorrelations, but the rate of decay depends on the properties of the process.

The autocorrelation function has been widely used for analysis of hydrologic time series, including for identification of relevant antecedent conditions and temporal persistence governing the process (e.g. Matalas 1963a; Salas 1993); see also Salas et al. (1995). However, autocorrelations are not characteristic enough to distinguish random from deterministic chaotic signals; see Chap. 6 for additional details.

3.4.6 *Power Spectrum*

The spectral analysis is most useful in isolating periodicities in a time series, which are often best delineated by analyzing the data in the frequency domain. The power spectrum is a widely used tool for studying the oscillations of a time series. It is defined as the square of the coefficients in a Fourier series representation of the time series. It shows the variance of the function at different frequencies. For a discrete time series X_i, $i = 1, 2, \ldots, N$, if the power spectrum $P(f)$ obeys a power law form

$$P(f) \propto f^{-\beta} \quad (3.6)$$

where f is the frequency β is the spectral exponent, this is an indication of the absence of characteristic timescale in the range of the power law. In such a case, a fractal or scale-invariant behavior may be assumed to hold (e.g. Fraedrich and Larnder 1993).

For a purely random process, the power spectrum oscillates randomly about a constant value, indicating that no frequency explains any more of the variance of the sequence than any other frequency. Periodic or quasi-periodic signals show sharp spectral lines; measurement noise adds a continuous floor to the spectrum. Thus, in the spectrum, signal and noise are readily distinguished.

Spectral analysis, and especially the power spectrum, has been widely applied to study hydrologic time series, including for identification of temporal persistence and scale-invariant behavior (e.g. Roesner and Yevjevich 1966; Quimpo 1967; Fraedrich and Larnder 1993; Olsson et al. 1993; Tessier et al. 1996; Menabde et al. 1997; Pelletier and Turcotte 1997; Sivakumar 2000; Mathevet et al. 2004). However, power spectrum has limited ability in distinguishing between noise and chaotic signals, since the latter can also have sharp spectral lines but even in the absence of noise can have broadband spectrum; see Chap. 6 for additional details.

Basically, both the autocorrelation function and the spectral density function contain the same information (since the spectral density is defined as the Fourier transform of the autocorrelation function). The difference is that this information is presented in the time (or space) domain by the autocorrelation function and in the frequency domain by the spectral density function.

3.5 Parametric Methods

In parametric methods, the *structure* of the model is specified a priori, and the number and nature of the parameters are generally fixed in advance. A large number of parametric methods have been developed in the literature. These methods include autoregressive (AR), moving average (MA), autoregressive moving average (ARMA), autoregressive integrated moving average (ARIMA), gamma autorgressive (GAR), periodic counterparts of AR, ARMA, and GAR, disaggregation, fractional Gaussian noise (FGN), point process, Markov chain process, scaling, and others. Extensive details of these models are available in Box and Jenkins (1970), Brillinger (1975), Chatfield (1989), and Brockwell and Davis (1991), among others. These methods have been extensively applied in hydrology for studying a wide variety of time series and associated problems; see Clarke (1994), Haan (1994), Hipel and McLeod (1994), and Salas et al. (1995) for details. Here, a brief account of some of these methods and their applications in hydrology is presented.

3.5.1 *Autoregressive (AR) Models*

In an autoregressive (AR) model, the present outcome is considered to be a linear combination of the signal in the past (with a finite memory) plus additive noise. For a time series, X_i, an autoregressive model of order p, i.e. AR(p) model or simply the AR model, is given by (e.g. Box and Jenkins 1970; Salas et al. 1995):

$$X_i = \mu + \phi_1(X_{i-1} - \mu) + \phi_2(X_{i-2} - \mu) + \cdots + \phi_p(X_{i-p} - \mu) + \varepsilon_i \tag{3.7}$$

where μ is the mean value of the series, ϕ is the regression coefficient, and ε_i is the error or noise, which is usually assumed to be an uncorrelated normal random variable with mean zero and variance σ_ε^2 (i.e. white Gaussian noise). The error ε_i is also uncorrelated with $X_{i-1}, X_{i-2}, \ldots, X_{i-p}$. Since ε_i is normally distributed, X_i is also normally distributed. In order to determine the order p of the autoregression required to describe the persistence adequately, it is necessary to estimate $p + 2$ parameters: $\phi_1, \phi_2, \ldots, \phi_p, \mu$, and σ_ε^2. Several methods have been proposed in the literature for an efficient estimation of these parameters; see Jenkins and Watts (1968) and Kendall and Stuart (1968) for some early studies.

The mean, variance, and autocorrelation function of the AR(p) process are (e.g. Box and Jenkins 1970; Salas et al. 1995)

$$E(X) = \mu \tag{3.8a}$$

$$Var(X) = \sigma^2 = \frac{\sigma_\varepsilon^2}{1 - \sum_{j=1}^{p} \phi_j \rho_j} \tag{3.8b}$$

$$\rho_\tau = \phi_1 \rho_{\tau-1} + \phi_1 \rho_{\tau-2} + \cdots + \phi_p \rho_{\tau-p} \tag{3.8c}$$

respectively. Equation (3.8c) is also known as the Yule–Walker equation. Given the model parameters, these three equations are useful for determining the properties of a model. Similarly, given a set of observations $X_1, X_2, \ldots, X_N$, they are useful for estimating the parameters of the model.

The simplest form of the AR model when $p = 1$, i.e. AR(1) model, is given by:

$$X_i = \mu + \phi_1 (X_{i-1} - \mu) + \varepsilon_i \tag{3.9}$$

Equation (3.9) is popularly known as the first-order Markov model. Such a model states that the value of X in one time period is dependent only on the value of X in the preceding time period plus a random component (with the random component independent of X). The AR(1) model has three parameters to be estimated: μ, ϕ_1, and σ_ε^2. For this model, Eqs. (3.8b) and (3.8c) become

$$\sigma^2 = \frac{\sigma_\varepsilon^2}{1 - \phi_1^2} \tag{3.10a}$$

$$\rho_\tau = \phi_1 \rho_{\tau-1} = \phi_1^\tau \tag{3.10b}$$

Autoregressive models have been widely used in hydrology. In particular, the first-order Markov model and other low-order AR models have been used for modeling annual hydrologic time series as well as seasonal and daily time series after standardization; see, for example, Thomas and Fiering (1962), Yevjevich (1964, 1975), Matalas (1967), and Quimpo (1967) for some early studies.

3.5.2 Moving Average (MA) Models

The moving average (MA) is used to smooth various types of time series. The moving average process used in the stochastic generation of hydrologic time series is somewhat different. In this, the moving average process describes the deviations of a sequence of events from their mean value. For a time series, X_i, a moving average model of order q, i.e. MA(q) model or simply the MA model, is given by (e.g. Box and Jenkins 1970):

$$X_i = \mu + \varepsilon_i - \theta_1 \varepsilon_{i-1} - \theta_2 \varepsilon_{i-2} - \cdots - \theta_q \varepsilon_{i-q} \tag{3.11}$$

The simplest form of the MA model when $q = 1$, i.e. MA(1) model, is given by:

$$X_i = \mu + \varepsilon_i - \theta_1 \varepsilon_{i-1} \tag{3.12}$$

For this model,

$$\sigma^2 = \sigma_\varepsilon^2 \left(1 - \theta_1^2\right) \tag{3.13a}$$

$$\rho_\tau = \frac{-\theta_1}{1 - \theta_1^2} \tag{3.13b}$$

The moving average model has been applied in many early studies on hydrologic time series analysis, especially to study annual series. For instance, Matalas (1963b) used the MA model to relate the effective annual precipitation and the annual runoff. Yevjevich (1963) used the MA model to relate the mean annual runoff to the annual effective precipitation.

Although useful, the MA model has not been particularly effective in the analysis of many hydrologic time series when applied independently. On the other hand, the model has been found to be very useful when combined with some other models, such as the autoregressive model, as described below.

3.5.3 Autoregressive Moving Average (ARMA) Models

The autoregressive moving average (ARMA) models combine any direct autocorrelation properties of a time series with the smoothing effects of an updated running mean through the series. The ARMA(p,q) model or simply the ARMA model is defined as (e.g. Box and Jenkins 1970; Salas et al. 1995):

$$X_i = \mu + \phi_1 (X_{i-1} - \mu) + \cdots + \phi_p \left(X_{i-p} - \mu\right) + \varepsilon_i - \theta_1 \varepsilon_{i-1} - \cdots - \theta_q \varepsilon_{i-q} \tag{3.14}$$

with p autoregressive parameters $\phi_1, \phi_2, \ldots, \phi_p$, and q moving average parameters $\theta_1, \theta_2, \ldots, \theta_q$. As above, the noise ε_i is an uncorrelated normal process with mean zero and variance σ_ε^2 and is also uncorrelated with $X_{i-1}, X_{i-2}, \ldots, X_{i-p}$. An ARMA $(p,0)$ model is the same as an AR(p) model and an ARMA($0,q$) model is the same as the MA(q) model.

One of the merits of the ARMA process is that, in general, it is possible to fit a model with a small number of parameters, i.e. $p + q$. This number is generally smaller than the number of parameters that would be necessary using either an AR model or an MA model. This principle is called the parsimony of parameters.

A simple form of the ARMA model with $p = 1$ and $q = 1$, i.e. the ARMA(1,1) model, is given by:

$$X_i = \mu + \phi_1(X_{i-1} - \mu) + \varepsilon_i - \theta_1\varepsilon_{i-1} \tag{3.15}$$

with $-1 < \phi_1 < 1$ and $-1 < \theta_1 < 1$. The variance and lag-1 autocorrelation coefficient of the ARMA(1,1) model are:

$$\sigma^2 = \frac{1 - 2\phi_1\theta_1 + \theta_1^2}{1 - \phi_1^2}\sigma_\varepsilon^2 \tag{3.16a}$$

$$\rho_1 = \frac{(1 - \phi_1\theta_1)(\phi_1 - \theta_1)}{1 - 2\phi_1\theta_1 + \theta_1^2} \tag{3.16b}$$

respectively. Furthermore, the autocorrelation function is:

$$\rho_\tau = \phi_1\rho_{\tau-1} = \rho_1\phi_1^{\tau-1} \quad \tau > 1 \tag{3.16c}$$

Comparing Eqs. (3.16c) and (3.10b), one may observe that ρ_τ of the AR(1) process is less flexible than that of the ARMA(1,1) process, since the former depends on the sole parameter ϕ_1, while the latter depends on ϕ_1 and θ_1. In general, AR processes are short memory processes, while ARMA processes are long-memory processes (Salas et al. 1979; Salas 1993).

The ARMA(1,1) model and other low-order ARMA models are very useful in hydrology. They have found widespread applications in hydrology; see Carlson et al. (1970), McKerchar and Delleur (1974), Moss and Bryson (1974), Tao and Delleur (1976), Hipel et al. (1977), Lettenmaier and Burges (1977), McLeod et al. (1977), Delleur and Kavvas (1978), Hipel and McLeod (1978a), Cooper and Wood (1982), Salas and Obeysekera (1982), and Stedinger and Taylor (1982) for some early applications. The applications have been on annual hydrologic time series, seasonal time series after seasonal standardization, daily time series either after seasonal standardization or by separating the year into several seasons and applying different models to the daily series in each season, and many others.

3.5.4 Gamma Autoregressive (GAR) Models

Since the AR, MA, and ARMA models are based on the assumption that the error ε_i, and, hence, the time series X_i are normally distributed, they cannot be directly applied to skewed hydrologic time series. Their application to skewed time series requires transformation of the time series into Normal processes. The gamma autoregressive (GAR) model does not require such a transformation and, thus, offers a direct modeling approach for skewed time series. The GAR model is based on the assumption that the underlying series have a gamma marginal distribution. The GAR model is defined as (Lawrance and Lewis 1981):

$$X_i = \phi X_{i-1} + \varepsilon_i \tag{3.17}$$

where ϕ is the autoregressive coefficient, ε_i is a random component, and X_i has a three-parameter gamma marginal distribution given by the function:

$$f_x(x) = \frac{\alpha^\beta (x - \lambda)^{\beta - 1} \exp[-\alpha(x - \lambda)]}{\Gamma(\beta)} \tag{3.18}$$

where λ, α, and β are the location, scale, and shape parameters, respectively.

There are two ways to obtain the variable ε_i. For integer values of β, ε is given by (Gaver and Lewis 1980):

$$\varepsilon = \frac{\lambda(1 - \phi)}{\beta} + \sum_{j=1}^{\beta} \eta_j \tag{3.19}$$

where $\eta_j = 0$, with probability ϕ; $\eta_j = \exp(\alpha)$, with probability $(1 - \phi)$; and $\exp(\alpha)$ is an exponentially distributed random variable with expected value $1/\alpha$. This approach is valid for skewness coefficient less than or equal to 2.0. For non-integer values of β, ε is given by (Lawrance 1982):

$$\varepsilon = \lambda(1 - \phi) + \eta \tag{3.20}$$

where

$$\eta = 1 \quad if\ M = 0 \tag{3.21a}$$

$$\eta = \sum_{j=1}^{M} E_j \phi^{U_j} \quad if\ M > 0 \tag{3.21b}$$

where M is an integer random variable with Poisson distribution of mean value $-\beta\ln(\phi)$. The set $U_1,\ldots, U_M$ are independent and identically distributed (iid) random variables with uniform distribution (0,1) and the set $E_1,\ldots, E_M$ are iid random variables exponentially distributed with mean $1/\alpha$ (Salas 1993).

The GAR model has been applied in many hydrologic studies, including for modeling annual streamflow and rainfall disaggregation; see Fernandez and Salas (1990), Cigizoglu and Bayazit (1998), and Koutsoyiannis and Onof (2001), among others.

3.5.5 Periodic Models: PAR, PARMA, and PGAR Models

To incorporate periodicity in processes (e.g. seasonality in rainfall and streamflow), a number of periodic models have been suggested in the literature; see Box and Jenkins (1970), Brockwell and Davis (1991), and Salas et al. (1995) for extensive details. Of particular significance are the periodic counterparts of AR, ARMA, and GAR models. In what follows, periodic AR (PAR), periodic ARMA (PARMA) and multiplicative PARMA, and periodic GAR (PGAR) models are described, as these models have found widespread applications in hydrology; see Loucks et al. (1981), Salas (1993), Hipel and McLeod (1994), and Salas et al. (1995) for some comprehensive accounts.

3.5.5.1 Periodic AR (PAR) Model

Let us consider a periodic time series represented by $X_{y,s}$, $y = 1, 2, \ldots, N$; $s = 1, 2, \ldots, \omega$, where y is the year and s is the season, N is the number of years of record, and ω is the number of seasons in a year. Here, s can represent a day, week, month, or season. If s represents a month, then $\omega = 12$. The periodic autoregressive model of order p, i.e. PAR(p) model, for such a time series is defined as:

$$X_{y,s} = \mu_s + \sum_{j=1}^{p} \phi_{j,s}\left(X_{y,s-j} - \mu_{s-j}\right) + \varepsilon_{y,s} \tag{3.22}$$

where $\varepsilon_{y,s}$ is an uncorrelated normal variable with mean zero and variance $\sigma_s^2(\varepsilon)$, and it is also uncorrelated with $X_{y,s-1}, X_{y,s-2}, \ldots, X_{y,s-p}$. The model parameters are μ_s, $\phi_{1,s}$, $\phi_{2,s}$, $\ldots$, $\phi_{p,s}$ and $\sigma_s^2(\varepsilon)$ for $s = 1, 2, \ldots, \omega$. In Eq. (3.22), if $s - j \leq 0$, then $X_{y,s-j}$ becomes $X_{y-1,\omega+s-j}$ and μ_{s-j} becomes $\mu_{\omega+s-j}$. The simplest PAR(p) model, i.e. PAR(1) model, can be written as:

$$X_{y,s} = \mu_s + \phi_{1,s}\left(X_{y,s-1} - \mu_{s-1}\right) + \varepsilon_{y,s} \tag{3.23}$$

The low-order PAR models have been widely used in hydrology, especially for monthly (and seasonal) rainfall and streamflow simulations; see Hannan (1955), Thomas and Fiering (1962), Fiering and Jackson (1971), Delleur et al. (1976), Salas and Abdelmohsen (1993), and Shahjahan Mondal and Wasimi (2006), among others.

3.5.5.2 Periodic ARMA (PARMA) Model

In a similar manner, the periodic autoregressive moving average model, i.e. PARMA(p,q) model, for the above time series $X_{y,s}$ is defined, by extending the PAR(p) model, as:

$$X_{y,s} = \mu_s + \sum_{j=1}^{p} \phi_{j,s}\left(X_{y,s-j} - \mu_{s-j}\right) + \varepsilon_{y,s} - \sum_{j=1}^{q} \theta_{j,s}\varepsilon_{y,s-j} \tag{3.24}$$

The model parameters are μ_s, $\phi_{1,s}$, $\phi_{2,s}$, …, $\phi_{p,s}$, $\theta_{1,s}$, $\theta_{2,s}$, …, $\theta_{p,s}$, and $\sigma_s^2(\varepsilon)$ for $s = 1, 2, \ldots, \omega$. When $q = 0$, the model becomes the well-known PARMA(p,0) or PAR(q) model. The simplest PARMA(p,q) model, i.e. PARMA(1,1) model, can be written as:

$$X_{y,s} = \mu_s + \phi_{1,s}\left(X_{y,s-1} - \mu_{s-1}\right) + \varepsilon_{y,s} - \theta_{1,s}\varepsilon_{y,s-1} \tag{3.25}$$

Low-order PARMA models, such as PARMA(1,0) and PARMA(1,1), have been widely used for modeling seasonal hydrologic processes, especially for simulating monthly and weekly flows; see Delleur and Kavvas (1978), Hirsch (1979), Vecchia (1985), Ula (1990), Salas and Obeysekera (1992), Bartolini and Salas (1993), Salas (1993), Salas and Abdelmohsen (1993), Rasmussen et al. (1996), and Tesfaye et al. (2006), among others. In such studies, the method of moments is typically used for estimation of the model parameters.

Physically-based or conceptual arguments of the hydrologic cycle of a watershed or river basin justify the applicability of the PARMA models. For instance, Salas and Obeysekera (1992) showed that assuming that the precipitation input is an uncorrelated periodic-stochastic process and under some linear reservoir considerations for the groundwater storage, the stochastic model for seasonal streamflow becomes a PARMA(1,1) process. Some studies have also suggested a constant parameter ARMA(2,2) model with periodic independent residuals (e.g. Claps et al. 1993; Murrone et al. 1997).

3.5.5.3 Multiplicative PARMA Model

It is obvious that preservation of both seasonal and annual statistics should be a desirable property of periodic stochastic models. However, such dual preservation of statistics is also difficult to achieve with low-order PAR models, such as the PAR(1) and PAR(2). While PARMA models offer the possibility of preserving both seasonal and annual statistics, due to their more flexible correlation structure, there are some concerns that such models have too many parameters (although the number may be reduced by keeping some of the parameters constant). To overcome these problems, the family of multiplicative models was proposed as an alternative (Box and Jenkins 1970).

Multiplicative models have the characteristic of linking the variable $X_{y,s}$ with $X_{y,s-1}$ and $X_{y-1,s}$. However, without periodic parameters, such models often cannot produce the seasonality in the covariance structure of the process, as was shown by McKerchar and Delleur (1974) in their study on simulation and forecasting of monthly streamflows. A model, with periodic parameters, that can overcome these limitations is the multiplicative PARMA model. The multiplicative PARMA(1,1) × $(1,1)_\omega$ model is given by:

$$X_{y,s} = \mu_s + \Phi_{1,s}\left(X_{y-1,s} - \mu_s\right) + \phi_{1,s}\left(X_{y,s-1} - \mu_{s-1}\right) - \Phi_{1,s}\phi_{1,s}\left(X_{y-1,s-1} - \mu_{s-1}\right) + \varepsilon_{y,s} - \Theta_{1,s}\varepsilon_{y-1,s} - \theta_{1,s}\varepsilon_{y,s-1} + \Theta_{1,s}\theta_{1,s}\varepsilon_{y-1,s-1} \tag{3.26}$$

where $\Phi_{1,s}$, $\Theta_{1,s}$, $\theta_{1,s}$, and $\sigma_s(\varepsilon)$ are the model parameters.

3.5.5.4 Periodic GAR (PGAR) Model

The above PARMA and multiplicative PARMA models for modeling periodic time series require transformation of the time series into Normal. An alternative that can overcome this problem is the periodic GAR model of order 1, i.e. PGAR(1) model. This model has periodic correlation structure and periodic gamma marginal distribution (e.g. Fernandez and Salas 1986). Let us consider that $X_{y,s}$ is a periodic correlated variable with gamma marginal distribution with location λ_s, scale α_s, and shape β_s parameters varying with s, and $s = 1, 2, \ldots, \omega$. Then, the variable $Z_{y,s} = X_{y,s} - \lambda_s$ is a two-parameter gamma that can be represented by the model

$$X_{y,s} = \lambda_s + \phi_s\left(X_{y,s-1} - \lambda_{s-1}\right) + \left(X_{y,s-1} - \lambda_{s-1}\right)^{\delta_s} W_{y,s} \tag{3.27}$$

where ϕ_s is the periodic autoregressive coefficient, δ_s is the periodic autoregressive exponent, and $W_{y,s}$ is the noise process. This model has a periodic correlation structure equivalent to that of the PAR(1) process. An early application of this model in hydrology was made by Fernandez and Salas (1986) for modeling the weekly streamflows in several rivers in the United States. The model favorably compared with respect to the Normal-based models (e.g. the PAR model after logarithmic transformation) in reproducing the basic statistics usually considered for streamflow simulation.

3.5.6 *Extension of AR, ARMA, PAR, and PARMA Models to Multiple Variables*

3.5.6.1 Multivariate AR and Multivariate ARMA Models

Let us consider multiple time series $X_i^{(1)}$, $X_i^{(2)}$, ..., $X_i^{(n)}$, with n representing the number of variables or sites. This multiple time series can be represented as a column vector $\mathbf{X}_i$, with elements $X_i^{(1)}$, $X_i^{(2)}$, ..., $X_i^{(n)}$. With this, the simplest multivariate AR model, i.e. multivariate AR(1) model, can be written as (Matalas 1967):

$$\boldsymbol{X}_i = \boldsymbol{\mu} + \boldsymbol{A}_1(\boldsymbol{X}_{i-1} - \boldsymbol{\mu}) + \boldsymbol{B}\boldsymbol{\varepsilon}_i \tag{3.28}$$

where $\boldsymbol{\mu}$ is a column parameter vector with elements $\mu_i^{(1)}$, $\mu_i^{(2)}$, ..., $\mu_i^{(n)}$; and $\mathbf{A}_1$ and $\mathbf{B}$ are $n \times n$ parameter matrices. The noise term $\boldsymbol{\varepsilon}_i$ is also a column vector of noises $\varepsilon_i^{(1)}$, $\varepsilon_i^{(2)}$, ..., $\varepsilon_i^{(n)}$, each with zero mean such that $\mathbf{E}(\boldsymbol{\varepsilon}_i\boldsymbol{\varepsilon}_i^T) = \mathbf{I}$, where $\mathbf{T}$ denotes the transpose of the matrix and $\mathbf{I}$ is the identity matrix, and $\mathbf{E}(\boldsymbol{\varepsilon}_i\boldsymbol{\varepsilon}_{i-\tau}^T) = 0$ for $\tau \neq 0$. In addition, $\boldsymbol{\varepsilon}_i$ is uncorrelated with $\mathbf{X}_{i-1}$ and $\boldsymbol{\varepsilon}_i$ is also normally distributed.

In a similar manner, the simplest multivariate ARMA model, i.e. multivariate ARMA(1,1) model, is given by:

$$\boldsymbol{X}_i = \boldsymbol{\mu} + \boldsymbol{A}_1(\boldsymbol{X}_{i-1} - \boldsymbol{\mu}) + \boldsymbol{B}\boldsymbol{\varepsilon}_i - \boldsymbol{C}_1\boldsymbol{\varepsilon}_{i-1} \tag{3.29}$$

where $\mathbf{C}_1$ is an additional $n \times n$ parameter matrix.

The multivariate AR and multivariate ARMA models have found a number of applications in hydrology; see Pegram and James (1972), Ledolter (1978), Cooper and Wood (1982), Stedinger et al. (1985), and Chaleeraktrakoon (1999, 2009), among others; see also Salas et al. (1995) for a comprehensive account. A particular limitation in using the full multivariate AR and multivariate ARMA models is that they often lead to complex parameter estimation. To overcome this issue, simplifications have been suggested. One such simplification is the assumption of a diagonal matrix for $\mathbf{A}_1$ in the multivariate AR model (Eq. 3.28) (e.g. Matalas 1967) and for $\mathbf{A}_1$ and $\mathbf{C}_1$ in the multivariate ARMA model (Eq. 3.29) (e.g. Salas et al. 1995). These simplified models are called *contemporaneous* models, to imply that only the dependence of concurrent values of the X's are considered important. The diagonalization of the parameter matrices allows model decoupling into component univariate models, so that the model parameters do not have to be estimated jointly and that univariate modeling procedures can be employed. The contemporaneous AR model is termed as the CAR model and the contemporaneous ARMA model is popularly known as the CARMA model.

3.5.6.2 Multivariate PAR and Multivariate PARMA Models

The multivariate AR and multivariate ARMA models can be modified for periodic series, similar to the modifications presented earlier for the univariate periodic series. Let us consider a periodic multiple time series $X_{y,s}^{(1)}$, $X_{y,s}^{(2)}$, …, $X_{y,s}^{(n)}$, where y is the year, s is the season, and n is the number of variables or sites. These time series can be represented as a column vector $\mathbf{X}_{y,s}$, with elements $X_{y,s}^{(1)}$, $X_{y,s}^{(2)}$, …, $X_{y,s}^{(n)}$. With this, the simplest multivariate periodic AR model, i.e. multivariate PAR(1) model, can be written as:

$$\boldsymbol{X}_{y,s} = \boldsymbol{\mu}_s + \boldsymbol{A}_s\left(\boldsymbol{X}_{y,s-1} - \boldsymbol{\mu}_s\right) + \boldsymbol{B}_s\boldsymbol{\varepsilon}_{y,s} \tag{3.30}$$

where $\boldsymbol{\mu}$ is a column parameter vector with elements $\mu_s^{(1)}$, $\mu_s^{(2)}$, …, $\mu_s^{(n)}$; and $\mathbf{A}_s$ and $\mathbf{B}_s$ are $n \times n$ parameter matrices. All these three parameters are periodic. The noise term $\boldsymbol{\varepsilon}_{y,s}$ is also a column vector of noises $\varepsilon_{y,s}^{(1)}$, $\varepsilon_{y,s}^{(2)}$, …, $\varepsilon_{y,s}^{(n)}$, each with zero mean such that $\mathbf{E}(\boldsymbol{\varepsilon}_{y,s}\boldsymbol{\varepsilon}_{y,s}^T) = \mathbf{I}$, and $\mathbf{E}(\boldsymbol{\varepsilon}_{y,s}\boldsymbol{\varepsilon}_{y,s-\tau}^T) = 0$ for $\tau \neq 0$. In addition, $\boldsymbol{\varepsilon}_{y,s}$ is uncorrelated with $\mathbf{X}_{y,s-1}$ and $\boldsymbol{\varepsilon}_{y,s}$ is also normally distributed.

In a similar manner, the simplest multivariate periodic ARMA model, i.e. multivariate PARMA(1,1) model, is given by:

$$\boldsymbol{X}_{y,s} = \boldsymbol{\mu}_s + \boldsymbol{A}_s\left(\boldsymbol{X}_{y,s-1} - \boldsymbol{\mu}_s\right) + \boldsymbol{B}_s\boldsymbol{\varepsilon}_{y,s} - \boldsymbol{C}_s\boldsymbol{\varepsilon}_{y,s-1} \tag{3.31}$$

where $\mathbf{C}_s$ is an additional $n \times n$ periodic parameter matrix. Simplifications of these models can be made through diagonalization. These simplified models are called contemporaneous periodic AR model and contemporaneous periodic ARMA model, respectively, for PAR and PARMA models.

A number of studies have used the multivariate PAR and multivariate PARMA and their contemporaneous versions for modeling hydrologic time series; see Vecchia (1985), Bartolini et al. (1988), Ula (1990), Salas and Abdelmohsen (1993), and Rasmussen et al. (1996), among others.

3.5.7 *Disaggregation Models*

Many hydrologic design and operational problems often require data at much finer scales than that are commonly available through measurements. For instance, studies on floods and design of urban drainage structures require rainfall data at hourly scale or at even finer scales. However, rainfall data are widely available only at the daily scale. Therefore, it becomes necessary to disaggregate the available daily rainfall data to hourly or finer-scale data. For streamflow, the problem may be to obtain monthly flows from annual flows, since simulations of annual flows are much more accurate than those of monthly flows. These data issues are applicable not only to time but also to space and, consequently, to space-time. In view of these, development and applications of disaggregation models have been an important part of stochastic hydrology. Such disaggregation models can be broadly categorized into temporal disaggregation models, spatial disaggregation models, and space-time disaggregation models. Here, a brief account of such models is presented, with a description of the Valencia-Schaake model (Valencia and Schaake 1973) for disaggregation of streamflows from the annual scale to the seasonal scale (e.g. monthly), as this model has been widely used in hydrology, especially in streamflow studies.

The Valencia-Schaake model to disaggregate streamflows from the annual scale to the seasonal scale (with number of seasons s) at n sites is written as:

$$\boldsymbol{Y} = \boldsymbol{AX} + \boldsymbol{B\varepsilon} \tag{3.32}$$

where $\mathbf{Y}$ is an ns vector of disaggregated seasonal values (monthly streamflows), X is an n vector of aggregate annual values (annual streamflows), $\mathbf{A}$ and $\mathbf{B}$ are $ns \times ns$ parameter matrices, and $\boldsymbol{\varepsilon}$ is an ns vector of independent standard normal variables. The matrix $\mathbf{A}$ is estimated to reproduce the correlation between aggregate and disaggregate values, whereas the matrix $\mathbf{B}$ is estimated to produce the correlation between individual disaggregate components. Parameter estimation, based on the method of moments, generally leads to the preservation of the first- and second-order moments at all levels of aggregation. In this model, the derivation of the monthly streamflow values is accomplished in two or more steps. First, the aggregate annual values are modeled so as to reproduce the desired annual statistics (e.g. based on the ARMA(1,1) model). Then, synthetic annual values are generated and subsequently disaggregated into the seasonal values by means of the model parameters in Eq. (3.32).

The Valencia-Schaake model generally performs well for disaggregation, as the variance-covariance properties of the seasonal data are preserved and the generated seasonal values also add up to the annual values. However, it also suffers from the following shortcomings, among others: (1) the model does not preserve the covariances of the first season of a year and any preceding season; (2) the model is compatible mainly with data that exhibit Gaussian distributions, but some hydrologic data (including daily and sub-daily rainfall data as well as monthly streamflow data) are seldom normally distributed. Therefore, it becomes necessary to apply some kind of normalizing transformations to the data before its application. However, it is often difficult to find a general normalizing transformation and retain the statistical properties of the process; (3) the model involves an excessive number of parameters; and (4) the linear nature of the model limits it from representing any nonlinearity in the dependence structure between variables, except through the normalizing transformation used.

Some of these limitations have been addressed by certain modifications to the model. For instance, Mejia and Rouselle (1976) presented a modification to preserve the covariances of the first season of a year and any preceding season. Their model is given by:

$$\boldsymbol{Y} = \boldsymbol{AX} + \boldsymbol{B\varepsilon} + \boldsymbol{CZ} \tag{3.33}$$

where **C** is an additional parameter matrix and **Z** is a vector of seasonal values from the previous year (usually only the last season of the previous year). The issues regarding summability (i.e. the requirement that disaggregate variables should add up to the aggregate quantity), Gaussian assumption, and number of parameters have been addressed by many studies through other modifications (e.g. Tao and Delleur, 1976; Hoshi and Burges 1979; Lane 1979; Todini 1980; Stedinger and Vogel 1984; Stedinger et al. 1985; Grygier and Stedinger 1988; Koutsoyiannis 1992; Santos and Salas 1992; Salas 1993). These limitations also led to the proposal of new disaggregation models that do not necessarily exhibit the generality of the Valencia-Schaake model. Such models are based on non-dimensionalized Markov process (e.g. Woolhiser and Osborne 1985; Hershenhorn and Woolhiser 1987), dynamic step-wise disaggregation (e.g. Koutsoyiannis and Xanthopoulos 1990; Koutsoyiannis 2001), Bartlett-Lewis rectangular pulses (e.g. Bo et al. 1994; Glasbey et al. 1995), Poisson cluster process (e.g. Connolly et al. 1998; Koutsoyiannis and Onof 2001), scaling (Perica and Foufoula-Georgiou 1996; Olsson 1998), and more general forms (e.g. Koutsoyiannis 2000). Furthermore, to overcome the issues relating to a priori definition of the model structure in these parametric models, nonparametric and semi-parametric approaches for disaggregation have also been proposed (e.g. Tarboton et al. 1998; Nagesh Kumar et al. 2000; Srinivas and Srinivasan 2005; Prairie et al. 2007; Nowak et al. 2010); see also Sect. 3.6 for a discussion of nonparametric approaches.

3.5.8 Markov Chain Models

A Markov chain is a discrete-time stochastic process that undergoes transitions from one state to another on a state space. Let us consider a discrete-time process $X(i)$ developing through time i. Let us denote the values this process takes as X_i, $i = 1, 2, \ldots, N$. With this, given the entire history of the process, the probability of the process being equal to X_i at time i can be written as:

$$P[X(i) = X_i | X(1) = X_1, X(2) = X_2, \ldots, X(i-1) = X_{i-1}] \tag{3.34}$$

Simplification of Eq. (3.34) as:

$$P[X(i) = X_i | X(i-1) = X_{i-1}] \tag{3.35}$$

indicates that the outcome of the process at time i can be defined by using only the outcome at time $i-1$. A process with this property is known as the first-order Markov chain or a simple Markov chain.

Let us consider that there is a total of u number of states of the process and that any given state is denoted by l. With this, instead of using X_i, one can use $X(i)$ to represent the state of the process, such as $X(i) = l$, $l = 1, 2, \ldots, u$, which means that $X(i)$ is at state l. For instance, for rainfall modeling, a total of two states (i.e. $u = 2$) is normally considered: $l = 1$ for a dry state (i.e. no rain) and $l = 2$ for a wet state.

The first-order Markov chain is defined by its transition probability matrix $\mathbf{P}(i)$. The matrix $\mathbf{P}(i)$ is a square matrix with elements $p_{kl}(i)$ and is given by:

$$p_{kl}(i) = P[X(i) = l | X(i-1) = k] \tag{3.36}$$

for all k, l pairs. Therefore, one can write that

$$\sum_{l=1}^{u} p_{kl}(i) = 1 \quad k = 1, \ldots, u \tag{3.37}$$

If the transition probability matrix $\mathbf{P}(i)$ does not depend on time i, then the Markov chain is a homogeneous chain or a stationary chain. In such a case, the transition probability matrix is denoted as P and the elements are denoted as p_{kl}. The probabilities that are often useful in hydrology, especially in rainfall, are the r-step transition probability $p_{kl}^{(r)}$ (assuming that the chain changes from state k to state l in r steps), the marginal distribution $q_l(i)$ given the distribution $q_l(1)$ (i.e. for the initial state), and the steady-state probability vector $\mathbf{q}^*$; see Parzen (1962) for details.

The Markov chain models have been widely used in hydrology, including for rainfall (e.g. Gabriel and Neumann 1962; Haan et al. 1976; Buishand 1977; Chin 1977; Katz 1977; Roldan and Woolhiser 1982; Chang et al. 1984; Mimikou 1984; Foufoula-Georgiou and Lettenmaier 1987; Bardossy and Plate 1991; Zucchini and Guttorp 1991; Katz and Parlange 1995; Rajagopalan et al. 1996; Wilks 1998; Hughes et al. 1999; Thyer and Kuczera 2000; Robertson et al. 2004; Lennartsson et al. 2008), streamflow (e.g. Yevjevich 1972; Şen 1976, 1990; Chung and Salas 2000; Cancelliere and Salas 2004, 2010; Bayazit and Önöz 2005; Akyuz et al. 2012; Sharma and Panu 2014), and water storage (e.g. Moran 1954; Lloyd 1963; Gani 1969), among others. In recent years, use of Markov chains for parameter estimation and uncertainty in hydrologic models has also been gaining considerable attention; see Bates and Campbell (2001), Marshall et al. (2004), Vrugt et al. (2008), and Vrugt (2016) for some accounts.

3.5.9 Point Process Models

The theory of point processes (e.g. Cox 1958; Neyman and Scott 1958; Bartlett 1963) has been one of the earliest tools for modeling rainfall (e.g. Le Cam 1958). A fundamental assumption in rainfall models based on point processes is that the occurrence of rain storms is a Poisson process. Assuming that the storm arrivals are governed by a Poisson process, the number of storms N_i in a time interval (1,i) arriving at a location is Poisson-distributed with a parameter λ_i (storm arrival rate).

$$P[N_i = n] = (\lambda_i)^n \exp(-\lambda_i) \quad n = 1, 2, \ldots \tag{3.38}$$

If n storms arrived in the interval (1,i) at times $i_1, \ldots, i_n$, then the number of storms in any time interval I (= i/n) is also Poisson-distributed with parameter λ_I. It is further assumed that the rainfall amount (R) associated with a storm arrival is white noise (e.g. gamma-distributed) and that the number of storms N_i and amount R are assumed to be independent. Thus, rainfall amounts $r_1, \ldots, r_n$ correspond to storms occurring at times $i_1, \ldots, i_n$. Such a rainfall generating process is called Poisson white noise (PWN) and is a simple example of a point process.

Under this formulation, the cumulative rainfall in the interval (1,i) is given by:

$$Z_i = \sum_{j=1}^{N_i} R_j \tag{3.39}$$

The cumulative rainfall Z_i is called a compound Poisson process.

In the PWN model, the rainfall is assumed to occur instantaneously, so the storms have zero duration, which is also unrealistic. Modifications to the PWN model consider rainfall as an occurrence with a random duration E and intensity Y, called the Poisson rectangular pulse (PRP) model (Rodriguez-Iturbe et al. 1984). Commonly, E and Y are assumed to be independent and exponentially distributed. In this formulation, n storms may occur at times $i_1, \ldots, i_n$ with associated durations and intensities (y_1, e_1), …, (y_n, e_n), and the storms may overlap, so that the aggregated process X_t becomes autocorrelated. The PRP model is better conceptualized than the PWN model, but it is still limited when applied to rainfall data.

In view of the limitations of the simple point process models, such as the PWN and PRP, several alternative models, often with greater complexity and sophistication, have also been suggested. These include models based on Cox processes, renewal processes, cluster processes, and others; see Le Cam (1958), Buishand (1977), Kavvas and Delleur (1981), Smith and Carr (1983), Rodriguez-Iturbe et al. (1984, 1986, 1987), Ramirez and Bras (1985), Foufoula-Georgiou and Guttorp (1986), and Foufoula-Georgiou and Lettenmaier (1987) for some early applications of these concepts for modeling rainfall. Comprehensive reviews and comparisons of different models are also available in the literature; see, for instance, Onof et al. (2000) for some details. Among these, the cluster-based models have been widely used, and in particular the Neyman-Scott model and the Bartlett-Lewis model. A brief account of these two models is presented next.

3.5.9.1 Neyman-Scott Model

The Neyman-Scott model for rainfall (Rodriguez-Iturbe et al. 1987) was based on the cluster point process model of Neyman and Scott (1958). The model assumes that there exists a generating mechanism called the "storm origin" in any storm event which may be passing fronts or some other criteria for convection storms from which rain cells develop. The Neyman-Scott model is described by three independent elementary stochastic processes: (1) a process that sets the origins of the storms; (2) a process that determines the number of rain cells generated by each storm; and (3) a process that defines the origin of the cells. The origins of the storms (L) are governed by a Poisson process with rate parameter λ. At a point on the ground, the storm is conceptualized as a random number of rain cells C, which are Poisson or geometrically distributed. The cell origins and positions (B) are independently separated from the storm origin by distances which are exponentially distributed with parameter β. No cell origins are assumed to be located at the storm origin. A rectangular pulse is associated independently with each cell origin with random duration, E, and with random intensity, Y. The duration and intensity are assumed to be exponentially distributed with parameters η and $1/\mu_x$, respectively, and are independent of each other. With these, the basic Neyman-Scott model consists of five variables (origin of storms—L, number of cells—C, position of cells —B, duration of cells—E, and intensity of cells—Y) and five parameters (λ, $1/\mu_c$, β, η, $1/\mu_x$). Two storms and cells may overlap, and the total rainfall intensity at any point in time i, i.e. X_i, is given by the sum of the intensities of the individual cells active at time i. If the rainfall cell is described by an instantaneous random rainfall depth, the resulting rainfall process is known as Neyman-Scott white noise (NSWN). On the other hand, if the rainfall cell is a rectangular pulse, the rainfall process is known as Neyman-Scott rectangular pulse (NSRP).

Parameter estimation of Neyman-Scott models has been extensively studied in the hydrologic literature using the method of moments and other approaches (e.g. Kavvas and Delleur 1981; Obeysekera et al. 1987; Entekhabi et al. 1989; Islam et al. 1990; Cowpertwait 1995). A major problem with parameter estimation in Neyman-Scott models is that parameters estimated based on data for one level of aggregation (e.g. hourly) may be significantly different from those estimated from data for another level of aggregation (e.g. daily). Weighted moments estimates of various timescales in a least squares fashion have been proposed as an alternative (e.g. Entekhabi et al. 1989). Constraints may be set on the parameters based on the physical understanding of the process that can improve parameter estimation (Cowpertwait and O'Connell 1997), as shown in a space-time cluster model (Waymire et al. 1984). However, the difficulty in estimating the parameters even when using physical considerations persists.

Since the introduction of the Neyman-Scott model for rainfall modeling, numerous studies have applied the model and its variants for rainfall modeling and for other subsequent analysis. Details of such applications can be seen in Cowpertwait (1991), Puente et al. (1993), Cowpertwait et al. (1996a, b, 2002, 2007), Calenda and Napolitano (1999), Favre et al. (2002), Evin and Favre (2008), Leonard et al. (2008), and Burton et al. (2010), among others.

3.5.9.2 Bartlett-Lewis Model

The Bartlett-Lewis model for rainfall was originally proposed by Rodriguez-Iturbe et al. (1987). The model works in a somewhat similar manner to that of the Neyman-Scott model. Similar to the Neyman-Scott model, the Bartlett-Lewis model represents the arrival of rain storms as a Poisson process with rate λ, with each storm generating a cluster of cell arrivals. However, the clustering mechanism assumes that the time intervals between successive cells (rather than the temporal distances of the cells from their storm origin) are independent and identically distributed random variables. The intervals between successive cells are assumed to be exponentially distributed, so that cell arrivals constitute a secondary Poisson process of rate β. Each cell is associated with a rectangular pulse of rain, of random duration, E, and with random intensity, Y. In the simplest version of the model, these are both assumed to be exponentially distributed with parameters η and $1/\mu_x$, respectively, and are independent of each other. The cell origin process terminates after a time that is also exponentially distributed with rate γ. This basic version, therefore, has five parameters in total: λ, β, η, $1/\mu_x$, and γ.

Since the original model of Rodriguez-Iturbe et al. (1987), several modifications have been suggested to the Bartlett-Lewis rainfall model. One early modification was proposed by Rodriguez-Iturbe et al. (1988), which involves randomization of the cell duration parameter and related temporal storm characteristics to enable variation between storms. This model, called the Bartlett-Lewis Random Parameter (BLRPR) model, extends the basic model by allowing parameter η, that specifies the duration of cells, to vary randomly between storms. This is achieved by assuming that the η values for different storms are independent and identically-distributed random variables from a gamma distribution with index (shape) α and rate parameter (scale) ν. This way, the model is re-parameterized in such a way that, rather than keeping β (the cell arrival rate) and γ (the storm termination rate) constant for each storm, the ratio of both these parameters to η is kept constant. This means that, for higher η (i.e. typically shorter cell durations), there are correspondingly shorter storm durations and shorter cell inter-arrival times. This model improves the simulation of dry spell lengths. Other modifications to the BLRP model include the Bartlett-Lewis rectangular pulse gamma model (Onof and Wheater 1994), hybrid-based model (Gyasi-Agyei and Willgoose 1997), spatial-temporal model based on Bartlett-Lewis process (Northrop 1998), the Bartlett-Lewis Instantaneous Pulse (BLIP) model (Copertwait et al. 2007), and the random parameter Bartlett-Lewis Instantaneous Pulse (BLIPR) model (Kaczmarska et al. 2014), among others.

The original Bartlett-Lewis model and its modified versions have been applied in numerous rainfall studies. Extensive details of such applications, including reviews and comparisons of different stochastic methods, are available in Islam et al. (1990), Onof and Wheater (1993), Copertwait (1995, 1998, 2004, 2010), Khaliq and Cunnane (1996), Verhoest et al. (1997, 2010), Cameron et al. (2000), Onof et al. (2000), Koutsoyiannis and Onof (2001), Smithers et al. (2002), Marani and Zanetti (2007), Copertwait et al. (2011), Pui et al. (2012), Pham et al. (2013), and Kaczmarska et al. (2014), among others.

3.5.10 Other Models

In addition to the models discussed above, a number of other stochastic parametric models also exist for time series analysis. Such models include those that are either variants/extensions of the models discussed above or based on very different concepts. Examples of such models are autoregressive integrated moving average (ARIMA) models, fractional Gaussian noise (FGN) models, broken line (BL) models, and scaling models. All these models have found important applications in hydrology; see Mandelbrot and Van Ness (1968), Mandelbrot and Wallis (1969), Carlson et al. (1970), Mandelbrot (1972), Mejia et al. (1972, 1974), Delleur and Kavvas (1978), Hipel and McLeod (1978b), Salas et al. (1982), Gupta and Waymire (1990), Ahn and Salas (1997), Montanari et al. (1997), Venugopal et al. (1999), Deidda (2000), and Seed et al. (2000) for some earlier applications.

Among these models, the scaling-based models have been gaining far more attention over the past two decades or so. The scaling-based models are generally based on self-similar or scale-invariant structure of the underlying process, and originate from turbulence theory (e.g. Mandelbrot 1974; Meneveau and Sreenivasan 1987). Comprehensive accounts of the relevance of scaling-based concepts in hydrology and their applications can be found in Mandelbrot (1983), Gupta et al. (1986), Rodriguez-Iturbe and Rinaldo (1997), and Sposito (2008), among others. Scaling theories are viewed in many different ways and, consequently, many different types of models have been developed and applied in hydrology. For instance, these models can be categorized into canonical and microcanonical, bounded and unbounded, simple-scaling and multi-scaling. Extensive details of such models and their applications in hydrology can be found in Schertzer and Lovejoy (1987), Lovejoy and Schertzer (1990), Gupta and Waymire (1993), Olsson et al. (1993), Over and Gupta (1996), Carsteanu and Foufoula-Georgiou (1996), Gupta et al. (1994, 1996), Perica and Foufoula-Georgiou (1996), Menabde et al. (1997), Venugopal et al. (1999), Veneziano et al. (2000, 2006), Ferraris et al. (2003), Molnar and Burlando (2005), Dodov and Foufoula-Georgiou (2005), Marani and Zanetti (2007), Pui et al. (2012), and Markonis and Koutsoyiannis (2015), among others.

A more recent comprehensive account of scaling-based methods and their applications in hydrology can be found in Veneziano and Langousis (2010).

3.5.11 Remarks

While the commonly used parametric models, such as the ones discussed above, are indeed useful for modeling hydrologic time series, they have certain important limitations. For instance: (1) since single linear model is often fit to the entire data, the significance of local neighborhood is not given due consideration; (2) since the *structure* of the model is specified a priori and the number and nature of the parameters are generally fixed in advance, there is little flexibility; (3) the parameter estimation procedure is often complicated, especially when the number of parameters is large, which is also often the case; and (4) the models are often not able to capture several properties of hydrologic time series, including asymmetric and/or multimodal conditional and marginal probability distributions, persistent large amplitude variations at irregular time intervals, amplitude-frequency dependence, apparent long memory, nonlinear dependence between X_i versus $X_{i-\tau}$ for lag τ, and time irreversibility, among others; see Yakowitz (1973), Jackson (1975), Kendall and Dracup (1991), Lall and Sharma (1996) for some additional details. These limitations led to the development and applications of nonparametric methods in hydrology; see Yakowitz (1979, 1985, 1993), Adamowski (1985), Karlsson and Yakowitz (1987), Bardsley (1989), Kendall and Dracup (1991), Smith (1991), Smith et al. (1992), Lall and Sharma (1996), and Lall et al. (1996) for some early studies. The next section presents an overview of nonparametric methods and their hydrologic applications.

3.6 Nonparametric Methods

The nonparametric methods make no prior assumptions on the model structure. Instead, the model structure is determined from the data. Although nonparametric methods may also involve parameters, the number and nature of the parameters are not fixed in advance and, thus, are flexible. The nonparametric methods approximate the conditional and marginal distributions of a time series and simulate from these. For instance, for a time series X_i, $i = 1, 2, \ldots, N$, an order p model can be expressed as a simulation from a conditional probability density function

$$f(X_i|X_{i-1}, X_{i-2}, \ldots, X_{i-p}) = \frac{f(X_i, X_{i-1}, X_{i-2}, \ldots, X_{i-p})}{\int f(X_i, X_{i-1}, X_{i-2}, \ldots, X_{i-p})\,dX_i} \tag{3.40}$$

This approach is different from the one adopted in parametric models, which essentially simulate from a Gaussian conditional distribution because they assume the data is normally distributed.

There exist many nonparametric methods in the literature; see Efron and Tibishirani (1993), Higgins (2003), Sprent and Smeeton (2007), and Hollander et al. (2013) for details. Among the nonparametric methods that have found applications in hydrology are the nearest neighbor resampling, block bootstrap resampling, kernel density estimator (KDE), *k*-nearest neighbor bootstrap resampling (KNNR), *k*-nearest neighbor with local polynomial regression (LPK), nonparametric order *p* simulation with long-term dependence (NPL), and hybrid models. Details of these methods and applications in hydrology are available in Yakowitz (1979), Lall and Sharma (1996), Vogel and Shallcross (1996), Sharma et al. (1997), Tarboton et al. (1998), Rajagopalan and Lall (1999), Sharma and O'Neill (2002), Srinivas and Srinivasan (2005, 2006), and Prairie et al. (2006), among others. Some of these methods are briefly described next.

3.6.1 Bootstrap and Block Bootstrap

The bootstrap (Efron 1979; Efron and Tibishirani 1993) is perhaps the simplest nonparametric technique for time series analysis. The bootstrap is a statistical method that involves resampling the original time series (with replacement) to estimate the distribution of a statistic (e.g. mean, variance, correlation). The classical idea is to resample the original time series (with replacement) to generate B bootstrap samples, from which one can simulate B estimates of a given statistic, leading to an empirical probability distribution of the statistic. Let us consider a time series X_i, $i = 1, 2, \ldots N$, denoted as x and the task is to estimate the empirical probability distribution of a statistic $\hat{\theta}_i$. Each observation X_i is resampled (with replacement) with an equal probability of $1/N$. The sample x continues to be resampled with replacement B times, until B bootstrap samples x_i, $i = 1, 2, \ldots, B$ are obtained. Each bootstrap sample x_i yields a bootstrap estimate of the statistic θ leading to the B bootstrap estimates $\hat{\theta}_i, i = 1, 2, \ldots, B$.

By simply resampling (with replacement) from the original time series, the bootstrap can be used as a nonparametric time series model. A particular challenge in such a model, however, is to resample the time series to preserve the temporal correlation ρ (and spatial covariance) structure of the original time series. This is because the classic bootstrap assumes that the data are independent and identically distributed and resamples from each prior data point with equal probability. One way to address this problem is by resampling λ-year blocks, so that the resulting sequence will be approximately independent and, thus, can preserve the serial correlation structure of the time series. This kind of resampling is known as the moving-blocks bootstrap (e.g. Künsch 1989; Efron and Tibshirani 1993). Here, a block length $\lambda \approx N/K$, where N is the length of the time series and K is the number of blocks to resample, is chosen, as opposed to a single observation, in the bootstrap. Sampling blocks of length λ allows one to retain the original correlation ρ among the observations within each block, and yet adjacent blocks are uncorrelated. The basic idea is to choose a large enough block length λ so that observations more than λ time units apart will be nearly independent.

Numerous studies have applied the bootstrapping, block bootstrapping, and moving block bootstrapping, and related techniques in hydrology; see Labadie et al. (1987), Zucchini and Adamson (1989), Kendall and Dracup (1991), Vogel and Shallcross (1996), Ouarda et al. (1997), Kundzewicz and Robson (2004), Yue and Pilon (2004), Noguchi et al. (2011), Önöz and Bayazit (2012), Sonali and Nagesh Kumar (2013), and Hirsch et al. (2015), among others.

3.6.2 *Kernel Density Estimate*

Kernel density estimation entails a weighted moving average of the empirical frequency distribution of the time series. Most nonparametric density estimators can be expressed as kernel density estimators (Scott 1992).

For a univariate time series of observations X_i, $i = 1, 2, \ldots, N$, the kernel probability density estimator at any point X is written as:

$$\hat{f}(X) = \sum_{i=1}^{N} \frac{1}{N\lambda_i} K\left(\frac{X - X_i}{\lambda_i}\right) \tag{3.41}$$

where K(.) is a kernel function centered on the observation X_i that is usually taken to be a symmetric, positive, probability density function with finite variance, and λ is a bandwidth or "scale" parameter of the kernel centered at X_i. A fixed kernel density estimator uses a constant bandwidth, λ, irrespective of the location of X. Such a fixed estimate is formed by summing kernels with bandwidth λ centered at each observation X_i, as given by:

$$\hat{f}(X) = \sum_{i=1}^{N} \frac{1}{N\lambda} K\left(\frac{X - X_i}{\lambda}\right) \tag{3.42}$$

The kernel K(.) is a symmetric function centered on the observation X_i, that is positive, integrates to unity, has first moment equal to zero and finite variance. This is similar to the histogram construction, where individual observations contribute to the density by placing a rectangular box (analogous to the kernel function) in the prespecified bin the observation lies in (Sharma et al. 1997). The histogram is sensitive to the position and size of each bin. Use of smooth kernel functions makes the kernel density estimate in Eq. (3.41) smooth and continuous.

There are many possible kernel functions, including uniform, triangular, Normal, Epanechnikov, Bisquare, and others; see Silverman (1986) and Scott (1992) for details. However, the Gaussian kernel function is widely used. It is given by:

$$K(X) = \frac{1}{(2\pi)^{1/2}} \exp\left(-\frac{X^2}{2}\right) \tag{3.43}$$

The univariate kernel density estimate in Eq. (3.42) can be easily extended to a multivariate one with a dimension d. For instance, using a Gaussian kernel function, the multivariate kernel probability density $\hat{f}(X)$ of a d-dimensional variable set $\mathbf{X}$ is estimated as:

$$\hat{f}(\mathbf{X}) = \frac{1}{N}\sum_{i=1}^{N} \frac{1}{(2\pi)^{d/2}\lambda^d \det(\mathbf{S})^{1/2}} \exp\left(-\frac{(\mathbf{X}-\mathbf{X}_i)^T \mathbf{S}^{-1}(\mathbf{X}-\mathbf{X}_i)}{2\lambda^2}\right) \tag{3.44}$$

where $\mathbf{X}_i$ is the ith multivariate data point for a sample of size N, $\mathbf{S}$ is the sample covariance of the variable set $\mathbf{X}$.

The bandwidth λ is key to an accurate estimate of the probability density. A large value of λ results in an oversmoothed probability density, with subdued models and overenhanced tails. A small value, on the other hand, can lead to density estimates overly influenced by individual data points, with noticeable bumps in the tails of the probability density. Several operational rules for choosing optimal values of λ are available in the literature; see Silverman (1986), Adamowski and Feluch (1991), Scott (1992), Sain et al. (1994), and Rajagopalan et al. (1997) for some details. One of the widely used guidelines is the least squares cross validation (LSCV).

The kernel density estimation-based approach started finding its applications in hydrology in the 1980s; see Adamowski (1985), Schuster and Yakowitz (1985), Bardsley (1989), Adamowski and Feluch (1990), Guo (1991), Lall et al. (1993, 1996), Moon and Lall (1994), Lall (1995), and Rajagopalan and Lall (1995) for some early studies. However, the real impetus came in the latter part of the 1990s.

Sharma et al. (1997) presented a multivariate kernel density estimate as a nonparametric alternative to the lag-p autoregressive model for monthly streamflow simulation. Their model, called the NP_p model, specifically considered the lag-1 situation (i.e. p = 1), (i.e. NP1 model), which means that the problem is bivariate kernel density estimate. They used kernel density estimators with Gaussian kernels, and selected the bandwidth λ using least squares cross validation (LSCV). Their procedure for nonparametric streamflow simulation is as follows: (1) Form bivariate sample set $\mathbf{X}_i = (X_i, X_{i-1})$; (2) Estimate bandwidth λ using least squares cross validation and estimate $\mathbf{S}$; (3) Initialize $i = 0$, and obtain $x_{i=0}$ (The initialization can be done in two ways, either by sampling from the marginal or by using warm-up); (4) From the given value $x_{i-1} = X_{i-1}$, select one of the observations X_i according to weight w_i; (5) Simulate x_i from ith kernel slice, and correct for negative simulations; and (6) Repeat steps (1) to (5) until the desired length of data is simulated. Further details about the procedure, including the use of the LSCV method for bandwidth selection and its sensitivity analysis for streamflow simulation, are available in Sharma et al. (1998).

The nonparametric kernel density estimator by Sharma et al. (1997) has been modified to incorporate other characteristics of hydrologic time series. For instance, Sharma and O'Neill (2002) developed a model to incorporate the long-term interannual variability for monthly streamflow simulation (NPL model). They particularly focused on order 1, and called the model as NPL1 model. Salas and Lee (2010) suggested other modifications. In the development of a *k*-nearest neighbor resampling algorithm with gamma kernel perturbation model (KGK model), they used an aggregate variable (KGKA model) and a pilot variable (KGKP model) to the kernel density estimator to lead the generation of the seasonal flows. They compared the proposed models with another nonparametric model that considers the reproduction of the interannual variability.

Kernel density estimator-based nonparametric models have found a number of applications in hydrology over the last two decades, including for rainfall and streamflow simulation and disaggregation (or downscaling). Extensive details of such applications are available in Rajagopalan et al. (1997), Tarboton et al. (1998), Rajagopalan and Lall (1999), Harrold et al. (2003a, b), Kim and Valdés (2005), Srikanthan et al. (2005), Ghosh and Mujumdar (2007), Block et al. (2009), Mehrotra and Sharma (2007a, b, 2010), Mehrotra et al. (2012), Li et al. (2013), Mirhosseini et al. (2015), and Viola et al. (2016), among others.

Although the kernel-based methods have been found to be useful for a number of hydrologic applications, they also possess certain limitations. For instance, they often have problems at the boundaries of the variables (e.g. zero in rainfall and streamflow), thus resulting in bias (e.g. Lall and Sharma 1996), which gets exaggerated at higher dimensions. The methods can also simulate negative values, although this is to a lesser degree when compared to parametric models. The methods are also very difficult to use in higher dimensions, since estimation of optimal bandwidths at such dimensions is not trivial. These limitations led to other nonparametric methods. Among these, the methods based on the concept of nearest neighbors have been of particular significance and, therefore, are discussed next.

3.6.3 k-Nearest Neighbor Resampling (KNNR)

In the *k*-nearest neighbor resampling (e.g. Lall and Sharma 1996), the conditional PDF is approximated using *k*-nearest neighbors of the current value X_i in the time series and one of the neighbors is selected as the value for the next time step. A brief description of the *k*-nearest neighbor resampling method for a single variable time series is as follows:

1. Formulate a conditioning 'feature vector' based on the current value X_i of the time series and the past values (i.e. X_{i-1}, X_{i-2}, …), with the number of past values identified based on serial correlation;
2. Identify the k nearest neighbors of the feature vector based on Euclidean distance metric;
3. Assign weights to these k neighbors based on their distances, with the nearest neighbor getting the largest weight and the farthest neighbor getting the smallest weight;
4. Normalize these weights to create a probability mass function. This is termed as the 'weight metric;'
5. Resample one of the neighbors using this 'weight metric;'
6. The successor of the resampled neighbor becomes the simulated value for the next time step, X_{i+1}.
7. Repeat Steps (1)–(6) to generate several simulations.

This approach is simple and robust, as it can easily simulate from conditional PDF of any dimension. The simulations are generally insensitive to the choice of the weight function, as long as the selected weight function weighs the nearest neighbor the most relative to the farthest.

The number of neighbors k can be thought of as a smoothing parameter. For smaller k, the PDF approximation is based on only a few points and, therefore, has the ability to capture the local features. A larger k, on the other hand, can smooth out local features, but can capture global features. Therefore, an appropriate selection of k is key to the performance of the k-nearest neighbor resampling method. There are several methods for selecting an appropriate number for k, including heuristic and objective methods (e.g. Generalized Cross Validation (GCV)). In general, the choice $k = \sqrt{N}$ is quite robust.

Although k-NN technique has been found to perform reasonably well for a number of applications in hydrology, it has an important drawback in that the values not seen in the historical record cannot be simulated. This means, there may not be enough variety in simulations, while the use of kernel density estimators can alleviate this problem (e.g. Sharma et al. 1997; Tarboton et al. 1998). Another shortcoming is that the variance may be underestimated, especially when the historical time series are correlated. This problem may be overcome by using gamma kernel perturbation, as has been done by Salas and Lee (2010), mentioned earlier.

The nonparametric k-NN method and their modifications (see also Sect. 3.6.4 for another example), including for multiple sites and multiple variables have found numerous applications in hydrology. For instance, Rajagopalan and Lall (1999) implemented a lag-1 multivariate resampling model to simulate daily rainfall and other weather variables at a single site. Yates et al. (2003) extended this model to simulate daily rainfall and weather variables at multiple sites simultaneously. A semi-parametric approach to rainfall and weather generation was proposed. In this, a Markov chain model was fitted for each month separately to simulate the rainfall occurrence and then k-NN lag-1 resampling was used to generate the vector of weather variables conditionally on the transitional state of the rainfall occurrence. For instance, if the simulated current period's occurrence was wet and the following period dry, then the neighbors are obtained from the historical time series that have the same transition (Apipattanavis et al. 2007). Souza Filho and Lall (2003) applied the k-NN method for multisite seasonal ensemble streamflow forecasts by formulating the feature vector as a set of predictors for streamflow. Still other studies that have applied the k-NN method in hydrology include those by Karlsson and Yakowitz (1987), Buishand and Brandsma (2001), Brath et al. (2002), Mehrotra and Sharma (2006), Prairie et al. (2007), Towler et al. (2009), Eum et al. (2010), Gangopadhyay et al. (2005, 2009), Goyal et al. (2012), Lee and Jeong (2014), Lu and Xin (2014), and Sharifazari and Araghinejad (2015), among others.

3.6.4 k-Nearest Neighbors with Local Polynomial Regression

As mentioned earlier, the k-NN bootstrap method has an important limitation in that it cannot simulate the values that are not seen in the historical record. This limitation can be addressed by making certain modifications to the method. One such modification for streamflow simulation was presented by Prairie et al. (2006) based on local polynomial regression (Loader 1999). In this method, with the value of the current period available, local regression is used to obtain the mean value of a future period. Then, k neighbors are computed from the data for the current period and residuals from the regression at these k neighbors are resampled using the k-NN approach (see Sect. 3.6.3) and added to the mean value. Therefore, in this method, instead of resampling the historical values, residuals are resampled from the neighborhood. A brief description of this method is as follows (Prairie et al. 2006):

1. Fit a local polynomial for each timestep i dependent on the previous timestep $i - 1$

$$X_i = g(X_{i-1}) + \varepsilon_i \tag{3.45}$$

where $g(X_{i-1})$ is the local polynomial;

2. Save the residuals (ε_i) from the fit;
3. Once the simulated value of the current period (i.e. X^*_{i-1}) is obtained, estimate the mean flow of the next period $\widehat{X}^*_i$ from Eq. (3.45), not including the residual;
4. Obtain k-NN of X^*_{i-1};
5. Assign weights to these k neighbors based on their distances, with the nearest neighbor getting the largest weight and the farthest neighbor getting the smallest;

6. Normalize these weights to create a probability mass function. This is termed as the 'weight metric;'
7. Resample one of the neighbors using this 'weight metric.' Its residual (ε_i^*) is added to the mean estimate $\widehat{X}_i^*$. Thus, the simulated value for the next timestep becomes $X_i^* = \widehat{X}_i^* + \varepsilon_i^*$;
8. Repeat Steps (1)–(7) for other time periods to obtain an ensemble of simulations.

The local polynomial based *k*-NN method has the following advantages over the traditional *k*-NN method: (1) it provides simulations even for values not seen in the historical record; (2) residual resampling captures the local variability more effectively; and (3) the local regression fit has the ability to capture any arbitrary (linear or nonlinear) relationship in the time series and also extrapolate beyond the range of observations (Prairie et al. 2006).

The local polynomial *k*-NN method and its further extensions has been successfully applied to study different problems in hydrology, including for rainfall and streamflow simulation, forecasting, and downscaling. Extensive details of such applications can be seen in Regonda et al. (2005, 2006a, b), Singhrattna et al. (2005), Grantz et al. (2005), Block and Rajagopalan (2007), Bracken et al. (2010), Li and Singh (2014), and Verdin et al. (2016), among others.

3.6.5 Others

There exist many other stochastic nonparametric methods that have also found important applications in hydrology. In some cases, nonparametric models have been combined with parametric models to serve as hybrid models. For instance, Srinivas and Srinivasan (2001) proposed a hybrid moving block bootstrap (HMBB) model for multi-season streamflow simulation, by combining a parametric periodic model and a nonparametric block bootstrap. Srinivas and Srinivasan (2005, 2006) presented further modifications and improvements to this hybrid model. Srivatsav et al. (2011) used the hybrid model of Srinivas and Srinivasan (2006) as a simulation engine to develop an efficient simulation-optimization-based hybrid stochastic modeling framework. Srivatsav and Simonovic (2014) used the maximum entropy bootstrap model to provide analytical procedures for multi-site, multi-season weather generation and streamflow generation; see also Cook and Buckley (2009), and Cook et al. (2013) for maximum entropy bootstrap model applications in hydrology. Still other recent studies of interest on nonparametric or hybrid models in hydrology are Lambert et al. (2003), Mehrotra and Sharma (2007a), Wong et al. (2007), Basinger et al. (2010), Lee et al. (2010), Kalra and Ahmad (2011), Keylock (2012), Stagge and Moglen (2013), Haerter et al. (2015), Mehrotra et al. (2015), and Langousis et al. (2016), among others.

3.7 Summary

Over the past half a century or so, stochastic time series methods have been an important part of hydrologic studies. A large number of parametric and nonparametric (as well as hybrid or semi-parametric) methods have been developed and applied in almost all areas of hydrology, including for simulation, forecasting, and disaggregation (or downscaling) of rainfall and streamflow. This chapter has presented an overview of several parametric and nonparametric methods and their hydrologic applications. It is clear that stochastic time series methods have dominated hydrologic studies in recent decades, and there is no question that this trend will continue to grow in the future. In this regard, an area that is gaining particular attention is the application of stochastic methods for downscaling outputs from global climate models (GCMs), especially with the clear recognition of the impacts of climate change on our water resources at the global scale as well as at the regional and local scales. While stochastic time series methods have been dominating hydrologic research in recent decades, a large number of other approaches, especially those that address the nonlinear and related properties of time series, have also been developed and extensively applied in hydrology. Some of these methods that have found widespread applications in hydrology are discussed in the next chapter.

References

Adamowski K (1985) Nonparametric kernel estimation of flood frequencies. Water Resour Res 21 (11):1585–1590

Adamowski K, Feluch W (1990) Nonparametric flood-frequency analysis with historical information. J Hydraul Eng 116(8):1035–1047

Adamowski K, Feluch W (1991) Application of nonparametric regression to ground water level prediction. Can J Civ Eng 18:600–606

Ahn H, Salas JD (1997) Groundwater head sampling based on stochastic analysis. Water Resour Res 33(12):2769–2780

Apipattanavis S, Podestá G, Rajagopalan B, Katz RW (2007) A semiparametric multivariate and multisite weather generator. Water Resour Res 43:W11401. doi:10.1029/2006WR005714

Aykuz DE, Bayazit M, Önöz B (2012) Markov chain models for hydrological drought characteristics. J Hydrometeorol 13(1):298–309

Bardossy A, Plate EJ (1991) Space-time model for daily rainfall using atmospheric circulation patterns. Water Resour Res 28(5):1247–1259

Bardsley WE (1989) Using historical data in nonparametric flood estimation. J Hydrol 108:249–255

Bartlett MS (1955) An introduction to stochastic processes. Cambridge University Press, Cambridge

Bartlett MS (1963) The spectral analysis of point processes. J R Stat Soc Ser B 25(2):264–296

Bartolini P, Salas JD (1993) Modeling of streamflow processes at different time scales. Water Resour Res 29(8):2573–2587

Bartolini P, Salas JD, Obeysekera JTB (1988) Multivariate periodic ARMA(1,1) processes. Water Resour Res 24(8):1237–1246

Basinger M, Montalto F, Lall U (2010) A rainwater harvesting system reliability model based on nonparametric stochastic rainfall generator. J Hydrol 392:105–118
Bates BC, Campbell EP (2001) A Markov chain Monte Carlo scheme for parameter estimation and inference in conceptual rainfall-runoff modeling. Water Resour Res 37(4):937–947
Bayazit M, Önöz B (2005) Probabilities and return periods of multisite droughts. Hydrol Sci J 50:605–615
Berndstein S (1938) Equations différentielles stochastiques. Actualités Sci Ind No 138:5–31
Block P, Rajagopalan B (2007) Interannual variability and ensemble forecast of Upper Blue Nile Basin Kiremt season precipitation. J Hydrometeorol 8:327–343
Block PJ, Souza Filho FA, Sun L, Kwon H-H (2009) A streamflow forecasting framework using multiple climate and hydrological models. J Ame Wat Resour Assoc 45(4):828–843
Bo Z, Islam S, Eltahir EAB (1994) Aggregation-disaggregation properties of a stochastic rainfall model. Water Resour Res 30(12):3423–3435
Bochner S (1949) Diffusion equations and stochastic processes. Proc Natl Acad Sci USA 35 (7):368–370
Bochner S (1955) Harmonic analysis and the theory of probability. University of California Press, Berkeley
Bordoy R, Burlando P (2014) Stochastic downscaling of climate model precipitation outputs in orographically complex regions: 2. Downscaling methodology. Water Resour Res 50(1):562–579
Box G, Jenkins G (1970) Time series analysis, forecasting, and control. Holden-Day, San Francisco
Bracken C, Rajagopalan B, Prairie J (2010) A multisite seasonal ensemble streamflow forecasting technique. Water Resour Res 46(3):W03532. doi:10.1029/2009WR007965
Bras RL, Rodriguez-Iturbe I (1985) Random functions and hydrology. Addison-Wesley, MA
Brath A, Montanari A, Toth E (2002) Neural networks and non-parametric methods for improving real-time forecasting through conceptual hydrological models. Hydrol Earth Syst Sci 6(4):627–640
Brillinger DR (1975) Time series: data analysis and theory. Holt, Rinehart & Winston, New York
Brockwell PJ, Davis RA (1991) Time series: theory and methods. Springer-Verlag, New York
Buishand TA (1977) Stochastic modeling of daily rainfall sequences. Veenman and Zonen, Wageningen, The Netherlands
Buishand TA, Brandsma T (2001) Multisite simulation of daily precipitation and temperature in the Rhine basin by nearest-neighbor resampling. Water Resour Res 37(11):2761–2776
Burton A, Fowler HJ, Blenkinsop S, Kilsby CG (2010) Downscaling transient climate change using a Neyman-Scott Rectangular Pulses stochastic rainfall model. J Hydrol 381:18–32
Calenda G, Napolitano F (1999) Parameter estimation of Neyman-Scott processes for temporal point rainfall simulation. J Hydrol 225:45–66
Cameron D, Beven KJ, Tawn J (2000) Modelling extreme rainfalls using a modified random pulse Bartlett-Lewis stochastic rainfall model (with uncertainty). Adv Water Resour 24:203–211
Cancelliere A, Salas JD (2004) Drought length probabilities for periodic-stochastic hydrologic data. Water Resour Res 40:W02503. doi:10.1029/2002WR001750
Cancelliere A, Salas JD (2010) Drought probabilities and return period for annual streamflow series. J Hydrol 391:77–89
Carlson RF, MacCormick AJA, Watts DG (1970) Application of linear models to four annual streamflow series. Water Resour Res 6:1070–1078
Carsteanu A, Foufoula-Georgiou E (1996) Assessing dependence of weights in a multiplicative cascade model of temporal rainfall. J Geophys Res 101:26363–26370
Chaleeraktrakoon (1999) Stochastic procedure for generating seasonal flows. J Hydrol Eng 4 (4):337–343
Chaleeraktrakoon C (2009) Parsimonious SVD/MAR(1) procedure for generating multisite multiseason flows. J Hydrol Eng 14(5):516–527
Chang TJ, Kavvas ML, Delleur JW (1984) Daily precipitation modeling by discrete autoregressive moving average processes. Water Resour Res 20(5):565–580

Chatfield C (1989) The analysis of time series: An introduction. CRC Press, USA
Chin EH (1977) Modeling daily precipitation occurrence process with Markov chains. Water Resour Res 13(6):849–956
Chung C, Salas JD (2000) Drought occurrence probabilities and risk of dependent hydrological process. J Hydrol Eng 5:259–268
Cigizoglu HK, Bayazit M (1998) Application of gamma autoregressive model to analysis of dry periods. J Hydrol Eng 3(3):218–221
Claps P, Rossi F, Vitale C (1993) Conceptual-stochastic modeling of seasonal runoff using autoregressive moving average models and different time scales of aggregation. Water Resour Res 29(8):2545–2559
Clarke RT (1994) Statistical modelling in hydrology. John Wiley & Sons, Chichester, UK
Connolly RD, Schirmer J, Dunn PK (1998) A daily rainfall disaggregation model. Agric For Meteorol 92:105–117
Cook BI, Buckley BM (2009) Objective determination of monsoon season onset, withdrawal, and length. J Geophys Res 114(D23):D23109. doi:10.1029/2009JD012795
Cook ER, Palmer JG, Ahmed M, Woodhouse CA, Fenwick P, Zafar MU, Wahab M, Khan N (2013) Five centuries of Upper Indus River flow from tree rings. J Hydrol 486:365–375
Cooper DM, Wood EF (1982) Identification of multivariate time series and multivariate input-output models. Water Resour Res 18(4):937–946
Cowpertwait PSP (1991) Further developments of the Neyman-Scott clustered point process for modeling rainfall. Water Resour Res 27:1431–1438
Cowpertwait PSP (1995) A generalized spatial-temporal model of rainfall based on a clustered point process. Proc R Soc London Ser A 450:163–175
Cowpertwait PSP (1998) A Poisson-cluster model of rainfall: high-order moments and extreme values. Proc R Soc London Ser A 454:885–898
Cowpertwait PSP (2004) Mixed rectangular pulses models of rainfall. Hydrol Earth Syst Sci 8 (5):993–1000
Cowpertwait PSP (2010) A spatial-temporal point process model with a continuous distribution of storm types. Water Resour Res 46(12):W12507. doi:10.1029/2010WR009728
Cowpertwait PSP, O'Connell PE (1997) A regionalised Neyman-Scott model of rainfall with convective and stratiform cells. Hydrol Earth Syst Sci 1:71–80
Cowpertwait PSP, O'Connell PE, Metcalfe AV, Mawdsley J (1996a) Stochastic point process modelling of rainfall, I. Single-site fitting and validation. J Hydrol 175(1–4):17–46
Cowpertwait PSP, O'Connell PE, Metcalfe AV, Mawdsley J (1996b) Stochastic point process modelling of rainfall, II. Regionalisation and disaggregation. J Hydrol 175(1–4):47–65
Cowpertwait PSP, Kilsby CG, O'Connell PE (2002) A space-time Neyman-Scott model of rainfall: empirical analysis of extremes. Water Resour Res 38(8):1–14
Cowpertwait PSP, Isham V, Onof C (2007) Point process models of rainfall: developments for fine-scale structure. Proc R Society London Ser A 463(2086):2569–2587
Cowpertwait PSP, Xie G, Isham V, Onof C, Walsh DCI (2011) A fine-scale point process model of rainfall with dependent pulse depths within cells. Hydrol Sci J 56(7):1110–1117
Cox DR (1958) Some problems connected with statistical inference. Ann Math Statist 29(2):357–372
D'Onofrio D, Palazzi E, von Hardenberg J, Provenzale A, Calmanti S (2014) Statistical rainfall downscaling of climate models. J Hydrometeorol 15:830–843
Deidda R (2000) Rainfall downscaling in a space-time multifractal framework. Water Resour Res 36(7):1779–1784
Delleur JW, Kavvas ML (1978) Stochastic models for monthly rainfall forecasting and synthetic generation. J Appl Meteorol 17:1528–1536
Delleur JW, Tao PC, Kavvas ML (1976) An evaluation of the practicality and complexity of some rainfall and runoff time series models. Water Resour Res 12(5):953–970
Dodov B, Foufoula-Georgiou E (2005) Fluvial processes and streamflow variability: Interplay in the scale-frequency continuum and implications for scaling. Water Resour Res 41:W05005. doi:10.1029/2004WR003408

Doob JL (1934) Stochastic processes and statistics. Proc Natl Acad Sci USA 20:376–379
Doob JL (1938) Stochastic processes with an integer-valued parameter. Trans Amer Math Soc 44:87–150
Doob JL (1940) Regularity properties of certain families of chance variables. Trans Amer Math Soc 47:455–486
Doob JL (1945) Markov chains–denumerable case. Trans Amer Math Soc 58:455–473
Doob JL (1953) Stochastic processes. John Wiley & Sons Inc, New York
Dynkin EB (1959) Foundations of the theory of Markov processes. Fizmatgiz, Moscow
Efron B (1979) Bootstrap methods: Another look at the Jackknife. Ann Stat 7:1–26
Efron B, Tibishirani R (1993) An introduction to the bootstrap. Chapman and Hall, New York
Entekhabi D, Rodriguez-Iturbe I, Eagleson PS (1989) Probabilistic representation of the temporal rainfall process by the modified Neyman-Scott rectangular pulses model: parameter estimation and validation. Water Resour Res 25(2):295–302
Eubank RL (1988) Spline smoothing and nonparametric regression. Marcel Dekker Inc, New York
Eum H, Simonovic S, Kim Y (2010) Climate change impact assessment using k-nearest neighbor weather generator: case study of the Nakdong River Basin in Korea. J Hydrol Eng 15(10):772–785
Evin G, Favre A-C (2008) A new rainfall model based on the Neyman-Scott process using cubic copulas. Water Resour Res 44:W03433. doi:10.1029/2007WR006054
Favre A-C, Musy A, Morgenthaler S (2002) Two-site modeling of rainfall based on the Neyman-Scott process. Water Resour Res 38(12):1307. doi:10.1029/2002WR001343
Feller W (1950) An introduction to probability theory and its applications. John Wiley, New York
Fernandez B, Salas JD (1986) Periodic gamma-autoregressive processes for operational hydrology. Water Resour Res 22(10):1385–1396
Fernandez B, Salas JD (1990) Gamma-autoregressive models for stream-flow simulation. J Hydraul Eng 116(11):1403–1414
Ferraris L, Gabellani S, Parodi U, Rebora N, von Hardenberg J, Provenzale A (2003) Revisiting multifractality in rainfall fields. J Hydrometeorol 4:544–551
Fiering M (1967) Streamflow synthesis. Hard University Press, Cambridge, MA
Fiering MB, Jackson BB (1971) Synthetic streamflows. Am Geophys Union Water Resour, Monograph 1:1–98
Foufoula-Georgiou E, Guttorp P (1986) Compatibility of continuous rainfall occurrence models with discrete rainfall observations. Water Resour Res 22(8):1316–1322
Foufoula-Georgiou E, Lettenmaier DP (1987) A Markov renewal model of rainfall occurrences. Water Resour Res 23(5):875–884
Fowler HJ, Blenkinsop S, Tebaldi C (2007) Linking climate change modeling to impacts studies: recent advances in downscaling techniques for hydrological modeling. Int J Climatol 27(12):1547–1578
Fraedrich K, Larnder C (1993) Scaling regimes of composite rainfall time series. Tellus 45A:289–298
Gabriel KR, Neumann J (1962) A Markov chain model for daily rainfall occurrence at Tel Aviv. Q J R Meteorol Soc 88:90–95
Gangopadhyay S, Clark M, Rajagopalan B (2005) Statistical downscaling using K-nearest neighbors. Water Resour Res 41:W02024. doi:10.1029/2004WR003444
Gangopadhyay S, Harding BL, Rajagopalan B, Lukas JJ, Fulp TJ (2009) A nonparametric approach for paleohydrologic reconstruction of annual streamflow ensembles. Water Resour Res 45:W06417. doi:10.1029/2008WR007201
Gani J (1969) Recent advances in storage and flooding theory. Adv Appl Prob 1:90–110
Gardiner CW (2009) Stochastic methods: a handbook for the natural and social sciences. Springer-Verlag, Berlin and Heidelberg
Gaver DP, Lewis PAW (1980) First-order autoregressive gamma sequences and point processes. Adv Appl Prob 12:727–745
Gelhar LW (1993) Stochastic subsurface hydrology. Prentice-Hall, Englewood Cliffs, New Jersey

Ghosh S, Mujumdar PP (2007) Nonparametric methods for modeling GCM and scenario uncertainty in drougth assessment. Water Resour Res 43:W07405. doi:10.1029/2006WR005351

Glasbey CA, Cooper C, McGechan MB (1995) Disaggregation of daily rainfall by conditional simulation from a point-process model. J Hydrol 165:1–9

Govindaraju RS (ed) (2002) Stochastic methods in subsurface contaminant hydrology. ASCE, New York

Goyal MK, Ojha CSP, Burn DH (2012) Nonparametric statistical downscaling of temperature, precipitation, and evaporation in a semiarid region in India. J Hydrol Eng 17(5):615–627

Grantz K, Rajagopalan B, Clark M, Zagona E (2005) A technique for incorporating large-scale climate information in basin-scale ensemble streamflow forecasts. Water Resour Res 41: W10410. doi:10.1029/2004WR003467

Grillakis MG, Koutroulis AG, Tsanis IK (2013) Multisegment statistical bias correction of daily GCM precipitation output. J Geophys Res 118(8):3150–3162

Grygier JC, Stedinger JR (1988) Condensed disaggregation procedures and conservation corrections for stochastic hydrology. Water Resour Res 24(10):1574–1584

Guo SL (1991) Nonparametric variable kernel estimation with historical floods and paleoflood information. Water Resour Res 27(1):91–98

Gupta VK, Waymire E (1981) The mathematical structure of rainfall representations: 3. Some applications of the point process theory to rainfall processes. Water Resour Res 17(5):1287–1294

Gupta VK, Waymire E (1990) Multiscaling properties of spatial rainfall and river flow distributions. J Geophys Res 95(D3):1999–2009

Gupta VK, Waymire E (1993) A statistical analysis of mesoscale rainfall as a random cascade. J Appl Meteorol 32:251–267

Gupta VK, Rodríguez-Iturbe I, Wood EF (eds) (1986) Scale problems in hydrology: runoff generation and basin response. Reidel Publishing Company, FD, p 244

Gupta VK, Mesa OJ, Dawdy DR (1994) Multiscaling theory of flood peaks: Regional quantile analyis. Water Resour Res 30(12):3405–3421

Gupta VK, Castro S, Over TM (1996) On scaling exponents of spatial peak flows from rainfall and river network geometry. J Hydrol 187(1–2):81–104

Gyasi-Agyei Y, Willgoose G (1997) A hybrid model for point rainfall modeling. Water Resour Res 33:1699–1706

Haan CT (1994) Statistical methods in hydrology. Iowa University Press, Iowa

Haan CT, Allen DM, Street JD (1976) A Markov chain model of daily rainfall. Water Resour Res 12(3):443–449

Haerter JO, Eggert B, Moseley C, Piani C, Berg P (2015) Statistical precipitation bias correction of gridded model data using point measurements. Geophys Res Lett 42:1919–1929. doi:10.1002/2015GL063188

Hannan EJ (1955) A test for singularities in Sydney rainfall. Austr Jour Phys 8(2):289–297

Härdle W, Bowman AW (1988) Bootstrapping in nonparametric regression: local adaptive smoothing and confidence bands. J Am Stat Assoc 83:102–110

Harms AA, Campbell TH (1967) An extension to the Thomas-Fiering model for the sequential generation of streamflow. Water Resour Res 3(3):653–661

Harrold TI, Sharma A, Sheather SJ (2003a) A nonparametric model for stochastic generation of daily rainfall occurrence. Water Resour Res 39(10):1300. doi:10.1029/2003WR002182

Harrold TI, Sharma A, Sheather SJ (2003b) A nonparametric model for stochastic generation of daily rainfall amounts. Water Resour Res 39(12):1343. doi:10.1029/2003WR002570

Hershenhorn J, Woolhiser DA (1987) Disaggregation of daily rainfall. J Hydrol 95:299–322

Higgins JJ (2003) Introduction to modern nonparametric statistics. Duxbury Press, North Scituate, Massachusetts

Hipel KW, McLeod AI (1978a) Preservation of the rescaled adjusted range. 2, Simulation studies using Box-Jenkins models. Water Resour Res 14(3):509–516

Hipel KW, McLeod AI (1978b) Preservation of the rescaled adjusted range. 3, Fractional Gaussian noise algorithms. Water Resour Res 14(3):517–518
Hipel KW, McLeod AI (1994) Time series modeling of water resources and environmental systems. Elsevier Science, Amsterdam
Hipel KW, McLeod AI, Lennox WC (1977) Advances in Box-Jenkins modeling, 1 Model construction. Water Resour Res 13(3):567–575
Hirsch RM (1979) Synthetic hydrology and water supply reliability. Water Resour Res 15 (6):1603–1615
Hirsch RM, Archfield SA, De Cicco LA (2015) A bootstrap method for estimating uncertainty of water quality trends. Environ Modell Softw 73:148–166
Hollander M, Wolfe DA, Chicken E (2013) Nonparametric statistical methods. John Wiley & Sons, Hoboken, New Jersey
Hoshi K, Burges SJ (1979) Disaggregatio of streamflow volumes. J Hydraul Div ASCE 195 (HY1):27–41
Hughes J, Guttorp P, Charles S (1999) A non-homogeneous hidden Markov model for precipitation occurrence. Appl Stat 48(1):15–30
Hurst HE (1951) Long-term storage capacity of reservoirs. Trans Am Soc Civil Eng 116:770–808
Hurst HE (1956) Methods of using long-term storage in reservoirs. Proc Inst Civil Eng 1:519–543
Islam S, Entekhabi D, Bras RL, Rodriguez-Iturbe I (1990) Parameter estimation and sensitivity analysis for the modified Bartlett-Lewis rectangular pulses model of rainfall. J Geophys Res 95:2093–2100
Itô K (1944) Stochastic integral. Proc Imp Acad Tokyo 20:519–524
Itô (1946) On a stochastic integral equation. Proc Japan Acad Tokyo 22:32–35
Jackson BB (1975) Markov mixture models for drought lengths. Water Resour Res 11(1):75–95
Jazwinski A (1970) Stochastic processes and filtering theory. Academic Press, New York
Jenkins GM, Watts DG (1968) Spectral analysis and its applications. Holden-Day, San Francisco
Kaczmarska J, Isham V, Onof C (2014) Point process models for finite-resolution rainfall. Hydrol Sci J 59(11):1972–1991
Kalra A, Ahmad S (2011) Evaluating changes and estimating seasonal precipitation for Colorado River Basin using stochastic nonparametric disaggregation technique. Water Resour Res 47: W05555. doi:10.1029/2010WR009118
Karlin S, Taylor HM (1975) A first course in stochastic processes. Academic Press, San Diego, California
Karlsson M, Yakowitz S (1987) Nearest-neighbor methods for nonparametric rainfall-runoff forecasting. Water Resour Res 23(7):1300–1308
Katz RW (1977) Precipitation as a chain-dependent process. J Appl Meteor 16:671–676
Katz RW, Parlange MB (1995) Generalizations of chain-dependent processes: application to hourly precipitation. Water Resour Res 31:1331–1341
Kavvas ML, Delleur JW (1981) A stochastic cluster model of daily rainfall sequences. Water Resour Res 17(4):1151–1160
Kendall DR, Dracup JA (1991) A comparison of index-sequential and AR(1) generated hydrological sequences. J Hydrol 122:335–352
Kendall MG, Stuart A (1968) Advanced theory of statistics. Wiley, New York
Keylock KJ (2012) A resampling method for generating synthetic hydrological time series with preservation of cross-correlative structure and higher-order properties. Water Resour Res 48: W12521. doi:10.1029/2012WR011923
Khaliq MN, Cunnane C (1996) Modelling point rainfall occurrences with the modified Bartlett-Lewis rectangular pulses model. J Hydrol 180(1–4):109–138
Kim T, Valdés B (2005) Synthetic generation of hydrologic time series based on nonparametric random generation. J Hydrol Eng 10(5):395–404
Klemeš V (1978) Physically based stochastic hydrologic analysis. Adv Hydrosci 11:285–352
Kottegoda NT (1980) Stochastic water resources technology. The Macmillan Press, London
Koutsoyiannis D (1992) A nonlinear disaggregation method with a reduced parameter set for simulation of hydrologic series. Water Resour Res 28(12):3175–3191

Koutsoyiannis D (2000) A generalized mathematical framework for stochastic simulation and forecast of hydrologic time series. Water Resour Res 36:1519–1534
Koutsoyiannis D (2001) Coupling stochastic models of different timescales. Water Resour Res 37 (2):379–391
Koutsoyiannis D, Onof C (2001) Rainfall disaggregation using adjusting procedures on a Poisson cluster model. J Hydrol 246(1–4):109–122
Koutsoyiannis D, Xanthopoulos Th (1990) A dynamic model for short-scale rainfall disaggregation. Hydrol Sci J 35(3):303–322
Kundzewicz ZW, Robson AJ (2004) Change detection in hydrological records—a review of the methodology. Hydrol Sci J 49(1):7–19
Künsch HR (1989) The jackknife and the bootstrap for general stationary observations. Ann Stat 17:1217–1241
Labadie JW, Fontane D, Tabios GQ, Chou NF (1987) Stochastic analysis of dependable hydropower capacity. J Water Resour Plann Manage 113(3):422–477
Lall U (1995) Nonparametric function estimation: recent hydrologic applications. US Natl Rep Int Union Geod Geophys 1991–1994. Rev Geophys 33:1093–1102
Lall U, Sharma A (1996) A nearest neighbor bootstrap for resampling hydrologic time series. Water Resour Res 32(3):679–693
Lall U, Moon Y-I, Bosworth K (1993) Kernel flood frequency estimators: bandwidth selection and kernel choice. Water Resour Res 29(4):1003–1016
Lall U, Rajagopalan B, Tarboton DG (1996) A nonparametric wet/dry spell model for resampling daily precipitation. Water Resour Res 32:2803–2823
Lambert MF, Whiting JP, Metcalfe AV (2003) A non-parametric hidden Markov model for climate change identification. Hydrol Earth Syst Sci 7(5):652–667
Lane WL (1979) Applied stochastic techniques. User Manual. Division of Planning Technical Services, US Bur Reclam, Denver, Colo
Langousis A, Mamalakis A, Puliga M, Deidda R (2016) Threshold detection for the generalized Pareto distribution: Review of representative methods and application to the NOAA NCDC daily rainfall database. Water Resour Res 52:2659–2681
Lawrance AJ (1982) The innovation distribution of a gamma distributed autoregressive process. Scand J Statist 9:234–236
Lawrance AJ, Kottegoda NT (1977) Stochastic modeling of river flow time series. J R Stat Soc Ser A 140:1–47
Lawrance AJ, Lewis PAW (1981) A new autoregressive time series model in exponential variables [NEAR(1)]. Adv Appl Prob 13(4):826–845
Le Cam LA (1958) Un théorème sur la division d'un intervalle par des points pris au hasard. Publ Inst Statist Univ Paris 7:7–16
Le Cam LA (1961) A stochastic description of precipitation. In: Newman J (ed) Proc 4th Berkeley symp mathematics, statistics, and probability. University of California Press, Berkeley, pp 165–186
Ledolter J (1978) The analysis of multivariate time series applied to problems in hydrology. J Hydrol 36:327–352
Lee T, Jeong C (2014) Nonparametric statistical temporal downscaling of daily precipitation to hourly precipitation and implications for climate change scenarios…
Lee T, Salas JD, Prairie J (2010) An enhanced nonparametric streamflow disaggregation model with genetic algorithm. Water Resour Res 46:W08545. doi:10.1029/2009WR007761
Lennartsson J, Baxevani A, Chen D (2008) Modelling precipitation in Sweden using multiple step Markov chains and a composite model. J Hydrol 363(1):42–59
Leonard M, Lambert MF, Metcalfe AV, Cowpertwait PSP (2008) A space-time Neyman-Scott rainfall model with defined storm extent. Water Resour Res 44:W09402. doi:10.1029/2007WR006110
Lettenmaier DP, Burges SJ (1977) Operational assessment of hydrologic models of long-term persistence. Water Resour Res 13(1):113–124
Lévy P (1925) Calcul des probabilités. Gauthier-Villars, Paris

Lévy P (1937) Théorie de l'addition des variables aléatoires. Gauthier-Villars, Paris
Lévy (1948) Processus stochastiques et mouvement Brownien. Gauthier-Villars, Paris
Li C, Singh VP (2014) A multimodel regression-sampling algorithm for generating rich monthly streamflow scenarios. Water Resour Res 50:5958–5979
Li C, Singh VP, Mishra AK (2013) A bivariate mixed distribution with a heavy-tailed component and its application to single-site daily rainfall simulation. Water Resour Res 49:767–789
Lloyd EH (1963) A probability theory of reservoirs with serially correlated inputs. J Hydrol 1:99–128
Loader CR (1999) Bandwidth selection: classical or plug-in? Ann Statist 27(2):415–438
Loève M (1955) Probability theory. D van Nostrand Company Inc, New York
Loucks DP, Stedinger JR, Haith DA (1981) Water resources systems planning and analysis. Prentice-Hall, Englewood Cliffs, New Jersey
Lovejoy S, Schertzer D (1990) Multifractals, universality classes and satellite and radar measurements of clouds and rain fields. J Geophys Res 95(D3):2021–2034
Lu Y, Xin X (2014) Multisite rainfall downscaling and disaggregation in a tropical urban area. J Hydrol 509:55–65
Mandelbrot BB (1967) How long is the coast of Britain? Statistical self-similarity and fractional dimension. Science 156:636–638
Mandelbrot BB (1972) Broken line process derived as an approximation to fractional noise. Water Resour Res 8(5):1354–1356
Mandelbrot BB (1974) Intermittent turbulence in self-similar cascades: divergence of high moments and dimension of the carrier. J Fluid Mech 62(2):331–358
Mandelbrot BB (1977) Fractals: form, chance and dimension. W.H. Freeman and Co, New York
Mandelbrot BB (1983) The fractal geometry of nature. Freeman, New York
Mandelbrot BB, Van Ness JW (1968) Fractional Brownian motions, fractional noises and applications. SIAM Soc Ind Appl Math Rev 10(4):422–437
Mandelbrot BB, Wallis JR (1968) Noah, Joseph and operational hydrology. Water Resour Res 4 (5):909–918
Mandelbrot BB, Wallis JR (1969) Computer experiments with fractional Gaussian noises, 1. Averages and variances. Water Resour Res 5(1):228–267
Mann HB (1945) Nonparametric tests against trend. Econometrica 13:245–259
Marani M, Zanetti S (2007) Downscaling rainfall temporal variability. Water Resour Res 43: W09415. doi:10.1029/2006WR005505
Maraun D, Wetterhall F, Ireson AM, Chandler RE, Kendon EJ, Widmann M, Brienen S, Rust HW, Sauter T, Themeßl M, Venema VKC, Chun KP, Goodess CM, Jones RG, Onof C, Vrac M, Thiele-Eich I (2010) Precipitation downscaling under climate change: Recent developments to bridge the gap between dynamical models and the end user. Rev Geophys 48:RG3003. doi:10.1029/2009RG000314
Markonis Y, Koutsoyiannis D (2015) Scale-dependence of persistence in precipitation records. Nature Climate Change 6:399–401
Marshall L, Nott D, Sharma A (2004) A comparative study of Markov chain Monte Carlo methods for conceptual rainfall-runoff modeling. Water Resour Res 40:W02501. doi:10.1029/2003WR002378
Matalas NC (1963a) Autocorrelation of rainfall and streamflow minimums. US Geol Surv Prof Paper 434-B, Virginia
Matalas NC (1963b) Statistics of a runoff-precipitation relation. US Geol Surv Prof Paper 434-D, Virginia
Matalas NC (1967) Mathematical assessment of synthetic hydrology. Water Resour Res 3(4):937–945
Mathevet T, Lepiller M, Mangin A (2004) Application of time-series analyses to the hydrological functioning of an Alpine karstic system: the case of Bange-L'Eau-Morte. Hydrol Earth Syst Sci 8(6):1051–1064
McCuen RH (2003) Hydrologic analysis and design. Prentice-Hall Inc., New Jersey

McKean HP (1969) Stochastic integrals. American Mathematical Society, Academic Press, New York
McKerchar AI, Delleur JW (1974) Application of seasonal parametric linear stochastic models to monthly flow data. Water Resour Res 10(2):246–255
McLeod AI, Hipel KW, Lennox WC (1977) Advances in Box-Jenkins modeling. 2, Applications. Water Resour Res 13(3):577–586
Mehrotra R, Sharma A (2006) Conditional resampling of hydrologic time series using multiple predictor variables: A K-nearest neighbour approach. Adv Water Resour 29(7):987–999
Mehrotra R, Sharma A (2007a) A semi-parametric model for stochastic generation of multi-site rainfall exhibiting low-frequency variability. J Hydrol 335:180–193
Mehrotra R, Sharma A (2007b) Preserving low-frequency variability in generated daily rainfall sequences. J Hydrol 345:102–120
Mehrotra R, Sharma A (2010) Development and application of a multisite rainfall stochastic downscaling framework for climate change impact assessment. Water Resour Res 46:W07526. doi:10.1029/2009WR008423
Mehrotra R, Sharma A (2012) An improved standardization procedure to remove systematic low frequency variability biases in GCM simulations. Water Resour Res 48:W12601. doi:10.1029/WR012446
Mehrotra R, Westra S, Sharma A, Srikanthan R (2012) Continuous rainfall simulation: 2. A regionalized daily rainfall generation approach. Water Resour Res 48:W01536. doi:10.1029/2011WR010490
Mehrotra R, Li J, Westra S, Sharma A (2015) A programming tool to generate multi-site daily rainfall using a two-stage semi parametric model. Environ Modell Softw 63:230–239
Mejia JM, Rouselle J (1976) Disaggregation models in hydrology revisited. Water Resour Res 12(2):185–186
Mejia JM, Rodriguez-Iturbe I, Dawdy DR (1972) Streamflow simulation. 2, The broken line process as a potential model for hydrologic simulation. Water Resour Res 8(4):931–941
Mejia JM, Dawdy DR, Nordin CF (1974) Streamflow simulation. 3, The broken line process and operational hydrology. Water Resour Res 10(2):242–245
Menabde M, Harris D, Seed A, Austin G, Stow D (1997) Multiscaling properties of rainfall and bounded random cascades. Water Resour Res 33(12):2823–2830
Meneveau C, Sreenivasan KR (1987) Simple multifractal cascade model for fully developed turbulence. Phys Rev Lett 59:1424–1427
Mimikou (1984) A study for improving precipitation occurrences modelling with a Markov chain. J Hydrol 70(1–4):25–33
Mirhosseini G, Srivastava P, Sharifi A (2015) Developing probability-based IDF curves using kernel density estimator. J Hydrol Eng 20(9):04015002. doi:10.1061/(ASCE)HE.1943-5584.0001160
Molnar P, Burlando P (2005) Preservation of rainfall properties in stochastic disaggregation by a simple random cascade model. Atmos Res 77(1–4):137–151
Montanari A, Rosso R, Taqqu MS (1997) Fractionally differenced ARIMA models applied to hydrologic time series: identification, estimation, and simulation. Water Resour Res 33(5):1035–1044
Moon Y-I, Lall U (1994) A kernel quantile function estimation for flood frequency analysis. Water Resour Res 30(11):3095–3103
Moran PAP (1954) A probability theory of dams and storage systems. Aust J Appl Sci 5:116–124
Moss ME, Bryson MC (1974) Autocorrelation structure of monthly streamflows. Water Resour Res 10(4):737–744
Murrone F, Rossi F, Claps P (1997) Conceptually-based shot noise modeling of streamflows at short time interval. Stoch Hydrol Hydraul 11(6):483–510
Nagesh Kumar D, Lall L, Peterson MR (2000) Multisite disaggregation of monthly to daily streamflow. Water Resour Res 36(7):1823–1833
Neyman J, Scott EL (1958) Statistical approach to problems of cosmology. J R Stat Soc Ser B 20:1–29

Noguchi K, Gel YR, Duguay CR (2011) Bootstrap-based tests for trends in hydrological time series, with application to ice phenology data. J Hydrol 410:150–161

Northrop P (1998) A clustered spatial-temporal model of rainfall. Proc R Soc London Ser A 454:1875–1888

Nowak K, Prairie J, Rajagopalan B, Lall U (2010) A non-parametric stochastic approach for multisite disaggregation of annual to daily streamflow. Water Resour Res 46:W08529. doi:10.1029/2009WR008530

Obeysekera JTB, Tabios G, Salas JD (1987) On parameter estimation of temporal rainfall models. Water Resour Res 23(10):1837–1850

Olsson J (1998) Evaluation of a scaling cascade model for temporal rainfall disaggregation. Hydrol Earth Syst Sci 2(1):19–30

Olsson J, Niemczynowicz J, Berndtsson R (1993) Fractal analysis of high-resolution rainfall time series. J Geophys Res 98(D12):23265–23274

Onof C, Wheater HS (1993) Modelling of British rainfall using a random parameter Bartlett-Lewis rectangular pulse model. J Hydrol 149:67–95

Onof C, Wheater H (1994) Improvements to the modelling of British rainfall using a modified random parameter Bartlett-Lewis rectangular pulse model. J Hydrol 157:177–195

Onof C, Chandler RE, Kakou A, Northrop P, Wheater HS, Isham V (2000) Rainfall modelling using Poisson-cluster processes: a review of developments. Stoch Environ Res Risk Assess 14(6):384–411

Önöz B, Bayazit M (2012) Block bootstrap for Mann-Kendall trend test of serially dependent data. Hydrol Process 26(23):3552–3560

Ouarda TBMJ, Labadie JW, Fontane DG (1997) Indexed sequential hydrologic modeling for hydropower capacity estimation. J Am Water Resour Assoc 33(6):1337–1349

Over TM, Gupta VK (1996) Statistical analysis of mesoscale rainfall: dependence of a random cascade generator on large-scale forcing. J Appl Meteorol 33:1526–1542

Parzen E (1962) On estimation of a probability density function and mode. Ann Math Statist 33:1065–1076

Pegram GGS, James W (1972) Multilag multivariate autoregressive model for the generation of operational hydrology. Water Resour Res 8(4):1074–1076

Pelletier JD, Turcotte DL (1997) Long-range persistence in climatological and hydrological time series: analysis, modeling and application to drought hazard assessment. J Hydrol 203:198–208

Perica S, Foufoula-Georgiou E (1996) Model for multiscale disaggregation of spatial rainfall based on coupling meteorological and scaling descriptions. J Geophys Res 101:26347–26361

Pham MT, Vanhaute WJ, Vandenberghe S, De Baets B, Verhoest NEC (2013) An assessment of the ability of Bartlett-Lewis type of rainfall models to reproduce drought statistics. Hydrol Earth Syst Sci 17(12):5167–5183

Prairie JR, Rajagopalan B, Fulp TJ, Zagona EA (2006) Modified K-NN model for stochastic streamflow simulation. J Hydrol Eng 11(4):371–378

Prairie J, Rajagopalan R, Lall U, Fulp T (2007) A stochastic nonparametric technique for space-time disaggregation of streamflows. Water Resour Res 43:W03432. doi:10.1029/2005WR004721

Puente CE, Bierkens MFP, Diaz-Granados MA, Dik PE, Lopez MM (1993) Practical use of analytically derived runoff models based on rainfall point processes. Water Resour Res 29:3551–3560

Pui A, Sharma A, Mehrotra R, Sivakumar B, Jeremiah E (2012) A comparison of alternatives for daily to sub-daily rainfall disaggregation. J Hydrol 470–471:138–157

Quimpo RG (1967) Stochastic model of daily river flow sequences. Hydrology paper 18, Colorado State University, Fort Collins, Colorado

Rajagopalan B, Lall U (1995) A kernel estimator for discrete distributions. J Nonparametric Statist 4:406–426

Rajagopalan B, Lall U (1999) A k-nearest-neighbor simulator for daily precipitation and other weather variables. Water Resour Res 35(10):3089–3101

Rajagopalan B, Lall U, Tarboton DG (1996) Nonhomogeneous Markov model for daily precipitation. J Hydrol Eng 1(1):33–40

Rajagopalan B, Lall U, Tarboton DG (1997) Evaluation of kernel density estimation methods for daily precipitation resampling. Stoch Hydrol Hydraul 11:523–547

Ramirez J, Bras RL (1985) Conditional distributions of Neyman-Scott models for storm arrivals and their use in irrigation scheduling. Water Resour Res 21(3):317–330

Rao SG, Rao RA (1984) Nonlinear stochastic model of rainfall runoff process. Water Resour Res 20(2):297–309

Rasmussen PF, Salas JD, Fagherazzi L, Bobée B (1996) Estimation and validation of contemporaneous PARMA models for streamflow simulation. Water Resour Res 32 (10):3151–3160

Regonda SK, Rajagopalan B, Lall U, Clark M, Moon Y-I (2005) Local polynomial method for ensemble forecast of time series. Nonlinear Proc Geophys 12:397–406

Regonda SK, Rajagopalan B, Clark M, Zagona E (2006a) A multimodel ensemble forecast framework: Application to spring seasonal flows in the Gunnison River Basin. Water Resour Res 42:W09404. doi:10.1029/2005WR004653

Regonda SK, Rajagopalan B, Clark M (2006b) A new method to produce categorical streamflow forecasts. Water Resour Res 42:W09501. doi:10.1029/2006WR004984

Robertson A, Kirshner S, Smyth P (2004) Downscaling of daily rainfall occurrence over northeast Brazil using a hidden Markov model. J Climate 17(22):4407–4424

Rodriguez-Iturbe I, Rinaldo A (1997) Fractal river basins: chance and self-organization. Cambridge University Press, Cambridge

Rodriguez-Iturbe I, Gupta VK, Waymire E (1984) Scale considerations in the modeling of temporal rainfall. Water Resour Res 20(11):1611–1619

Rodriguez-Iturbe I, Cox DR, Eagleson PS (1986) Spatial modeling of total storm rainfall. Proc R Soc Lond A 403:27–50

Rodriguez-Iturbe I, Cox DR, Isham V (1987) Some models for rainfall based on stochastic point processes. Proc R Soc London Ser A 410:269–288

Rodriguez-Iturbe I, Cox DR, Isham V (1988) A point process model for rainfall: further developments. Proc R Soc London Ser A 417:283–298

Roesner LA, Yevjevich VM (1966) Mathematical models for time series of monthly precipitation and monthly runoff. Colo State Univ Hydol Pap 15

Roldan J, Woolhiser DA (1982) Stochastic daily precipitation models: 1. A comparison of occurrence processes. Water Resour Res 18:1451–1459

Ross (1996) Stochastic processes. John Wiley & Sons, New York

Sain SR, Baggerly KA, Scott DW (1994) Cross-validation of multivariate densities. J Am Stat Assoc 89:807–817

Salas JD (1993) Analysis and modeling of hydrologic time series. In: Maidment DR (ed) Handbook of hydrology. McGraw-Hill, New York, pp 19.1–19.72

Salas JD, Abdelmohsen MW (1993) Initialization for generating single-site and multisite low-order periodic autoregressive and moving average processes. Water Resour Res 29 (6):1771–1776

Salas JD, Lee T (2010) Nonparametric simulation of single site seasonal streamflow. J Hydrol Eng 15(4):284–296

Salas JD, Obeysekera JTB (1982) ARMA model identification of hydrologic time series. Water Resour Res 18(4):1011–1021

Salas JD, Obeysekera JTB (1992) Conceptual basis of seasonal streamflow time series models. J Hydraul Eng 118(8):1186–1194

Salas JD, Smith RA (1981) Physical basis of stochastic models of annual flows. Water Resour Res 17(2):428–430

Salas JD, Boes DC, Yevjevich V, Pegram GGS (1979) Hurst phenomenon as a pre-asymptotic behavior. J Hydrol 44(1):1–15

Salas JD, Boes DC, Smith RA (1982) Estimation of ARMA models with seasonal parameters. Water Resour Res 18(4):1006–1010

Salas JD, Delleur JR, Yevjevich VM, Lane WL (1995) Applied modeling of hydrologic time series. Water Resources Publications, Littleton, Colorado

Santos EG, Salas JD (1992) Stepwise disaggregation scheme for synthetic hydrology. J Hydraul Eng 118(5):765–784

Schertzer D, Lovejoy S (1987) Physical modeling and analysis of rain and clouds by anisotropic scaling multiplicative processes. J Geophys Res 92(D8):9693–9714

Schuster E, Yakowitz S (1985) Parametric/nonparametric mixture density estimation with application to flood-frequency analysis. Water Resour Bull 21(5):797–803

Scott DW (1992) Multivariate density estimation: theory, practice and visualization. John Wiley and Sons Inc, New York

Seed AW, Draper C, Srikanthan R, Menabde M (2000) A multiplicative broken-line model for time series of mean areal rainfall. Water Resour Res 36(8):2395–2399

Şen Z (1976) Wet and dry periods of annual flow series. J Hydraul Div Amer Soc Civ Eng 102:1503–1514

Şen Z (1990) Critical drought analysis by second-order Markov chain. J Hydrol 120:183–202

Shahjahan Mondal S, Wasimi SA (2006) Generating and forecasting monthly flows of the Ganges river with PAR model. J Hydrol 323:41–56

Sharifazari S, Araghinejad S (2015) Development of a nonparametric model for multivariate hydrological monthly series simulation considering climate change impacts. Water Resour Manage 29(14):5309–5322

Sharma A, O'Neill R (2002) A nonparametric approach for representing interannual dependence in monthly streamflow sequences. Water Resour Res 38(7):1100. doi:10.1029/2001WR000953

Sharma A, Tarboton DG, Lall U (1997) Streamflow simulation: a nonparametric approach. Water Resour Res 33(2):291–308

Sharma A, Lall U, Tarboton DG (1998) Kernel bandwidth selection for a first order nonparametric streamflow simulation model. Stoch Hydrol Hydraul 12:33–52

Sharma TC, Panu US (2014) A simplified model for predicting drought magnitudes: a case of streamflow droughts in Canadian Prairies. Water Resour Manage 28:1597–1611

Sikorska AE, Montanari A, Koutsoyiannis D (2015) Estimating the uncertainty of hydrological predictions through data-driven resampling techniques. J Hydrol Eng 20(1):A4014009

Silverman BW (1986) Density estimation for statistics and data analysis. Chapman & Hall, New York

Singhrattna N, Rajagopalan B, Clark M, Krishna Kumar K (2005) Forecasting Thailand summer monsoon rainfall. Int J Climatol 25(5):649–664

Sivakumar B (2000) Fractal analysis of rainfall observed in two different climatic regions. Hydrol Sci J 45(5):727–738

Smith JA (1991) Long-range streamflow forecasting using nonparametric regression. Water Resour Bull 27(1):39–46

Smith JA, Carr AF (1983) A point process model of summer season rainfall occurrences. Water Resour Res 19(1):95–103

Smith JA, Day GN, Kane MD (1992) Nonparametric framework for long range streamflow forecasting. J Water Res Plng Mgmt 118(1):82–92

Smithers JC, Pegram GGS, Schulze RE (2002) Design rainfall estimation in South Africa using Bartlett-Lewis rectangular pulse rainfall models. J Hydrol 258(1–4):83–99

Sonali P, Kumar Nagesh (2013) Review of trend detection methods and their application to detect temperature changes In India. J Hydrol 476:212–227

Souza Filho FA, Lall U (2003) Seasonal to interannual ensemble streamflow forecasts for Ceara, Brazil: Applications of a multivariate semiparametric algorithm. Water Resour Res 39 (11):1307. doi:10.1029/2002WR001373

Spitzer F (1964) Principles of random walk. Springer, New York

Sposito G (2008) Scale dependence and scale invariance in hydrology. Cambridge University Press, Cambridge, UK

Sprent P, Smeeton NC (2007) Applied nonparametric statistical methods. Chapman & Hall, Boca Raton

Srikanthan R, McMahon TA (1983) Stochastic simulation of daily rainfall for Australian stations. Trans ASAE 26:754–759
Srikanthan R, Harrold TI, Sharma A, McMahon TA (2005) Comparison of two approaches for generation of daily rainfall data. Stoch Environ Res Risk Assess 19:215–226
Srinivas VV, Srinivasan K (2001) A hybrid stochastic model for multiseason streamflow simulation. Water Resour Res 37(10):2537–2549
Srinivas VV, Srinivasan K (2005) Hybrid moving block bootstrap for stochastic simulation of multi-site multi-season streamflows. J Hydrol 302(1–4):307–330
Srinivas VV, Srinivasan K (2006) Hybrid matched-block bootstrap for stochastic simulation of multiseason streamflows. J Hydrol 329(1–2):1–15
Srivatsav RK, Simonovic SP (2014) Multi-site, multivariate weather generator using maximum entropy bootstrap. Clim Dyn 44:3431–3448
Srivatsav RK, Srinivasan K, Sudheer KP (2011) Simulation-optimization framework for multi-season hybrid stochastic models. J Hydrol 404(3):209–225
Stagge JH, Moglen GE (2013) A nonparametric stochastic method for generating daily climate-adjusted streamflows. Water Resour Res 49:6179–6193
Stedinger JR, Taylor MR (1982) Synthetic streamflow generation: 1. Model verification and validation. Water Resour Res 18(4):909–918
Stedinger JR, Vogel RM (1984) Disaggregation procedures for generating serially correlated flow vectors. Water Resour Res 20(1):47–56
Stedinger JR, Lettenmaier DP, Vogel RM (1985) Multisite ARMA(1,1) and disaggregation models for annual streamflow generation. Water Resour Res 21(4):497–509
Tao PC, Delluer JW (1976) Multistation, multiyear synthesis of hydrologic time series by disaggregation. Water Resour Res 12(6):1303–1312
Tarboton DG, Sharma A, Lall U (1998) Disaggregation procedures for stochastic hydrology based on nonparametric density estimation. Water Resour Res 34(1):107–119
Tesfaye YG, Meerschaert MM, Anderson PL (2006) Identification of periodic autoregressive moving average models and their application to the modeling of river flows. Water Resour Res 42:W01419. doi:10.1029/2004WR003772
Tessier Y, Lovejoy S, Hubert P, Schertzer D, Pecknold S (1996) Multifractal analysis and modeling of rainfall and river flows and scaling, causal transfer functions. J Geophys Res 101 (D21):26427–26440
Thomas HA, Fiering MB (1962) Mathematical synthesis of streamflow sequences for the analysis of river basins by simulation. In: Mass A et al (eds) Design of water resource systems. Harvard University Press, Cambridge, Massachusetts, pp 459–493
Thyer MA, Kuczera G (2000) Modelling long-term persistence in hydro-climatic time series using a hidden state Markov model. Water Resour Res 36:3301–3310
Todini E (1980) The preservation of skewness in linear disaggregation schemes. J Hydrol 47:199–214
Towler E, Rajagopalan B, Seidel C, Summers RS (2009) Simulating ensembles of source water quality using a k-nearest neighbor resampling approach. Environ Sci Technol 43:1407–1411
Ula TA (1990) Periodic covariance stationarity of multivariate periodic autoregressive moving average processes. Water Resour Res 26(5):855–861
Valencia DR, Schaake JC (1973) Disaggregation processes in stochastic hydrology. Water Resour Res 9(3):580–585
Vecchia AV (1985) Periodic autoregressive-moving average (PARMA) modeling with applications to water resources. Water Resour Bull 21(5):721–730
Veneziano D, Langousis A (2010) Scaling and fractals in hydrology. In: Sivakumar B, Berndsson R (eds) Advances in data-based approaches for hydrologic modeling and forecasting. World Scientific, Singapore, pp 107–243
Veneziano D, Moglen G, Furcolo P, Iacobellis V (2000) Stochastic model of the width function. Water Resour Res 36(4):1143–1157
Veneziano D, Langousis A, Furcolo P (2006) Multifractality and rainfall extremes: A review. Water Resour Res 42:W06D15. doi:10.1029/2005WR004716

Venugopal V, Foufoula-Georgiou E, Sapozhnikov V (1999) A space-time downscaling model for rainfall. J Geophys Res-Atmos 104(D16):19705–19721

Verdin A, Funk C, Rajagopalan B, Kleiber W (2016) Kriging and local polynomial methods for blending satellite-derived and gauge precipitation estimates to support hydrologic early warning systems. IEEE Trans Geoscience and Remote Sensing 54(5):2552–2562

Verhoest N, Troch PA, De Troch FP (1997) On the applicability of Bartlett-Lewis rectangular pulses models in the modeling of design storms at a point. J Hydrol 202(1–4):108–120

Verhoest NEC, Vandenberghe S, Cabus P, Onof C, Meca-Figueras T, Jameleddine S (2010) Are stochastic point rainfall models able to preserve extreme flood statistics? Hydrol Process 24:3439–3445

Viola F, Francipane A, Caracciolo D, Pumo D, La Loggia G, Noto LV (2016) Co-evolution of hydrological components under climate change scenarios in the Mediterranean area. Sci Total Environ 544:515–524

Vogel RM, Shallcross AL (1996) The moving blocks bootstrap versus parametric time series models. Water Resour Res 32(6):1875–1882

Vrugt JA (2016) Markov chain Monte Carlo simulation using the DREAM software package: theory, concepts, and MATLAB implementation. Environ Modell Softw 75:273–316

Vrugt JA, ter Braak CJF, Clark MP, Hyman JM, Robinson BA (2008) Treatment of input uncertainty in hydrologic modeling: doing hydrology backward with Markov chain Monte Carlo simulation. Water Resour Res 44:W00B09. doi:10.1029/2007WR006720

Wasko C, Pui A, Sharma A, Mehrotra R, Jeremiah E (2015) Representing low-frequency variability in continuous rainfall simulations: a hierarchical random Bartlett Lewis continuous rainfall generation model. Water Resour Res 51(12):9995–10007

Waymire E, Gupta VK, Rodriguez-Iturbe I (1984) A spectral theory of rainfall intensity at the meso-β scale. Water Resour Res 20(10):1453–1465

Wiener N (1949) Extrapolation, interpolation, and smoothing of stationary time series. MIT Press

Wilks DL (1998) Multisite generalisation of a daily stochastic precipitation generation model. J Hydrol 210:178–191

Wilks DS (2010) Use of stochastic weather generators for precipitation downscaling. Wiley Interdiscip Rev 1:809–907

Wong H, Ip W-C, Zhang R, Xia J (2007) Nonparametric time series models for hydrologic forecasting. J Hydrol 332:337–347

Woolhiser DA, Osborne HB (1985) A stochastic model of dimensionless thunderstorm rainfall. Water Resour Res 21(4):511–522

Yakowitz S (1973) A stochastic model for daily river flows in an arid region. Water Resour Res 9 (5):1271–1285

Yakowitz S (1979) A nonparametric Markov model for daily river flow. Water Resour Res 15 (5):1035–1043

Yakowitz S (1985) Markov flow models and the flood warning problem. Water Resour Res 21 (1):81–88

Yakowitz (1993) Nearest-neighbor regression estimation for null-recurrent Markov time series. Stochastic Processes Their Appl 48:311–318

Yates D, Gangopadhyay S, Rajagopalan B, Strzepek K (2003) A technique for generating regional climate scenarios using a nearest neighbor bootstrap. Water Resour Res 39(7):1199. doi:10.1029/2002WR001769

Yevjevich VM (1963) Fluctuations of wet and dry years—Part I, Research data assembly and mathematical models. Hydrology paper 1, Colorado State University, Fort Collins, Colorado

Yevjevich VM (1964) Fluctuations of wet and dry years—Part II, analysis by serial correlation. Hydrology paper 4, Colorado State University, Fort Collins, Colorado

Yevjevich VM (1972) Stochastic processes in hydrology. Water Resour Publ, Fort Collins, Colorado

Yevjevich VM (1975) Generation of hydrologic time series. Hydrology paper 56, Colorado State University, Fort Collins, CO

Yue S, Pilon P (2004) A comparison of the power of the t test, Mann-Kendall and bootstrap tests for trend detection. Hydrol Sci J 49(1):21–37

Zucchini W, Adamson PT (1989) Bootstrap confidence intervals for design storms from exceedence series. Hydrol Sci J 34:41–48

Zucchini W, Guttorp P (1991) A hidden Markov model for space-time precipitation. Water Resour Res 27:1917–1923

Chapter 4
Modern Nonlinear Time Series Methods

Abstract Advances in computational power, scientific concepts, and data measurements have led to the development of numerous nonlinear methods to study complex systems normally encountered in various scientific fields. These nonlinear methods often have very different conceptual bases and levels of sophistication and have been found suitable for studying many different types of systems and associated problems. Their relevance to hydrologic systems and ability to model and predict the salient characteristics of hydrologic systems have led to their extensive applications in hydrology over the past three decades or so. This chapter presents an overview of some of the very popular nonlinear methods that have found widespread applications in hydrology. The methods include: nonlinear stochastic methods, data-based mechanistic models, artificial neural networks, support vector machines, wavelets, evolutionary computing, fuzzy logic, entropy-based techniques, and chaos theory. For each method, the presentation includes a description of the conceptual basis and examples of applications in hydrology.

4.1 Introduction

The nonlinear nature of hydrologic systems has been known for many decades (e.g. Minshall 1960; Jacoby 1966; Amorocho 1967, 1973; Dooge 1967; Amorocho and Brandstetter 1971; Bidwell 1971; Singh 1979). It is evident in various ways and at almost all spatial and temporal scales. The hydrologic cycle itself is an example of a system exhibiting nonlinear behavior, with almost all of the individual components themselves exhibiting nonlinear behavior as well. The climatic inputs and landscape characteristics are changing in a highly nonlinear fashion, and so are the outputs, often in unknown ways. The rainfall-runoff process is nonlinear, almost regardless of the basin area, land uses, rainfall intensity, and other influencing factors; see Singh (1988) for a comprehensive review of earlier black-box and conceptual models of nonlinear rainfall-runoff processes. In fact, the effects of nonlinearity can be tremendous, especially when the system is sensitively dependent on initial conditions. This means, even small changes in the inputs may result in large changes in the outputs (and large changes in the inputs may turn out to cause only small changes in

B. Sivakumar, *Chaos in Hydrology*, DOI 10.1007/978-90-481-2552-4_4

the outputs), a situation popularly termed as 'chaos' in the nonlinear science literature (e.g. Lorenz 1963).

Although the nonlinear nature of hydrologic systems has been known for a long time now, much of the early research in hydrologic systems (particularly during 1960s–1980s), including development and applications of time series methods, essentially resorted to linear stochastic approaches (e.g. Thomas and Fiering 1962; Harms and Campbell 1967; Yevjevich 1972; Valencia and Schaake 1973; Klemeš 1978; Beaumont 1979; Kavvas and Delleur 1981; Salas and Smith 1981; Srikanthan and McMahon 1983; Bras and Rodriguez-Iturbe 1985; Salas et al. 1995), which continue to be prevalent in hydrology; see also Chap. 3 for details. One of the important factors that contributed to, or necessitated, the use of linear approaches was the lack of computational power to develop the (perhaps more complex) nonlinear mathematical models.

However, significant developments in computational power during the past three decades or so, and also major advances in measurement technology and mathematical concepts, have facilitated formulation of nonlinear approaches as viable alternatives for complex systems. This subsequently led to applications of nonlinear approaches to study the nonlinear and related properties of hydrologic systems. These applications started to gain attention in the 1990s, and have skyrocketed in recent years.

The nonlinear approaches that are popular in hydrology include: nonlinear stochastic methods, data-based mechanistic models, artificial neural networks, support vector machines, evolutionary computing, fuzzy logic, wavelets, entropy-based techniques, and deterministic chaos theory, among others. The outcomes of the applications of these approaches for hydrologic modeling and forecasting are certainly encouraging, especially considering the fact that we are still in the 'exploratory stage' in regards to these approaches, as opposed to the much more established linear stochastic approaches. Details of applications of these nonlinear approaches in hydrology can be found in Young and Beven (1994), Kumar and Foufoula-Georgiou (1997), Singh (1997, 1998, 2013), ASCE Task Committee (2000a, b), Govindaraju and Rao (2000), Sivakumar (2000, 2004a, 2009), Dibike et al. (2001), Kavvas (2003), Gupta et al. (2007), Şen (2009), Young and Ratto (2009), Abrahart et al. (2010); see also Sivakumar and Berndtsson (2010a) for a compilation of applications of many of these approaches in hydrology. This chapter presents an overview of each of these nonlinear methods and their applications in hydrology. Since the linear stochastic methods and also the methods that make no prior assumptions regarding linearity/nonlinearity have already been extensively discussed in Chap. 3, only a very brief account of the nonlinear stochastic methods is presented.

4.2 Nonlinear Stochastic Methods

The inadequacy of many of the linear stochastic time series methods, discussed in Chap. 3, for studying natural and physical systems was realized as early as in the 1950s. This led to the beginning of the development of nonlinear stochastic methods in the late 1960s (e.g. Klein and Preston 1969; Caughey 1971), which intensified further in the 1980s (e.g. Ozaki 1980; Tong 1983; Priestley 1988). As the name suggests, the nonlinear stochastic methods are, in essence, designed to capture the salient properties of systems/time series that deviate from linearity and associated properties. Consequently, the nonlinear stochastic methods also model the nonlinear, nonperiodic, nonstationary, heteroscedastic properties of systems or their combinations and, therefore, are basically certain variants or extensions of linear stochastic methods (discussed in Chap. 3). Extensive details of nonlinear stochastic methods and their applications can be found in Tong (1983), Tuma and Hannon (1984), Priestley (1988), Seber and Wild (2003), and Pázman (2010), among others.

While studies on the nonlinear nature of hydrologic processes actually began in the 1960s (e.g. Minshall 1960; Jacoby 1966; Amorocho 1967; Dooge 1967), nonlinear stochastic methods, in their specific context, started to find their applications in hydrology only about two decades later. This development was also the result of advances in linear stochastic methods and their applications in hydrology during 1960s–1980s (e.g. Thomas and Fiering 1962; Harms and Campbell 1967; Yevjevich 1972; Valencia and Schaake 1973; Klemeš 1978; Beaumont 1979; Kavvas and Delleur 1981; Salas and Smith 1981; Srikanthan and McMahon 1983), especially for analysis of rainfall and streamflow time series; see Chap. 3 for details. Since 1980s, however, nonlinear stochastic methods have found extensive applications in almost all areas of hydrology (e.g. Rao and Rao 1984; Koutsoyiannis and Xanthopoulos 1990; Rodriguez-Iturbe et al. 1991; Koutsoyiannis 1992; Lall 1995; Serrano 1995; Lall and Sharma 1996; Govindaraju 2002; Kavvas 2003; Cayar and Kavvas 2009), keeping pace with the theoretical developments in nonlinear stochastic concepts. Among the hydrologic problems studied using nonlinear stochastic methods are simulation and disaggregation of rainfall and streamflow, rainfall-runoff modeling, groundwater flow and contaminant transport, and soil moisture dynamics. Indeed, many of the nonparametric stochastic methods may also belong to the class of nonlinear stochastic methods, although the assumption of nonlinearity in their development is not explicitly stated; see Chap. 3 for details of the nonparametric methods and their hydrologic applications.

4.3 Data-based Mechanistic Models

The term 'data-based mechanistic' (DBM) was first used only in the 1990s (Young and Lees 1993), but the basic concepts of this approach to modeling dynamic systems had been developed over many years before that. For example, the concepts were first introduced in the early 1970s (Young 1974). Since then, the DBM concepts have been strengthened further and also applied to many different systems in diverse areas, including hydrology; see Young (1998, 2006, 2010a, 2013) for some reviews. A recent and comprehensive account of DBM concepts and their applications can be found in Young (2011).

The DBM concepts have particularly become popular in hydrologic and environmental system studies, since they take into account many of the salient characteristics of such systems and address important challenges in modeling, including nonlinearity (e.g. Young 1993; Young and Beven 1994; McIntyre et al. 2011) and simplicity in complexity (e.g. Young et al. 1996; Young and Parkinson 2002). The DBM modeling may also offer, in its own way, a unified view of real systems, through combining the deductive and inductive modeling philosophies (e.g. Young 1992, 2013; Young and Ratto 2009).

The DBM approach involves seven major stages:

(1) Definition of the objectives, consideration of the types of most appropriate models, and specification of scale and likely data availability;
(2) Conversion of deterministic simulation equations to a stochastic form. This involves the assumption that the associated parameters and inputs are inherently uncertain and can only be characterized in some suitable stochastic form, such as a probability distribution function (pdf) for the parameters and a time series model for the inputs;
(3) Application of dominant mode analysis (DMA) to enhance the understanding of the relative importance of different parts of the simulation model in explaining the dominant behavioral mechanisms;
(4) Derivation of more complete understanding of the links between the high-order simulation model and its reduced order representation (Stage 3) through performing multiple DMA analysis over a user-specified range of simulation model parameter values. State-dependent parameter regression (SDR) analysis is then applied to these DMA results for estimating the parametric mapping and obtaining a full dynamic emulation model;
(5) Identification of an appropriate model structure and order for experimental time series through a process of statistical inference applied directly to the time series data and based on a generic class of dynamic models;
(6) Reconciliation of the data-based model (Stage 5) with the dynamic emulation version of the simulation model (Stage 4); and
(7) Model validation

It must be noted that, although these seven stages are general stages in DBM modeling, the actual (number of) stages required is usually application-specific. Therefore, the above stages should simply be considered as 'tools' to be used at the discretion of the modeler (Young and Ratto 2009). Note also that the computational algorithms developed for DBM modeling are available in the CAPTAIN Toolbox, for use in the Matlab software environment (see http://captaintoolbox.co.uk/Captain_Toolbox.html). These algorithms also allow for modeling directly in continuous-time terms and the advantages of such an approach are discussed, for example, in Young and Garnier (2006).

Over the past three decades, the DBM modeling approach has found extensive applications in a wide range of problems associated with hydrologic (and environmental) systems. These include, among others, rainfall-runoff modeling, water quality modeling, water level forecasting, flood routing and forecasting, and model simplifcation (Young and Beck 1974; Young and Beven 1994; Lees 2000; Young 2001, 2002, 2003, 2010b, 2013; Chappell et al. 2006; Romanowicz et al. 2006, 2008; Young et al. 2004, 2007; Ratto et al. 2007; Ochieng and Otieno 2009; McIntyre et al. 2011; Beven et al. 2012).

4.4 Artificial Neural Networks

An artificial neural network (ANN) is a massively parallel-distributed information-processing system that has certain performance characteristics resembling biological neural networks of the human brain, where knowledge is acquired through a learning process and finding optimum weights for the different connections between the individual nerve cells (Haykin 1994). A particular advantage of the ANN is that even when no prior knowledge of the actual physical process and the exact relationship between sets of input and output data is available, the network can be 'trained' to 'learn' such a relationship through a transformation function, also called activation or transfer or threshold function. It is this ability of the ANN to 'train' and 'learn' the output from a given input makes it capable of explaining large-scale arbitrarily complex nonlinear problems, such as those encountered in hydrologic systems.

An ANN is generally characterized by (a) its architecture that represents the pattern of connection between nodes; (b) its method of determining the connection weights; and (c) the activation function (see Fausett 1994 for details). A typical ANN consists of a number of nodes that are organized according to a particular arrangement. One way of characterizing ANNs is by the number of layers (e.g. single-layer, bi-layer, multi-layer). Another way of characterizing ANNs is based on the direction of information flow and processing, e.g. feed-forward (where the information flows through the nodes from the input to the output side) and recurrent (where the information flows through the nodes in both directions). Among the several combinations of ANNs, the multi-layer feedforward networks, also popularly known as multi-layer perceptrons (MLPs), trained with a back-propagation (BP) learning algorithm have been the most widely used, since they have been found to provide the best performance with regard to input–output function approximation, such as in forecasting applications.

An MLP can have many layers. Figure 4.1 shows the structure of a typical MLP with just one hidden layer. The first layer connects with the input variables and is, thus, called the input layer. The last layer connects to the output variables and is called the output layer. The layer inbetween the input layer and the output layer is called the hidden layer. While it is very common to have only one hidden layer in an MLP, multiple hidden layers can be included as well, thus making the network four-, five-, and higher-layer ones. The processing elements in each layer are called nodes (or units). Each of these nodes is connected to the nodes of the neighboring layers. The parameters associated with each of these connections are called weights.

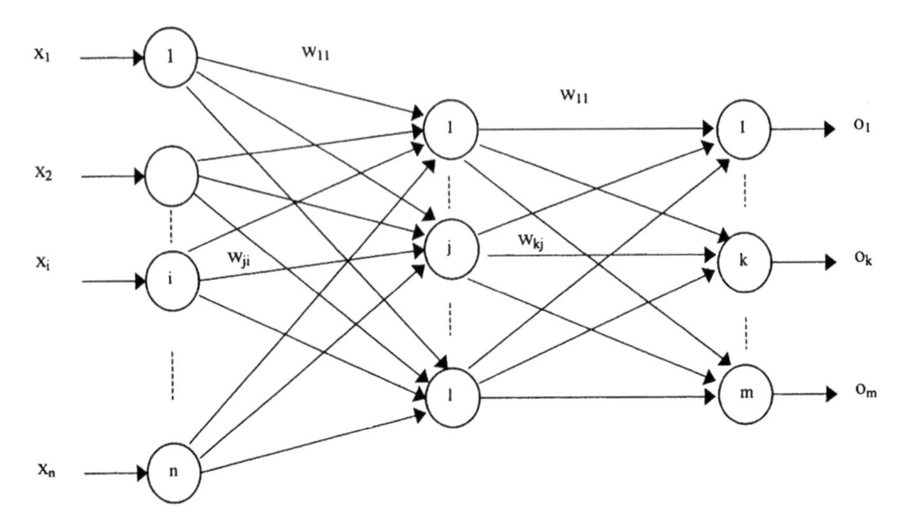

Fig. 4.1 Structure of a typical multi-layer perceptron with one hidden layer (*source* Sivakumar et al. (2002b))

Figure 4.2 shows the structure of a typical node in the hidden layer or output layer. The node j (i.e. each node in the corresponding layer) receives incoming signals from every node i in the previous layer. A weight w_{ji} is associated with each incoming signal x_i. The weighted sum of all the incoming signals to node j is then the effective incoming signal s_j to node j, given by:

$$s_j = \sum_{i=0}^{n} w_{ji} x_i \tag{4.1}$$

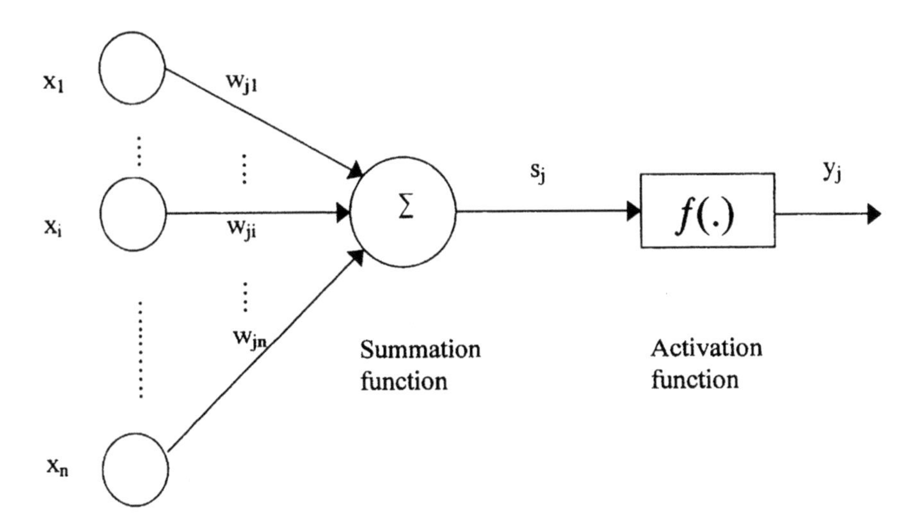

Fig. 4.2 Structure of a typical node in the hidden layer or output layer (*source* Sivakumar et al. (2002b))

This effective incoming signal s_j is passed through a nonlinear activation function to produce the outgoing signal y_j of node j.

In an MLP trained with back-propagation algorithm, the most commonly used activation function is the sigmoid function. The salient characteristics of the sigmoid function are: (a) it is bounded above and below; (b) it is monotonically increasing; and (c) it is continuous and differentiable everywhere (see Hecht-Nielsen 1990 for details). The sigmoid function most often used for ANNs is the logistic function:

$$f(s_j) = \frac{1}{1 + \exp^{-s_j}} \tag{4.2}$$

in which s_j can vary in the range $-\infty$ to $+\infty$, whereas y_j is bounded between 0 and 1.

Applications of ANNs in hydrology started in the early 1990s (e.g. French et al. 1992; Hsu et al. 1995; Maier and Dandy 1996; Minns and Hall 1996; Shamseldin 1997), and have skyrocketed since then (e.g. ASCE Task Committee 2000a, b; Govindaraju and Rao 2000; Maier and Dandy 2000; See and Openshaw 2000a, b; Abrahart and White 2001; Coulibaly et al. 2001, 2005; Dawson and Wilby 2001; Khalil et al. 2001; Sivakumar et al. 2002b; Wilby et al. 2003; Abrahart et al. 2004; Jain et al. 2004; Dawson et al. 2006; Kişi 2007; Jothiprakash and Garg 2009). In addition to MLPs, other types of networks, such as the Generalized Regression Neural Networks (GRNN) and Radial Basis Neural Networks (RBNN), have also been employed (e.g. Jayawardena and Fernando 1998; Cigizoglu 2005; Kişi 2006; Chang et al. 2009; Fernando and Shamseldin 2009). ANNs have been applied for numerous purposes in hydrology, including forecasting or simulation or estimation (rainfall, river flow, river stage, groundwater table depth, sediment, water quality, evapotranspiration), data infilling or missing data estimation, and disaggregation or downscaling (including GCM outputs), among others. Recent years have also witnessed the applications of 'hybrid' multi-models, i.e. combining neural networks with one or more other models (e.g. fuzzy logic, wavelets, genetic algorithm), and other advanced concepts (e.g. neuroemulation, self-organizing maps) (e.g. Hsu et al. 2002; Jain and Srinivasulu 2004; Moradkhani et al. 2004; Chang et al. 2007; Chidthong et al. 2009; Adamowski and Sun 2010; Alvisi and Franchini 2011; Abrahart et al. 2012a).

Despite the advantages of ANNs and the encouraging outcomes of their applications in hydrology, there have and continue to be strong criticisms on their use. The reasons for such criticisms are wide-ranging, such as ANNs are weak extrapolators, they lack physical explanation, they are unable to provide information on input/parameter selection, and they are affected by overparameterization problems; see, for instance, ASCE Task Committee (2000b), Gaume and Gosset (2003), and Koutsoyiannis (2007) for further details. Many studies have addressed these (and many other) concerns in ANN applications and offered strength to their usefulness and effectivness for modeling and forecasting hydrologic time series (e.g. Abrahart et al. 1999; Wilby et al. 2003; Jain et al. 2004; Sudheer and Jain 2004; Chen and Adams 2006). However, there is no question that a lot more still needs to be done to allay such concerns. Recent comprehensive accounts of ANN applications in hydrology during the last two decades can be found in Abrahart et al. (2010, 2012b).

4.5 Support Vector Machines

Support vector machines (SVMs) are a range of classification and regression algorithms that have been formulated from the principles of statistical learning theory (Cortes and Vapnik 1995; Vapnik 1998). The formulation follows the structural risk minimization (SRM) principle, which seeks to minimize an upper bound of the generalization error, rather than minimize the empirical error as is done in traditional empirical risk minimization (ERM) principle employed by conventional neural networks and other techniques. It is this difference that equips SVM with greater ability to generalize, which is the goal in statistical learning. In addition, SVM is equivalent to solving a linear constrained quadratic programming problem and so it achieves a network structure that is always unique and globally optimal. Support vector machines can be applied for both classification and regression problems. The regression problem is far more prevalent in time series analysis and, therefore, a brief description of support vector regression (SVR) is presented here. For further details about support vector machines, see Cristianini and Shawe-Tayler (2000), Hamel (2009), and Steinwart and Christmann (2008), among others.

In support vector regression, the aim is basically to estimate a functional dependency $f(\vec{x})$ between a given set of sampled points $X = \{\vec{x_1}, \vec{x_2}, \ldots, \vec{x_l}\}$ taken from R^n and target values $Y = \{y_1, y_2, \ldots, y_l\}$, with $y_i \in R$. If these samples are assumed to have been generated independently from an unknown probability distribution function $P(\vec{x}, y)$ and a class of functions (Vapnik 1998), according to:

$$F = \{f | f(\vec{x}) = (\vec{w}, \vec{x}) + B : \vec{w} \in R^n, R^n \to R\} \tag{4.3}$$

where $\vec{w}$ and B are coefficients to be estimated from the input data, then the basic problem here is to find a function $f(\vec{x}) \in F$ that minimizes a risk function:

$$\left[R[f(\vec{x})] = \int l(y - f(\vec{x}), \vec{x}) dP(\vec{x}, y)\right] \tag{4.4}$$

where l is a loss function used to measure the deviation between the target values, y, and estimated values, $f(\vec{x})$. Since the probability distribution function $P(\vec{x}, y)$ is unknown, one cannot simply minimize $R[f(\vec{x})]$ directly but can only compute the empirical risk function as:

$$R_{emp}[f(\vec{x})] = \frac{1}{N} \sum_{i=1}^{N} l(y_i - f\vec{x}_i)) \tag{4.5}$$

This traditional empirical risk minimization, however, is not advisable without any means of structural control or regularization. Therefore, a regularized risk function with the smallest steepness among the functions that minimize the empirical risk function can be used as:

$$R_{reg}[f(\vec{x})] = C_c \sum_{x_i \in X} l_\varepsilon(y_i - f(\vec{x}_i)) + \frac{1}{2}\|\vec{w}\|^2 \tag{4.6}$$

where C_c is a positive constant (i.e. an additional capacity control parameter) to be chosen beforehand. The constant C_c that influences a trade-off between an approximation error and the regression (weight) vector $\|\vec{w}\|$ is a design parameter. The loss function here, which is called 'ε-insensitive loss function,' has a particular advantage that not all the input data are needed for describing the regression vector $\vec{w}$ and, thus, can be expressed as:

$$l_\varepsilon(y_i - f(\vec{x}_i)) = \begin{cases} 0 & \textit{for } |y_i - f(\vec{x}_i)| < \varepsilon \\ |y_i - f(\vec{x}_i)| & \textit{otherwise} \end{cases} \tag{4.7}$$

This function serves as a biased estimator when combined with a regularization term $\left(\gamma\|\vec{w}\|^2\right)$. If the difference between the predicted $f(\vec{x}_i)$ and the measured value y_i is less than ε, then the loss is equal to zero. The choice of the value of ε is generally easier than the choice of C_c, and it is often given as a desired percentage of the output values y_i. Hence, nonlinear regression function is given by a function that minimizes the regularized risk function (Eq. (4.6)) subject to the loss function (Eq. (4.7)) as given by (Vapnik 1998; Çimen 2008):

$$f(x) = \sum_{i=1}^{N} (\alpha_i^* - \alpha_i) K(x, x_i) + B \tag{4.8}$$

where α_i, $\alpha_i^* \geq 0$ are the Lagrange multipliers, B is a bias term, and $K(x, x_i)$ is the kernel function. The data are often assumed to have zero mean, and so the bias term can be dropped. The kernel function is primarily to enable operations to be performed in the input space, rather than the potentially high-dimensional feature space. Hence, an inner product in the feature space has an equivalent kernel in the input space. In general, the kernel functions created by the support vector regression are the functions with the polynomial, exponential radial basis, Gaussian radial basis, multilayer perceptron, and splines, among others.

Applications of support vector machines in hydrology started only around the beginning of this century, but have grown enormously since then (e.g. Dibike et al. 2001; Liong and Sivapragasam 2002; Choy and Chan 2003; Asefa et al. 2004, 2005; Bray and Han 2004; Yu et al. 2004; Sivaprakasam and Muttil 2005; Khan and Coulibaly 2006; Tripathi et al. 2006; Yu and Liong 2007; Anandhi et al. 2008; Çimen 2008; Lamorski et al. 2008; Wu et al. 2008, 2009; Karamouz et al. 2009; Kişi and Çimen 2009, 2011; Lin et al. 2009a, b, 2013; Chen et al. 2010; Maity et al.

2010; Samsudin et al. 2011). The applications include forecasting of flows/floods, downscaling of precipitation, estimation of suspended sediment, downscaling and forecasting of evapotranspiration, reservoir operation, prediction of lake water level, development of pedotransfer functions for water retention of soils, estimation of removal efficiency for settling basins, and many others. In some of these studies, support vector machines have also been coupled with other techniques. While the outcomes of these studies are certainly encouraging, establishing connections between SVMs and hydrologic systems/processes and interpreting the outcomes remain challenging, as is the case in almost all data-based approaches.

4.6 Wavelets

A wavelet is a mathematical function that cuts up data into different frequency components and then studies each component with a resolution matched to its scale. A wavelet transform is the representation of a function by wavelets.

Wavelet transforms have advantages over traditional Fourier transforms in analyzing physical systems (e.g. identification of temporal localization of dominant events) where the signal contains discontinuities and sharp spikes, influenced by non-periodic and/or nonstationary events. Furthermore, statistical significance tests for the application of wavelet transform (e.g. Torrence and Compo 1998) also provide a quantitative measure of variance change. Wavelet transforms are broadly classified into discrete wavelet transforms (DWTs) and continuous wavelet transforms (CWTs). Both DWTs and CWTs are continuous-time (analog) transforms, and can be used to represent continuous-time (analog) signals. While CWTs operate over every possible scale and transition, DWTs use a specific subset of scale and translation values or representation grid.

A large number of wavelets have beeen developed on the basis of DWTs and CWTs, and are accordingly called discrete wavelets and continuous wavelets. Examples of discrete wavelets are Daubechies wavelet, Haar wavelet, Legendre wavelet, and Mathieu wavelet, while Beta wavelet, Hermitian wavelet, Mexican hat wavelet, Morlet wavelet, and Shannon wavelet are examples of continuous wavelets. Details on these wavelets and related issues can be found in Daubechies (1988, 1990, 1992), Heil and Walnut (1989), Mallat (1989), Chui (1992), Farge (1992), Roques and Meyer (1993), Jawerth and Sweldens (1994), Torrence and Compo (1998), and Labat (2005, 2010a), among others.

Mathematically, a wavelet transform decomposes a time series X_t with a set of functions $\psi(t, s)$ (called "daughter wavelets") derived from the dilations (s) and translations (t) of a "mother wavelet" $\psi_0(t)$:

$$\psi(t, s) = \frac{1}{s^{1/2}} \psi_0\left(\frac{t' - t}{s}\right) \tag{4.9}$$

where the dilation parameter s (> 0) corresponds to scale or temporal period and, hence, connects the wavelet size to the resolutions of particular frequencies; and the translation parameter t controls the locations of wavelet in the time domain. The term $s^{1/2}$ is the energy normalization factor to keep the energy of daughter wavelets the same as the energy of the mother wavelet (Lau and Weng 1995). The continuous wavelet transform of a time series X_t with respect to the analyzing wavelet $\psi_0(t)$ is defined by the convolution of the two, given by:

$$W(t,s) = \frac{1}{s^{1/2}} \int \psi^* \left(\frac{t' - t}{s} \right) X_t dt \tag{4.10}$$

where ψ^* indicates the complex conjugate of ψ defined on the time and scale and $W(t, s)$ is the generated wavelet coefficient. One of the most commonly used continuous wavelets is the Mexican hat wavelet. The Mexican hat mother wavelet is given by:

$$\psi_0(t) = \frac{2}{\sqrt{3}} \pi^{-1/4} (1 - t^2) \exp(-t^2/2) \tag{4.11}$$

It is the second derivative of a Gaussian function. The shape of the wavelet is shown in Fig. 4.3a.

Figure 4.3 illustrates the processes involved in the application of the wavelet transform to a time series. The example shown is the application of the Mexican hat wavelet to the annual precipitation time series from the Pearl River basin in South China (Niu 2010). The transform calculation is undertaken at different dilation scales and locally around time positions, and the wavelet coefficient map is created (see Fig. 4.3d). The map value indicates the correlation intensity (Gaucherel 2002) of the precipitation time series and the wavelet shapes at different time and dilation scales. The small circle spot on the map of Fig. 4.3d indicates a short fluctuation that can be related to a small-scale wavelet, and the large circle spot indicates a relatively gentle trend. The isolines of the wavelet coefficient present stage features. The solid lines with positive value indicate a relatively wet period, as compared to the average level of precipitation amount over the period of 1951–2000; the dashed lines with negative value indicate a relative dry period; and the zero isoline indicates a transition at large scales or an abrupt change at small scales. From Fig. 4.3d, it can be seen that at a ten-year scale, the precipitation variation is relative dry, relative wet, relative dry, relative wet. Also, the minimum and maximum of the wavelet coefficient at small scales correspond to the severest drought in 1963 and the whole basin flood in 1994, respectively (Niu 2010).

The time position for discrete wavelet transform is not continuous (e.g. dyadic positions). It is a process of time series decomposition for high-frequency and low-frequency parts. The high-frequency part accounts for details, while the low-frequency part accounts for approximation. After one time decomposition, the wavelet coefficients will be half the number of the previous data points.

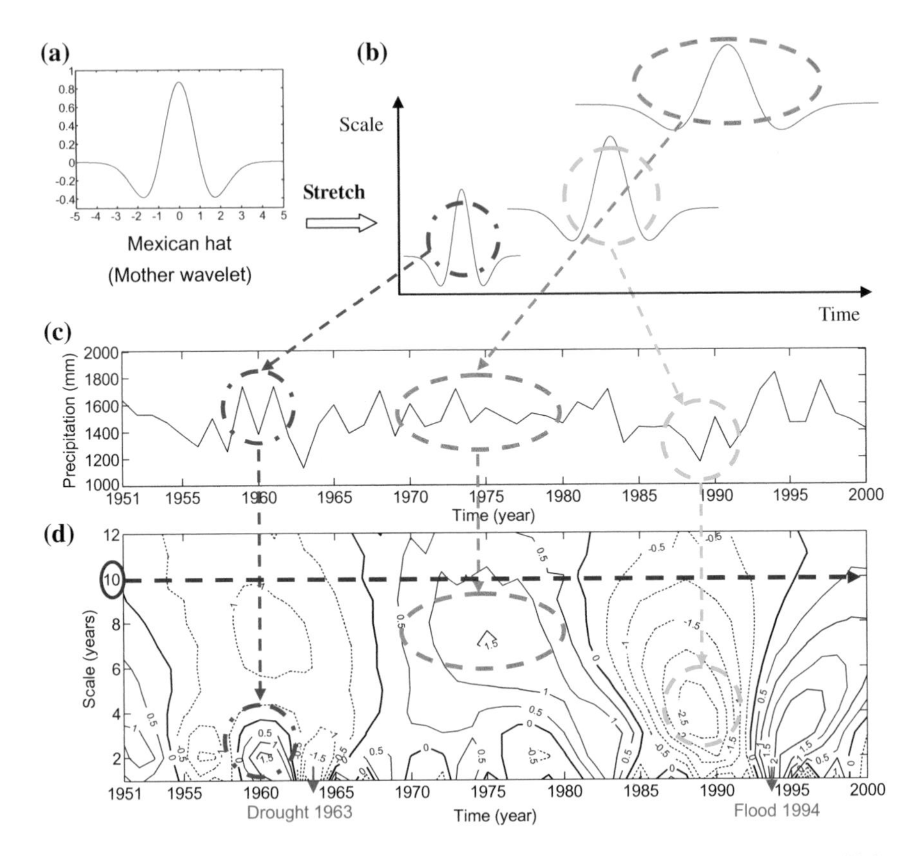

Fig. 4.3 Illustration of the wavelet transform process of **a** Mexican hat mother wavelet; and **b** its daughter wavelet for **c** annual precipitation of the whole Pearl River basin of South China, resulting in **d** wavelet coefficient map (*source* Jun Niu, personal communication)

The Haar wavelet is the simplest wavelet function for the discrete wavelet transform, which is given as:

$$\psi_0(t) = \begin{cases} 1 & 0 \le t < \frac{1}{2} \\ -1 & \frac{1}{2} \le t < 1 \\ 0 & otherwise \end{cases} \tag{4.12}$$

It has an advantage in that it does not produce any edge effects for finite samples of length that are multiples of 2 (Saco and Kumar 2000). In the daughter wavelet $\psi_{m,n}(t)$ (Daubechies 1988; Mallat 1989), 2^m is the scale index and n is time location index.

$$\psi_{m,n}(t) = 2^{-m/2}\psi(2^{-m}t - n) = \frac{1}{\sqrt{2^m}}\psi\left(\frac{t - n2^m}{2^m}\right) \tag{4.13}$$

For a time series $f(t)$, the wavelet coefficient ($d_{m,n}$) can be calculated from:

$$d_{m,n} = \int f(t)\psi_{m,n}(t)dt \tag{4.14}$$

Application of the Haar wavelet transform to hydrologic time series (e.g. streamflow) can be found in Saco and Kumar (2000).

Applications of wavelets in hydrology started in the 1990s (see Foufoula-Georgiou and Kumar (1995) for an earlier review), and have been continuing at a very fast pace since then. Wavelets have been applied for studies on precipitation fields and variability, river flow forecasting, streamflow simulation, rainfall-runoff relations, drought forecasting, suspended sediment discharge, and water quality, among others (e.g. Kumar and Foufoula-Georgiou 1993, 1997; Kumar 1996; Venugopal and Foufoula-Georgiou 1996; Smith et al. 1998; Labat et al. 2000, 2005; Bayazit et al. 2001; Gaucherel 2002; Kim and Valdés 2003; Aksoy et al. 2004; Sujono et al. 2004; Coulibaly and Burn 2006; Gan et al. 2007; Kang and Lin 2007; Lane 2007; Schaefli et al. 2007; Adamowski 2008; Labat 2008, 2010b; Niu 2012; Niu and Sivakumar 2013).

The outcomes of these studies clearly indicate the utility of wavelets for analyzing hydrologic signals as well as their superiority over some traditional signal processing methods (e.g. Fourier transforms). However, serious concerns on the lack of physical interpretation of the results from wavelet analysis and other issues remain (e.g. Maruan and Kurths 2004). Development of wavelet-based models that take into account the intrinsic multi-scale nature of physical relationships of hydrologic processes would help address these concerns. Another way to allay the concerns may be by coupling wavelets with other methods that can represent, at least to a certain degree, the physical relationships. Current studies on wavelets in hydrology provide good indications as to the positive direction in which we are moving; see Labat (2010a) for an excellent recent review of wavelets and their applications in hydrology and in the broader field of Earth sciences.

4.7 Evolutionary Computing

Evolutionary computing generally refers to computation based on principles of Darwinian theory of biological evolution (Darwin 1859) for studying complex systems and associated problems. There are currently various methods under the umbrella of evolutionary computing, each formulated for a certain purpose(s) and interprets/uses the evolution principles in a certain way(s). Broadly, evolutionary computing includes, among others: (1) evolutionary algorithms (e.g. evolutionary programming, evolution strategy, genetic algorithm, genetic programming, differential evolution, eagle strategy); (2) swarm intelligence (e.g. ant colony optimization, particle swarm optimization, bees algorithm, cuckoo search); (3) self-organization (e.g. self-organizing map, growing neural gas, and competitive learning); and others (e.g. cultural algorithm, firefly algorithm, harmony search, parallel simulated annealing). Details of these methods are available in Fogel et al. (1966), Holland (1975), Schwefel (1981), Goldberg (1989), and Koza (1992), among others.

Figure 4.4 presents a schematic representation of flow of information in a typical evolutionary computing algorithm. The evolution in artificial media (e.g. computer) begins, just as in natural evolution, through creation of an initial set (generation) of contending or competing solutions (population in the form of mathematical equations, set of rules or sequences of numbers/patterns) for the problem of interest (e.g. model design, parameter estimation, pattern recognition, optimization). The initial set may be generated either by randomly creating a population of initial solutions or by utilizing the available knowledge about the problem. The 'offsprings' are then generated from the 'parent' solutions of a given generation by means of 'reproduction.' A new population is then produced through 'crossover operation' (in a way similar to sexual reproduction), by exchanging parts (chromosomes) from any two existing parent solutions. Then, by randomly replacing a part of the individual parent solution with a randomly generated new structure, the 'mutation' operation (asexual reproduction) builds a member for the new generation. The 'permutation' operation randomly switches two 'components' (genes) within the individual 'parent.' The resulting new generation 'offspring' solutions are then evaluated for their effectiveness (measured in terms of prediction error, pattern recognition accuracy, optimality, or other evaluation criteria that are relevant to the problem) in

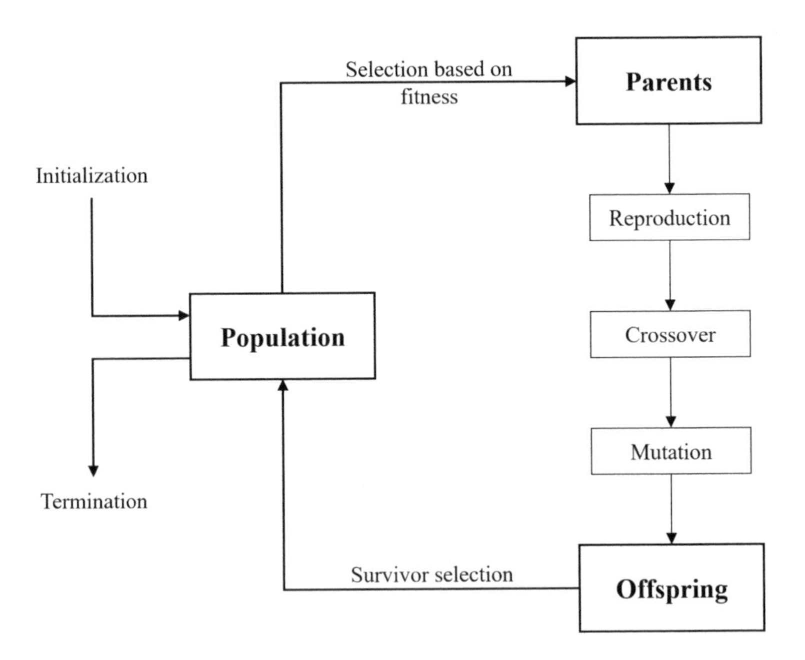

Fig. 4.4 Schematic representation of flow of information in a typical evolutionary computing algorithm

solving the problem using a 'fitness criterion' that tests the ability to reproduce the known behavior. The population in the new generation undergoes selection based on the 'survival of the fittest' criterion (Darwin 1859) and the best set of solutions that satisfies the criterion is chosen as parents for the next round of improvement (through subsequent reproduction). The evolutionary computation algorithm is normally run either for a pre-specified number of generations or until a desired performance is attained by the best solution.

Applications of evolutionary computing methods in hydrology started in the early 1990s, and they have grown enormously since then. So far, evolutionary computing methods have found applications in diverse areas and associated problems in hydrology, including: (a) hydrologic processes, including rainfall, evapotranspiration, soil moisture, and sediment transport (e.g. Babovic 2000; Şen and Oztopel 2001; Makkeasorn et al. 2006; Aytek and Kişi 2007; Parasuraman et al. 2007); (b) rainfall-runoff simulation and modeling, including catchment model calibration, hybrid modeling with artificial neural networks and other data-based methods, and process modeling (e.g. Wang 1991; Franchini 1996; Balascio et al. 1998; Abrahart et al. 1999; Savic et al. 1999; Liong et al. 2002; Tayfur and Moramarco 2008; Zhang et al. 2009); (c) groundwater system, including groundwater remediation, monitoring, sampling network design, and parameter estimation (e.g. McKinney and Lin 1994; Ritzel et al. 1994; Cieniawski et al. 1995; Aly and Peralta 1999; Aral et al. 2001; Karpouzos et al. 2001; Erickson et al. 2002); (d) water quality issues, including water pollution control, waste load allocation, and health risks (e.g. Chen and Chang 1998; Mulligan and Brown 1998; Vasques et al. 2000; Burn and Yulianti 2001; Whigham and Rechnagel 2001; Zou et al. 2007); (e) urban water systems, including water distribution systems, urban drainage systems, and wastewater systems (e.g. Dandy et al. 1996; Gupta et al. 1999; Montesinos et al. 1999; Rauch and Harremoes 1999; Vairavamoorthy and Ali 2000; Hong and Bhamidimarri 2003); and (f) reservoir control and operations, including reservoir planning and operations, irrigation, and control of floods and droughts (e.g. Oliveira and Loucks 1998; Chang and Chen 1998; Wardlaw and Sharif 1999; Chen 2003; Chang et al. 2003). Reviews of applications of evolutionary computing techniques in hydrology can be found in Babovic (1996), Savic and Khu (2005), and Babovic and Rao (2010), among others.

Despite these advances, evolutionary computing techniques and their applications in hydrology are still an emerging area of research and, thus, our knowledge of these methods remains very limited. Furthermore, there are also certain important issues or difficulties pertaining to hydrologic systems/data (and real and complex systems in general) that may constrain the success of these methods, such as selection of data sets (for training and validation), parameter settings, model structure, and effects of noise (see, for example, Babovic and Rao (2010) for some details). As a result, we have not been able to considerably explore the methods in hydrology, to the extent we would like, until now. This situation, however, will certainly change in the near future, as we continue our research in this direction both vigorously and rigorously. The increasing availability of hydrologic data and the computing power should help in this regard.

4.8 Fuzzy Logic

Fuzzy logic is based on fuzzy set theory, and is appropriate for systems where empirical relationships are not well-defined or impractical to model. The foundations of fuzzy set theory, to deal specifically with non-statistical uncertainties, were first developed by Zadeh (1965). A fuzzy logic model is a logical-mathematical procedure based on a "IF-THEN" rule system that allows for the reproduction of the human way of thinking in computational form; see Zadeh (1968) for an early study. In general, a fuzzy rule system consists of four basic components, as shown in Fig. 4.5:

1. Fuzzification of variables—process that transforms the "crisp" variable into a "fuzzy" variable. For each variable, input (e.g. rainfall) or output (e.g. runoff),

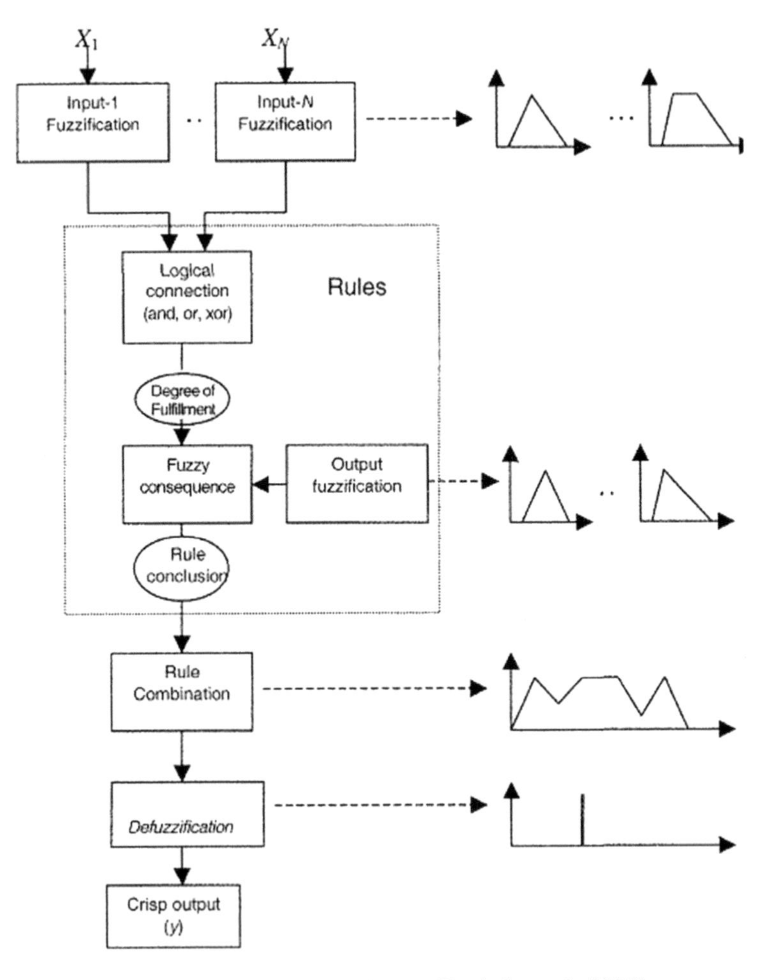

Fig. 4.5 Components of a fuzzy rule system (*source* Hundecha et al. (2001))

a set of membership functions is defined. A membership function basically defines the degree to which the value of a variable belongs to the group and is usually a linguistic term (e.g. high, low).
2. Construction of fuzzy rules—process that links the input and output variables. Membership functions of each variable are related to the output normally through a series of IF-THEN statements or rules. For example, one rule would be: IF the rainfall amount is low (linguistic term represented by a membership function), THEN the runoff amount is low (linguistic term represented by a membership function).
3. Fuzzy inference—process that elaborates and combines rule outputs. The rules are mathematically evaluated and the results are combined. Each rule is evaluated through a process called implication, and the results of all of the rules are combined in a process called aggregation.
4. Defuzzification of output—process that transforms the fuzzy output into a crisp output.

There exist a variety of methodologies for developing fuzzy rule systems, including those proposed by Mamdani and Assilian (1975), Tsukamoto (1979), and Takagi and Sugeno (1985). Some of these typically follow the scheme just mentioned above, whereas others use a composite procedure for fuzzy inference and output fuzzification.

Fuzzy logic models have several advantages over many statistical and other methods. They: (1) are conceptually easy to understand; (2) are flexible; (3) are tolerant of imprecise data; (4) can model nonlinear functions of arbitrary complexity; (5) can be built on top of the experience of experts; (6) can be blended with conventional control techniques; and (7) do not necessarily replace conventional control methods, but often augment them and simplify their implementation. Since real systems, such as hydrologic systems, typically present grey or fuzzy information and are also often present data limitations, particularly on extreme events, fuzzy logic models present an excellent opportunity and useful tool for such systems. Consequently, fuzzy logic has been applied in various scientific and engineering fields. Extensive details about fuzzy set theory, fuzzy logic, and fuzzy rule systems, as well as their applications in different fields can be found in Dubois (1980), Kosko (1993), Bárdossy and Duckstein (1995), Ross (1995), Zadeh et al. (1996), Klir et al. (1997), Bogardi et al. (2003), and Şen (2009), among others.

Applications of fuzzy logic in hydrology started in the 1990s, and have continued to intensify since then (e.g. Bárdossy and Disse 1993; Bárdossy et al. 1995, 2006; Özelkan et al. 1996; Pesti et al. 1996; Fontane et al. 1997; Mujumdar and Sasikumar 1999; Pongracz et al. 1999; Abebe et al. 2000; Hundecha et al. 2001; Özelkan and Duckstein 2001; Xiong et al. 2001; Bárdossy and Samaniego 2002; Cheng et al. 2002; Mahabir et al. 2003; Tayfur et al. 2003; Nayak et al. 2004; Chang et al. 2005; Alvisi et al. 2006; Kişi et al. 2006; Lohani et al. 2007; Simonovic 2009). The applications include rainfall-runoff, runoff forecasting, river-level monitoring and forecasting systems, water quality, stage-discharge-sediment concentration relationships, reservoir operation, simulation of actual component hydrologic processes, relations between atmospheric circulation patterns and precipitation, infiltration, classification of atmospheric patterns, droughts, and others. Many studies have combined fuzzy logic with other data-driven approaches as well. For comprehensive reviews of fuzzy logic applications in hydrology, see Bárdossy and Duckstein (1995), Bogardi et al. (2003), and Şen (2009), among others.

4.9 Entropy-based Models

The concept of entropy was originally formulated by Shannon (1948). Entropy can be regarded as a measure of information or disorder or uncertainty. Uncertainty about an event suggests that the event may take on different values, and information is gained by observing the event only if there is uncertainty about it. If an event occurs with a high probability, it conveys less information and vice versa. On the other hand, more information is needed to characterize less probable or more uncertain events or reduce uncertainty about the occurrence of such an event. In a similar vein, if an event is more certain to occur, its occurrence or observation conveys less information and less information is needed to characterize it. This suggests that a more uncertain event transmits more information or that more information is needed to characterize it. This means that there is a connection between entropy, information, and uncertainty.

Almost a decade after Shannon's introduction of entropy theory, Jaynes (1957a, b) developed the principle of maximum entropy (POME) for deriving a least biased probability distribution subject to given information expressed mathematically in terms of constraints. Subsequently, Jaynes (1979) also developed the theorem of concentration for hypothesis testing. In addition to the Shannon entropy, there are also other types of entropies, including Kolmogorov entropy (Kolmogorov 1956, 1958), Rényi's entropy (Rényi 1961), Tsallis entropy (Tsallis 1988), epsilon entropy (Rosenthal and Binia 1988), algorithmic entropy (Zurek 1989), Kapur entropy (Kapur 1989), and exponential entropy (Pal and Pal 1991), among others.

During the past few decades, the concept of entropy has found extensive applications in hydrology and related fields, including hydraulics, geomorphology, and meteorology. Leopold and Langbein (1962) were probably the first to employ the entropy concept in hydrology with their study on landscape evolution. However, the real impetus to entropy-based modeling was provided in the early 1970s with the works of Amorocho and Espildora (1973) and Sonuga (1972, 1976), among others. Since then, numerous studies have and continue to employ the concept to address a wide variety of hydrologic and water resources systems and problems (e.g. Chiu 1987, 1991; Jowitt 1991; Fiorentino et al. 1993; Singh and Guo 1995a, b; Cao and Knight 1997; Koutsoyiannis 2005a, b; de Araujo 2007; Singh

2010a, b, c). Comprehensive reviews of such studies can be found in Harmancioglu et al. (1992), Singh and Fiorentino (1992), and Singh (1997, 1998, 2013).

Applications of the entropy concept in hydrology may be broadly grouped into three classes: (1) physical; (2) statistical; and (3) mixed. In the first class, applications involve certain physical law(s) in the form of a flux-concentration relation and a hypothesis on the cumulative probability distribution of either flux or concentration, depending on the problem. Rainfall-runoff modeling, infiltration capacity estimation, movement of soil moisture, distribution of velocity in water courses, hydraulic geometry, channel cross-section shape, sediment concentration and discharge, sediment yield, longitudinal river bed profile, and rating curve are examples for this type of applications (e.g. Chiu 1987; Cao and Knight 1997; Singh 2010a, b; Singh and Zhang 2008a, b; Singh et al. 2003a, b). The general objective in this class of problems is the derivation of the design variable as a function of space or time.

The second class of applications does not directly invoke any physical law and is entirely statistical, even though physics may appear indirectly through the specification of constraints. Examples of such applications are derivation of frequency distributions for given constraints, estimation of frequency distribution parameters in terms of given constraints, evaluation and design of monitoring networks in space and/or time, flow forecasting, spatial and inverse spatial analysis, grain-size distribution, complexity analysis, and clustering analysis (e.g. Krstanovic 1991a, b, 1993a, b; Krasovskaia and Gottschalk 1992; Krasovskaia 1997; Singh 1998).

The third class of applications involves deriving relations between entropy and design variables and then establishing relations between design variables and system characteristics. Geomorphologic relations for elevation, slope, and fall, and evaluation of water distribution systems fall in this class (Yang 1971; Fiorentino et al. 1993).

A more recent review of entropy-based modeling in hydrology is presented in Singh (2011), with particular focus on hydrologic synthesis. The review: (1) revisits the definition of entropy and entropy theory; (2) presents a general methodology for the application of the theory; (3) shows how entropy theory couples statistical information with physical laws and how it can be employed to derive useful physical constructs in space and/or space; and (4) provides a review of physical applications of the entropy theory for hydrologic synthesis. Singh (2013) presents entropy theory applications in an even broader context of environmental and water engineering.

4.10 Nonlinear Dynamics and Chaos

Hydrologic systems are not only nonlinear and interdependent, but also possess hidden determinism and order. Their 'complex' and 'random-looking' behaviors need not always be the outcome of 'random' systems but can also arise from simple deterministic systems with sensitive dependence on initial conditions, called 'chaos;' see Lorenz (1963) for an introduction of 'chaos.' Although the discovery of 'chaos theory' in the 1960s brought about a noticeable change in our perception of 'complex' systems, the theory did not find applications in hydrology for over two decades or so, due mainly to the absence of powerful computers and nonlinear mathematical tools. However, with computational advances in the 1970s and development of new nonlinear methods in the 1980s (e.g. Grassberger and Procaccia 1983; Wolf et al. 1985; Farmer and Sidorowich 1987), applications of nonlinear dynamics and chaos concepts in hydrology started during the late 1980s–early 1990s (e.g. Hense 1987; Rodriguez-Iturbe et al. 1989; Sharifi et al. 1990; Wilcox et al. 1991; Berndtsson et al. 1994; Jayawardena and Lai 1994). Since then, there has been an enormous growth in chaos studies in hydrology (e.g. Abarbanel and Lall 1996; Puente and Obregon 1996; Porporato and Ridolfi 1997; Liu et al. 1998; Wang and Gan 1998; Lambrakis et al. 2000; Sivakumar et al. 1999, 2001a, b, 2005, 2007; Elshorbagy et al. 2002; Faybishenko 2002; Sivakumar 2002; Sivakumar and Jayawardena 2002; Zhou et al. 2002; Manzoni et al. 2004; Regonda et al. 2004; Dodov and Foufoula-Georgiou 2005; Salas et al. 2005; Hossain and Sivakumar 2006; Dhanya and Nagesh Kumar 2010, 2011; Kyoung et al. 2011; Sivakumar and Singh 2012). Comprehensive accounts of such applications are presented in Sivakumar (2000, 2004a, 2009) and Sivakumar and Berndtsson (2010b). The application areas and problems include rainfall, river flow, rainfall-runoff, sediment transport, groundwater contaminant transport, modeling, prediction, noise reduction, scaling, disaggregation, missing data estimation, reconstruction of system equations, parameter estimation, and catchment classification, among others. Despite the criticisms and suspicions on such studies and the reported outcomes due to various reasons (e.g. Schertzer et al. 2002; Koutsoyiannis 2006), these studies and their outcomes certainly provide different perspectives and new avenues to study hydrologic systems and processes; see Sivakumar et al. (2002a, c) and Sivakumar (2004b, 2005a) for some details. In fact, arguments as to the potential of chaos theory to serve as a bridge between our traditional and dominant deterministic and stochastic theories have also been put forward (e.g. Sivakumar 2004a, 2009). The discussion of chaos theory and its applications in hydrology are the focus of the rest of the book.

4.11 Summary

Since the 1980s, along with advances in computational power and measurement techniques, nonlinear time series methods, such as the ones described above, have found extensive applications in hydrology. Such applications have allowed us to study hydrologic data in far more detail than ever before and have significantly advanced our understanding of hydrologic systems and processes. Despite this progress, however, there are also growing concerns in regards to two important issues, among others: (1) our tendency to 'specialize' in individual scientific theories and mathematical methods, rather than finding ways to integrate them to better address the larger hydrologic issues; and (2) the lack of ‘physical’ explanation of these concepts and the parameters involved in the methods to real catchments and their salient properties (see Sivakumar 2005b, 2008a, b for some details). Although there have certainly been some efforts in advancing research in this direction (e.g. Wilby et al. 2003; Jain et al. 2004; Sivakumar 2004b; Sudheer and Jain 2004; Hill et al. 2008), there is clearly far more that needs to be done, to realize the true potential of modern nonlinear time series methods for studying hydrology. These issues will be discussed further in Part C, as relevant in the context of applications of nonlinear dynamic and chaos concepts, and in Part D, with a look into the future.

References

Abarbanel HDI, Lall U (1996) Nonlinear dynamics of the Great Salt Lake: system identification and prediction. Climate Dyn 12:287–297

Abebe AJ, Solomatine DP, Venneker RGW (2000) Application of adaptive fuzzy rule-based models for reconstruction of missing precipitation events. Hydrol Sci J 45(3):425–436

Abrahart RJ, White SM (2001) Modelling sediment transfer in Malawi: comparing backpropagation neural network solutions against a multiple linear regression benchmark using small data sets. Phys Chem Earth Part B: Hydrol Oceans Atmos 26(1):19–24

Abrahart RJ, See LM, Kneale PE (1999) Using pruning algorithms and genetic algorithms to optimise network architectures and forecasting inputs in a neural network rainfall-runoff model. J Hydroinfor 1(2):103–114

Abrahart RJ, Kneale PE, See LM (eds) (2004) Neural networks for hydrological modelling. A.A. Balkema Publishers, Rotterdam, The Netherlands

Abrahart RJ, See LM, Dawson CW, Shamseldin AY, Wilby RL (2010) Nearly two decades of neural network hydrologic modeling. In: Sivakumar B, Berndtsson R (eds) Advances in data-based approaches for hydrologic modeling and forecasting. World Scientific Publishing Company, Singapore, pp 267–346

Abrahart RJ, Mount NJ, Shamseldin AY (2012a) Neuroemulation: definition and key benefits for water resources research. Hydrol Sci J 57:407–423

Abrahart RJ, Anctil F, Coulibaly P, Dawson CW, Mount NJ, See LM, Shamseldin AY, Solomatine DP, Toth E, Wilby RL (2012b) Two decades of anarchy? Emerging themes and outstanding challenges for neural network river forecasting. Prog Phys Geogr 36(4):480–513

Adamkowski (2008) River flow forecasting using wavelet and cross-wavelet transform models. Hydrol Process 22(25):4877–4891

Adamowski J, Sun K (2010) Development of a coupled wavelet transform and neural network method for flow forecasting of non-perennial rivers in semi-arid watersheds. J Hydrol 390:85–91

Aksoy H, Akar T, Unal NE (2004) Wavelet analysis for modeling suspended sediment discharge. Nord Hydrol 35(2):165–174

Alvisi S, Franchini M (2011) Fuzzy neural networks for water level and discharge forecasting with uncertainty. Environ Modell Softw 26:523–537

Alvisi S, Mascellani G, Franchini M, Bárdossy A (2006) Water level forecasting through fuzzy logic and artificial neural network approaches. Hydrol Earth Syst Sci 10:1–17

Aly AH, Peralta RC (1999) Optimal design of aquifer cleanup systems under uncertainty using a neural network and a genetic algorithm. Water Resour Res 35(8):2523–2532

Amorocho J (1967) The nonlinear prediction problems in the study of the runoff cycle. Water Resour Res 3(3):861–880

Amorocho J (1973) Nonlinear hydrologic analysis. Adv Hydrosci 9:203–251

Amorocho J, Brandstetter A (1971) Determination of nonlinear functional response functions in rainfall-runoff processes. Water Resour Res 7(5):1087–1101
Amorocho J, Espildora B (1973) Entropy in the assessment of uncertainty in hydrologic systems and models. Water Resour Res 9(6):1511–1522
Anandhi A, Srinivas VV, Nanjundiah RS, Nagesh Kumar D (2008) Downscaling precipitation to river basin in India for IPCC SRES scenarios using support vector machine. Int J Climatol 28 (3):401–420
Aral MM, Guan J, Maslia ML (2001) Identification of contaminant source location and release history in aquifers. J Hydraul Eng ASCE 6(3):225–234
ASCE Task Committee (2000a) Artificial neural networks in hydrology. I: preliminary concepts. J Hydrol Engg. 5(2):115–123
ASCE Task Committee (2000b) Artificial neural networks in hydrology. II: hydrologic applications. J Hydrol Engg. 5(2):124–137
Asefa T, Kemblowski MW, Urroz G, McKee M, Khalil A (2004) Support vectors-based groundwater head observation networks design. Water Resour Res 40:W11509. doi:10.1029/2004WR003304
Asefa T, Kemblowski M, Lall U, Urroz G (2005) Support vector machines for nonlinear state space reconstruction: application to the Great Salt Lake time series. Water Resour Res 41: W12422. doi:10.1029/2004WR003785
Aytek A, Kişi Ö (2007) A genetic programming approach to suspended sediment modelling. J Hydrol 351(3–4):288–298
Babovic V (1996) Emergence evolution intelligence Hydroinformatics—a study of distributed and decentralised computing using intelligent agents. A. A. Balkema Publishers, Rotterdam, Holland
Babovic V (2000) Data mining and knowledge discovery in sediment transport. Comp Aid Civil Infrastr Engg 15(5):383–389
Babovic V, Rao R (2010) Evolutionary computing in hydrology. In: Sivakumar B, Berndtsson R (eds) Advances in data-based approaches for hydrologic modeling and forecasting. World Scientific Publishing Company, Singapore, pp 347–369
Balascio CC, Palmeri DJ, Gao H (1998) Use of a genetic algorithm and multi-objective programming for calibration of a hydrologic model. Trans ASAE 44(3):615–619
Bárdossy A, Disse M (1993) Fuzzy rule-based models for infiltration. Water Resour Res 29 (2):373–382
Bárdossy A, Duckstein L (1995) Fuzzy rule-based modeling in geophysical, economic, biological and engineering systems. CRS Press, Boca Raton, FL
Bárdossy A, Samaniego L (2002) Fuzzy rule-based classification of remotely sensed imagery. IEEE Trans on Geosci Rem Sens 40:362–374
Bárdossy A, Duckstein L, Bogardi I (1995) Fuzzy rule-based classification of atmospheric circulation patterns. Int J Climatol 5:1087–1097
Bárdossy A, Mascellani G, Franchini M (2006) Fuzzy unit hydrograph. Water Resour Res 42: W02401. doi:10.1029/2004WR003751
Bayazit M, Önöz B, Aksoy H (2001) Nonparametric streamflow simulation by wavelet or Fourier analysis. Hydrol Sci 46(4):623–634
Beaumont C (1979) Stochastic models in hydrology. Prog Phys Geogr 3:363–391
Berndtsson R, Jinno K, Kawamura A, Olsson J, Xu S (1994) Dynamical systems theory applied to long-term temperature and precipitation time series. Trends Hydrol 1:291–297
Beven KJ, Leedal DT, Smith PJ, Young PC (2012) Identification and representation of state dependent non-linearities in flood forecasting using the DBM methodology. In: Wang L, Garnier H (eds) System identification, environmetric modelling and control. Springer, Berlin, pp 341–366
Bidwell VJ (1971) Regression analysis of nonlinear catchment systems. Water Resour Res 7:1118–1126
Bogardi I, Bárdossy A, Duckstein L, Pongracz R (2003) Fuzzy logic in hydrology and water resources. Fuzzy logic in geology. Elsevier Science, Amsterdam, pp 153–190

Bras RL, Rodriguez-Iturbe I (1985) Random functions and hydrology. Addison-Wesley, Reading, Massachusetts
Bray M, Han D (2004) Identification of support vector machines for runoff modelling. J Hydroinf 6(4):265–280
Burn DH, Yulianti JS (2001) Waste-load allocation using genetic algorithms. J Wat Resour Plan Manage ASCE 127(2):121–129
Cao S, Knight DW (1997) Entropy-based approach of threshold alluvial channels. J Hydraul Res 35(4):505–524
Caughey TK (1971) Nonlinear theory of random vibrations. Advances in Applied Mechanics, vol 11. Academic Press, New York
Cayar M, Kavvas ML (2009) Ensemble average and ensemble variance behavior of unsteady, one-dimensional groundwater flow in unconfined, heterogeneous aquifers: an exact second-order model. Stoch Environ Res Risk Assess 23:947–956
Chang F-J, Chen L (1998) Real-coded genetic algorithm for rule-based flood control reservoir management. Wat Resour Manage 12:185–198
Chang F-J, Lai J-S, Kao L-S (2003) Optimisation of operation rule curves and flushing schedule in a reservoir. Hydrol Process 17:1623–1640
Chang L-C, Chang F-J, Tsai Y-H (2005) Fuzzy exemplar-based inference system for flood forecasting. Water Resour Res 41:W02005. doi:10.1029/2004W003037
Chang F-J, Chang L-C, Wang Y-S (2007) Enforced self-organizing map neural networks for river flood forecasting. Hydrol Process 21:741–749
Chang L-C, Chang F-J, Wang Y-P (2009) Autoconfiguring radial basis function networks for chaotic time series and flood forecasting. Hydrol Process 23:2450–2459
Chappell NA, Tych W, Chotai A, Bidin K, Sinun W, Thang HC (2006) BARUMODEL: combined data based mechanistic models of runoff response in a managed rainforest catchment. Forest Ecol Manage 224:58–80
Chen HS, Chang N-B (1998) Water pollution control in the river basin by fuzzy genetic algorithm-based multiobjective programming modeling. Wat Sci Tech 37(8):55–63
Chen J, Adams BJ (2006) Integration of artificial neural networks with conceptual models in rainfall-runoff modeling. J Hydrol 318(1–4):232–249
Chen L (2003) Real coded genetic algorithm optimization of long term reservoir operation. J Ame Water Resour Assoc 39(5):1157–1165
Chen ST, Yu PS, Tang YH (2010) Statistical downscaling of daily precipitation using support vector machines and multivariate analysis. J Hydrol 47(5):721–738
Cheng CT, Ou CP, Chau KW (2002) Combining a fuzzy optimal model with a genetic algorithm to solve multi-objective rainfall-runoff model calibration. J Hydrol 268:72–86
Choy KY, Chan CW (2003) Modelling of river discharges and rainfall using radial basis function networks based on support vector regression. Int J Syst Sci 34(14–15):763–773
Chidthong Y, Tanaka H, Supharatid S (2009) Developing a hybrid multi-model for peak flood forecasting. Hydrol Process 23:1725–1738
Chiu CL (1987) Entropy and probability concepts in hydraulics. J Hydraul Eng 113(5):583–600
Chiu CL (1991) Application of entropy concept in open channel flow study. J Hydraul Eng 117 (5):615–628
Chui CK (1992) An introduction to wavelets. Academic Press, Boston
Cieniawski SE, Eheart JW, Ranjithan S (1995) Using genetic algorithms to solve a multiobjective groundwater monitoring problem. Water Resour Res 31(2):399–409
Cigizoglu H (2005) Application of generalized regression neural networks to intermittent flow forecasting and estimation. J Hydrol Eng 10(4):336–341
Cortes C, Vapnik V (1995) Support-vector machines. Mach Learn 20:273–297
Coulibaly P, Burn DH (2006) Wavelet analysis of variability in annual Canadian streamflows. Water Resour Res 40(3):W03105. doi:10.1029/2003WR002667
Coulibaly P, Anctil F, Aravena R, Bobée B (2001) Artificial neural network modeling of water table depth fluctuation. Water Resour Res 37(4):885–896

Coulibaly P, Dibike YB, Anctil F (2005) Downscaling precipitation and temperature with temporal neural networks. J Hydrometeor 6(4):483–496
Cristianini N, Shawe-Taylor J (2000) An introduction to support vector machines and other kernel-based learning methods. Cambridge University Press, 204 pp
Çimen M (2008) Estimation of daily suspended sediments using support vector machines. Hydrol Sci J 53(3):656–666
Dandy GC, Simpson AR, Murphy LJ (1996) An improved genetic algorithm for pipe network optimisation. Water Resour Res 32(2):449–458
Darwin C (1859) On the origin of species by means of natural selection, or the preservation of favoured races in the struggle for life. John Murray, London
Daubechies I (1988) Orthonormal bases of compactly supported wavelets. Commun Pure Appl Math XLI:901–996
Daubechies I (1990) The wavelet transform, time-frequency localization and signal analysis. IEEE Trans Inform Theory 36(5):961–1005
Daubechies I (1992) Ten lectures on Wavelets. CSBM-NSF Series Appli Math, SIAM Publications, 357 pp
Dawson CW, Wilby RL (2001) Hydrological modelling using artificial neural networks. Prog Phys Geog 25(1):80–108
Dawson CW, Abrahart RJ, Shamseldin AY, Wilby RL (2006) Flood estimation at ungauged sites using artificial neural networks. J Hydrol 319(1–4):391–409
de Araujo JC (2007) Entropy-based equation to assess hillslope sediment production. Earth Surf Process Landf 32:2005–2018
Dhanya CT, Nagesh Kumar D (2010) Nonlinear ensemble prediction of chaotic daily rainfall. Adv Water Resour 33:327–347
Dhanya CT, Nagesh Kumar D (2011) Multivariate nonlinear ensemble prediction of daily chaotic rainfall with climate inputs. J Hydrol 403:292–306
Dibike YB, Velickov S, Slomatine D, Abbott MB (2001) Model induction with support vector machines: introduction and applications. J Comp Civil Eng 15(3):208–216
Dodov B, Foufoula-Georgiou E (2005) Incorporating the spatio-temporal distribution of rainfall and basin geomorphology into nonlinear analysis of streamflow dynamics. Adv Water Resour 28(7):711–728
Dooge JCI (1967) A new approach to nonlinear problems in surface water hydrology: hydrologic systems with uniform nonlinearity. Int Assoc Sci HydrolPubl 76:409–413
Dubois D (1980) Fuzzy sets and systems, theory and applications. Academic Press, New York
Elshorbagy A, Simonovic SP, Panu US (2002) Estimation of missing streamflow data using principles of chaos theory. JHydrol 255:123–133
Erickson M, Mayer A, Horn J (2002) Multi-objective optimal design of groundwater remediation systems: application of the niched Pareto genetic algorithm (NPGA). Adv Water Resour 25 (1):51–65
Farge M (1992) Wavelet transforms and their applications to turbulence. Annu Rev Fluid Mech 24:395–457
Farmer DJ, Sidorowich JJ (1987) Predicting chaotic time series. Phys Rev Lett 59:845–848
Fausett L (1994) Fundamentals of neural networks. Prentice Hall, Englewood, NJ
Faybishenko B (2002) Chaotic dynamics in flow through unsaturatedfractured media. Adv Water Resour 25(7):793–816
Fernando DAK, Shamseldin AY (2009) Investigation of internal functioning of the radial-basis-function neural network river flow forecasting models. ASCE J Hydrol Eng 14:286–292
Fiorentino M, Claps P, Singh VP (1993) An entropy-based morphological analysis of river-basin networks. Water Resour Res 29(4):1215–1224
Fogel LJ, Owens AJ, Walsh MJ (1966) Artificial intelligence through simulated evolution. John Wiley, New York
Fontane DG, Gates TK, Moncada E (1997) Planning reservoir operations with imprecise objectives. J Water Resour Plann Manage 123(3):154–162

Foufoula-Georgiou E, Kumar P (eds) (1995) Wavelets in Geophysics. Academic Press, New York
Franchini M (1996) Using a genetic algorithm combined with a local search method for the automatic calibration of conceptual rainfall-runoff models. Hydrol Sci J 41(1):21–40
French MN, Krajewski WF, Cuykendall RR (1992) Rainfall forecasting in space and time using a neural network. J Hydrol 137(1–4):1–31
Gan TY, Gobena AK, Wang Q (2007) Precipitation of southwestern Canada: Wavelet, scaling, multifractal analysis, and teleconnection to climate anomalies. J Geophys Res 112(D10110). doi:10.1029/2006JD007157
Gaucherel C (2002) Use of wavelet transform for temporal characterisation of remote watersheds. J Hydrol 269(3–4):101–121
Gaume E, Gosset R (2003) Over-parameterisation, a major obstacle to the use of artificial neural networks in hydrology? Hydrol Earth Syst Sci 7(5):693–706
Goldberg DE (1989) Genetic algorithms in search, optimisation, and machine learning. Addison-Wesley, Reading
Govindaraju RS (2002) Stochastic methods in subsurface contaminant hydrology. American Society of Civil Engineers, New York
Govindaraju RS, Rao AR (eds) (2000) Artificial neural networks in hydrology. Kluwer Academic Publishers, Dordrecht, The Netherlands
Grassberger P, Procaccia I (1983) Measuring the strangeness of strange attractors. Physica D 9:189–208
Gupta I, Gupta A, Khanna P (1999) Genetic algorithm for optimisation of water distribution systems. Environ Modell Softw 14:437–446
Gupta VK, Troutman BM, Dawdy DR (2007) Towards a nonlinear geophysical theory of floods in river networks: an overview of 20 years of progress. In: Tsonis AA, Elsner JB (eds) Twenty years of nonlinear dynamics in geosciences. Springer Verlag
Hamel L (2009) Knowledge discovery with support vector machines. Wiley-Interscience, USA 246 pp
Harmancioglu NB, Singh VP, Alpaslan N (1992) Versatile uses of the entropy concept in water resources. In: Singh VP, Fiorentino M (eds) Entropy and energy dissipation in water resources. Kluwer, Dordrecht, Netherlands, pp 91–118
Harms AA, Campbell TH (1967) An extension to the Thomas-Fiering model for the sequential generation of streamflow. Water Resour Res 3(3):653–661
Haykin S (1994) Neural networks: a comprehensive foundation. MacMillan, New York, USA
Hecht-Nielsen R (1990) Neurocomputing. Addison-Wesley, Reading, Massachusetts, USA
Heil CE, Walnut DF (1989) Continuous and discrete wavelet transforms. SIAM Review 31 (4):628–666
Hense A (1987) On the possible existence of a strange attractor for the southern oscillation. Beitr Phys Atmos 60(1):34–47
Hill J, Hossain F, Sivakumar B (2008) Is correlation dimension a reliable proxy for the number of dominant influencing variables for modeling risk of arsenic contamination in groundwater? Stoch Environ Res Risk Assess 22(1):47–55
Holland JH (1975) Adaptation in natural and artificial systems. University of Michigan Press, Ann Arbor, MI
Hong Y-S, Bhamidimarri R (2003) Evolutionary self-organising modelling of a municipal wastewater treatment plant. Wat Res 37:1199–1212
Hossain F, Sivakumar B (2006) Spatial pattern of arsenic contaminationin shallow wells of Bangladesh: regional geology and nonlinear dynamics. Stoch Environ Res Risk Assess 20(1–2):66–76
Hsu K-L, Gupta HV, Sorooshian S (1995) Artificial neural network modeling of the rainfall-runoff process. Water Resour Res 31(10):2517–2530
Hsu K-L, Gupta HV, Gao X, Sorooshian S, Imam B (2002) Self-organizing linear output map (SOLO): an artificial neural network suitable for hydrologic modeling and analysis. Water Resour Res 38(12):W01302. doi:10.1029/2001WR000795

Hundecha Y, Bárdossy A, Theisen HW (2001) Development of a fuzzy logic-based rainfall-runoff model. Hydrol Sci J 46(3):363–376

Jacoby SLS (1966) A mathematical model for nonlinear hydrologic systems. J Geophys Res 71 (20):4811–4824

Jain A, Srinivasulu S (2004) Development of effective and efficient rainfall-runoff models using integration of deterministic, real-coded genetic algorithms and artificial neural network techniques. Water Resour Res 40:W04302

Jain A, Sudheer KP, Srinivasulu S (2004) Identification of physical processes inherent in artificial neural network rainfall runoff models. Hydrol Process 18(3):571–581

Jawerth B, Sweldens W (1994) An overview of wavelet based multiresolution analyses. SIAM review 36(3):377–412

Jayawardena AW, Fernando DAK (1998) Use of radial basis function type artificial neural networks for runoff simulation. Comput-Aided Civ Infrastruct Engng 13:91–99

Jayawardena AW, Lai F (1994) Analysis and prediction of chaos in rainfall and stream flow time series. J Hydrol 153:23–52

Jaynes ET (1957a) Information theory and statistical mechanics, I. Phys Rev 106:620–630

Jaynes ET (1957b) Information theory and statistical mechanics, II. Phys Rev 108:171–190

Jaynes ET (1979) Concentration of distributions at entropy maxima. Paper presented at the 19th NBER-NSF Seminar on Bayesian Statistics, Montreal, October 1979. In: Jaynes ET, Papers on Probability, Statistics, and Statistical Physics, edited by Rosenkratz RD, Reidel Publishing Company, Boston, 1983, pp 315–336

Jothiprakash V, Garg V (2009) Reservoir sedimentation estimation using artificial neural network. J Hydrol Eng 14(9):1035–1040

Jowitt PW (1991) A maximum entropy view of probability-distributed catchment models. Hydrol Sci J 36(2):123–134

Kang S, Lin H (2007) Wavelet analysis of hydrological and water quality signals in an agricultural watershed. J Hydrol 338:1–14

Kapur JN (1989) Maximum entropy models in science and engineering. Wiley, New Delhi, India

Karamouz M, Ahmadi A, Moridi A (2009) Probabilistic reservoir operation using Bayesian stochastic model and support vector machine. Adv Water Resour 32(11):1588–1600

Karpouzos DK, Delay F, Katsifarakis KL, de Marsily G (2001) A multipopulation genetic algorithm to solve the inverse problem in hydrogeology. Water Resour Res 37(9):2291–2302

Kavvas ML (2003) Nonlinear hydrologic processes: conservation equations for determining their means and probability distributions. ASCE J Hydrol Eng 8(2):44–53

Kavvas ML, Delleur JW (1981) A stochastic cluster model of daily rainfall sequences. Water Resour Res 17(4):1151–1160

Khalil M, Panu US, Lennox WC (2001) Groups and neural networks based streamflow data infilling procedures. J Hydrol 241(3–4):153–176

Khan MS, Coulibaly P (2006) Application of support vector machine in lake water level prediction. J Hydrol Eng 11(3):199–205

Kim TW, Valdés JB (2003) A nonlinear model for drought forecasting based on conjunction of wavelet transforms and neural networks. J Hydrol Engg 8:319–328

Kişi Ö (2006) Generalized regression neural networks for evapotranspiration modeling. Hydrol Sci J 51(6):1092–1105

Kişi Ö (2007) Evapotranspiration modelling from climatic data using a neural computing technique. Hydrol Process 21(14):1925–1934

Kişi Ö, Çimen M (2009) Evapotranspiration modelling using support vector machines. Hydrol Sci J 54(5):918–928

Kişi Ö, Çimen M (2011) A wavelet-support vector machine conjunction model for monthly streamflow forecasting. J Hydrol 399:132–140

Kişi Ö, Karahan ME, Şen Z (2006) River suspended sediment modeling using a fuzzy logic approach. Hydrol Process 20:4351–4362

Klein LR, Preston RS (1969) Stochastic nonlinear models. Econometrica 37(1):95–106

Klemeš V (1978) Physically based stochastic hydrologic analysis. Adv Hydrosci 11:285–352

Klir GJ, St Clair UH, Yuan B (1997) Fuzzy set theory foundations and applications. Prentice Hall, New Jersey, 245 pp

Kolmogorov AN (1956) Aysmptotic characteristics of some completely bounded metric spaces. Dokl Akad Nauk SSSR 108:585–589

Kolmogorov AN (1958) New metric invariant of transitive dynamical systems and endomorphisms of Lebesgue spaces. Dokl Akad Nauk SSSR 119(N5):861–864

Kosko B (1993) Fuzzy thinking: the new science of fuzzy logic. Hyperion, NY

Koutsoyiannis D (1992) A nonlinear disaggregation method with a reduced parameter set for simulation of hydrologic series. Water Resour Res 28(12):3175–3191

Koutsoyiannis D (2005a) Uncertainty, entropy, scaling and hydrological stochastics. 1. Marginal distributional properties of hydrological processes and state scaling. Hydrol Sci J 50(3): 381–404

Koutsoyiannis D (2005b) Uncertainty, entropy, scaling and hydrological stochastics. 2. Time dependence of hydrological processes and time scaling. Hydrol Sci J 50(3):405–426

Koutsoyiannis D (2006) On the quest for chaotic attractors in hydrological processes. Hydrol Sci J 51(6):1065–1091

Koutsoyiannis D (2007) Discussion of "Generalized regression neural networks for evapotranspiration modelling" by O. Kişi. Hydrol Sci J 52(4):832–835

Koutsoyiannis D, Xanthopoulos T (1990) A dynamic model for short-scale rainfall disaggregation. Hydrol Sci J 35(3):303–322

Koza JR (1992) Genetic programming: on the programming of computers by means of natural selection. MIT Press, Cambridge, MA

Krasovskaia I (1997) Entropy-based grouping of river flow regimes. J Hydrol 202:173–191

Krasovskaia I, Gottschalk L (1992) Stability of river flow regimes. Nord Hydrol 23:137–154

Krstanovic PF, Singh VP (1991a) A univariate model for longterm streamflow forecasting: I. Development. Stochastic Hydrol Hydraul 5:173–188

Krstanovic PF, Singh VP (1991b) A univariate model for longterm streamflow forecasting: II. Application. Stochastic Hydrol Hydraul 5:189–205

Krstanovic PF, Singh VP (1993a) A real-time flood forecasting model based on maximum entropy spectral analysis: I. Development. Water Resour Manage 7:109–129

Krstanovic PF, Singh VP (1993b) A real-time flood forecasting model based on maximum entropy spectral analysis: II. Application. Water Resour Manage 7:131–151

Kumar P (1996) Role of coherent structure in the stochastic dynamic variability of precipitation. J Geophys Res 101(26):393–404

Kumar P, Foufoula-Georgiou E (1993) A multicomponent decomposition of spatial rainfall fields. Segregation of large and small scale features using wavelet transform. Water Resour Res 29 (8):2515–2532

Kumar P, Foufoula-Georgiou E (1997) Wavelet analysis for geophysical applications. Rev Geophys 35(4):385–412

Kyoung MS, Kim HS, Sivakumar B, Singh VP, Ahn KS (2011) Dynamic characteristics of monthly rainfall in the Korean peninsula under climate change. Stoch Environ Res Risk Assess 25(4):613–625

Labat D (2005) Recent advances in wavelet analyses: Part 1. A review of concepts. J Hydrol 314:275–288

Labat D (2008) Wavelet analysis of annual discharge records of the world's largest rivers. Adv Water Resour 31:109–117

Labat D (2010a) Wavelet analysis in hydrology. In: Sivakumar B, Berndtsson R (eds) Advances in data-based approaches for hydrologic modeling and forecasting. World Scientific Publishing Company, Singapore, pp 371–410

Labat D (2010b) Cross wavelet analyses of annual continental freshwater discharge and selected climate indices. J Hydrol 385:269–278

Labat D, Ababou R, Mangin A (2000) Rainfall-runoff relations for karstic springs - Part II: Continuous wavelet and discrete orthogonal multiresolution analyses. J Hydrol 238:149–178

Labat D, Ronchail J, Guyot J-L (2005) Recent advances in wavelet analyses: Part 2—Amazon, Parana, Orinoco and Congo discharges time scale variability. J Hydrol 314:289–311
Lall U (1995) Recent advances in nonparametric function estimation: hydraulic applications. U.S. National Report for International Union of Geodesy and Geophysics 1991–1994. Rev Geophys 33:1092–1102
Lall U, Sharma A (1996) A nearest neighbor bootstrap for resampling hydrologic time series. Water Resour Res 32(3):679–693
Lambrakis N, Andreou AS, Polydoropoulos P, Georgopoulos E, Bountis T (2000) Nonlinear analysis and forecasting of a brackish karstic spring. Water Resour Res 36(4):875–884
Lamorski K, Pachepsky Y, Slawihski C, Walczak RT (2008) Using support vector machines to develop pedotransfer functions for water retention of soils in Poland. Soil Sci Soc Am J 72 (5):1243–1247
Lane SN (2007) Assessment of rainfall–runoff models based upon wavelet analysis. Hydrol Process 21:586–607
Lau KM, Weng HY (1995) Climate signal detection using wavelet transform: how to make a time series sing. Bull Am Meteorol Soc 76:2391–2402
Lees MJ (2000) Data-based mechanistic modelling and forecasting of hydrological systems. J Hydroinform 2:15–34
Leopold LB, Langbein WB (1962) The concept of entropy in landscape evolution. Geol Surv Prof Pap 500-A, USGS, U.S. Department of the Interior, Washington, DC, 1–55
Lin G-F, Chen G-R, Wu M-C, Chou Y-C (2009a) Effective forecasting of hourly typhoon rainfall using support vector machines. Water Resour Res 45:W08440. doi:10.1029/2009WR007911
Lin G-F, Chen G-R, Huang P-Y, Chou Y-C (2009b) Support vector machine-based models for hourly reservoir inflow forecasting during typhoon-warning periods. J Hydrol 372(1–4):17–29
Lin G-F, Lin H-Y, Wu M-C (2013) Development of support-vector machine-based model for daily pan evaporation estimation. Hydrol Process 27:3115–3127
Liong SY, Sivapragasam C (2002) Flood stage forecasting with support vector machines. J Am Water Res Assoc 38(1):173–186
Liong SY, Gautam TR, Khu ST, Babovic V, Muttil N (2002) Genetic programming: A new paradigm in rainfall-runoff modelling. J Am Water Resour Assoc 38(3):705–718
Liu Q, Islam S, Rodriguez-Iturbe I, Le Y (1998) Phase-space analysis of daily streamflow: characterization and prediction. Adv Water Resour 21:463–475
Lohani AK, Goel NK, Bhatia KKS (2007) Deriving stage-discharge-sediment concentration relationships using fuzzy logic. Hydrol Sci J 52(4):793–807
Lorenz EN (1963) Deterministic nonperiodic flow. J Atmos Sci 20:130–141
Mahabir C, Hicks FE, Robinson Fayek A (2003) Application of fuzzy logic to forecast seasonal runoff. Hydrol Process 17(18):3749–3762
Maier HR, Dandy GC (1996) The use of artificial neural networks for the prediction of water quality parameters. Water Resour Res 32(4):1013–1022
Maier HR, Dandy GC (2000) Neural networks for the prediction and forecasting of water resources variables: a review of modelling issues and applications. Environ Model Softw 15 (1):101–123
Maity R, Bhagwat PP, Bhatnagar A (2010) Potential of support vector regression for prediction of monthly streamflow using endogenous property. Hydrol Process 24:917–923
Makkeasorn A, Chang N-B, Beaman M, Wyatt C, Slater C (2006) Soil moisture estimation in a semiarid watershed using RADARSAT-1 satellite imagery and genetic programming. Water Resour Res 42(W09401), doi:10.1029/2005WR004033
Mallat S (1989) A theory for multiresolution signal decomposition: The wavelet representation. IEEE Tran Pattern Anal Mach Intel 11(7):674–693
Mamdani EH, Assilian S (1975) An experiment in linguistic synthesis with a fuzzy logic controller. Int J Man-Machine Studies 7(1):1–13
Manzoni S, Porporato A, D'Odorico P, Laio F, Rodriguez-Iturbe I (2004) Soil nutrient cycles as a nonlinear dynamical system. Nonlinear Processes Geophys 11:589–598

Maruan D, Kurths J (2004) Cross wavelet analysis: significance testing and pitfalls. Nonlinear Process Geophys 11:505–514

McIntyre N, Young PC, Orellana B, Marshall M, Reynolds B, Wheater H (2011) Identification of nonlinearity in rainfall-flow response using data-based mechanistic modeling. Water Resour Res 47:W03515. doi:10.1029/2010WR009851

McKinney DC, Lin MD (1994) Genetic algorithm solution of groundwater management models. Water Resour Res 30(6):1897–1906

Minns AW, Hall MJ (1996) Artificial neural networks as rainfall-runoff models. Hydrol Sci J 41 (3):399–417

Minshall NE (1960) Predicting storm runoff on small experimental watersheds. J Hydraul Div Am Soc Eng. 86(HYB):17–38

Montesinos P, Garcia-Guzman A, Ayuso JL (1999) Water distribution network optimisation using a modified genetic algorithm. Water Resour Res 35(11):3467–3473

Moradkhani H, Hsu K-L, Gupta HV, Sorooshian S (2004) Improved streamflow forecasting using self-organizing radial basis fucntion artificial neural networks. J Hydrol 295:246–262

Mujumdar PP, Sasikumar K (1999) A fuzzy risk approach for seasonal water quality management in a river system. Water Resour Res 38(1):1–9

Mulligan AE, Brown LC (1998) Genetic algorithms for calibrating water quality models. J Environ Eng ASCE 124(3):202–211

Nayak PC, Sudheer KP, Rangan DM, Ramasastri KS (2004) A neuro-fuzzy computing technique for modeling hydrological time series. J Hydrol 291:52–66

Niu J (2010) A comprehensive analysis of terrestrial hydrological processes over the Pearl River basin in South China. Ph.D. thesis, 202 pp, University of Hong Kong, Hong Kong

Niu J (2012) Precipitation in the Pearl River basin, South China: scaling, regional patterns, and influence of large-scale climate anomalies. Stoch Environ Res Risk Assess 27(5):1253–1268

Niu J, Sivakumar (2013) Scale-dependent synthetic streamflow generation using a continuous wavelet transform. J Hydrol 496:71–78

Ochieng G, Otieno F (2009) Data-based mechanistic modelling of stochastic rainfall-flow processes by state dependent parameter estimation. Environ Modell Softw 24:279–284

Oliveira R, Loucks DP (1997) Operating rules for multi-reservoir systems. Water Resour Res 33 (4):839–852

Ozaki T (1980) Nonlinear time series models for nonlinear random vibrations. J Appl Probab 17:84–93

Özelkan EC, Duckstein L (2001) Fuzzy conceptual rainfall-runoff models. J Hydrol 253:41–68

Özelkan EC, Ni F, Duckstein L (1996) Relationship between monthly atmospheric circulation patterns and precipitation: fuzzy logic and regression approaches. Water Resour Res 32:2097–2103

Pal NR, Pal SK (1991) Entropy: a new definition and its applications. IEEE Trans Syst Man Cyber 21(5):1260–1270

Parasuraman K, Elshorbagy A, Carey S (2007) Modeling the dynamics of evapotranspiration process using genetic programming. Hydrol Sci J 52(3):563–578

Pázman A (2010) Nonlinear statistical models. Springer, Mathematics and its applications series 259 pp

Pesti G, Shrestha BP, Duckstein L, Bogardi I (1996) A fuzzy rule-based approach to drought assessment. Water Resour Res 32:1741–1747

Pongracz R, Bogardi I, Duckstein L (1999) Application of fuzzy rule-based modeling technique to regional drought. J Hydrol 224:100–114

Porporato A, Ridolfi R (1997) Nonlinear analysis of river flow time sequences. Water Resour Res 33(6):1353–1367

Priestley MB (1988) Non-linear and non-stationary time series analysis. Academic Press, London

Puente CE, Obregon N (1996) A deterministic geometric representation of temporal rainfall. Results for a storm in Boston. Water Resour Res 32(9):2825–2839

Rao SG, Rao RA (1984) Nonlinear stochastic model of rainfall runoff process. Water Resour Res 20(2):297–309

Ratto N, Young PC, Romanowicz R, Pappenberger F, Saltelli A, Pagano A (2007) Uncertainty, sensitivity analysis and the role of data based mechanistic modelling in hydrology. Hydrol Earth Syst Sci 11:1249–1266

Rauch W, Harremoes P (1999) Genetic algorithms in real time control applied to minimise transient pollution from urban wastewater systems. Water Res 33(5):1265–1277

Regonda S, Sivakumar B, Jain A (2004) Temporal scaling in river flow: can it be chaotic? Hydrol Sci J 49(3):373–385

Rényi A (1961) On measures of entropy and information. Proc 4th Berkeley Symposium on Math Stat Prob, vol 1. Berkeley, CA, pp 547–561

Ritzel BJ, Wayland Eheart J, Ranjithan S (1994) Using genetic algorithms to solve a multiple objective groundwater pollution containment problem. Water Resour Res 30(5):1589–1603

Rodriguez-Iturbe I, De Power FB, Sharifi MB, Georgakakos KP (1989) Chaos in rainfall. Water Resour Res 25(7):1667–1675

Rodriguez-Iturbe I, Entekhabi D, Bras RL (1991) Nonlinear dynamics of soil moisture at climate scales, 1. Stochastic analysis. Water Resour Res 27(8):1899–1906

Romanowicz RJ, Young PC, Beven KJ (2006) Data assimilation and adaptive forecasting of water levels in the River Severn catchment. Water Resour Res 42(W06407). doi:10.1029/2005WR005373

Romanowicz RJ, Young PC, Beven KJ, Pappenberger F (2008) A data based mechanistic approach to nonlinear flood routing and adaptive flood level forecasting. Adv Water Resour 31:1048–1056

Roques S, Meyer Y (eds) (1993) Progress in wavelet analysis and applications. Edit Frontières, 785 pp

Rosenthal H, Binia J (1988) On the epsilon entropy of mixed random variables. IEEE Trans Inf Theory 34(5):1110–1114

Ross JT (1995) Fuzzy logic with engineering applications. McGraw-Hill, NY, 593 pp

Saco P, Kumar P (2000) Coherent modes in multiscale variability of streamflow over the United States. Water Resour Res 36(4):1049–1067

Salas JD, Smith RA (1981) Physical basis of stochastic models of annual flows. Water Resour Res 17(2):428–430

Salas JD, Delleur JW, Yevjevich V, Lane WL (1995) Applied modeling of hydrologic time series. Water Resources Publications, Littleton, Colorado

Salas JD, Kim HS, Eykholt R, Burlando P, Green TR (2005) Aggregation and sampling in deterministic chaos: implications for chaos identification in hydrological processes. Nonlinear Processes Geophys 12:557–567

Samsudin R, Saad P, Shabri A (2011) River flow time series using least squares support vector machines. Hydrol Earth Syst Sci 15:1835–1852

Savic DA, Khu ST (2005) Evolutionary computing in hydrological sciences. In: Anderson MG (ed) Encyclopedia of hydrological sciences. John Wiley & Sons Ltd, London, pp 2–18

Savic DA, Walters GA, Davidson GW (1999) A genetic programming approach to rainfall-runoff modeling. Water Resour Manage 13:219–231

Schaefli B, Maraun D, Holschneider M (2007) What drives high flow events in the Swiss Alps? Recent developments in wavelet spectral analysis and their application to hydrology. Adv Water Resour 30:2511–2525

Schertzer D, Tchiguirinskaia I, Lovejoy S, Hubert P, Bendjoudi H (2002) Which chaos in the rainfall-runoff process? A discussion on 'Evidence of chaos in the rainfall-runoff process' by Sivakumar et al. Hydrol Sci J 47(1):139–147

Schwefel HP (1981) Numerical optimisation of computer models. John Wiley, Chichester

Seber GAF, Wild CJ (2003) Nonlinear regression. Wiley-Interscience, New Jersey

See LM, Openshaw S (2000a) A hybrid multi-model approach to river level forecasting. Hydrol Sci J 45(4):523–536

See L, Openshaw S (2000b) Applying soft computing approaches to river level forecasting. Hydrol Sci J 44(5):763–779

Serrano SE (1995) Analytical solutions of the nonlinear groundwater flow equation in unconfined aquifers and the effect of heterogeneity. Water Resour Res 31(11):2733–2742
Shamseldin AY (1997) Application of a neural network technique to rainfall-runoff modelling. J Hydrol 199(3–4):272–294
Shannon CE (1948) The mathematical theory of communications, I and II. Bell Syst Tech J 27:379–423
Sharifi MB, Georgakakos KP, Rodriguez-Iturbe I (1990) Evidence of deterministic chaos in the pulse of storm rainfall. J Atmos Sci 47:888–893
Simonovic SP (2009) Managing water resources: methods and tools for a systems approach. UNESCO, Paris
Singh VP (1979) A uniformly nonlinear hydrologic cascade model. Irrigation Power 36(3):301–317
Singh VP (1988) Hydrologic systems: rainfall-runoff modelling, vol 1. Prentice Hall, Englewood Cliffs, NJ
Singh VP (1997) The use of entropy in hydrology and water resources. Hydrol Process 11:587–626
Singh VP (1998) Entropy-based parameter estimation in hydrology. Kluwer, Boston 365 pp
Singh VP (2010a) Entropy theory for derivation of infiltration equations. Water Resour Res 46: W03527
Singh VP (2010b) Entropy theory for movement of moisture in soils. Water Resour Res 46: W03516
Singh VP (2010c) Tsallis entropy theory for derivation of infiltration equations. Trans ASABE 53 (2):447–463
Singh VP (2011) Hydrologic synthesis using entropy theory: review. J Hydrol Eng 16(5):421–433
Singh VP (2013) Entropy theory and its application in environmental and water engineering. Wiley, Oxford, UK
Singh VP, Fiorentino M (1992) A historical perspective of entropy applications in water resources. In: Singh VP, Fiorentino M (eds) Entropy and energy dissipation in water resources. Kluwer, Dordrecht, Netherlands, pp 21–61
Singh VP, Guo H (1995a) Parameter estimation for 2-parameter Pareto distribution by POME. Water Resour Manage 9:81–93
Singh VP, Guo H (1995b) Parameter estimation for 3-parameter generalized Pareto distribution by the principle of maximum entropy (POME). Hydrol Sci J 40(2):165–181
Singh VP, Zhang L (2008a) At-a-station hydraulic geometry: I. Theoretical development. Hydrol Process 22:189–215
Singh VP, Zhang L (2008b) At-a-station hydraulic geometry: II. Calibration and testing. Hydrol Process 22:216–228
Singh VP, Yang CT, Deng ZQ (2003a) Downstream hydraulic geometry relations: 1. Theoretical development. Water Resour Res 39(12):1–15
Singh VP, Yang CT, Deng ZQ (2003b) Downstream hydraulic geometry relations: 2. Calibration and testing. Water Resour Res 39(12):1–10
Sivakumar B (2000) Chaos theory in hydrology: important issues and interpretations. J Hydrol 227 (1–4):1–20
Sivakumar B (2002) A phase-space reconstruction approach to prediction of suspended sediment concentration in rivers. J Hydrol 258:149–162
Sivakumar B (2004a) Chaos theory in geophysics: past, present and future. Chaos Soliton Fract 19 (2):441–462
Sivakumar B (2004b) Dominant processes concept in hydrology: moving forward. Hydrol Process 18(12):2349–2353
Sivakumar B (2005a) Correlation dimension estimation of hydrologic series and data size requirement: myth and reality. Hydrol Sci J 50(4):591–604
Sivakumar B (2005b) Hydrologic modeling and forecasting: role of thresholds. Environ Model Softw 20(5):515–519

Sivakumar B (2008a) Dominant processes concept, model simplification and classification framework in catchment hydrology. Stoch Env Res Risk Assess 22(6):737–748

Sivakumar B (2008b) The more things change, the more they stay the same: the state of hydrologic modeling. Hydrol Process 22:4333–4337

Sivakumar B (2009) Nonlinear dynamics and chaos in hydrologic systems:latest developments and a look forward. Stoch Environ Res Risk Assess 23:1027–1036

Sivakumar B, Berndtsson R (2010a) Advances in data-based approaches for hydrologic modeling and forecasting. World Scientific Publishing Company, Singapore

Sivakumar B, Berndtsson R (2010b) Nonlinear dynamics and chaos in hydrology. In: Sivakumar B, Berndtsson R (eds) Advances in data-based approaches for hydrologic modeling and forecasting. World Scientific Publishing Company, Singapore, pp 411–461

Sivakumar B, Jayawardena AW (2002) An investigation of the presence of low-dimensional chaotic behavior in the sediment transport phenomenon. Hydrol Sci J 47(3):405–416

Sivakumar B, Singh VP (2012) Hydrologic system complexity and nonlinear dynamic concepts for a catchment classification framework. Hydrol Earth Syst Sci 16:4119–4131

Sivakumar B, Phoon KK, Liong SY, Liaw CY (1999) A systematic approach to noise reduction in chaotic hydrological time series. J Hydrol 219(3–4):103–135

Sivakumar B, Berndttson R, Olsson J, Jinno K (2001a) Evidence of chaos in the rainfall-runoff process. Hydrol Sci J 46(1):131–145

Sivakumar B, Sorooshian S, Gupta HV, Gao X (2001b) A chaotic approach to rainfall disaggregation. Water Resour Res 37(1):61–72

Sivakumar B, Berndtsson R, Olsson J, Jinno K (2002a) Reply to 'which chaos in the rainfall-runoff process?' by Schertzer et al. Hydrol Sci J 47(1):149–158

Sivakumar B, Jayawardena AW, Fernando TMGH (2002b) River flow forecasting: Use of phase-space reconstruction and artificial neural networks approaches. J Hydrol 265(1–4):225–245

Sivakumar B, Persson M, Berndtsson R, Uvo CB (2002c) Is correlation dimension a reliable indicator of low-dimensional chaos in short hydrological time series? Water Resour Res 38(2). doi:10.1029/2001WR000333

Sivakumar B, Harter T, Zhang H (2005) Solute transport in a heterogeneous aquifer: a search for nonlinear deterministic dynamics. Nonlinear Processes Geophys 12:211–218

Sivakumar B, Jayawardena AW, Li WK (2007) Hydrologic complexity and classification: a simple data reconstruction approach. Hydrol Process 21(20):2713–2728

Sivaprakasam C, Muttil N (2005) Discharge rating curve extension: a new approach. Water Resour Manage 19(5):505–520

Smith LC, Turcotte D, Isacks BL (1998) Stream flow characterization and feature detection using a discrete wavelet transform. Hydrol Process 12:233–249

Sonuga JO (1972) Principle of maximum entropy in hydrologic frequency analysis. J Hydrol 17 (3):177–219

Sonuga JO (1976) Entropy principle applied to the rainfall-runoff process. J Hydrol 30:81–94

Srikanthan R, McMahon TA (1983) Stochastic simulation of daily rainfall for Australian stations. Trans ASAE:754–766

Steinwart I, Christmann A (2008) Support vector machines. Springer, Germany, 602 pp

Sudheer KP, Jain A (2004) Explaining the internal behaviour of artificial neural network river flow models. Hydrol Process 18(4):833–844

Sujono J, Shikasho S, Hiramatsu K (2004) A comparison of techniques for hydrograph recession analysis. Hydrol Process 18:403–413

Şen Z (2009) Fuzzy logic and hydrologic modeling. CRC Press, Boca Raton, FL

Şen Z, Oztopal A (2001) Genetic algorithms for the classification and prediction of precipitation occurrence. Hydrol Sci J 46(2):255–267

Takagi T, Sugeno M (1985) Fuzzy identification of systems and its application to modeling and control. IEEE Trans Syst Man Cyber 15(1):116–132

Tayfur G, Moramarco T (2008) Predicting hourly-based flow discharge hydrographs from level data using genetic algorithms. J Hydrol 352(1–2):77–93

Tayfur G, Ozdemir S, Singh VP (2003) Fuzzy logic algorithm for runoff-induced sediment transport from bare soil surfaces. Adv Water Resour 26:1249–1256

Thomas HA, Fiering MB (1962) Mathematical synthesis of streamflow sequences for the analysis of river basins by simulation. In: Mass A et al (eds) Design of water resource systems. Harvard University Press, Cambridge, Massachusetts, pp 459–493

Tong H (1983) Threshold models in non-linear time series analysis. Springer-Verlag

Torrence C, Compo GP (1998) A practical guide to wavelet analysis. Bull Am Meteorol Soc 79 (1):62–78

Tripathi S, Srinivas VV, Nanjundian RS (2006) Downscaling of precipitation for climate change scenarios: a support vector machine approach. J Hydrol 330(3–4):621–640

Tsallis C (1988) Possible generalization of Boltzmann-Gibbs statistics. J Stat Phys 52(1–2):479–487

Tsukamoto Y (1979) An approach to fuzzy reasoning method. In: Gupta MM, Ragade RK, Yager RR (eds) Advances in fuzzy set theory and application. North-Holland, Amsterdam, pp 137–149

Tuma NB, Hannan MT (1984) Social dynamics: models and methods. Academic Press, Orlando, FL, p 602

Vairavamoorthy K, Ali M (2000) Optimal design of water distribution systems using genetic algorithms. Comp Aided Civil Infrastructure Eng 15:374–382

Valencia DR, Schaake JL (1973) Disaggregation processes in stochastic hydrology. Water Resour Res 9(3):211–219

Vapnik V (1998) The nature of statistical learning theory. Springer Verlag, New York, USA

Vasques JA, Maier HR, Lence BJ, Tolson BA, Foschi RO (2000) Achieving water quality system reliability using genetic algorithms. J Environ Eng ASCE 126(10):954–962

Venugopal V, Foufoula-Georgiou E (1996) Energy decomposition of rainfall in the time-frequency-scale domain using wavelet packets. J Hydrol 187:3–27

Wang QJ (1991) The genetic algorithm and its application to calibrating conceptual rainfall-runoff models. Water Resour Res 27(9):2467–2471

Wang Q, Gan TY (1998) Biases of correlation dimension estimates of streamflow data in the Canadian prairies. Water Resour Res 34(9):2329–2339

Wardlaw R, Sharif M (1999) Evaluation of genetic algorithms for optimal reservoir system operation. J Wat Resour Plan Manage ASCE 125(1):25–33

Whigham PA, Rechnagel F (2001) Predicting chlorophyll-a in freshwater lakes by hybridising process-based models and genetic algorithms. Ecol Model 146:243–251

Wilby RL, Abrahart RJ, Dawson CW (2003) Detection of conceptual model rainfall-runoff processes inside an artificial neural network. Hydrol Sci J 48(2):163–181

Wilcox BP, Seyfried MS, Matison TM (1991) Searching for chaotic dynamics in snowmelt runoff. Water Resour Res 27(6):1005–1010

Wolf A, Swift JB, Swinney HL, Vastano A (1985) Determining Lyapunov exponents from a time series. Physica D 16:285–317

Wu CL, Chau KW, Li YS (2008) River stage prediction based on a distributed support vector regression. J Hydrol 358(1–2):96–111

Wu CL, Chau KW, Li YS (2009) Predicting monthly streamflow using data-driven models coupled with data-preprocessing techniques. Water Resour Res 45:W08432. doi:10.1029/2007WR006737

Xiong LH, Shamseldin AY, O'Connor KM (2001) A nonlinear combination of the forecasts of rainfall-runoff models by the first order Takagi-Sugeno fuzzy system. J Hydrol 245(1–4):196–217

Yang CT (1971) Potential energy and stream morphology. Water Resour Res 2(2):311–322

Yevjevich VM (1972) Stochastic processes in hydrology. Water Resources Publications, Fort Collins, Colorado

Young PC (1974) Recursive approaches to time-series analysis. Bull Inst Math Appl 10:209–224

Young (1992) Parellel processes in hydrology and water: a unified time series approach. J Inst Water Environ Manage 6:598–612

Young PC (1993) Time variable and state dependent modelling of nonstationary and nonlinear time series. In: Subba Rao T (ed) Developments in time series analysis, Chapman and Hall, London, pp 374–413
Young PC (1998) Data-based mechanistic modeling of environmental, ecological, economic and engineering systems. Environ Model Softw 13:105–122
Young PC (2001) Data-based mechanistic modelling and validation of rainfall-flow processes. In: Anderson MG, Bates PD (eds) Model validation: perspectives in hydrological science. Wiley, Chichester, pp 117–161
Young PC (2002) Advances in real-time flood forecasting. Philos Trans R Soc, Phys Eng Sci 360 (9):1433–1450
Young PC (2003) Top-down and data-based mechanistic modelling of rainfall-flow dynamics at the catchment scale. Hydrol Process 17:2195–2217
Young PC (2006) The data-based mechanistic approach to the modelling, forecasting and control of environmental systems. Annu Rev Control 30:169–182
Young PC (2010a) Gauss, Kalman and advances in recursive parameter estimation. J Forecast 30:104–146
Young PC (2010b) Real-time updating in flood forecasting and warning. In: Pender GJ, Faulkner H (eds) Flood Risk Science and Management. Wiley-Blackwell, Oxford, UK, pp 163–195
Young PC (2011) Recursive estimation and time-series analysis: an introduction for the student and practitioner. Springer, Berlin
Young PC (2013) Hypothetico-inductive data-based mechanistic modeling of hydrological systems. Water Resour Res 49:915–935. doi:10.1992/wrcr.20068
Young PC, Beck MB (1974) The modelling and control of water quality in a river system. Automatica 10:455–468
Young PC, Beven KJ (1994) Data-based mechanistic modeling and rainfall-flow non-linearity. Environmetrics 5(3):335–363
Young PC, Garnier H (2006) Identification and estimation of continuous-time, data-based mechanistic models for environmental systems. Environ Model Softw 21:1055–1072
Young PC, Lees MJ (1993) The active mixing volume: a new concept in modelling environmental systems. In: Barnett V, Turkman K (eds) Statistics for the Environment. Wiley, Chichester, pp 3–43
Young PC, Parkinson S (2002) Simplicity out of complexity. In: Beck MB (ed) Environmental foresight and models: a manifesto. Elservier, Oxford, pp 251–294
Young PC, Ratto M (2009) A unified approach to environmental systems modeling. Stoch Environ Res Risk Assess 23:1037–1057
Young PC, Parkinson SD, Lees MJ (1996) Simplicity out of complexity: Occam's razor revisited. J Appl Stat 23:165–210
Young PC, Chotai A, Beven KJ (2004) Data-based mechanistic modelling and the simplification of environmental systems. In: Wainwright J, Mullgan M (eds) Environmental modelling: finding simplicity in complexity. Wiley, Chichester, pp 371–388
Young PC, Castelletti A, Pianosi F (2007) The data-based mechanistic approach in hydrological modelling. In: Castelletti A, Sessa RS (eds) Topics on system analysis and integrated water resource management. Elsevier, Amsterdam, pp 27–48
Yu X, Liong SY (2007) Forecasting of hydrologic time series with ridge regression in feature space. J Hydrol 332(3–4):290–302
Yu XY, Liong SY, Babovic V (2004) EC-SVM approach for real-time hydrologic forecasting. J Hydroinf 6(3):209–233
Zadeh LA (1965) Fuzzy sets. Inform. Control 8:338–353
Zadeh LA (1968) Fuzzy algorithms. Inf. Control 12:94–102
Zadeh LA, Klir GJ, Yuan B (1996) Fuzzy sets, fuzzy logic and fuzzy systems. World Scientific Publishers, Singapore
Zhang X, Srinivasan R, Bosch D (2009) Calibration and uncertainty analysis of the SWAT model using genetic algorithms and Bayesian model averaging. J Hydrol 374(3–4):307–317

Zhou Y, Ma Z, Wang L (2002) Chaotic dynamics of the flood series in the Huaihe River Basin for the last 500 years. J Hydrol 258:100–110

Zou R, Lung W-S, Wu J (2007) An adaptive neural network embedded GA approach for inverse water quality modeling. Water Resour Res 43:W08427. doi:10.1029/2006WR005158

Zurek WH (1989) Algorithmic randomness and physical entropy. Phys Rev A 40(8):4731–4751

Part II
Nonlinear Dynamics and Chaos

Chapter 5
Fundamentals of Chaos Theory

Abstract Almost all natural, physical, and socio-economic systems are inherently nonlinear. Nonlinear systems display a very broad range of characteristics. The property of "chaos" refers to the combined existence of nonlinear interdependence, determinism and order, and sensitive dependence in systems. Chaotic systems typically have a 'random-looking' structure. However, their determinism allows accurate predictions in the short term, although long-term predictions are not possible. Since 'random-looking' structures are a common encounter in numerous systems, the concepts of chaos theory have gained considerable attention in various scientific fields. This chapter discusses the fundamentals of chaos theory. First, a brief account of the definition and history of the development of chaos theory is presented. Next, several basic properties and concepts of chaotic systems are described, including attractors, bifurcations, interaction and interdependence, state phase and phase space, and fractals. Finally, four examples of chaotic dynamic systems are presented to illustrate how simple nonlinear deterministic equations can generate highly complex and random-looking structures.

5.1 Introduction

Nonlinear dynamics is the study of the evolution of nonlinear systems. In nonlinear systems, the relationship between cause and effect is not proportional and determinate but rather vague and difficult to discern. Nonlinear systems may be characterized by periods of both linear and nonlinear interactions between variables. This means that the dynamic behavior may reveal linear continuity at certain time periods, while the relationships between variables may change, resulting in dramatic structural and behavioral change, during other periods. The dramatic change from one qualitative behavior to another is referred to as a "bifurcation." Consequently, nonlinear systems are capable of generating very complex behavior over time. Studies on nonlinear systems evidence three types of temporal behavior: (1) stable (a mathematical equilibrium or fixed point); (2) oscillation between mathematical points in a stable, smooth, and periodic manner; or (3) seemingly

B. Sivakumar, *Chaos in Hydrology*, DOI 10.1007/978-90-481-2552-4_5

random, devoid of pattern (or non-periodic behavior) where uncertainty dominates and predictability breaks down. These behaviors may occur intermittently throughout the "life" of a nonlinear system. One regime may dominate for some periods, while other regimes dominate at other times. It is the potential for a variety of behaviors that represents the dynamics of nonlinear systems.

These discoveries gave rise to a new science of chaos in the 1960s. During the last half a century or so, the new science has found applications in numerous fields, including meteorology, biology, ecology, economics, engineering, environment, finance, politics, and social sciences. The chaos paradigm has profound implications for the previously (and still largely) dominant Newtonian view of a mechanistic and predictable universe. While a Newtonian universe was founded on the basis of linearity, stability, and order and, thus, certainty and predictability, chaos theory reveals that nonlinearity, instability, and disorder and, hence, uncertainty and unpredictability are not only widespread in nature but also essential to the evolution of complexity in the universe. Thus, chaos theory, as relativity theory and quantum theory before it, presents another strike against a singular commitment to the determinism of a Newtonian view of the natural realm.

With the focus of chaos theory on nonlinearity, instability, and uncertainty, the application of this theory to hydrology was a predictable eventuality, since hydrologic systems (and the Earth system at large) are inherently clearly nonlinear, where cause and effect are often a puzzling maze. The obvious value in applying chaos theory to hydrologic systems has served as an impetus for the emergence of the application of this theory to hydrologic phenomena, as discussed in detail in Part C. Time series analysis is essential to these efforts, as researchers strive to examine how nonlinear and chaotic behavior occurs and changes over time. As this book is about chaos in hydrology, hydrologic examples are given priority here to explain the relevance of the basic concepts of nonlinear dynamics and chaos.

5.2 Definition of Chaos

In common parlance, the word 'chaos,' derived from the Ancient Greek word Χάος, typically means a state lacking order or predictability; in other words, chaos is synonymous to 'randomness.' In modern dynamic systems science literature, however, the term 'chaos' is used to refer to situations where complex and 'random-looking' behaviors arise from simple deterministic systems with sensitive dependence on initial conditions; therefore, chaos and randomness are quite different. This latter definition has important implications for system modeling and prediction: randomness is irreproducible and unpredictable, while chaos is reproducible and predictable in the short term (due to determinism) but irreproducible and unpredictable only in the long term (due to sensitivity to initial conditions).

The three fundamental properties inherent in the definition of chaos, namely (a) nonlinear interdependence; (b) hidden determinism and order; and (c) sensitivity to initial conditions, are highly relevant in almost all real systems and the associated processes. In hydrology, for instance: (a) nonlinear interactions are dominant (albeit by varying degrees) among the components and mechanisms in the hydrologic cycle; (b) determinism and order are prevalent in daily temperature and annual river flow; and (c) contaminant transport in surface and sub-surface waters is highly sensitive to the time (e.g. rainy or dry season) at which the contaminants were released. The first property represents the 'general' nature of system processes, whereas the second and third represent their 'deterministic' and 'stochastic' natures, respectively. Furthermore, despite their complexity and random-looking behavior, hydrologic processes may be governed only by a few degrees of freedom (e.g. runoff in a well-developed urban catchment depends essentially on rainfall), another basic idea of chaos theory (e.g. Sivakumar 2004). All these properties make chaos theory a viable candidate for a balanced middle-ground approach between our dominant extreme-view deterministic and stochastic approaches. Further details on this will be discussed in Parts C and D.

5.3 Brief History of the Development of Chaos Theory

Although dynamics is an interdisciplinary subject today, it was originally a branch of physics. The subject began in the 1600s, when Newton invented (Newton 1687) differential equations, discovered his laws of motion and universal gravitation, and combined them to explain Kepler's laws of planetary motion. Specifically, Newton solved the two-body problem, originally studied by Kepler (1609)—the problem of calculating the motion of the Earth around the sun, given the inverse-square law of gravitational attraction between them. Subsequent generations of mathematicians and physicists tried to extend Newton's analytical methods to the three-body problem (e.g. sun, Earth, and moon), but curiously this problem turned out to be much more difficult to solve (e.g. Euler 1767; Lagrange 1772). After decades of effort, it was eventually realized that the three-body problem was essentially *impossible* to solve, in the sense of obtaining explicit formulas for the motions of the three bodies. At this point, the situation seemed hopeless.

The breakthrough came with the bifurcation theory studies of Henri Poincaré in the late 1800s, which were also the roots of chaos theory. Poincaré introduced (Poincaré 1890, 1896) a new point of view that emphasized qualitative rather than quantitative questions. For example, instead of asking for the exact positions of the planets at all times, he asked "is the solar system stable forever, or will some planets eventually fly off to infinity?" He also developed a powerful *geometric* approach to analyze such questions, and found that there can be orbits which are non-periodic (and yet not forever increasing nor approaching a fixed point) with sensitively dependent on initial conditions, thereby rendering long-term prediction impossible. This geometric approach has flowered into the modern subject of dynamics, with applications reaching far beyond celestial mechanics.

Despite the interesting findings by Poincaré, chaos theory remained in the background during the entire first half of the twentieth century, most likely due to lack of computational power; instead, dynamics was largely concerned with nonlinear oscillators and their applications in physics and engineering. Nonlinear oscillators played a vital role in the development of such technologies as radio, radar, phase-locked loops, and lasers. On the theoretical side, nonlinear oscillators also stimulated the invention of new mathematical techniques—pioneering works in this area include van der Pol (1927), Andronov (1929), Cartwright (1935), Levinson (1943), Smale (1960), and Littlewood (1966). Meanwhile, in a separate development, Poincaré's geometric methods were being extended to yield a much deeper understanding of classical mechanics, thanks to the work of Birkhoff (1927) and later Kolmogorov (1954), Arnol'd (1964), and Moser (1967).

The invention of high-speed computers in the 1950s changed the situation for chaos theory for the better, as computers allowed experimentation with equations in a way that was impossible before, especially the process of repeated iteration of mathematical formulas to study nonlinear dynamic systems. Such experiments led to Edward Lorenz's discovery, in 1963, of chaotic motion on a 'strange attractor' (Lorenz 1963). Lorenz studied a simplified model of convection rolls in the atmosphere to gain insight into the notorious unpredictability of the weather. He found that the solutions to his equations never settled down to equilibrium or to a periodic state; instead, they continued to oscillate in an irregular, aperiodic fashion. Moreover, when the simulations were started from two slightly different initial conditions, the resulting behaviors became totally different. The implication was that the system was inherently unpredictable—tiny errors in measuring the current state of the atmosphere (or any other chaotic system) would be amplified rapidly. But Lorenz also showed that there was structure (in the chaos)—when plotted in three dimensions, the solutions to his equations fell onto a butterfly-shaped set of points (see Sect. 5.11 for details).

The main developments in chaos theory were witnessed in the 1970s. Ruelle and Takens (1971) proposed a new theory for the onset of turbulence in fluids, based on abstract consideration about 'strange attractors.' A few years later, May (1976) found examples of chaos in iterated mappings arising in population biology, popularly known as the logistic equation (see Sect. 5.11 for details), and emphasized on the pedagogical importance of studying simple nonlinear systems, to counterbalance the often misleading linear intuition fostered by traditional education. The hidden beauty of chaos was also revealed through study of other simple nonlinear mathematical models, such as the Henon map (Henon 1976) and the Rössler system (Rössler 1976) (see Sect. 5.11 for details). Beautiful 'strange attractors' that described the final states of these systems were produced and studied, and routes that lead a dynamic system to chaos were discovered. Feigenbaum (1978) discovered that there are certain universal laws governing the transition from regular to chaotic behavior; for instance, completely different systems can go chaotic in the same way. Feigenbaum's work offered a link between chaos and phase transitions.

During the late 1970s–1980s, the study of chaos moved to the laboratory. Ingenious experiments were set up and chaotic behavior was studied in fluids, mechanical oscillators, semiconductors, and many others (e.g. Swinney and Gollub 1978; Linsay 1981; Teitsworth and Westervelt 1984; Mishina et al. 1985; Meissner and Schmidt 1986; Tufillario and Albano 1986; Briggs 1987; Su et al. 1987). Such experiments significantly enhanced the practical nature of chaos studies and elevated chaos theory from being just a mathematical curiosity and established it as a physical reality.

The purpose and nature of these laboratory experiments and the positive and interesting outcomes about the dynamics of the systems studied encouraged search for chaos also in systems outside the 'controlled' space—in Nature. However, the investigations also presented an enormous challenge, since the mathematical formulation for such 'uncontrolled' systems was not always known accurately. Despite this difficulty, advances in computational power and measurement technology facilitated development, in the 1980s and early 1990s, of a new set of mathematical techniques for chaos identification and prediction. Understandably, most of these techniques were based on or designed for time series. In these techniques, some earlier concepts were revisited and some new ones were developed. Among the concepts are data reconstruction, nonlinearity, dimensionality, entropy, predictability (e.g. Packard et al. 1980; Takens 1981; Grassberger and Procaccia 1983a, b, c; Wolf et al. 1985; Farmer and Sidorowich 1987; Casdagli 1989, 1992; Kennel et al. 1992; Theiler et al. 1992).

Since their developments, these techniques have been employed for identification and prediction of chaos in many real systems, including those encountered in the fields of atmosphere, biology, ecology, economics, engineering, environment, finance, politics, and society. Chaos theory has indeed become a widely applied scientific concept, including its use in such larger constructs as complexity theory, complex systems theory, synergetics, and nonlinear dynamics (e.g. Haken 1983; Nicolis and Prigogine 1989; Abarbanel 1996). The number of studies are already in hundreds of thousands, if not millions, and continues to grow every day. Examples of some early notable books on chaos theory and its applications are those by Schuster (1988), Ruelle (1989), Tong (1990), Tsonis (1992), Ott (1993), Hilborn (1994), Strogatz (1994), Kaplan and Glass (1995), Abarbanel (1996), Kiel and Elliott (1996), Williams (1997), and Kantz and Schreiber (2004). For a more general and non-mathematical description of chaos theory, the reader is referred to Gleick (1987) and, to some extent, Goerner (1994), among others.

5.4 Dynamical Systems and Stability Analysis

Any system whose evolution from some initial state is dictated by a set of rules is called a *dynamical system*. When these rules are a set of differential equations, the system is called a *flow*, because their solution is continuous in time. When the rules are a set of discrete difference equations, the system is referred to as a *map* (or *iterated map*). The evolution of a dynamical system is best described in its state space or phase space (see Sect. 5.9 for details), a coordinate system whose coordinates are all the variables that enter the mathematical formulation of the system (i.e. the variables necessary to completely describe the state of the system at any moment). To each possible state of the system, there corresponds a point in the state space or phase space.

5.5 Attractors

The term *attractor* is difficult to define in a rigorous way, and several different definitions exist. For instance, Baumol and Behabib (1989) define an attractor as "a set of points toward which complicated time paths starting in its neighborhood are attracted," while Pool (1989) defines an attractor as "the set of points in a phase space corresponding to all the different states of the system." Nevertheless, a workable definition may be this: an attractor is a geometric object that characterizes the long-term behavior of a system. In essence, an attractor functions as an abstract representation of the flow, or motion, of a system, by 'storing' information about a system's behavior over time. As mentioned above, the examination of an attractor is normally conducted by a mapping of the data onto a state space or phase space (see Sect. 5.9 for details).

Attractors can be used to obtain important qualitative and quantitative information about system evolution. A visual inspection of the attractor (e.g. shape, structure) often provides useful qualitative information on the nature of system dynamics; for instance, a perfectly-shaped and clearly-structured attractor is generally an indication of a deterministic system, whereas an imperfectly-shaped and scattered attractor is generally an indication of a stochastic system. On the other hand, estimation of certain suitable measures or *invariants* of the attractor (e.g. dimension, entropy) provides quantitative information on the extent of complexity of system dynamics; for instance, an attractor with a low dimension is generally an indication of a simple system, while an attractor with a high dimension is generally an indication of a complex system.

Studies of the attractors of numerous time series (synthetic and real) reveal that three common behavioral regimes emanating from nonlinear differential equations create uniquely-shaped attractors. A stable equilibrium generates a *point attractor*, in which the data are attracted to a single point on the mapping (Fig. 5.1a). A stable periodic oscillation generates a circular mapping, or *limit cycle*, as the data revolve back and forth between consistent mathematical points (Fig. 5.1b). The chaotic attractor is represented by a variety of unique shapes resulting in the labeling of such attractors as *strange attractors* (Fig. 5.1c). While point attractor and limit cycle are indeed observed in certain natural and physical systems, it is the strange attractor that is dominant in most systems. Investigations of various hydrologic time series also suggest that strange attractors are far more prevalent in hydrologic systems, when compared to point attractor and limit cycle.

A *strange attractor* is an attractor that exhibits sensitive dependence on initial conditions, i.e. small changes in the initial conditions may give rise to large effects in the final outcomes (and the inverse also applies) (see below and also Sect. 5.8). Strange attractors are called strange because they are often *fractal* sets (see Sect. 5.10 for details of the concepts of fractal and fractal dimension). However, fractality alone is not a sufficient condition for chaos. This is why distinction is made between a fractal attractor and a chaotic attractor; a fractal attractor is one that simply has a fractal dimension, while a chaotic attractor is one that has a fractal

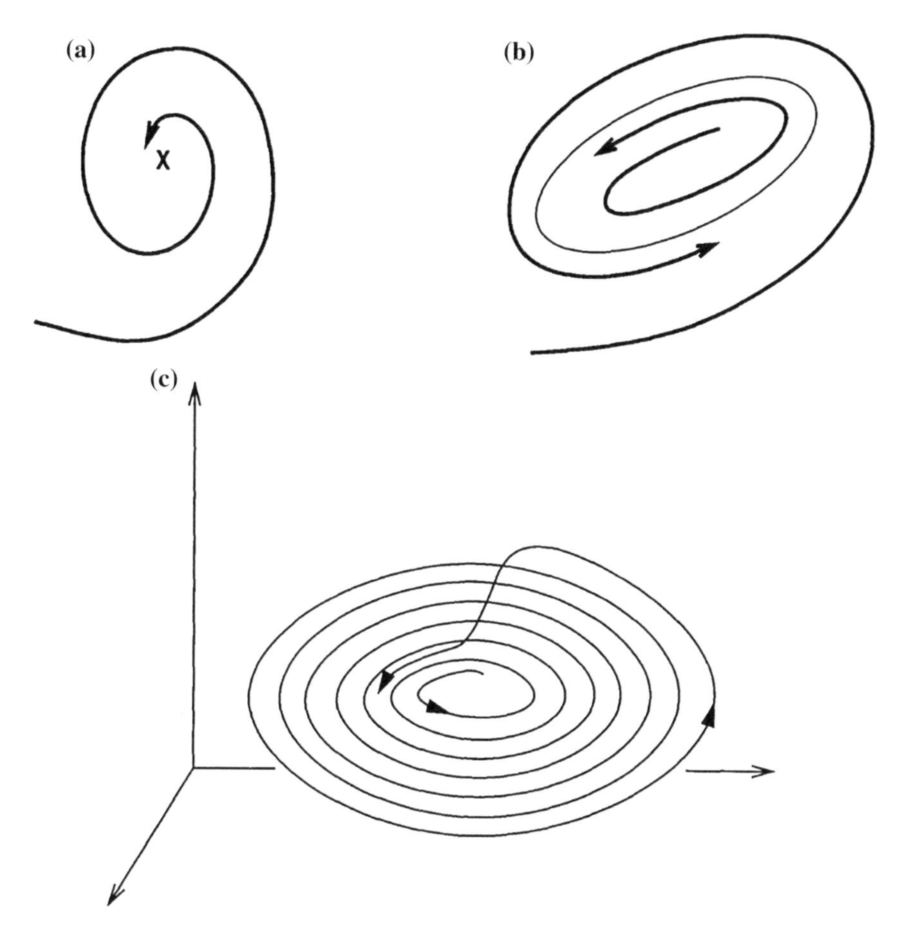

Fig. 5.1 Attractors: **a** fixed point; **b** limit cycle; and **c** strange (*source* Kantz and Schreiber (2004))

dimension and is also sensitively dependent on initial conditions. However, in studying real systems, the interest is often the combination of nonlinearity (including sensitivity to initial conditions) and dimensionality, especially in the context of identification of the appropriate type and complexity of models for reliable modeling and prediction (one of the main goals in chaos applications in hydrology).

Two mechanisms are responsible for the existence of a strange attractor: 'stretching' and 'compressing.' Stretching is responsible for a system's sensitive dependence on initial conditions. This means that two nearby points in state space or phase space, representing slightly different initial states of the system, will evolve along divergent trajectories and exhibit dramatically different states of the system after some finite time. This mechanism is responsible for the long-term unpredictability generally attributed to systems exhibiting chaos. Compressing is responsible for the recurrent behavior exhibited by all chaotic (as opposed to stochastic) systems. If the initial conditions were to be stretched apart indefinitely, the trajectories would not be confined to a bounded region of state space or phase space. To ensure that the trajectories do not run off to infinity, the flow must somehow be returned to a bounded region of state space or phase space. This mechanism is responsible for patterns which *almost* repeat themselves, and is at the heart of both the metric approaches (e.g. correlation dimension method) and the topological approaches (e.g. close returns plot) for the analysis of nonlinear time series (see Chap. 6 for various methods).

5.6 Bifurcations

A *bifurcation* is a transformation from one type of behavior to a *qualitatively* different type of behavior. A corollary to the concept of bifurcation is that a system may have more than one attractor, i.e. a single system may have more than one different form of behavior.

Bifurcation is an important property of nonlinear systems, since qualitative changes in structure and behavior of such systems are commonplace. Nonlinear systems are often characterized by periods of both linear and nonlinear interactions. While the system behavior may reveal linear continuity during some periods, the relationships between variables or parameters (called control parameters) may change during other periods, resulting in dramatic structural and behavioral changes. The points or parameter values at which bifurcations occur are called *bifurcation points*.

A *bifurcation diagram*, which is essentially a two-dimensional graph, provides an overview of how a system's behavior varies for different values of a control parameter, as shown in Fig. 5.2 for the population dynamic system represented by a first-order nonlinear difference equation, called the logistic equation (May 1974) (see Sect. 5.11 for further details about the logistic equation). In Fig. 5.2, the single line part of the graph represents attractors that repeat themselves after one cycle, the first parabola represents two-cycle attractors, the double parabola four-cycle attractors, and so on. The bifurcation diagram thus allows one to visualize where the system's behavior is essentially the same and where it undergoes qualitative transformations of behavior (i.e. bifurcation points).

The bifurcation diagram also produces a very different image of how behavioral change takes place. It shows a type of change called *punctuated equilibrium*, rather than the classical calculus-shape image of smooth, continuous, and traceable change. As the control parameter increases, the system goes through periods of stable sameness, punctuated by abrupt transitions to qualitatively different forms of behavior. Change can be sometimes smooth and sometimes discontinuous, and the effects of a particular perturbation cannot be tracked across a bifurcation.

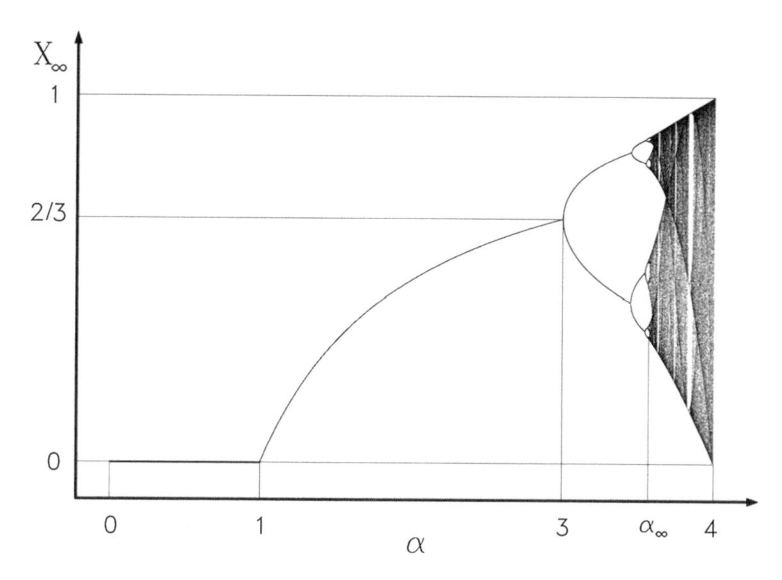

Fig. 5.2 Bifurcation diagram (*source* Carlos E. Puente, personal communication)

5.7 Interaction and Interdependence

Interaction, generally defined as "affecting and being affected," in a system means that it is not possible to separate the variables governing it. Interaction between the variables in a system essentially gives rise to their interdependence. In linear systems, the interaction between variables is often simple, linear, and one-directional and, thus, there is often dependence of one variable on another in one direction. In nonlinear systems, however, this process is complicated. Interaction between variables occurs in many different ways (often in feedback forms) and in varying degrees of nonlinearity, and so the variables are interdependent on each other, i.e. every variable is dependent on every other variable in a direct or indirect way. The hydrologic cycle is an excellent example of nonlinear interactions and interdependencies among variables, since every component in the hydrologic cycle is connected with every other component, either directly or indirectly (see Chaps. 1 and 2 for details of the hydrologic cycle and hydrologic system properties). These nonlinear interactions and interdependencies among variables are inherent characteristics of nonlinear dynamic systems, and chaotic systems in particular. They are why linear and reductionism-based approaches cannot work well for real systems, which are inherently nonlinear. They, thus, make the modeling and prediction of the evolution of chaotic systems difficult, especially their long-term evolution.

Interactive dynamics can create pull, in the form of an attractor, discussed above. Such dynamics can also be self-stabilizing. A self-stabilizing dynamic is said to have structural stability, which means it is resistant to change. However, the structure that interactive dynamics produce can also fall apart, and instability arises. In interactive dynamics, instability may lead to a new form. For example, bifurcation diagrams show where one dynamic flow pattern transforms itself into another, as discussed above. Thus, interactive dynamics have three different options: to exist, to not exist, or change forms.

It must be noted that, in interactive dynamics, a change of form does not mean that the underlying process or equation has changed. Rather, it only means that the same process has just reorganized into a different pattern. Multiple attractor systems provide a concrete model of how one process can create many forms.

Before the discovery of chaos, the concept of stability implied a single final state as in equilibrium or homeostatis or a repeated pattern (e.g. orbits of the solar system). However, it became known through chaos theory that the structural stability of a nonlinear (and chaotic) system can be a stability of a very different type. A system with structural stability may never repeat the same way twice (strange attractor) or may move back and forth between multiple distinct stable states. Such systems may even be in a locally stable pattern that is nevertheless part of an overall progression of states.

5.8 Sensitivity to Initial Conditions

Sensitivity to initial conditions is an inherent property of chaotic systems. This property refers to amplification (or propagation) of any small change in the initial conditions on the evolution of a system over a period of time. The implication of this property is that prediction of the behavior of a chaotic system in the long term is almost impossible, despite its fundamentally deterministic nature. This property is due to the fractal nature of the system attractor. The fractal nature of the (strange) attractor not only implies non-periodic orbits but also causes nearby trajectories to diverge. Trajectories that are initiated from (even slightly) different conditions will reach the attracting set after a certain time. However, two nearby trajectories do not stay close to each other; they diverge and follow totally different paths in the attractor. Therefore, the state of the system after some time can be anything, including randomness, despite the fact that the initial conditions were very close to each other.

In nonlinear dynamic systems, the effects of small disturbances are crucial. While steady state or periodic regime will damp such disturbances, chaotic regimes tend to generate positive feedback and amplify such disturbances. As a result, the system behavior may alter, change, and explode over time. This property of nonlinear dynamic systems generated the 'butterfly effect' metaphor: Can the flapping of a butterfly's wings in one place (e.g. New York) create a tornado in a far off place (e.g. Tokyo)?

The above findings have profound implications. If one knows exactly the initial conditions, then one can simply follow the trajectory that corresponds to the evolution of the system from those initial conditions and basically predict the evolution forever. The problem, however, is that we cannot have perfect knowledge of initial conditions. Our data (measurements) are only approximate and there always exist certain deviations from the actual initial conditions. The measured and actual conditions may be very close to each other, but they are not the same. Therefore, even if we completely know the physical laws that govern the systems, our predictions of the systems at a later time can be totally different from the actual values, essentially due to the nature of the underlying attractor (regardless of the sophistication of the methodology). Furthermore, systems with very *similar* starting conditions in their evolutions may diverge to very different systems and structure over time. This point has important implications for hydrologic and other real systems, since *virtually identical* systems can generate unique, and totally different, histories. In essence: (1) two seemingly totally different time series may have arisen from similar underlying system dynamics; and (2) two seemingly similar time series may have arisen from totally different underlying system dynamics.

5.9 State Space and Phase Space

The evolution of a nonlinear dynamic system (or any system, for that matter) can be represented through mapping of the governing variables (e.g. data) at different times in a two-, three-, or higher-dimensional space. A point in this space corresponds to a particular state of the system.

When the values of the actual variables governing the system are used to map the evolution of the system, then it is called *state space*. For instance, the use of rainfall (R) and temperature (T) for representation of the dynamics of streamflow (Q) is a state space. This, in essence, is a multi-variable representation in a multi-dimensional space. Figure 5.3a shows an example of a two-variable (rainfall and temperature) representation in a two-dimensional space for representing streamflow dynamics, assuming $Q = f(R, T)$ (ignoring other factors).

However, there are many situations where the values of the actual variables are not available. In such situations, 'proxy' variables are used to map the system evolution, and this is called *phase space*. This kind of representation is often made using values of an available single variable with a suitable delay time (τ), which can reliably represent the system dynamic changes. For instance, to study streamflow dynamics, streamflows observed at different times Q_t, where $t = 1, 2, \ldots, N$, can be treated as different variables by including a delay time (τ), such that the variables are $Q_{t+\tau}$, $Q_{t+2\tau}$, $Q_{t+3\tau}$, ..., $Q_{t+m\tau}$, where m is the total number of variables. This kind of representation is, in essence, a single-variable representation in a multi-dimensional space. Figure 5.3b shows an example of a single-variable (streamflow) representation in a two-dimensional space for representing streamflow dynamics (Q_t versus $Q_{t+\tau}$).

Phase space is a very useful concept for representing the evolution of real systems, since for such systems: (1) all of the actual governing variables are often not known a priori; and (2) even if the variables are known a priori, data corresponding to one or more variables are not always available. Therefore, studies investigating chaos in real systems mostly resort to system representation in phase

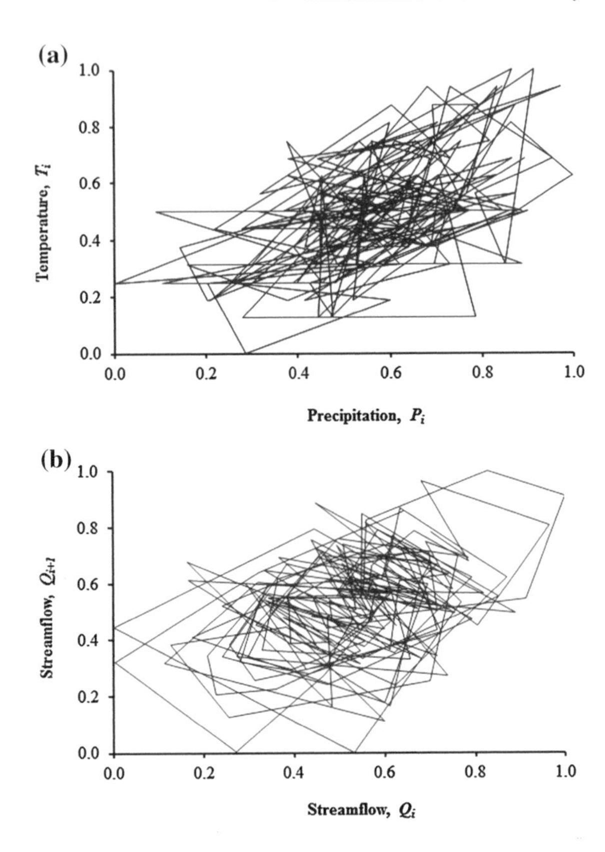

Fig. 5.3 Streamflow dynamics represented as: **a** state space (rainfall and temperature); and **b** phase space

space, and hydrology is no exception to this. Although phase space is a useful concept to obtain important qualitative information about the trajectory of the system evolution and nature and complexity of the attractor, caution must be exercised in constructing the phase space and interpreting the phase space diagram. One major reason for this is the absence of consensus regarding the selection of the delay time value τ. The selection of an appropriate τ is critical, since only an optimum τ gives the best separation of neighboring trajectories within the minimum embedding space. If τ is too small, then there is little new information contained in each subsequent datum and the reconstructed attractor is compressed along the identity line; this situation is termed as 'redundance' (Casdagli et al. 1991). On the other hand, if τ is too large, and the dynamics happen to be chaotic, then all relevant information for phase space reconstruction is lost, since neighboring trajectories diverge, and averaging in time and/or space is no longer useful (Sangoyomi et al. 1996); this situation is termed as 'irrelevance' (Casdagli et al. 1991). Further details regarding phase space reconstruction and the associated issues will be presented in Chaps. 6 and 7.

5.10 Fractal and Fractal Dimension

Fractals refer to a particular type of structure created by an iterative, self-referential process. Fractals are basically sets defined by the three related principles of self-similarity, scale-invariance, and power law relations. When these principles converge, fractal patterns form. The technical definition of fractals has to do with the *strange* fact that they have a fractional dimension (the reason behind the name *strange attractor*), as opposed to structures that have integer dimensions (e.g. point —dimension zero, line—dimension one, plane—dimension two, solid—dimension three). Indeed, fractals turn out to be an appropriate description of most naturally occurring forms, such as mountains, clouds, trees, rivers, and other structures; see Mandelbrot (1983) for a comprehensive account of fractals in Nature. Their ubiquity and connection to natural geometry validate the sense that nonlinear dynamics are more normative than linear dynamics. Fractal structure is an important characteristic of chaotic phenomena. However, neither is fractality alone a sufficient condition for chaos, nor can all systems that exhibit fractal structure be considered chaotic (see Abraham and Shaw 1984).

Since self-similarity, scale-invariance, and power law relations form important principles of fractals, they are basic diagnostics of fractals. However, it is the dimensionality of the sets that provides a quantitative measure of fractals. Generally speaking, the dimensionality of a fractal structure or attractor is called the *fractal dimension*. The fractal dimension is basically an invariant parameter that characterizes a fractal set, and is an index defining the complexity as a ratio of the change in details of patterns to the change in scale. There are many different measures of fractal dimension, and there are often many different ways to estimate such measures as well. In some cases, a particular measure is called by different names; in some others, minor differences between measures are ignored. Some of these measures and techniques are rather simple and easy to implement, while others are more complex and sophisticated. The measures for fractal dimension include: box-counting dimension, capacity dimension, correlation dimension, generalized dimension, Hausdorff dimension, information dimension, Kaplan–Yorke dimension, Lyapunov dimension, and Rényi dimension; for details, see Hausdorff (1918), Rényi (1959, 1971), Kaplan and Yorke (1979), and Grassberger and Procaccia (1983a, b), among others. The estimation of the correlation dimension, especially the Grassberger–Procaccia algorithm (Grassberger and Pracaccia 1983a, b), and its use in the identification of chaos will be discussed in detail in Chap. 6.

5.11 Examples of Chaotic Dynamic Equations

While nonlinear systems can take a wide array of behaviors over time, three behavioral regimes are commonly observed: (1) stable equilibrium; (2) periodic oscillation; and (3) chaos. These regimes have been identified based on analysis of time series representing numerous systems or equations. Popular among these systems are: (1) the logistic map; (2) the Henon map; (3) the Lorenz system; and (4) the Rössler system. These systems are briefly described here, for a better understanding of nonlinear systems and their behaviors.

5.11.1 Logistic Map

The logistic equation is a first-order nonlinear difference equation, and is widely used in the study of population dynamics (e.g. May 1974, 1976). It takes the form

$$x_{t+1} = kx_t(1 - x_t) \tag{5.1}$$

where x represents a variable (an animal population value), t is time, and k is a constant parameter or boundary value (representing the fertility rate). Depending upon the value of the parameter k and the initial condition of the variable x (i.e. x_0), different behavioral regimes can occur. It must be noted, however, that there are limits to the range of k ($0 < k < 4.0$) and x_0 ($0 < x_0 < 1.0$).

Figure 5.4a shows the three different behavioral regimes of the logistic map, depending upon the value of k.

1. When k is between 0 and 3, the logistic map converges to a stable equilibrium. For instance, the top row in Fig. 5.4a shows the time series obtained from the logistic map with $k = 2.827$ and $x_0 = 0.97$. It must be noted, however, that convergence to stability generally requires more iterations when k approaches 3.
2. Periodic behavior of the logistic map starts to occur when $k > 3$. Periodic behavior is cyclical or oscillatory behavior that repeats an identifiable pattern. This behavior initiates instability into the equation as the data start to oscillate. Such a change in the qualitative behavior is referred to as a bifurcation, as discussed earlier (see Fig. 5.2). Again, depending upon the value of k, different kinds of periodic behavior may occur: two-period cycle occurs when $3 < k < 3.5$, four-period cycle when k is approximately 3.5, and eight-period cycle when k is between 3.56 and 3.57. The middle row in Fig. 5.4a, for instance, shows the four-period cycle obtained with $k = 3.5$ and $x_0 = 0.97$. This process of cycles doubling in the number of alternating and continuous patterns of values is called *period doubling*. It is this continuous bifurcation of period doubling that eventuates in the "road to chaos" (Feigenbaum 1978). This process of period doubling continues, as k increases, until the onset of chaos.
3. Chaotic behavior occurs when k is between 3.8 and 4.0. The bottom row in Fig. 5.4a, for instance, shows the time series from the logistic equation when $k = 3.98$ and $x_0 = 0.90$. This regime represents another clear bifurcation or qualitative change in the system's behavior. What distinguishes chaos from the other regimes of behavior is the *apparent* lack of clear pattern, as can be seen

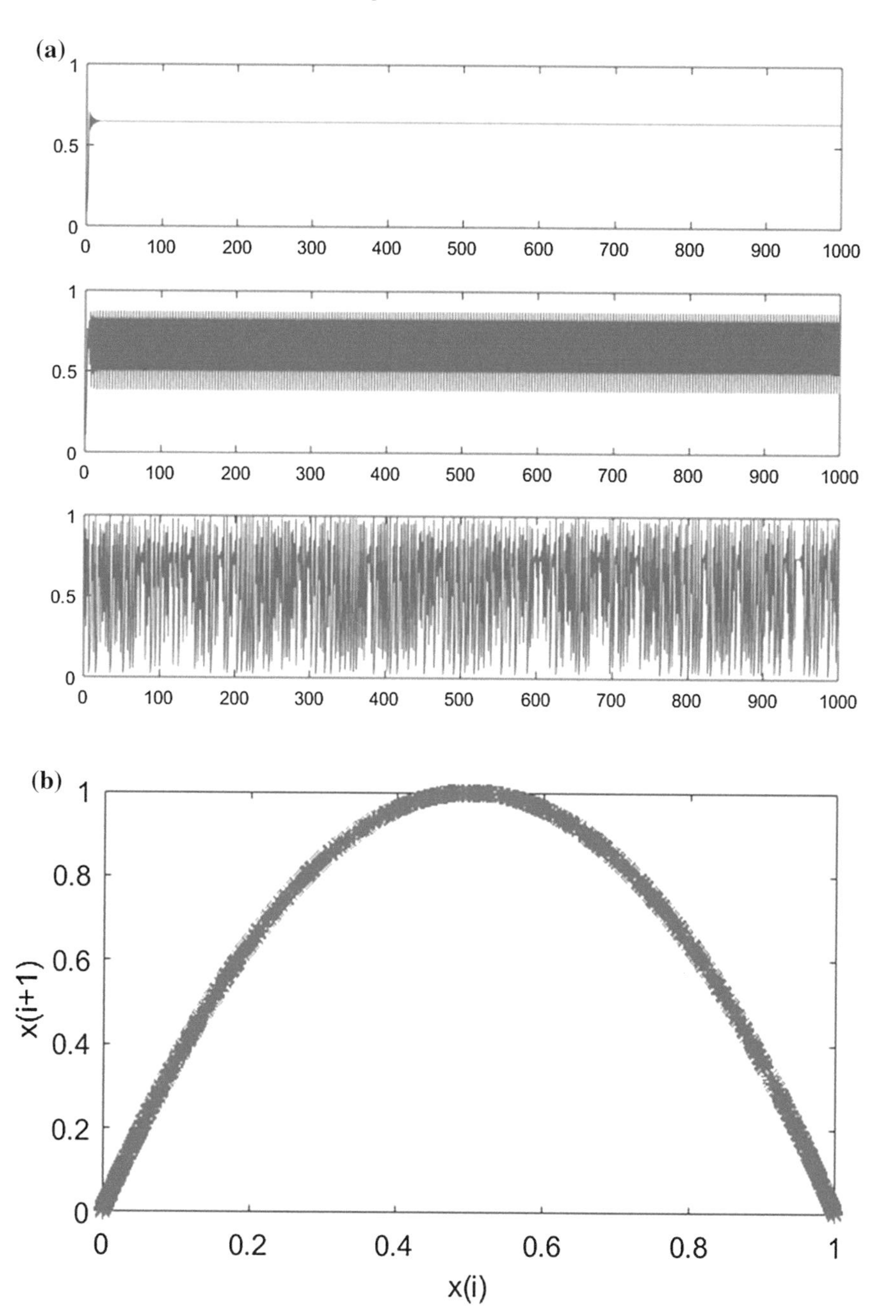

Fig. 5.4 Logistic map: **a** time series generated using $x_0 = 0.97$ and $k = 2.827$ (*top*), $x_0 = 0.97$ and $k = 3.50$ (*middle*), and $x_0 = 0.90$ and $k = 3.98$ (*bottom*); and **b** phase space diagram (*source* Hong-Bo Xie, personal communication)

from the bottom row of Fig. 5.4a (it is *aperiodic*), but nevertheless chaotic behavior remains within definable parameters. While this behavior appears random, it actually is not. It can be generated by a simple deterministic equation, as is clear from the logistic equation analysis.

Figure 5.4b shows the trajectory of the Logistic time series in two dimensions, data embedded in a two-dimensional phase space, with a delay time $\tau = 1$. As can be seen, despite its seemingly irregular structure, the time series still exhibits a clear attractor in a well-defined region in the phase space. The correlation dimension of this attractor is about 0.5.

5.11.2 Henon Map

One of the most celebrated simple dynamical systems that exhibit a strange attractor is the Henon map (Henon 1976). It is a simple deterministic two-dimensional map, given by:

$$X_{i+1} = a - X_i^2 + bY_i \tag{5.2a}$$

$$Y_{i+1} = X_i \tag{5.2b}$$

Depending upon the values of a and b, as well as the initial conditions for X (i.e. X_0) and Y (i.e. Y_0), the map yields a variety of behaviors, ranging from convergence to a periodic orbit to intermittence to chaotic dynamics. For $|b| \leq 1$, there exist initial conditions for which trajectories stay in a bounded region. When $a = 0.15$ and $b = 0.3$, a typical sequence of X_i has a stable periodic orbit as an attractor. However, when $a = 1.4$ and $b = 0.3$, a typical sequence of X_i (Fig. 5.5a) will not be periodic but chaotic. The initial values of X (i.e. X_0) and Y (i.e. Y_0) used for this data series are 0.13 and 0.50, respectively.

Figure 5.5b shows the trajectory of this time series in two dimensions. Similar to the observation for the Logistic series, the Henon time series, despite its seemingly irregular structure, exhibits a clear attractor in a well-defined region in the phase space. The dimensionality of this attractor is about 1.22. The identification of the dynamic behavior of the Henon map is discussed in more detail in Chap. 6.

5.11.3 Lorenz System

Arguably, the most popular system of equations in the context of chaos theory is the Lorenz equations. The Lorenz equations represent an early deterministic model of the weather formulated by Lorenz (1963), based on the convection equations of Saltzman (1962). The same equations also arise in models of lasers and dynamos, and they *exactly* describe the motion of a certain waterwheel.

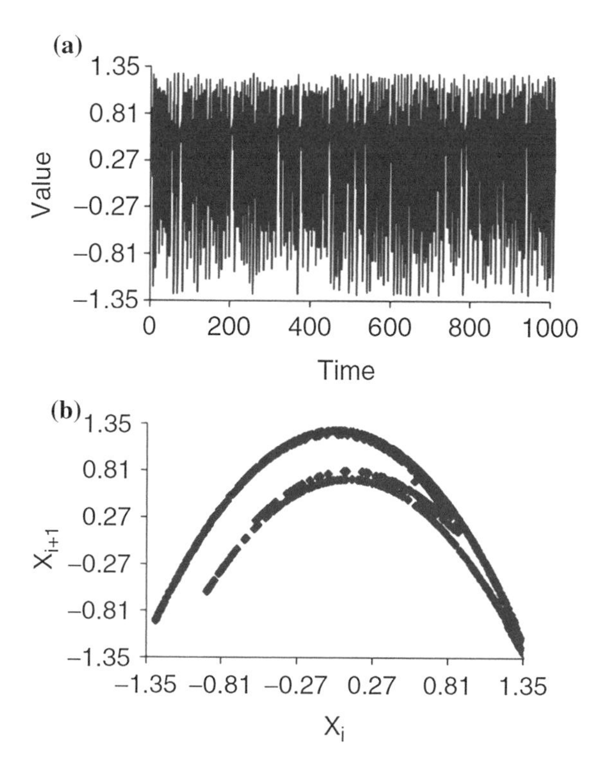

Fig. 5.5 Henon map: **a** time series; and **b** phase space diagram

The Lorenz equations present an approximate description of a fluid layer heated from below. The fluid at the bottom gets warmer and rises, creating convection. For a choice of the constants that correspond to sufficient heating, the convection may take place in an irregular and turbulent manner. The Lorenz equations are given by:

$$\dot{x} = \sigma(y - x) \tag{5.3a}$$

$$\dot{y} = rx - y - xz \tag{5.3b}$$

$$\dot{z} = xy - bz \tag{5.3c}$$

where x is proportional to the intensity of the convection motion, y is proportional to the horizontal temperature variation, z is proportional to the vertical temperature variation, and σ (Prandtl number), r (Rayleigh number), and b (related to the height of the fluid layer) are constants. The equations have only two nonlinearities, the quadratic terms xy and xz. Figure 5.6a shows the time series of the first component above, obtained using the parameter values of σ = 16.0, r = 45.92, and b = 4.0.

The Lorenz equations, despite their deterministic nature, could give rise to different behavioral regimes, ranging from simple determinism (stable fixed points, stable limit cycles) to extremely erratic dynamics. Over a wide range of parameters,

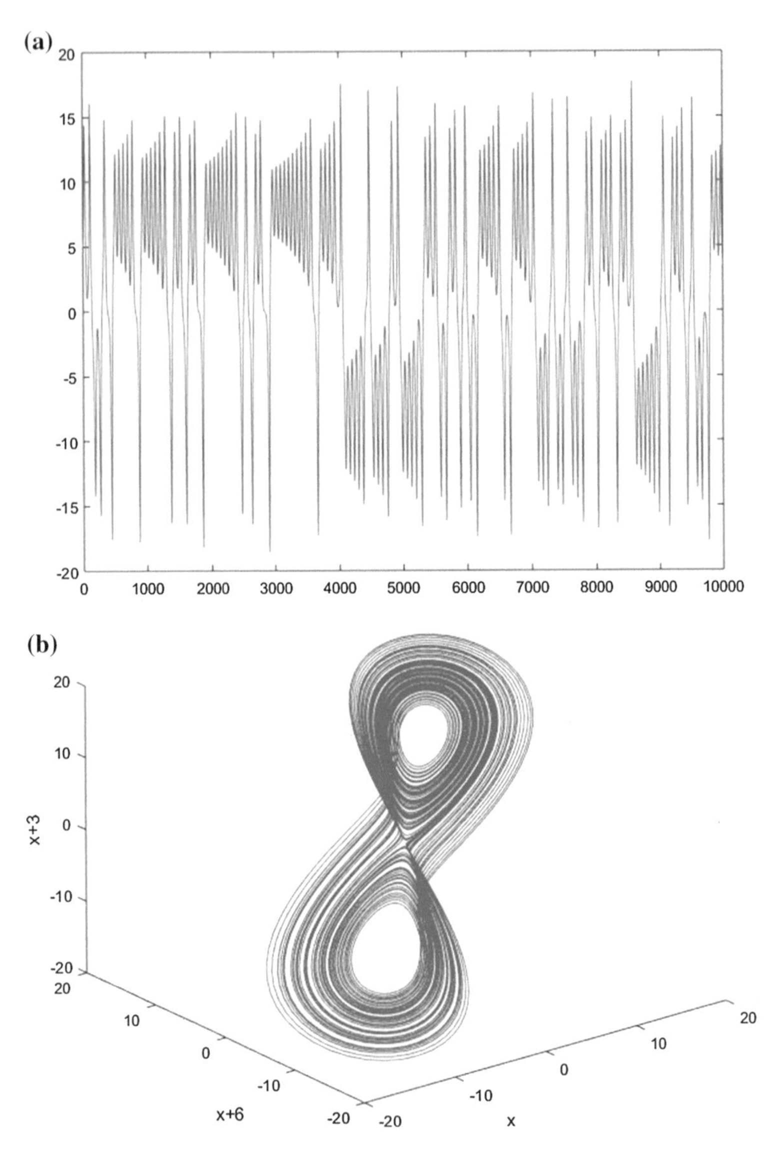

Fig. 5.6 Lorenz system: **a** time series (first component); and **b** phase space diagram (*source* Hong-Bo Xie, personal communication)

the solutions oscillate irregularly; they never exactly repeat but, at the same time, always remain in a bounded region. Wonderful structures can also emerge, which can be easily visualized from the trajectory in the state space or phase space (attractor). Figure 5.6b, for instance, shows one such structure, a clear butterfly pattern, obtained from plotting the trajectory of two of the system components (first versus third) in a two-dimensional space (state space). When plotted the trajectories in three dimensions, the solutions settle onto a complicated set.

Unlike stable fixed points and limit cycles, the Lorenz attractor is strictly aperiodic. The trajectory starts near the origin, then swings to the right, and then dives into the center of a spiral on the left. After a very slow spiral outward, the trajectory shoots back over to the right side, spirals around a few times, shoots over to the left, spirals around, and so on indefinitely. Extensive studies have shown that the fine structure of the Lorenz attractor is made up of infinitely nested layers (infinite area) that occupy zero volume. The number of circuits made on either side varies unpredictably from one cycle to the next. In fact, the sequence of the number of circuits has many of the characteristics of a *random* sequence. One may think of the Lorenz attractor as a Cantor-like set in a higher dimension. Its fractal dimension has been estimated to be about 2.06 (e.g. Grassberger and Procaccia 1983a, b).

5.11.4 Rössler System

The Rössler system is a system of three nonlinear differential equations, given by:

$$\dot{x} = -y - z \tag{5.4a}$$

$$\dot{y} = x + ay \tag{5.4b}$$

$$\dot{z} = bx + cz + xz \tag{5.4c}$$

The Rössler equations, originally studied by Rössler (1976), describe the spread of disease, and have been used effectively to model measles and whooping cough epidemics in children (Rössler 1976; Schaffer et al. 1986; Schaffer 1987). The Rössler equations were structured to have some similarity with the Lorenz equations but with an intention to make an easier qualitative analysis. In a somewhat similar way to the Lorenz attractor, an orbit within the Rössler attractor follows an outward spiral close to the x–y plane around an unstable fixed point. Once the graph spirals out sufficiently, a second fixed point influences the graph, causing a rise and twist in the z-dimension. Although each variable oscillates, in the time domain, within a fixed range of values, the oscillations are chaotic, such as the one obtained when $a = 0.1$, $b = 0.1$, and $c = 18$ (Fig. 5.7a), such as the attractor shown in Fig. 5.7b. The dimensionality of this attractor is about 2.01.

By keeping the values of a and b the same and by changing only the value of c (between 4 and 18, for example), a variety of behaviors can be observed, such as the period-1 orbit, period-2 orbit, sparse chaotic attractor, and filled-in chaotic attractor. For the same values of a and b, the Rössler equations also show some identifiable peaks at a variety of frequencies that appear above broad background noise, in contrast to the Lorenz equations, which show no identifiable structure other than high power at low frequency with a decay in power as the frequency increases.

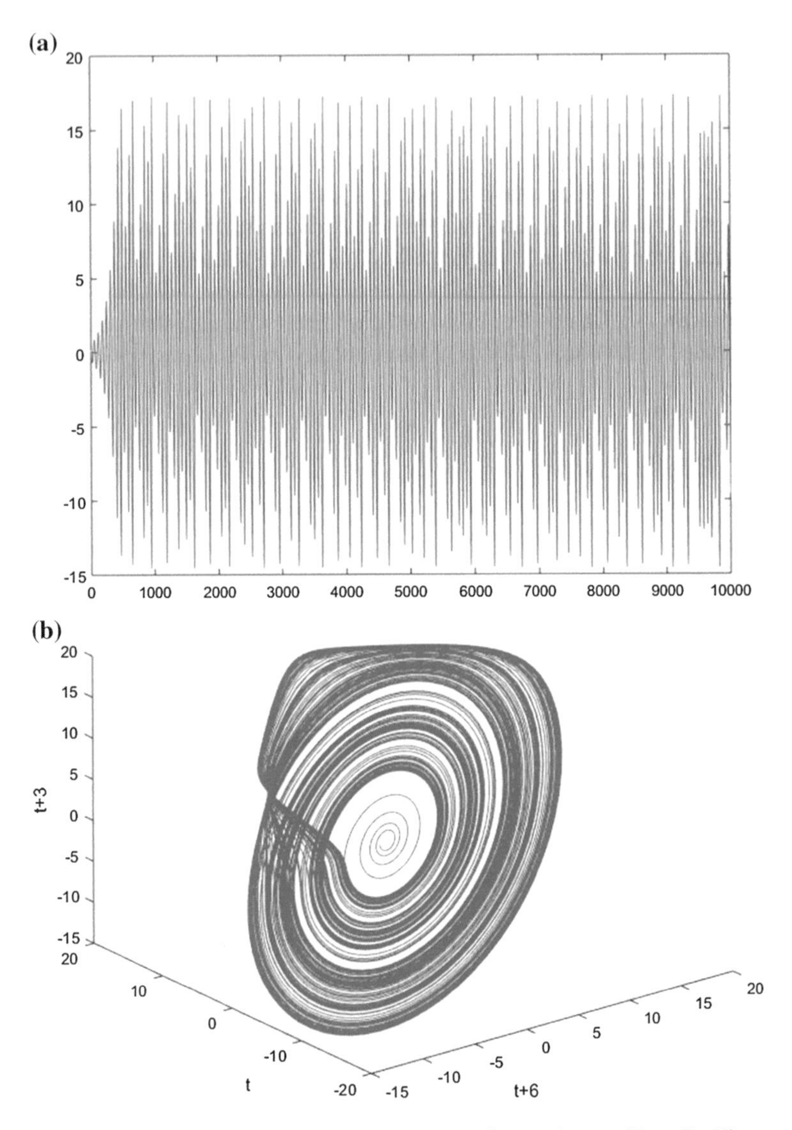

Fig. 5.7 Rössler system: **a** time series; and **b** phase space diagram (*source* Hong-Bo Xie, personal communication)

5.12 Summary

Nonlinearity is inherent in Nature. Nonlinear systems exhibit a wide range of characteristics. Among these, a particularly interesting characteristic is "chaos," which refers to situations where complex and 'random-looking' behaviors arise from simple deterministic systems with sensitive dependence on initial conditions.

Since its discovery in 1963 (Lorenz 1963), and especially with the development of various methods for its identification and prediction in the 1980s, the concept of chaos has attracted considerable attention in numerous fields. This chapter has discussed the fundamental concepts and properties of chaotic systems and also presented details of four popular synthetic chaotic systems for the purpose of illustration. Chapter 6 will present details of a number of methods for chaos identification and prediction, and Chap. 7 will discuss the important issues in the applications of chaos methods for real systems.

References

Abarbanel HDI (1996) Analysis of observed chaotic data. Springer-Verlag, New York

Abraham R, Shaw C (1984) Dynamics of the geometry and behavior, part two: stable and chaotic behavior. Aerial Press, Santa Cruz, CA

Andronov AA (1929) Les cycles limits de Poincaré et al théorie des oscillations autoentretenues. C R Acad Sci 189:559–561

Arnol'd VI (1964) Instability of dynamical systems with many degrees of freedom. Sov Math Dokl 5:581–585

Baumol WJ, Benhabib J (1989) Chaos: significance, mechanism, and economic applications. J Econ Pers 3:77–105

Birkhoff GD (1927) Dynamical systems. American Mathematical Society, New York

Briggs K (1987) Simple experiments in chaotic dynamics. Am J Phys 55:1083–1089

Cartwright ML (1935) Some inequalities in the theory of functions. Math Ann 111:98–118

Casdagli M (1989) Nonlinear prediction of chaotic time series. Physica D 35:335–356

Casdagli M (1992) Chaos and deterministic versus stochastic nonlinear modeling. J R Stat Soc B 54(2):303–328

Casdagli M, Eubank S, Farmer JD, Gibson J (1991) State space reconstruction in the presence of noise. Physica D 51:52–98

Euler L (1767) De motu rectilineo trium corporum se mutuo attrahentium, Novi commentarii academiæ scientarum Petropolitanæ 11, pp. 144–151 in Oeuvres, Seria Secunda tome XXV Commentationes Astronomicæ (p 286)

Farmer DJ, Sidorowich JJ (1987) Predicting chaotic time series. Phys Rev Lett 59:845–848

Feigenbaum MJ (1978) Quantitative universality for a class of nonlinear transformations. J Stat Phys 19:25–52

Gleick J (1987) Chaos: making a new science. Penguin Books, New York

Goerner SJ (1994) Chaos and the Evolving Ecological Universe. Gordon and Breach, Langhorne

Grassberger P, Procaccia I (1983a) Measuring the strangeness of strange attractors. Physica D 9:189–208

Grassberger P, Procaccia I (1983b) Characterisation of strange attractors. Phys Rev Lett 50 (5):346–349

Grassberger P, Procaccia I (1983c) Estimation of the Kolmogorov entropy from a chaotic signal. Phys Rev A 28:2591–2593

Haken H (1983) Advanced synergetics column instability hierarchies of self-organizing systems and devices. Springer-Verlag, Berlin

Hausdorff (1918) Dimension und äußeres Maß. Math Annalen 79(1–2):157–179

Henon M (1976) A two-dimensional mapping with a strange attractor. Commun Math Phys 50:69–77

Hilborn RC (1994) Chaos and nonlinear dynamics: an introduction for scientists and engineers. Oxford University Press, Oxford, UK

Kantz H, Schreiber T (2004) Nonlinear time series analysis. Cambridge University Press, Cambridge
Kaplan DT, Glass L (1995) Understanding nonlinear dynamics. Springer, New York
Kaplan JL, Yorke JA (1979) Chaotic behavior of multidimensional difference equations. In: Peitgen HO, Walther HO (eds) Functional differential equations and approximations of fixed points, Lecture notes in mathematics, Springer-Verlag, Berlin, pp 204
Kennel MB, Brown R, Abarbanel HDI (1992) Determining embedding dimension for phase space reconstruction using a geometric method. Phys Rev A 45:3403–3411
Kepler J (1609) Astronomia nova
Kiel LD, Elliott E (1996) Chaos theory in the social sciences: foundations and applications. The University of Michigan Press, Ann Arbor, USA, 349 pp
Kolmogorov AN (1954) On the conservation of conditionally periodic motions under small perturbation of the Hamiltonian function. Dokl Akad Nauk SSR 98:527–530
Lagrange JL (1772) Essai sur le Problème des Trois Corps, Prix de l'Académie Royale des Sciences de Paris, tome IX, in vol. 6 of Oeuvres (p 292)
Levinson N (1943) On a non-linear differential equation of the second order. J Math Phys 22:181–187
Linsay PS (1981) Period doubling and chaotic behaviour in a driven anharmonic oscillator. Phys Rev Lett 47:1349–1352
Littlewood JE (1966) Unbounded solutions of $\ddot{y} + g(y) = p(t)$. J London Math Soc 41:491–496
Lorenz EN (1963) Deterministic nonperiodic flow. J Atmos Sci 20:130–141
Mandelbrot BB (1983) The fractal geometry of nature. W. H. Freedman and Company, New York
May RM (1974) Stability and complexity in model ecosystems. Princeton University Press, Princeton, NJ
May RM (1976) Simple mathematical models with very complicated dynamics. Nature 261:459–467
Meissner H, Schmidt G (1986) A simple experiment for studying the transition from order to chaos. Am J Phys 54:800–804
Mishina T, Kohmoto T, Hashi T (1985) Simple electronic circuit for demonstration of chaotic phenomena. Am J Phys 53:332–334
Moser JK (1967) Convergent series expansions for quasi-periodic motions. Math Ann 169:136–176
Newton I (1687) Philosophiæ naturalis principia mathematica. Burndy Library, UK
Nicolis G, Prigogine I (1989) Exploring complexity: an introduction. W. H. Freeman and Company, New York
Ott E (1993) Chaos in dynamical systems. Cambridge University Press, Cambridge, UK
Packard NH, Crutchfield JP, Farmer JD, Shaw RS (1980) Geometry from a time series. Phys Rev Lett 45(9):712–716
Poincare H (1890) On the three-body problem and the equations of dynamics (Sur le Problem des trios crops ci les equations de dynamique). Acta Math 13:270
Poincaré H (1896) Sur les solutions périodiques et le principe de moindre action (On the three-body problem and the equations of dynamics. C R Acad Sci 123:915–918
Pool R (1989) Where strange attractors lurk. Science 243:1292
Rényi A (1959) On the dimension and entropy of probability distributions. Acta Math Hungarica 10(1–2)
Rényi A (1971) Probability theory. North Holland, Amsterdam
Rössler OE (1976) An equation for continuous chaos. Phys Lett A 57:397–398
Ruelle D (1989) Chaotic evolution and strange attractors: the statistical analysis of time series for deterministic nonlinear systems. Cambridge University Press, Cambridge, UK
Ruelle D, Takens F (1971) On the nature of tubulence. Comm Math Phys 20:167–192
Saltzman B (1962) Finite amplitude free convection as an initial value problem–I. J Atmos Sci 19:329–341
Sangoyomi TB, Lall U, Abarbanel HDI (1996) Nonlinear dynamics of the Great Salt Lake: dimension estimation. Water Resour Res 32(1):149–159

Schaffer W (1987) Chaos in ecology and epidemiology. In: Degn H, Holden AV, Olsen LF (eds) Chaos in biological systems. Plenum Press, New York, pp 366–381

Schaffer W, Ellner S, Kot M (1986) Effects of noise on some dynamical models in ecology. J Math Biol 24:479–523

Schuster HG (1988) Deterministic chaos: an introduction. Physik Verlag, Weinheim

Sivakumar B (2004) Chaos theory in geophysics: past, present and future. Chaos Soliton Fract 19 (2):441–462

Smale S (1960) More inequalities for a dynamical system. Bull Amer Math Soc 66:43–49

Strogatz SH (1994) Nonlinear dynamics and chaos: with applications to physics, biology, chemistry, and engineering. Perseus Books, Cambridge

Su Z, Rollins RW, Hunt ER (1987) Measurements of f(α) spectra of attractors at transitions to chaos in driven diode resonator systems. Phys Rev A 36:3515–3517

Swinney HL, Gollub JP (1978) Hydrodynamic instabilities and the transition to turbulence. Phys Today 31(8):41–49

Takens F (1981) Detecting strange attractors in turbulence. In: Rand DA, Young LS (eds) Dynamical systems and turbulence, lecture notes in mathematics 898, Springer, Berlin, pp 366–381

Teitsworth SW, Westervelt RM (1984) Chaos and broad-band noise in extrinsic photoconductors. Phys Rev Lett 53(27):2587–2590

Theiler J, Eubank S, Longtin A, Galdrikian B, Farmer JD (1992) Testing for nonlinearity in time series: the method of surrogate data. Physica D 58:77–94

Tong H (1990) Non-linear time series analysis. Oxford University Press, Oxford, UK

Tsonis AA (1992) Chaos: from theory to applications. Plenum Press, New York

Tufillario NB, Albano AM (1986) Chaotic dynamics of a bouncing ball. Am J Phys 54:939–944

van der Pol B (1927) On relaxation-oscillations. The London, Edinburgh and Dublin Phil Mag & J of Sci 2(7):978–992

Williams GP (1997) Chaos theory tamed. Taylor & Francis Publishers, London, UK

Wolf A, Swift JB, Swinney HL, Vastano A (1985) Determining Lyapunov exponents from a time series. Physica D 16:285–317

Chapter 6
Chaos Identification and Prediction Methods

Abstract Considerable interest in studying the chaotic behavior of natural, physical, and socio-economic systems have led to the development of many different methods for identification and prediction of chaos. An important commonality among almost all of these methods is the concept of phase space reconstruction. Other than this, the methods largely have different bases and approaches and often aim to identify different measures of chaos. All these methods have been successfully applied in many different scientific fields. This chapter describes some of the most popular methods for chaos identification and prediction, especially those that have found applications in hydrology. These methods include: phase space reconstruction, correlation dimension method, false nearest neighbor method, Lyapunov exponent method, Kolmogorov entropy method, surrogate data method, Poincaré maps, close returns plot, and nonlinear local approximation prediction method. To put the utility of these methods in a proper perspective in the identification of chaos, the superiority of two of these methods (phase space reconstruction and correlation dimension) over two commonly used linear tools for system identification (autocorrelation function and power spectrum) is also demonstrated. Further, as the correlation dimension method has been the most widely used method for chaos identification, it is discussed in far more detail.

6.1 Introduction

If the mathematical formulation of the system is given, recognizing chaotic behavior is as easy as producing the Fourier spectra of the evolution of one of the variables. Since the evolution is deterministic, broadband noise spectra would be sufficient to identify chaos. Furthermore, since the number of variables is known, the generation of the state space and the attractor (a geometric object that characterizes the long-term behavior of a system) as well as the estimation of the metric and topological properties are fairly straightforward. However, when dealing with controlled experiments, where one cannot record all the variables, and/or with observables from uncontrolled systems (like hydrologic systems), whose mathematical formulation

B. Sivakumar, *Chaos in Hydrology*, DOI 10.1007/978-90-481-2552-4_6

and total number of variables may not be known exactly, life becomes a little bit more complicated. Fourier analysis alone cannot be used for proof of chaos, since the observable might be a random (stochastic) variable. Another problem with the use of Fourier spectra for system identification is that it is not reliable for distinguishing between stochastic and chaotic signals, since both exhibit continuous part (broadband) of the spectrum; indeed, almost all other linear tools are susceptible to this kind of problem as well.

In view of these, additional evidence, in the form of nonlinear properties of the system (e.g. trajectories of evolution, dimensions, Lyapunov exponents, and other relevant metrics), must be provided. To accomplish this, one must first find a way to reconstruct the phase space of the underlying dynamic system from observables of one or more variables representing the system. It is also relevant to note that oftentimes there may not be sufficient information about all the variables influencing the system and, thus, one may be forced to perform analysis and make inferences about the system based on only a handful of recorded series.

The methods currently available for chaos identification in a time series may broadly be grouped under two categories: metric approach and topological approach. The metric approach is characterized by the study of distances between points on a strange attractor (in the phase space). The topological approach is characterized by the study of the organization of the strange attractor. This chapter details some of the popular methods for identification of chaos in a time series and its prediction, especially those that have found widespread applications in the study of hydrologic time series. These include: phase space reconstruction, correlation dimension method, false nearest neighbor algorithm, Lyapunov exponent method, Kolmogorov entropy method, surrogate data method, Poincaré map, close returns plot, and nonlinear local approximation prediction method. While the primary purpose of the nonlinear local approximation prediction method is prediction of time series that has been found to exhibit chaotic dynamics, the method can also be used for chaos identification through an inverse approach.

The reliability of phase space reconstruction, chaos identification, and nonlinear prediction methods to real complex systems, such as hydrologic systems, has been under considerable debate, in view of their potential limitations when applied to such systems and the associated time series. Much of the criticism has been directed at the correlation dimension method (Sect. 6.4), and in particular the Grassberger–Procaccia algorithm for dimension estimation (e.g. Grassberger and Procaccia 1983a, b). Extensive details of the relevant issues are already available in the literature; see Sivakumar (2000, 2001, 2005) and Sivakumar et al. (1999, 2002a, b) for some details specific to hydrologic data. Some of these issues will also be discussed in Chap. 7. To put the nonlinear methods and their utility in chaos studies in a proper perspective, a brief discussion is made here first on two basic and widely used linear tools (autocorrelation function and power spectrum) and their limitations in chaos identification in a time series.

6.2 Linear Tools and Limitations

In the analysis of time series for identification of system properties, some kind of 'data manipulation' or 'data reconstruction' becomes necessary. In this context, it is customary to use two basic linear tools: autocorrelation function (ACF) and power spectrum; see also Chap. 3 for some additional details.

The autocorrelation function is a normalized measure of the linear correlation among successive values in a time series. The use of the autocorrelation function in characterizing the behavior lies in its ability to determine the degree of dependence present in the values. For a periodic process, the autocorrelation function is also periodic, indicating the strong relation between values that repeat over and over again. For a purely random process, the autocorrelation function fluctuates about zero, indicating that the process at any certain instance has no 'memory' of the past at all. Other stochastic processes generally have decaying autocorrelations, but the rate of decay depends on the properties of the process. The autocorrelation function of signals from a chaotic process is also expected to decay exponentially with increasing lag, because the states of a chaotic process are neither completely dependent (i.e. deterministic) nor completely independent (i.e. random) of each other, although this is not always the case.

The power spectrum is particularly useful for characterizing the regularities/irregularities in observed signals (time series). In general, for a purely random process, the power spectrum oscillates randomly about a constant value, indicating that no frequency explains any more of the variance of the sequence than any other frequency. For a periodic or quasi-periodic sequence, only peaks at certain frequencies exist; measurement noise adds a continuous floor to the spectrum; thus, in the spectrum, signal and noise are readily distinguished. Chaotic signals may also have sharp spectral lines but even in the absence of noise there will be continuous part (broadband) of the spectrum. This is an immediate consequence of the exponentially decaying autocorrelation function. Therefore, without additional information, it is impossible to infer from the spectrum whether the continuous part is due to noise on top of a (quasi-)periodic signal or chaoticity.

While the autocorrelation function and power spectrum provide convincing distinctions between random and periodic (or quasi-periodic) systems, they are not reliable for distinguishing between random and chaotic signals. To demonstrate this, they are applied herein to two artificially generated time series (Fig. 6.1a, b) that look very much alike (both look 'complex' and 'random') but are the outcomes of systems (equations) possessing significantly different dynamic characteristics. The first series (Fig. 6.1a) is the outcome of a pseudo random number generation function:

$$X_i = rand() \tag{6.1}$$

which yields independent and identically distributed numbers (generated between 0 and 1 herein). The second (Fig. 6.1b), however, is the outcome of a fully deterministic simple two-dimensional map (Henon 1976):

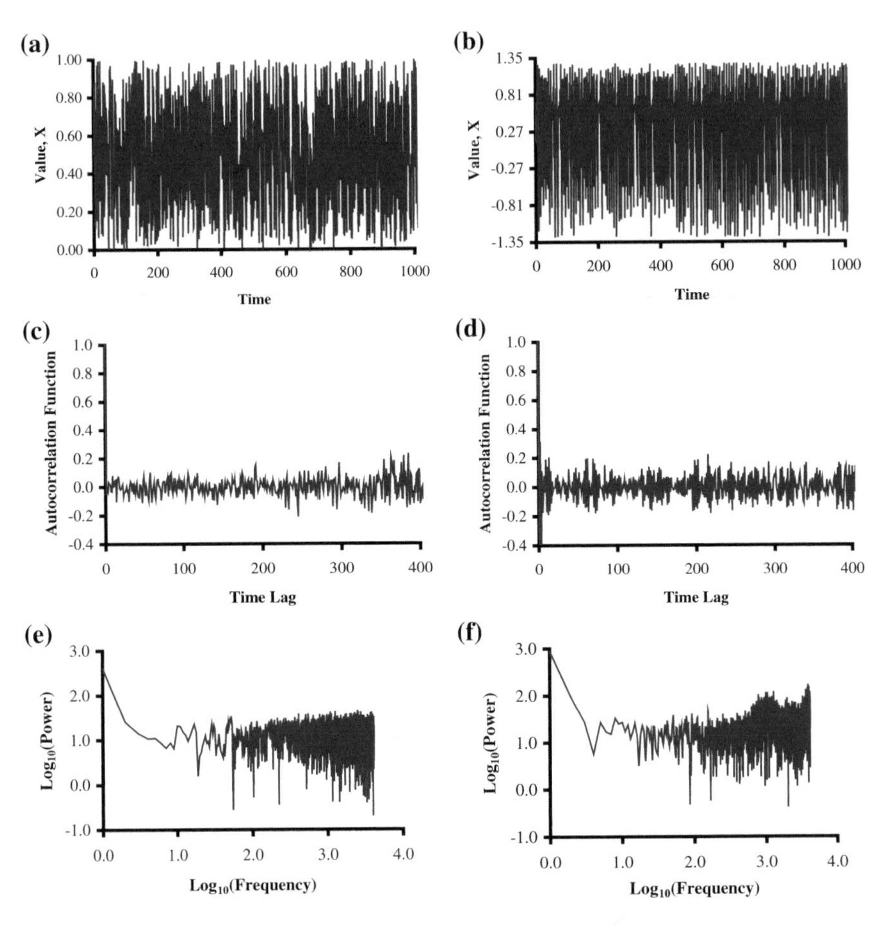

Fig. 6.1 Stochastic system (*left*) versus chaotic system (*right*): **a**, **b** time series; **c**, **d** autocorrelation function; and **e**, **f** power spectrum (*source* Sivakumar et al. (2007))

$$X_{i+1} = a - X_i^2 + bY_i \quad Y_{i+1} = X_i \tag{6.2}$$

which yields irregular solutions for many choices of a and b, but for $a = 1.4$ and $b = 0.3$, a typical sequence of X_i is chaotic. The initial values of X and Y used for this data series are 0.13 and 0.50, respectively.

Figure 6.1c, d shows the autocorrelation functions for these two series, while Fig. 6.1e, f presents the power spectra. It is clear that both the tools fail to distinguish between the two series. The failure is not just in 'visual' or 'qualitative' terms, but also in quantitative terms: for instance, for both series, the time lag at which the autocorrelation function first crosses the zero line is equal to 1 (especially no exponential dacay for the chaotic series) and the power spectral exponent is equal to 0 (indicating pure randomness in the underlying dynamics of both).

Granted that real systems (e.g. hydrologic systems) may be neither purely random nor as simple as a two-dimensional map, it is still safe to say that (these) linear tools may not be sufficient for studies on natural system properties and characterization, particularly when the system also possesses nonlinear properties. Consequently, one may need tools that can also represent such nonlinear properties, since natural systems are inherently nonlinear.

6.3 Phase Space Reconstruction

In the 'data reconstruction' context, another useful tool for 'embedding' the data to represent the evolution of a system in time (or in space) is the concept of phase space (e.g. Packard et al. 1980). Phase space is essentially a graph or a co-ordinate diagram, whose co-ordinates represent the variables necessary to completely describe the state of the system at any moment (in other words, the variables that enter the mathematical formulation of the system). The trajectories of the phase space diagram describe the evolution of the system from some initial state, which is assumed to be known, and hence represent the history of the system. The 'region of attraction' of these trajectories in the phase space provides at least important qualitative information on the nature of the underlying system dynamics, such as 'extent of complexity.'

For a dynamic system with known partial differential equations (PDEs), the system can be studied by discretizing the PDEs, and the set of variables at all grid points constitutes a phase space. One difficulty in constructing the phase space for such a system is that the (initial) values of many of the variables may not be known. However, a time series of a single variable of the system may be available, which may allow the attractor to be reconstructed. The idea behind such a reconstruction is that a (nonlinear) system is characterized by self-interaction, so that a time series of a single variable can carry the information about the dynamics of the entire multi-variable system. It is relevant to note that phase space may also be reconstructed using multiple variables when available (i.e. state space) (e.g. Cao et al. 1998); see also Chap. 5, Sect. 5.9.

Various embedding theorems and methods have been developed for phase space (or state space) reconstruction from an available single-variable (or multi-variable) time series. These include: (1) Whitney's embedding theorem (Whitney 1936); (2) Takens' delay embedding theorem (Takens 1981); (3) Fractal delay embedding prevalence theorem (Sauer et al. 1991); and (4) Filtered delay embeddings, such as singular value decomposition or singular spectrum analysis or principal component analysis (Broomhead and King 1986; Vautard et al. 1992; Elsner and Tsonis 1996). However, the Takens' delay embedding theorem is the most widely used one, especially in chaos identification studies, and is described here.

Given a single variable series, X_i, where $i = 1, 2, \ldots, N$, a multi-dimensional phase space can be reconstructed according to the Takens' delay embedding theorem as follows:

$$Y_j = (X_j, X_{j+\tau}, X_{j+2\tau}, \ldots, X_{j+(m-1)\tau/\Delta t}) \tag{6.3}$$

where $j = 1, 2, \ldots, N - 1(m - 1)\tau/\Delta t$; m is the dimension of the vector Y_j, called embedding dimension; and τ is an appropriate delay time taken to be a suitable integer multiple of the sampling time Δt. A correct phase space reconstruction in a dimension m generally allows interpretation of the system dynamics (if the variable chosen to represent the system is appropriate) in the form of an m-dimensional map f_T, given by:

$$Y_{j+T} = f_T(Y_j) \tag{6.4}$$

where Y_j and Y_{j+T} are vectors of dimension m, describing the state of the system at times j (current state) and $j + T$ (future state), respectively. With Eq. (6.4), the task is basically to find an appropriate expression for f_T (e.g. F_T) to predict the future.

To demonstrate the utility of phase space diagram for system identification, Fig. 6.2a, b present the phase space plots for the above two series (Fig. 6.1a, b). These diagrams correspond to reconstruction in two dimensions ($m = 2$) with delay time $\tau = 1$, i.e. the projection of the attractor on the plane $\{X_i, X_{i+1}\}$. For the first set, the points (of trajectories) are scattered all over the phase space (i.e. absence of an attractor), a clear indication of a 'complex' and 'random' nature of the underlying dynamics and potentially of a high-dimensional (and possibly random) system. On the other hand, the projection for the second set yields a very clear attractor (in a well-defined region), indicating a 'simple' and 'deterministic' (yet non-repeating) nature of the underlying dynamics and potentially of a low-dimensional (and possibly chaotic) system.

It is relevant to note that the selection of the minimum (or optimum) m and an appropriate τ for phase space reconstruction has been under considerable debate. Consequently, various methods have been formulated and guidelines developed. The case of embedding dimension will be discussed in Sect. 6.4.3, appropriately in the context of the number of dominant governing variables. The case of delay time will be discussed extensively in Chap. 7.

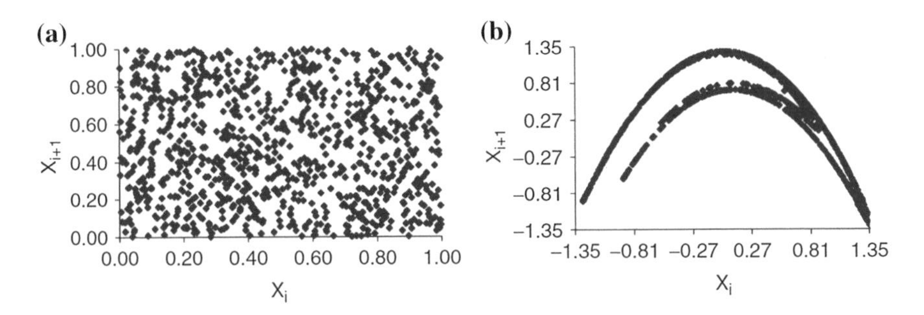

Fig. 6.2 Stochastic system (*left*) versus chaotic system (*right*): phase space diagram (*source* Sivakumar et al. (2007))

6.4 Correlation Dimension Method

6.4.1 *Basic Concept*

The word 'dimension' is derived from the Latin word 'dimensio,' which means measure. The dimension of a time series is, in a way, a representation of the number of dominant variables present in the evolution of the corresponding dynamic system. Dimension analysis also reveals the extent to which the variations in the time series are concentrated on a subset of the space of all possible variations. In the context of identification of chaos (more specifically, distinguishing between chaos and stochasticity), the central idea behind the application of the dimension approach is that systems whose dynamics are governed by stochastic processes are thought to have an infinite value for the dimension, whereas a finite, non-integer value of the dimension is considered to be an indication of the presence of chaos.

There are various forms of dimension, representing measures in many different ways. These include: box counting dimension, capacity dimension (Young 1984), correlation dimension (Grassberger and Procaccia 1983a, b), generalized or Rényi dimension (Rényi 1971; Grassberger 1983), Hausdorff dimension or Hausdorff-Besicovitch dimension (Hausdorff 1919), information dimension, Kaplan–Yorke dimension or Lyapunov dimension (Kaplan and Yorke 1979), local intrinsic dimension (Hediger et al. 1990), and others.

There are several ways to quantify the self-similarity of a geometric object by a dimension. However, we require the definition to coincide with the usual notion of dimension when applied to non-fractal objects: a finite collection of points is zero dimensional, lines have dimension one, surfaces two, and so on. For instance, let us propose a definition which is of particular interest in practical applications, such as in hydrologic applications, where the geometric object has to be reconstructed from a finite set of data points which are most likely to contain some errors as well. One such notion, called the correlation dimension, was introduced by Grassberger and Procaccia (1983a, b).

Correlation dimension is a measure of the extent to which the presence of a data point affects the position of the other points lying on the attractor in (a multi-dimensional) phase space. The correlation dimension method uses the correlation integral (or function) for determining the dimension of the attractor in the phase space and, hence, for distinguishing between low-dimensional and high-dimensional systems. The concept of the correlation integral is that a time series arising from deterministic dynamics will have a limited number of degrees of freedom equal to the smallest number of first-order differential equations that capture the most important features of the dynamics. Thus, when one constructs phase spaces of increasing dimension, a point will be reached where the dimension equals the number of degrees of freedom, beyond which increasing the phase space dimension will not have any significant effect on correlation dimension.

6.4.2 The Grassberger–Procaccia (G–P) Algorithm

Many algorithms have been formulated for the estimation of the correlation dimension of a time series (e.g. Grassberger and Procaccia 1983a, b; Theiler 1987; Grassberger 1990; Toledo et al. 1997; Carona 2000). Among these, the Grassberger–Procaccia algorithm (Grassberger and Procaccia 1983a, b) has and continues to be the most widely used one, especially in studies on hydrologic and geophysical systems. The algorithm uses the concept of phase space reconstruction for representing the dynamics of the system from an available single-variable (or multi-variable) time series, as presented in Eq. (6.3). For an m-dimensional phase space, the correlation function $C(r)$ is given by:

$$C(r) = \frac{2}{N(N-1)} \sum_{\substack{i,j \\ (1 \leq i < j \leq N)}} H\left(r - \|\mathbf{Y}_i - \mathbf{Y}_j\|\right) \tag{6.5}$$

where H is the Heaviside step function, with $H(u) = 1$ for $u > 0$, and $H(u) = 0$ for $u \leq 0$, where $u = r - \|\mathbf{Y}_i - \mathbf{Y}_j\|$, r is the vector norm (radius of sphere) centered on $\mathbf{Y}_i$ or $\mathbf{Y}_j$. If the time series is characterized by an attractor, then $C(r)$ and r are related according to:

$$\underset{\substack{r \to 0 \\ N \to \infty}}{C(r)} \approx \alpha r^{\vartheta} \tag{6.6}$$

where α is a constant and v is the correlation exponent or the slope of the Log $C(r)$ versus Log r plot given by:

$$\vartheta = \lim_{\substack{r \to 0 \\ N \to \infty}} \frac{\operatorname{Log} C(r)}{\operatorname{Log} r} \tag{6.7}$$

The slope is generally estimated by a least-squares fit of a straight line over a certain range of r (i.e. scaling regime) or through estimation of local slopes between r values.

The distinction between low-dimensional (and perhaps determinism) and high-dimensional (and perhaps stochastic) systems can be made using the v versus m plot. If v saturates after a certain m and the saturation value is low, then the system is generally considered to exhibit low-dimensional deterministic dynamics. The saturation value of v is defined as the correlation dimension (d) of the attractor. On the other hand, if v increases without bound with increase in m, the system under investigation is generally considered to exhibit high-dimensional stochastic dynamics.

6.4.3 *Identification of Number of System Variables*

The correlation dimension of an attractor provides information on the dimension of the phase space required for embedding the attractor, which, in turn, provides information on the number of variables governing the evolution of the corresponding dynamic system. According to the embedding theorem of Takens (Takens 1981), a dynamic system with an attractor dimension (e.g. correlation dimension) d can be adequately characterized in an ($m = 2d + 1$)-dimensional phase space, which is also consistent with the guideline by the Whitney's embedding theorem (Whitney 1936). However, Abarbanel et al. (1990) suggest that, in practice, $m > d$ would be sufficient. According to Fraedrich (1986), the nearest integer above d provides the minimum dimension of the phase space essential to embed the attractor, while the value of m at which the saturation of the correlation exponent occurs provides an upper bound on the dimension of the phase space sufficient to describe the motion of the attractor.

Considering these guidelines, it may be reasonable to assume that the nearest integer above the correlation dimension value is generally an indication of the number of variables dominantly governing the dynamics. Nevertheless, the identification of the number of dominant governing variables may also be open to interpretation, especially when it comes to real time series, where it is oftentimes hard to see a clear one-to-one correspondence between correlation exponent against embedding dimension (until the embedding dimension at which saturation of the correlation dimension occurs, if any); see Sivakumar et al. (2002a) for details.

6.4.4 *An Example*

To demonstrate the utility of the correlation dimension concept, Fig. 6.3a, c presents the correlation dimension results for the first set (Fig. 6.1a), whereas those for the second set (Fig. 6.1b) are shown in Fig. 6.3b, d. In each case, embedding dimensions from 1 to 10 are used for phase space reconstruction. The results indicate that the first set is the outcome of an infinite-dimensional system, i.e. absence of saturation in correlation exponent (Fig. 6.3c) and the presence of a large number of variables governing the system. However, the saturation of the correlation exponent observed for the second set (Fig. 6.3d), with a correlation dimension value of 1.22, indicates that the time series is the outcome of a low-dimensional system, whose systems are dominantly governed by just two variables. These results are consistent with the stochastic and chaotic systems (Eqs. (6.1) and (6.2)), thus indicating the utility and appropriateness of the correlation dimension for identification of chaos in a time series.

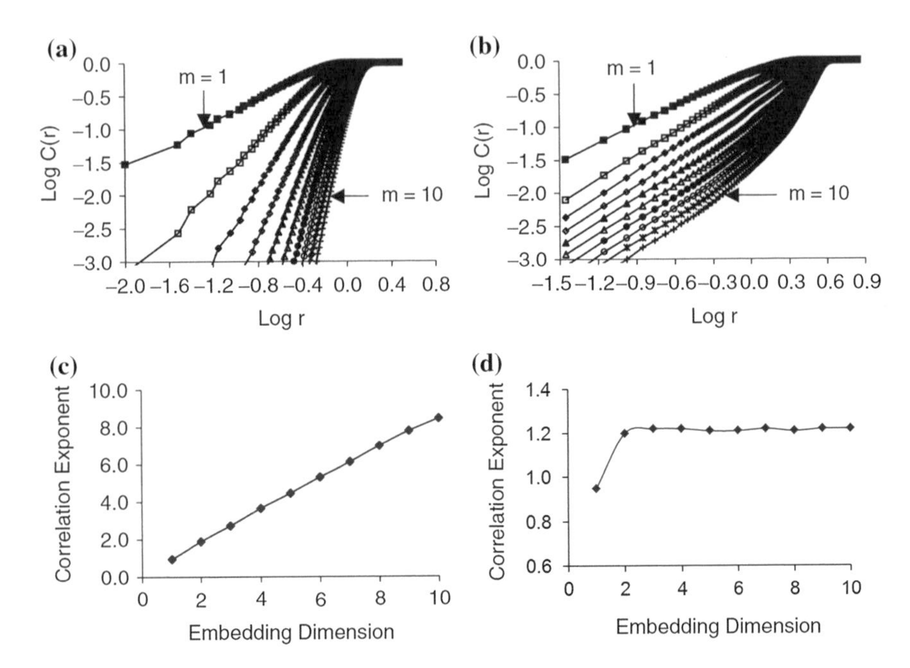

Fig. 6.3 Stochastic system (*left*) versus chaotic system (*right*): correlation dimension (*source* Sivakumar et al. (2007))

6.4.5 Improvements to the G–P Algorithm

There have been criticisms on the application of the Grassberger–Procaccia algorithm for correlation dimension estimation and chaos identification. These criticisms are in regards to temporal correlations (Theiler 1990), data size (e.g. Havstad and Ehlers 1989; Nerenberg and Essex 1990), data noise (Schreiber and Kantz 1996), presence of zeros (Tsonis et al. 1993, 1994), and even stochastic processes yielding low correlation dimensions (e.g. Osborne and Provenzale 1989). Details of these issues will be discussed in Chap. 7. In what follows, a brief account of the improvements made to the Grassberger–Procaccia algorithm in terms of computational efficiency is presented.

Direct application of the Grassberger–Procaccia algorithm requires computer time that scales as $O(N^2)$. This can make the algorithm computationally cumbersome, especially when we are dealing with (or require) large sample sizes. A more efficient algorithm for estimating the correlation dimension was developed by Theiler (1987). The philosophy behind Theiler's improvement is that it is not necessary to include in the calculations distances greater than a cutoff distance r_0, since they do not really define a scaling region. Theiler's algorithm (often referred to as the box-assisted algorithm) requires computer times that scale as $O(N \log N)$. Thus, depending on N, speedup factors up to 1000 over the usual Grassberger–Procaccia algorithm can be achieved. Another improvement was suggested by

Grassberger (1990), who proposed an optimized box-assisted algorithm that reduces the computer time by an additional factor of 2–4. A still further improvement to the neighbor searching algorithm for nonlinear time series analysis was proposed by Schreiber (1995), which requires computer times that scale as $O(N)$ without losing any of the precious pairs for correlation integral calculation.

6.5 False Nearest Neighbor (FNN) Algorithm

The false nearest neighbor (FNN) algorithm (Kennel et al. 1992) provides information on the minimum embedding dimension of the phase space (in other words, number of dominant variables) required for representing the system dynamics. It examines, in dimension m, the nearest neighbor $\boldsymbol{Y}_j^{NN}$ of every vector $\boldsymbol{Y}_j$, as it behaves in dimension $m + 1$. If the vector $\boldsymbol{Y}_j^{NN}$ is a true neighbor of $\boldsymbol{Y}_j$, then it comes to the neighborhood of $\boldsymbol{Y}_j$ through dynamic origins. On the other hand, if the vector $\boldsymbol{Y}_j^{NN}$ moves far away from vector $\boldsymbol{Y}_j$ as the dimension is increased, then it is declared a 'false nearest neighbor' as it arrived in the neighborhood of $\boldsymbol{Y}_j$ in dimension m by projection from a distant part of the attractor. When the percentage of these false nearest neighbors drops to zero, the geometric structure of the attractor has been unfolded and the orbits of the system are now distinct and do not cross (or overlap).

A key step in the false nearest neighbor algorithm is to determine how to decide upon increasing the embedding dimension that a nearest neighbor is false. Two criteria are generally used (Sangoyomi et al. 1996). These are:

- Loneliness tolerance: If $R_{m+1}(j) \geq 2R_A$, the jth vector has a false nearest neighbor, where $R_{m+1}(j)$ is the distance to the nearest neighbor of the jth vector (i.e., $\boldsymbol{Y}_j^{NN}$) in an embedding of dimension $(m + 1)$ and R_A is the standard deviation of the time series X_i, $i = 1, 2, \ldots, N$.
- Distance tolerance: If $[R_{m+1}(j) - R_m(j)] > \varepsilon R_m(j)$, the jth vector has a false nearest neighbor, where ε is a threshold factor (generally between 10 and 50), and the distance $R_{m+1}(j)$ is computed to the same neighbor that was identified with embedding m, but with the $(m + 1)$th coordinate (i.e., $X_{j-m\tau}$ appended to the jth vector and to its nearest neighbor with embedding m).

It is important to apply the first criterion because, with a finite and often short data set, as is generally the case in hydrologic series, under repeated embedding, the points may be stretched out far apart and yet cannot be moved any farther when the dimension is increased. The second criterion is used to check whether the nearest neighbors have moved far apart on increasing the dimension. The appropriate threshold ε is generally selected through experimentation.

The introduction of the false nearest neighbor concept (and other ad hoc tools) was partly a reaction to the finding that many results obtained for the genuine invariants, like the correlation dimension, have been spurious due to caveats of the estimation procedure. It turned out, however, that the false nearest neighbor algorithm also often suffers from the same problems (e.g. Fredkin and Rice 1995; Rhodes and Morari 1997). Several studies have addressed these problems and presented improvements. Fredkin and Rice (1995) showed that the original FNN algorithm can falsely indicate a stationary random process to be deterministic and also proposed remedial modifications. Rhodes and Morari (1997) showed that the original FNN algorithm could lead to incorrect estimation of the embedding dimension even in the presence of small amounts of noise, especially for large data sets. They also offered guidelines on the theoretically correct choice of the FNN threshold. Hegger and Kantz (1999) modified the original FNN algorithm to ensure a correct distinction between low-dimensional chaotic time series and noise. Observing that correlated noise processes can yield vanishing percentage of false nearest neighbors for rather low embedding dimensions and can be mistaken for deterministic signals, they advocated combining false nearest neighbor method with a surrogate data set. Aittokallio et al. (1999) introduced a graphical representation for the false nearest neighbor method, to take into account the distribution of neighboring points in the delay coordinates in addition to only the percentage of false nearest neighbors computed in the original FNN. Kennel and Abarbanel (2002) proposed an improved FNN method for estimation of the minimum necessary embedding dimension by correcting for systematic effects due to temporal oversampling, autocorrelation, and changing lag time. Ramdani et al. (2007) proposed a criterion, based on the estimation of a parameter defined by the averaged false nearest neighbor method, to detect determinism in short time series. They also investigated the robustness of this criterion in the case of deterministic time series corrupted by additive noise.

6.6 Lyapunov Exponent Method

Lyapunov exponents are the average exponential rates of divergence (expansion) or convergence (contraction) of nearby orbits in the phase space. Since nearby orbits correspond to nearly identical states, exponential orbital divergence means that systems whose initial differences that may not be possible to resolve will soon behave quite differently, i.e. predictive ability is rapidly lost. Any system containing at least one positive Lyapunov exponent is defined to be chaotic, with the magnitude of the exponent reflecting the timescale on which system dynamics become unpredictable. A negative Lyapunov exponent indicates that the orbit is stable and periodic. A zero Lyapunov exponent is an indication of a marginally or neutrally stable orbit, which often occurs near a point of bifurcation. An infinite Lyapunov exponent value is an indication of a stochastic system.

The Lyapunov exponent, λ_i, is closely related to several other measures of chaos. According to Ruelle (1983, 1989), the sum of all the positive λ_i is a measure of the Kolmogorov entropy (K entropy); this means that positive entropy exists when chaos exists. The λ_i is also linked to the information lost and gained during chaotic episodes (Ruelle 1980; Shaw 1981), and it is closely linked to the amount of information available for prediction. Kaplan and Yorke (1979) conjectured the λ_i to be related to the fractal character of the attractor; this was also subsequently supported by Russell et al. (1980) for some typical attractors. The λ_i itself has a fractal, non-continuous dimension, and that fractal quality is linked to the information available about the system (Kaplan and Yorke 1979; Young 1982; Ruelle 1989).

Many algorithms have been formulated for calculation of the Lyapunov exponents from a time series (e.g. Eckmann and Ruelle 1985; Sano and Sawada 1985; Wolf et al. 1985; Eckmann et al. 1986; Rosenstein et al. 1993; Kantz 1994). Among these, the algorithm by Wolf et al. (1985) has been the most widely used. The algorithm tracks a pair of arbitrarily close points over a trajectory to estimate the accumulated error per timestep. The points are separated in time by at least one orbit on the attractor. The trajectory is defined by the fiducial and test trajectories. They are tracked for a fixed time period or until the distance between the two components of the trajectory exceeds some specific value. In sequence, another test point near the fiducial trajectory is selected and the estimation proceeds. The end product is that the stretching and squeezing are averaged.

Figure 6.4 shows a representation of the Wolf et al. (1985) computation of a Lyapunov exponent; see also Vastano and Kostelich (1986). The initial data point, $\boldsymbol{Y}_1$ and its neighbor, $\boldsymbol{Z}_1$, are L_1 units apart. Over Δt, a series of timesteps from 1 to k, the two points $\boldsymbol{Y}$ and $\boldsymbol{Z}$ evolve until their distance, L'_1, is greater than some arbitrarily small ε. The $\boldsymbol{Y}$ value at k becomes $\boldsymbol{Y}_2$ and a new nearest neighbor, $\boldsymbol{Z}_2$, is selected. This procedure continues until the fiducial trajectory reaches the end of the time series. The replacement of the old point by its substitute point and the replacement of the error direction by a new directional vector constitute a renormalization of errors along the trajectory. The largest Lyapunov exponent, λ_1, is then given by:

$$\lambda_1 = \frac{1}{N\Delta t}\sum_{j=1}^{M} \log_2 \frac{L'_j}{L_j} \tag{6.8}$$

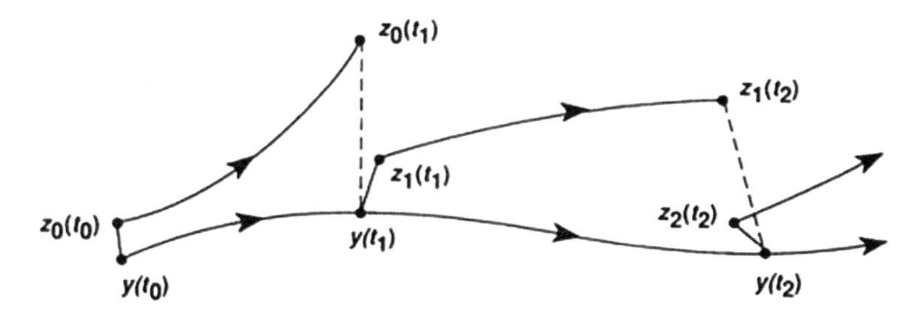

Fig. 6.4 Lyapunov exponent computation: Wolf et al. (1985) algorithm (*source* Vastano and Kostelich (1986))

where M is the number of replacement steps (where some arbitrarily small ε is exceeded) and N is the total number of timesteps that the fiducial trajectory progressed.

Although widely used, the algorithm by Wolf et al. (1985) has also been criticized on many grounds and its usefulness and efficiency questioned. An important concern is that the algorithm does not allow one to test for the presence of exponential divergence, but just assumes its existence and thus yields a finite exponent also for stochastic data, where the true exponent is infinite (Kantz and Scheriber 2004). Another concern is that the algorithm requires large amounts of data and long computing times.

In parallel with the algorithm by Wolf et al. (1985) and since then, many algorithms have been formulated for estimation of the Lyapunov exponents from a time series. The Eckmann–Ruelle algorithm (Eckmann and Ruelle 1985) offers some advantages over the Wolf et al. (1985) algorithm. It uses a least-squares approximation of the derivative matrices, an approach that was independently proposed by Sano and Sawada (1985). The algorithm also provides a useful way to calculate the entire range of Lyapunov exponent values, while the Wolf et al. (1985) algorithm will become substantially complicated if recovery of more than one positive Lyapunov exponent is attempted.

Furthermore, while the algorithm by Wolf et al. (1985) only uses a delay reconstruction of phase space, the algorithms by Sano and Sawada (1985) and Eckmann et al. (1986) involve the approximation of the underlying deterministic dynamics. This approach is highly efficient if the data allow for a good approximation of the dynamics. Another kind of algorithm has been introduced by Rosenstein et al. (1993) and Kantz (1994). It tests directly for the exponential divergence of nearby trajectories and, thus, facilitates decision on whether computation of a Lyapunov exponent for a time series really makes sense in the first place.

6.7 Kolmogorov Entropy Method

The Kolmogorov entropy, or K entropy, is the mean rate of information created by the system. It is important in characterizing the average predictability of a system of which it represents the sum of the positive Lyapunov exponents. The Kolmogorov entropy quantifies the average amount of new information on the system dynamics (or rate of uncoupling of correlations in phase space) brought by the measurement of a new value of the time series. In this sense, it measures the rate of information produced by the system. The value of $K = 0$ for periodic or quasi-periodic (i.e. completely predictable) time series; $K = \infty$ for white noise (i.e. unpredictable by definition); and $0 < K < \infty$ for chaotic system.

It is important to note that it is very difficult to directly estimate K entropy for a time series representing a system with greater than two degrees of freedom. Therefore, in the study of chaos, the only alternative is to estimate an approximation of the K entropy. This approximation is oftentimes the lower bound of the K entropy, which is the K_2 entropy. Grassberger and Procaccia (1983c) proposed the first algorithm for estimation of Kolmogorov entropy in the context of chaos analysis. Their algorithm uses the correlation sum or correlation integral, and the K_2 entropy is given as follows:

$$K_2(m) = \lim_{r \to 0}\left(\frac{1}{\tau}\{\log[C_m(r)] - \log[C_{m+1}(r)]\}\right) \tag{6.9}$$

and

$$K_2 = \lim_{m \to \infty}[K_2(m)] \tag{6.10}$$

where τ is the delay time, $C_m(r)$ is the value of $C(r)$ when the embedding dimension of phase space is m, $C_{m+1}(r)$ is the value of $C(r)$ when embedding dimension of phase space is $m + 1$. Again, the choice of τ and m is key to calculation of entropy. In practice, we need to consider dimension of embedding phase space as well as τ which has better simulating effect. The K_2 entropy and Kolmogorov entropy are thought to have the same qualitative behavior, with $K_2 = 0$ for periodic or quasi-periodic time series; $K_2 = \infty$ for white noise; and $0 < K_2 < \infty$ for chaotic system.

Since the correlation sum (or integral) is strongly affected by data size, data noise, and other factors, the estimation of Kolmogorov entropy is also often inaccurate. In view of this, a number of improvements have been made to obtain a better estimation of the Kolmogrov entropy. Important studies include Kantz and Schürmann (1996), Schürmann and Grassberger (1996), Diks (1999), Bonachela et al. (2008), and Jayawardena et al. (2010).

6.8 Surrogate Data Method

The surrogate data method (Theiler et al. 1992; Schreiber and Schmitz 1996) is actually not a chaos identification method, but rather a nonlinear detection method. It is essentially designed to answer such questions as follows: Is the apparent structure in the time series most likely due to nonlinearity or rather due to linear correlations? Is the irregularity (nonperiodicity) of the time series most likely due to nonlinear determinism or rather due to random inputs to the system or fluctuations in the parameters?

To address these, the surrogate data method makes use of the substitute data generated in accordance to the probabilistic structure underlying the original data. This means that the surrogate data possess some of the properties, such as the mean, the standard deviation, the cumulative distribution function, the power spectrum, etc., but are otherwise postulated as random, generated according to a specific null hypothesis. The null hypothesis here consists of a candidate linear process, and the goal is to reject the hypothesis that the original data have come from a linear stochastic process. The rejection of the null hypothesis can be made based on some discriminating statistics. If the discriminating statistics obtained for the surrogate data are significantly different from those of the original time series, then the null hypothesis can be rejected, and the original time series may be considered to have come from a nonlinear process. On the other hand, if the discriminating statistics obtained for the original data and surrogate data are not significantly different, then the null hypothesis cannot be rejected, and the original time series is considered to have come from a linear stochastic process.

The utility and effectiveness of the surrogate data method for nonlinearity detection critically depend on the following: (1) specification of null hypothesis and generation of surrogate data; (2) computation of discrimination statistics; and (3) calculation of measure of significance.

Some of the possibilities that can be used for specifying null hypothesis and generating surrogate data are temporally independent data, Ornstein–Uhlenbeck process, linearly autocorrelated Gaussian noise, static nonlinear transform of linear Gaussian noise, and nonlinear rescaling of a linear Gaussian process (e.g. Theiler et al. 1992; Schreiber and Schmitz 1996).

Once we have a number of surrogate data sets, the next step is to compute a discriminating statistic. In chaos studies, since the primary interest is to identify chaos in the time series, it would be desirable to use any of the statistics of 'invariants' used for chaos identification, such as correlation dimension, Lyapunov exponent, Kolmogorov entropy, and even prediction accuracy.

Let Q_{org} denote the statistic computed for the original time series and Q_{si} for the ith surrogate series generated under the null hypothesis. Let μ_s and σ_s denote, respectively, the mean and standard deviation of the distribution of Q_s. Then the measure of significance S is given by

$$S = \frac{|Q_{org} - \mu_s|}{\sigma_s} \tag{6.11}$$

An S value of ~ 2 cannot be considered very significant, whereas an S value of ~ 10 is highly significant (Theiler et al. 1992). The error on the significance value ΔS is estimated by:

$$\Delta S = \sqrt{(1 + 2\sigma_s^2/n)} \tag{6.12}$$

where n is the number of surrogate data sets.

Since its initial proposal by Theiler et al. (1992), numerous studies have examined the effectiveness of the surrogate data generation algorithms, nonlinear discriminating statistics, and related issues. Such studies have also offered important improvements to the surrogate data method and its application. Extensive details of these studies are available in Schreiber and Schmitz (1996, 1997, 2000), Schreiber (1998), Timmer (1998), Dolan and Spano (2001), Kugiumtzis (2001),

Engbert (2002), Breakspear et al. (2003), Keylock (2007), and Suzuki et al. (2007), among others. Prichard and Theiler (1994) proposed an extension to multivariate time series of the phase-randomized Fourier-transform algorithm for generating surrogate data.

6.9 Poincaré Maps

A Poincaré map (Poincaré 1899) is simply a 'slice' through the attractor in the state space, and is the intersection of a periodic orbit of a continuous dynamic system with a certain lower-dimensional sub-space transversal to the flow of the system. Therefore, a Poincaré map can be interpreted as a discrete dynamic system with a state space that is one dimension smaller than the original continuous dynamic system, but still providing all the information about the original system at specific moments in its evolution. Since it preserves many properties of periodic and quasi-periodic orbits of the original system and has a lower-dimensional state space, the Poincaré map is often used for analyzing the original system.

If the system evolution is n-periodic whose trajectory has a dimension one, then the Poincaré map consists of n points (dots) repeating indefinitely in the same order with a dimension zero. If the evolution is quasi-periodic whose trajectory has a dimension two (torus), then the Poincaré map consists of points that defines a closed limit cycle (dimension one). If the evolution is chaotic, then the Poincaré map is a collection of points that show interesting patterns with no obvious repetition, but often revealing the fractal nature of the underlying attractor.

The process of obtaining a Poincaré map corresponds to sampling the state of the system occasionally instead of continuously. In many cases, the appropriate sampling interval can be defined so that it corresponds to a physically meaningful measure of the dynamic system. For example, for a periodically forced oscillator, we may 'mark' the trajectory at times that are multiple integers of the forcing period. Then a sequence of strictly comparable points is accumulated.

6.10 Close Returns Plot

The close returns plot (CRP) (Gilmore 1993) is a topological method, and can be used to identify the presence of chaos in a time series without having to construct an embedding of the attractor. The CRP (or any other topology-based method) has several important advantages over the metric methods (e.g. correlation dimension, Lyapunov exponent, false nearest neighbor) in identifying and analyzing chaos in a time series: (1) it is reliable when applied to small data sets; (2) it is robust against noise; (3) it can be implemented with little computational effort; and (4) since it preserves the time-ordering of the data, it is able to provide additional information about the underlying system generating chaotic behavior.

A key feature of strange attractors (i.e. chaos) is that they are filled with unstable periodic orbits (of which unstable fixed points are a special case), i.e. they are 'dense.' Therefore, once a trajectory (i.e. system state) enters the neighborhood of one of these unstable orbits, it will remain near that orbit for a certain period of time until it diverges from that orbit by the exponential divergence. Because of the recurrence of trajectories and the bounded nature of the strange attractor, the system will eventually revisit the neighborhood of a previously visited location, i.e. region of 'close return.' This, however, is not the case in stochastic systems, where the trajectories move simply randomly and also do not necessarily have bounds even considering a short period of time. The topological approach essentially analyzes the way in which the mechanisms responsible for diverging and converging (i.e. stretching and compressing) the (strange) attractor act on the unstable periodic orbits, interwinding them in a very specific way.

The detection of the 'close returns' in a data set may be made by first computing the differences $|X_i - X_{i+t}|$ for all the values in the data set, then determining their extent of closeness, with respect to a particular distance, say, r (as is the case in the correlation dimension method, but only between individual data values, rather than between vectors), and finally plotting it on a color-coded graph. In the graph, the horizontal axis indicates the observation number i, where $i = 1, 2, \ldots, N$, and the vertical axis is designated as t, where $t = 1, 2, \ldots, N - i$. If the difference is less than r, it is coded black; if it is larger than r, then it is coded white. Close returns (i.e. intervals of periodicity) in the data set are indicated by horizontal line segments, which are key to identify chaos (or distinguishing chaos from other types of behavior). Periodic time series exhibit almost equally-spaced long horizontal (solid) black line segments running across the entire graph. Quasi-periodic time series also exhibit line segments, but patterns resembling a contour map. Stochastic time series exhibit a generally uniform array of black dots (without horizontal line segments). Chaotic time series exhibit a number of short horizontal line segments.

The appropriate range of r for a given time series can be identified as follows, similar to the one done in the case of correlation dimension method to compute the correlation function (sum). First, compute the maximum difference between any two observations in the set. Next, set r at some small fraction of that difference, and construct the close returns plot. If r is set too small, there will be an insufficient number of black points to identify a pattern that characterizes the data; as r is reduced below the standard deviation of additive noise, the pattern degrades. If r is too large, then the patterns will be hidden. Once an appropriate range for r is identified, there will be a sufficient array of points to allow determination of the type of pattern generated by the data. The level of r can be varied within that range without altering the qualitative nature of the pattern. The exact specification of r is, in most cases, *not* critical to the interpretation of the behavior of the data.

6.11 Nonlinear Prediction

An important purpose of identification of chaos in a time series is to attempt more reliable short-term predictions than those possible with other methods (e.g. stochastic methods), since the deterministic nature of chaotic systems allows reliable short-term predictions, although their sensitive dependence on initial conditions preclude long-term ones. An early method for prediction of chaotic time series was proposed by Farmer and Sidorowich (1987). Since then, there have been many advances (e.g. Casdagli 1989, 1992; Sugihara and May 1990; Tsonis and Elsner 1992; Abarbanel 1996), not only for prediction but also for chaos identification using the prediction results. All these methods are essentially based on 'local approximation,' and are generally put under the umbrella of 'nonlinear prediction' (NLP) methods.

It is recommended, and it is indeed a normal practice, to apply these nonlinear prediction methods only after chaos is identified in a time series using any of the chaos identification methods explained above, i.e. to use it mainly as a prediction tool. However, identification of chaos is not a necessary condition for the application of the prediction methods, since they can also help identify chaos in a time series through interpretation of the prediction results. This process is also termed as the "inverse approach." The inverse approach seems more reliable than the correlation dimension (or any other invariant) method for chaos identification, since it is essentially based on predictions, which is one of the primary purposes behind characterizing a system (stochastic or chaotic) towards appropriate model selection, among others. This is particularly the case when prediction accuracy against lead time is used as an indicator (e.g. Sugihara and May 1990), since a rapid decrease in prediction accuracy with increase in lead time is a typical characteristic of chaotic systems, i.e. sensitivity to initial conditions.

6.11.1 *Local Approximation Prediction*

Similar to the metric methods for chaos identification, explained above, the first step in chaos prediction method is phase space reconstruction of a time series according to Eq. (6.3) to reliably represent the underlying dynamics in the form of an m-dimensional map f_T (Eq. (6.4)). The problem then is to find an appropriate expression for f_T (e.g. F_T) to predict the future. There are several possible approaches for determining F_T. One promising approach is the "local approximation method" (Farmer and Sidorowich 1987), which uses only nearby states to make prediction. The basic idea in the local approximation method is to break up the domain F_T into local neighborhoods and fit parameters in each neighborhood separately. In this way, the underlying system dynamics are represented step by step locally in the phase space. To predict X_{j+T} based on $\boldsymbol{Y}_j$ (an m-dimensional vector) and past history, k nearest neighbors of $\boldsymbol{Y}_j$ are found on the basis of the minimum values of $\|Y_j - Y_j'\|$, with $j' < j$. If only one such neighbor is considered, the prediction of X_{j+T} would be X'_{j+T}. For k number of neighbors, the prediction of X_{j+T} could be taken as an average of the k values of X'_{j+T}. The value of k is determined by trial and error. It is also relevant to note that $k = 1$ for $m = 1$ is a very limited case of prediction, and is similar to the 'method of analogues,' originally proposed by Lorenz (1969). In general, however, both k and m are varied to find out the optimum predictions.

The prediction accuracy can be evaluated using a variety of measures, such as correlation coefficient, root mean square error, and coefficient of efficiency. In addition, direct time series plots and scatter diagrams can also be used to choose the best prediction results among a large combination of results achieved with different embedding dimensions (m) and number of neighbors (k), and also different delay times (τ).

Since the initial proposal by Farmer and Sidorowich (1987), numerous studies have contributed to the prediction of chaotic time series and also proposed significant improvements. These improvements have come in various forms, including local weighting, simplex method, local polynomials, and deterministic versus stochastic algorithm (e.g. Casdagli 1989, 1992; Sugihara and May 1990; Elsner and Tsonis 1992; Tsonis 1992; Tsonis and Elsner 1992; Abarbanel 1996). Which of these forms is the best for predictions is a difficult question to answer, as the results often strongly depend on the time series under investigation and also other considerations. For instance, the local averaging technique has an important advantage over the local polynomial technique in terms of computational efficiency, but the local polynomial technique seems to provide more accurate results. Many studies have also attempted to optimize the parameters involved in the prediction method (e.g. Phoon et al. 2002; Jayawardena et al. 2002).

6.11.2 Inverse Approach to Chaos Identification

While the local approximation method primarily serves as a prediction tool, the prediction results can also be used to distinguish between deterministic and stochastic systems, i.e. an inverse way to identify chaos. In general, a high prediction accuracy from the local approximation method may be considered as an indication of a deterministic system, whereas a low prediction accuracy is expected if the dynamics are stochastic. However, this is not a very effective way, since the predictions are strongly influenced by the two main parameters involved (m and k) as well as the lead time (T) considered. It should be noted that while the parameter m may be fixed a priori based on reliable available information (e.g. results from correlation dimension and false nearest neighbor methods), there is no reliable way to fix the parameter k in advance. Furthermore, even if m is reliably known a priori, it would still be helpful to make the predictions for different

m values, so that the optimum m can be compared and cross-verified against the dimension obtained from the other methods.

With results from the nonlinear prediction method at hand, the following guidelines may be adopted to distinguish between stochastic and chaotic systems:

1. Embedding dimension (m): If the time series exhibits deterministic chaos, then the prediction accuracy would increase to its best with the increase in the embedding dimension up to a certain point (low value of m), called the optimal embedding dimension (m_{opt}), and would remain close to its best for embedding dimensions higher than m_{opt}. For stochastic time series, there would be no increase in the prediction accuracy with an increase in the embedding dimension and the accuracy would remain the same for any value of m (e.g. Casdagli 1989);
2. Neighbors (k): Smaller number of neighbors would give the best predictions if the system dynamics are deterministic, whereas for stochastic systems the best predictions are achieved when the number of neighbors is large. This approach is also called the deterministic versus stochastic (DVS) algorithm (Casdagli 1992). The idea behind this approach is that since small k represents local models, it would be more appropriate for deterministic systems, while global models (large k) would be more appropriate for stochastic systems. However, if the best prediction is obtained using neither deterministic nor stochastic models but intermediate models (i.e. intermediate number of neighbors), then such a condition can be taken as an indication of chaotic behavior with some amount of noise in the data or chaos of moderate dimension (e.g. Casdagli 1992); and
3. Lead time (T): For a given embedding dimension and for a given number of neighbors, predictions in deterministic systems deteriorate considerably faster than in stochastic systems when the lead time is increased. This is due to the sensitivity of deterministic chaotic systems to initial conditions (Sugihara and May 1990).

6.12 Summary

Since the early 1980s, a number of methods have been developed for identification and prediction of chaos in time series and applied in many different scientific fields. Such methods are based on different concepts and aim to identify different measures of chaos, including dimensionality, entropy, and predictability. This chapter has described many popular methods for chaos identification and prediction, including through an example to demonstrate the utility of phase space reconstruction and correlation dimension methods for distinguishing between chaotic and stochastic time series. Although these methods generally work well, there are concerns regarding their appropriateness and reliability for real time series. These concerns are associated with the selection of parameters involved in the methods and the limitations of real time series. Chapter 7 offers a detailed discussion on some of these issues.

References

Abarbanel HDI (1996) Analysis of observed chaotic data. Springer-Verlag, New York, USA
Abarbanel HDI, Brown R, Kadtke JB (1990) Prediction in chaotic nonlinear systems: methods for time series with broadband Fourier spectra. Phys Rev A 41(4):1782–1807
Aittokallio T, Gyllenberg M, Hietarinta J, Kuusela T, Multamäki T (1999) Improving the false nearest neighbors method with graphical analysis. Phys Rev 60:416–421
Bonachela JA, Hinrichsen H, Munoz M (2008) Entropy estimations of small data set. J Phys A: Math Theor 41:202001
Breakspear M, Brammer M, Robinson PA (2003) Construction of multivariate surrogate sets from nonlinear data using the wavelet transform. Physica D 182:1–22
Broomhead DS, King GP (1986) Extracting qualitative dynamics from experimental data. Physica D 20:217–236
Cao L, Mees A, Judd K (1998) Dynamics from multivariate time series. Physica D 121:75–88
Carona A (2000) Adaptive box-assisted algorithm for correlation-dimension estimation. Phys Rev E 62:7872–7881
Casdagli M (1989) Nonlinear prediction of chaotic time series. Physica D 35:335–356
Casdagli M (1992) Chaos and deterministic versus stochastic nonlinear modeling. J Royal Stat Soc B 54(2):303–328
Diks C (1999) Nonlinear time series analysis, methods and applications. World Sci, Singapore
Dolan K, Spano ML (2001) Surrogates for nonlinear time series analysis. Phys Rev E 64:046128–1–046128-6
Eckmann JP, Ruelle D (1985) Ergodic theory of chaos and strange attractors. Rev Mod Phys 57 (3):617–656
Eckmann JP, Kamphorst SO, Ruelle D, Ciliberto S (1986) Lyapunov exponents from a time series. Phys Rev A 34:4971–4979
Elsner JB, Tsonis AA (1992) Nonlinear prediction, chaos and noise. Bull Amer Meteor Soc 73:49–60
Elsner JB, Tsonis AA (1996) Singular spectrum analysis: a new tool in time series analysis. Plenum Press, New York
Engbert R (2002) Testing for nonlinearity: the role of surrogate data. Chaos Soliton Fract 13 (1):79–84
Farmer DJ, Sidorowich JJ (1987) Predicting chaotic time series. Phys Rev Lett 59:845–848
Fraedrich K (1986) Estimating the dimensions of weather and climate attractors. J Atmos Sci 43:419–432
Fredkin DR, Rice JA (1995) Method of false nearest neighbors: a cautionary note. Phys Rev E 51 (4):2950–2954
Gilmore CG (1993) A new test for chaos. J Econ Behav Organ 22:209–237
Grassberger P (1983) Generalized dimension of strange attractors. Phys Lett A 97(6):227–230
Grassberger P (1990) An optimized box-assisted algorithm for fractal dimensions. Phys Lett A 148:63–68
Grassberger P, Procaccia I (1983a) Measuring the strangeness of strange attractors. Physica D 9:189–208
Grassberger P, Procaccia I (1983b) Characterisation of strange attractors. Phys Rev Lett 50 (5):346–349
Grassberger P, Procaccia I (1983c) Estimation of the Kolmogorov entropy from a chaotic signal. Phys Rev A 28:2591–2593
Hausdorff F (1919) Dimension und äußeres Maß. Math. Annalen 79(1–2):157–179
Havstad JW, Ehlers CL (1989) Attractor dimension of nonstationary dynamical systems from small data sets. Phys Rev A 39(2):845–853
Hediger T, Passamante A, Ferrell ME (1990) Characterizing attractors using local intrinsic dimensions calculated by singular-value decomposition and information-theoretic criteria. Phys Rev A 41:5325–5332

Hegger R, Kantz H (1999) Improved false nearest neighbor method to detect determinism in time series data. Phys Rev E 60(4):4970–4973
Henon M (1976) A two-dimensional mapping with a strange attractor. Commun Math Phys 50:69–77
Jayawardena AW, Li WK, Xu P (2002) Neighborhood selection for local modeling and prediction of hydrological time series. J Hydrol 258:40–57
Jayawardena AW, Xu P, Li WK (2010) Modified correlation entropy estimation for a noisy chaotic time series. Chaos 20:023104–1–023104-11
Kantz H (1994) A robust method to estimate the maximal Lyapunov exponent from a time series. Phys Lett A 185:77–87
Kantz H, Schreiber T (2004) Nonlinear time series analysis. Cambridge University Press, Cambridge
Kantz H, Schürmann T (1996) Enlarged scaling ranges for the KS-entropy and information dimension. Chaos 6:167–171
Kaplan JL, Yorke JA (1979) Chaotic behavior of multidimensional difference equations. In: Peitgen HO, Walther HO (eds) Functional differential equations and approximations of fixed points, Lecture Notes in Mathematics 730, Springer-Verlag, Berlin, pp 204–227
Kennel MB, Abarbanel HDI (2002) False nearest neighbors and false strands: a reliable minimum embedding dimension algorithm. Phys Rev E 66(026209):1–19
Kennel MB, Brown R, Abarbanel HDI (1992) Determining embedding dimension for phase space reconstruction using a geometric method. Phys Rev A 45:3403–3411
Keylock CJ (2007) A wavelet-based method for surrogate data generation. Physica D 225(2):219–228
Kugiumtzis (2001) On the reliability of the surrogate data test for nonlinearity in the analysis of noisy time series. Int J Bifurcation Chaos 11(7):1881–1896
Lorenz EN (1969) Atmospheric predictability as revealed by naturally occurring analogues. J Atmos Sci 26:636–646
Nerenberg MAH, Essex C (1990) Correlation dimension and systematic geometric effects. Phys Rev A 42(12):7065–7074
Osborne AR, Provenzale A (1989) Finite correlation dimension for stochastic systems with power-law spectra. Physica D 35:357–381
Packard NH, Crutchfield JP, Farmer JD, Shaw RS (1980) Geometry from a time series. Phys Rev Lett 45(9):712–716
Phoon KK, Islam MN, Liaw CY, Liong SY (2002) A practical inverse approach for forecasting of nonlinear time series. J Hydrol Engg 7(2):116–128
Poincaré H (1899) Les méthodes nouvelles de la méchanique céleste. Gauthier-Villars, Paris
Prichard D, Theiler J (1994) Generating surrogate data for time series with several simultaneously measured variables. Phys Rev Lett 73:951–954
Ramdani S, Bouchara F, Casties J-F (2007) Detecting determinism in short time series using a quantified averaged false nearest neighbors approach. Phys Rev E 76(3):036204, 1–14
Rényi A (1971) Probability theory. North Holland, Amsterdam
Rhodes C, Morari M (1997) False-nearest-neighbor algorithm and noise-corrupted time series. Phys Rev E 55:6162–6170
Rosenstein MT, Collins JJ, De Luca CJ (1993) A practical method for calculating largest Lyapunov exponents from small data sets. Physica D 65:117–134
Ruelle D (1980) Les attracteurs etranges. La Rech 108:132–144
Ruelle D (1983) Five turbulent problems. Physica D 7:40–44
Ruelle D (1989) Chaotic evolution and strange attractors. Cambridge University Press, New York
Russell DA, Hanson JD, Ott E (1980) Dimension of strange attractors. Phys Rev Lett 45: 1175–1178
Sangoyomi TB, Lall U, Abarbanel HDI (1996) Nonlinear dynamics of the Great Salt Lake: dimension estimation. Water Resour Res 32(1):149–159
Sano M, Sawada Y (1985) Measurement of the Lyapunov spectrum from a chaotic time series. Phys Rev Lett 55:1082–1085

Sauer T, Yorke J, Casdagli M (1991) Embeddology. J Stat Phys 65:579–616
Schreiber T (1995) Efficient neighbor searching in nonlinear time series analysis. Int J Bifurcat Chaos 5:349–358
Schreiber T (1998) Constrained randomization of time series data. Phys Rev Lett 80(10): 2105–2108
Schreiber T, Kantz H (1996) Observing and predicting chaotic signals: is 2% noise too much? In: Kadtke JB (ed) Kravtsov YuA. Predictability of complex dynamical systems, Springer Series in Synergetics, Springer, Berlin, pp 43–65
Schreiber T, Schmitz A (1996) Improved surrogate data for nonlinearity tests. Phys Rev Lett 77 (4):635–638
Schreiber T, Schmitz A (1997) Discrimination power of measures for nonlinearity in a time series. Phys Rev E 55(5):5443–5447
Schreiber T, Schmitz A (2000) Surrogate time series. Physica D 142(3–4):346–382
Schürmann T, Grassberger P (1996) Entropy estimation of symbol sequences. Chaos 6:414–427
Shaw RS (1981) Strange attractors, chaotic behavior, and information flow. Z Natürforsch A 36:80–112
Sivakumar B (2000) Chaos theory in hydrology: important issues and interpretations. J Hydrol 227 (1–4):1–20
Sivakumar B (2001) Rainfall dynamics at different temporal scales: a chaotic perspective. Hydrol Earth Syst Sci 5(4):645–651
Sivakumar B (2005) Correlation dimension estimation of hydrologic series and data size requirement: myth and reality. Hydrol Sci J 50(4):591–604
Sivakumar B, Phoon KK, Liong SY, Liaw CY (1999) A systematic approach to noise reduction in chaotic hydrological time series. J Hydrol 219(3–4):103–135
Sivakumar B, Berndtsson R, Olsson J, Jinno K (2002a) Reply to 'which chaos in the rainfall-runoff process?' by Schertzer et al. Hydrol Sci J 47(1):149–158
Sivakumar B, Persson M, Berndtsson R, Uvo CB (2002b) Is correlation dimension a reliable indicator of low-dimensional chaos in short hydrological time series? Water Resour Res 38(2). doi:10.1029/2001WR000333
Sivakumar B, Jayawardena AW, Li WK (2007) Hydrologic complexity and classification: a simple1433 data reconstruction approach. Hydrol Process 21(20):2713–2728
Sugihara G, May RM (1990) Nonlinear forecasting as a way of distinguishing chaos from measurement error in time series. Nature 344:734–741
Suzuki T, Ikeguchi T, Suzuki M (2007) Algorithms for generating surrogate data for sparsely quantized time series. Physica D 231(2):108–115
Takens F (1981) Detecting strange attractors in turbulence. In: Rand DA, Young LS (eds) Dynamical systems and turbulence, vol 898., Lecture Notes in MathematicsSpringer-Verlag, Berlin, Germany, pp 366–381
Theiler J (1987) Efficient algorithm for estimating the correlation dimension from a set of discrete points. Phys Rev A 36:4456–4462
Theiler J (1990) Estimating fractal dimension. J Opt Soc Am A: 7:1055–1073
Theiler J, Eubank S, Longtin A, Galdrikian B, Farmer JD (1992) Testing for nonlinearity in time series: the method of surrogate data. Physica D 58:77–94
Timmer J (1998) Power of surrogate data testing with respect to nonstationarity. Phys Rev E 58 (4):5153–5156
Toledo E, Toledo S, Almog Y, Akselrod S (1997) A vectorized algorithm for correlation dimension estimation. Phys Lett A 229(6):375–378
Tsonis AA (1992) Chaos: from theory to applications. Plenum Press, New York
Tsonis AA, Elsner JB (1992) Nonlinear prediction as a way of distinguishing chaos from random fractal sequences. Nature 358:217–220
Tsonis AA, Elsner JB, Georgakakos KP (1993) Estimating the dimension of weather and climate attractors: important issues about the procedure and interpretation. J Atmos Sci 50:2549–2555

Tsonis AA, Triantafyllou GN, Elsner JB, Holdzkom JJ II, Kirwan AD Jr (1994) An investigation on the ability of nonlinear methods toinfer dynamics from observables. Bull Amer Meteor Soc 75:1623–1633

Vastano JA, Kostelich EJ (1986) Comparison of algorithms for determining Lyapunov exponents from experimental data. In: Mayer-Kress G (ed) Dimenions and entropies in chaotic systems. Springer-Verlag, Berlin

Vautard R, Yiou P, Ghil M (1992) Singular-spectrum analysis: a toolkit for short, noisy chaotic signals. Physica D 58:95–126

Whitney H (1936) Differentiable manifolds. Ann Math 37:645–680

Wolf A, Swift JB, Swinney HL, Vastano A (1985) Determining Lyapunov exponents from a time series. Physica D 16:285–317

Young LS (1982) Dimension, entropy, and Lyapunov exponents. Ergodic Theory Dyn Syst 2:109–124

Young LS (1984) Dimension, entropy, and Lyapunov exponents in differentiable dynamical systems. Phys A 124:639–645

Chapter 7
Issues in Chaos Identification and Prediction

Abstract The existing methods for identification and prediction of chaos are generally based on the assumptions that the time series is infinite (or very long) and noise-free. There are also no clear-cut guidelines on the selection of parameters involved in the methods, especially in phase space reconstruction. Since data observed from real systems, such as hydrologic systems, are finite and often short and are always contaminated with noise (e.g. measurement error), there are concerns on the applications of chaos concepts and methods to real systems. Adding to this are complications that potentially arise due to issues that are specific to certain real systems, such as a large number of zeros in rainfall, runoff, and other hydrologic data. Therefore, it is important to study the issues related to methods and data that can potentially influence the outcomes of chaos studies. This chapter addresses four important issues in the applications of chaos methods to real time series, especially those that have particular relevance and gained considerable interest in hydrology: selection of delay time in phase space reconstruction, minimum data size required for correlation dimension estimation, influence of data noise, and influence of the presence of a large number of zeros in the data. Some specific examples are also presented in addressing the issues of data size and data noise.

7.1 Introduction

Since the initial development of methods for analysis of chaos in time series in the early 1980s (e.g. Packard et al. 1980; Takens 1981; Grassberger and Procaccia 1983a, b, c; Wolf et al. 1985), chaos theory has seen numerous theoretical and methodological advances as well as widespread practical applications in many different fields; see Chap. 5, Sect. 5.3 for some important references. However, the reliability of chaos theory-based methods, including those for phase space reconstruction, chaos identification, and chaos prediction, for application to real complex systems, such as hydrologic systems, has been under considerable debate. This is mainly due to some of the basic assumptions in the development of such methods and, therefore, their potential limitations when applied to real systems. For instance,

B. Sivakumar, *Chaos in Hydrology*, DOI 10.1007/978-90-481-2552-4_7

chaos theory-based methods inherently assume that the time series is infinite and noise-free, but real time series are always finite and often contaminated by noise. Much of the criticism on chaos analysis, however, has been directed at the correlation dimension method, and in particular the Grassberger–Procaccia algorithm (Grassberger and Procaccia 1983a, b) for dimension estimation. Part of the criticisms has been due to a number of studies carried out on data requirements for dimension estimation (e.g. Smith 1988; Havstad and Ehlers 1989; Nerenberg and Essex 1990), but the fact that the correlation dimension method has been the most widely used method for chaos detection has also contributed to such criticisms.

Among the important issues associated with chaos methods are: lack of information on the selection of parameters involved in the methods (e.g. delay time, embedding dimension, number of neighbors); and system characteristics and data constraints (e.g. scale, correlation, data size, data noise, presence of zeros). There is a plethora of literature, both in the nonlinear science field and in others (including hydrology), on these issues (e.g. Fraser and Swinney 1986; Holzfuss and Mayer-Kress 1986; Havstad and Ehlers 1989; Osborne and Provenzale 1989; Nerenberg and Essex 1990; Grassberger et al. 1991; Tsonis et al. 1993, 1994; Schreiber and Kantz 1996; Sivakumar 2000, 2001, 2005a; Sivakumar et al. 2002a, b). This chapter addresses a few of the major issues. As different issues may influence different methods in different ways, it is near impossible to offer a detailed account of each and every issue. Therefore, the discussion that follows focuses on issues that have significant influences on one or across a range of methods. Further, since the correlation dimension method has been the most criticized method, issues associated with this method are given more importance.

7.2 Delay Time

As explained in Chaps. 5 and 6, state space or phase space (e.g. Packard et al. 1980) is a very useful concept for representing the evolution of dynamic systems. Indeed, phase space reconstruction is a basic and necessary first step in almost all chaos identification and prediction methods. An appropriately constructed phase space is also crucial for a reliable estimation of invariants for chaos identification and also for reliable predictions. Delay embedding techniques (e.g. Takens 1981) have been widely used for the reconstruction of the phase space, although several other approaches also exist; see Sauer et al. (1991) for a more general account of embedology. The delay embedding techniques involve the use of a delay time, τ.

An appropriate delay time, τ, for the reconstruction of the phase space is necessary, because only an optimum τ gives the best separation of neighboring trajectories within the minimum embedding phase space (e.g. Frison 1994), whereas an inappropriate τ may lead to unreliable outcomes (e.g. underestimation or overestimation of invariants). For example, if τ is too small, then there is little new information contained in each subsequent datum and the reconstructed attractor is compressed along the identity line. This situation is termed as *redundance*

(Casdagli et al. 1991), and the result of which is an inaccurate estimation of the invariants, such as an underestimation of the correlation dimension (e.g. Havstad and Ehlers 1989). On the other hand, if τ is too large, and the dynamics happen to be chaotic, then all relevant information for the phase space reconstruction is lost, since neighboring trajectories diverge, and averaging in time and/or space is no longer useful (Sangoyomi et al. 1996). This situation is termed as *irrelevance* (Casdagli et al. 1991), and this may result in an inaccurate estimation of the invariants, such as an overestimation of the correlation dimension.

The obvious question now is: how to identify an appropriate delay time? Perhaps, the basic requirement in identifying a suitable τ is to make sure that successive data in the reconstructed vectors are not correlated. In other words, the choice of τ should be in terms of the decorrelation time of the time series under investigation. The question that comes next is: how to define the decorrelation time? Many researchers have addressed the selection of the decorrelation time and, hence, the selection of an appropriate τ for phase space reconstruction. A brief review of such studies is presented next.

7.2.1 Delay Time Selection

Since the proposal of the delay embedding theorem of Takens (Takens 1981) for chaos analysis, several methods and guidelines have been proposed for the selection of an appropriate delay time. These approaches are based on series correlation (e.g. autocorrelation, mutual information, high-order correlations), phase space extension (e.g. fill factor, wavering product, average displacement), and multiple autocorrelation and non-bias multiple autocorrelation (e.g. Fraser and Swinney 1986; Holzfuss and Mayer-Kress 1986; Liebert and Schuster 1989; Albano et al. 1991; Pfister and Buzug 1992a, b; Kembe and Fowler 1993; Rosenstein et al. 1994; Judd and Mees 1998; Lin et al. 1999). Three of the well-known methods among these are briefly discussed here: the autocorrelation function method, the mutual information method, and the correlation integral method.

The autocorrelation function method is the most commonly used method for delay time selection, for at least two reasons: (1) its computation is relatively simple; and (2) it is one of the most fundamental and standard statistical tools in any time series analysis. For a discrete time series, X_i, the autocorrelation function, r_τ, is determined according to:

$$r_\tau = \frac{\sum_{i=1}^{N-\tau} X_i X_{i+\tau} - \frac{1}{N-\tau}\sum_{i=1}^{N-\tau} X_{i+\tau}\sum_{i=1}^{N-\tau} X_i}{\left[\sum_{i=1}^{N-\tau} X_i^2 - \frac{1}{N-\tau}\left(\sum_{i=1}^{N-\tau} X_i\right)^2\right]^{1/2}\left[\sum_{i=1}^{N-\tau} X_{i+\tau}^2 - \frac{1}{N-\tau}\left(\sum_{i=1}^{N-\tau} X_{i+\tau}\right)^2\right]^{1/2}} \quad (7.1)$$

Within the autocorrelation function method, there are several guidelines for the selection of τ. For instance, Holzfuss and Mayer-Kress (1986) recommended using a value of lag time (or index lag) at which the autocorrelation function first crosses the zero line. Schuster (1988) suggested the use of the lag time at which the autocorrelation function attains 0.5, while Tsonis and Elsner (1988) suggested the selection of the lag time at which the autocorrelation function crosses 0.1.

Despite its widespread use, the appropriateness of the autocorrelation function method for the selection of τ has been seriously questioned. For example, Fraser and Swinney (1986) pointed out that the autocorrelation function method measures only the linear dependence between successive points and, thus, may not be appropriate for nonlinear dynamics. They suggested the use of the local minimum of the mutual information, which measures the general dependence, not just the linear dependence, between successive points. They reasoned that if τ is chosen to coincide with the first minimum of the mutual information, then the recovered state vector would consist of components that possess minimal mutual information between them, i.e., the successive values in the time series are statistically independent but (also) without any redundancy. For a discrete time series, with X_i and $X_{i-\tau}$ as successive values, for instance, the mutual information function, I_τ, is computed according to:

$$I_\tau = \sum_{i,i-\tau} P(X_i, X_{i-\tau}) \log_2 \left[\frac{P(X_i, X_{i-\tau})}{P(X_i)P(X_{i-\tau})} \right] \quad (7.2)$$

where $P(X_i)$ and $P(X_{i-\tau})$ are the individual probabilities of X_i and $X_{i-\tau}$, respectively, and $P(X_i, X_{i-\tau})$ is the joint probability density. The mutual information method is a more comprehensive method of determining proper delay time values (e.g. Tsonis 1992). However, the method has the disadvantage of requiring a large number of data, unless the dimension is small, and is computationally cumbersome.

An approach that is somewhat similar to the mutual information method but is based on the generalized correlation integral, known as the correlation integral method, to determine the delay time was proposed by Liebert and Schuster (1989). According to this method, the first minimum of the logarithm of the generalized correlation integral, $C(\tau, r, m)$, is considered to provide a proper choice of τ. For some radius r and embedding dimension m, one can calculate the correlation integral $C(r)$ as a function of τ. The logarithm of $C(\tau, r, m)$ is a measure of the averaged information content in the reconstructed vectors, and thus its minimum provides an easy way to define a proper τ.

For some attractors, it may not really matter which method is used for the selection of τ. For example, when applied to the Rössler system (Rössler 1976), the autocorrelation function, the mutual information, and the correlation integral methods all provide a value of τ approximately equal to one-fourth of the mean orbital period (Tsonis 1992). However, for some other attractors, the estimation of τ might depend strongly on the approach employed. An obvious way to have more confidence in the selection of τ may be to use different methods and check the consistency of the resulting τ values. However, this procedure may result in complications if different methods yield different τ values, and more so if the τ values are significantly different. Evidently, none of the aforementioned methods or rules has emerged as definitive for choosing τ. However, according to Tsonis (1992), the mutual information method is more comprehensive than the others and, therefore, may have an edge.

7.2.2 Delay Window Selection

A reliable alternative to address the issue of delay time selection is to try to fix the delay time window $\tau_w = \tau(m - 1)$, rather than just the delay time itself, since the delay time window is the one that is of actual interest at the end to represent the dynamics. An early attempt in this regard was made by Martinerie et al. (1992). Comparing the delay time window and delay times estimated using the autocorrelation function and mutual information methods, Martinerie et al. (1992) did not observe a consistent agreement between them. This is because, τ_w is basically the optimal time for representing the independence of the data, whereas autocorrelation function and mutual information methods determine only the first local optimal times in their estimation of τ. Kugiumtzis (1996) put emphasis on the relation between τ_w and dynamics of the underlying chaotic system and suggested to set $\tau_w > \tau_p$, the mean orbital period, with τ_p approximated from the oscillations of the time series.

Kim et al. (1998), through their analysis of time series generated from the Lorenz system, the Rabinovich–Fabrikant system, and the three-torus, showed that with an increase in the embedding dimension, the correlation dimension converges more rapidly for the case of τ_w held fixed than for the case of τ held fixed. Their study also revealed that such an outcome is especially the case for small data sets. Based on this distinction between τ and τ_w, Kim et al. (1999) subsequently developed a new technique to estimate both τ and τ_w. This technique, called the C–C method uses the Brock–Dechert–Scheinkman (BDS) statistic (Brock et al. 1991, 1996), which has its base on the correlation integral for testing nonlinearity in a time series. The main difference between the BDS method and the C–C method is in the inclusion of an additional parameter in the form of the lag time in the latter. The BDS method uses the statistic $S(m, N, r) = C(m, N, r) - C^m(1, N, r)$, where $C(m, N, r)$ is the correlation integral. The C–C method, on the other hand, uses the statistic $S(m, N, r, t) = C(m, N, r, t) - C^m(1, N, r, t)$ (the name 'C–C' comes from $C(m, N, r, t) - C^m(1, N, r, t)$. The C–C method was subsequently used by Kim et al. (2009) to estimate the general dependence and, hence, the nonlinear dynamic characteristics of rainfall, streamflow, and lake volume time series.

7.2.3 Remarks

Despite the numerous attempts over the past three decades or so, the issue of the selection of an appropriate delay time τ for phase space reconstruction continues to be challenging. This is clear from the many other suggestions made thus far, in addition to the methods and guidelines presented above. For instance, Wolf et al. (1985) suggested $\tau = T/m$, where T is the dominant periodicity (as revealed by Fourier analysis) and m is the embedding dimension. In this way, τ gives some measure of statistical independence of the data averaged over an orbit and, thus, is an appropriate approach if the autocorrelation function is periodic. Packard et al. (1980) suggested that τ should satisfy $\tau \ll I/\Lambda$, where I is the precision of measurement and Λ is the sum of all positive Lyapunov exponents of the flow. This ensures that the information-generating properties of the flow do not randomize information between successive sites on the recorded attractor. This approach, however, is not practical when we are dealing with an observable from an unknown dynamic system whose Lyapunov exponents are what we seek. Details of several other suggestions, including their advantages and limitations, can be found in Albano et al. (1991), Pfister and Buzug (1992a, b), Kembe and Fowler (1993), Rosenstein et al. (1994), Aguirre (1995), Kugiumtzis (1996), Judd and Mees (1998), Lin et al. (1999), Uzal et al. (2011), and Palit et al. (2013), among others.

The absence of a rigorous method or general guideline on the selection of delay time (and other parameters, such as embedding dimension) has motivated some researchers to search for a generic approach to phase space construction. For instance, Pecora et al. (2007) proposed an approach that views the issue of phase space reconstruction and choosing all the associated embedding parameters as being one and the same problem addressable using a single statistical test formulated directly from the reconstruction theorems. This view allows for varying delay times appropriate to the data and simultaneously helps decide on the embedding dimension. A second new statistic, undersampling, then acts as a check against overly long delay times and overly large embedding dimensions. The results of the application of this approach to a variety of time series (univariate, multivariate, data with multiple scales, and chaotic data) are encouraging. Nevertheless, there is still a long way to go in regards to resolving the issue of delay time, and the phase space reconstruction more broadly.

In the absence of clear-cut guidelines on the selection of τ, a practical approach is to experiment with different τ values to ascertain its effect, for example, on the estimation of invariants. Such an exercise is particularly fruitful for synthetic time series, since the dynamic properties (e.g. invariants) of such series are known a priori. However, one has to be careful in adopting this approach for real time series, since their dynamic properties are not known, and determination of which is indeed the task at hand. Nevertheless, the exercise can offer some important clues; see, for example, Sangoyomi et al. (1996) and Sivakumar et al. (1999a) for hydrologic time series.

Finally, in view of the difficulties with the delay time selection, it may be necessary to find alternative ways to reconstruct the phase space without having to define a proper delay time. One such approach may be to use not one but m observables of the same variable. Each sequence can be sampled independently of the other and, thus, could be used as an independent co-ordinate in an m-dimensional phase space. For example, instead of dealing with one time series representing a variable (e.g. rainfall) at a given point, we may measure the variable at m independent points, thus obtaining m such time series. Instead of sifting one time series to obtain the phase space co-ordinates, we can simply bring in one new time series at a time. Another approach may be to measure multiple variables from the system of interest and then reconstruct the phase space in a multi-variable sense (e.g. Cao et al. 1998; Porporato and Ridolfi 2001; Sivakumar et al. 2005; Hirata et al. 2006). However, this approach may also need the selection of a delay time depending upon the system under investigation; for instance, in the prediction of runoff, rainfall at the current time/previous time(s) as well as runoff at the previous time(s) may be needed to be included in the reconstruction vector, and the selection of the timestep(s) at which the variable(s) needs to be chosen can become very complicated; see Sivakumar et al. (2005) for details. In any case, both the above approaches may be difficult to implement in practice, since data at different points and of different variables may not be available and one may be forced to work with data available only at a single point and of a single variable, a situation commonly encountered in many different fields.

7.3 Data Size

The fiercest criticism on studies employing chaos theory-based methods to real time series has been with respect to the issue of data size (or length). One of the basic assumptions in the development of chaos identification and prediction methods is that the time series is infinite. However, since an 'infinite' time series simply does not exist, it precludes the reliability of any and all of the chaos identification and prediction methods! Therefore, a strict and inflexible adherence to the requirement of an 'infinite' time series for chaos analysis is not at all helpful. What would be useful, however, especially considering that real time series are often short (in hundreds or thousands in numbers), is to have an approximate estimate on the minimum length of data required for assessing the effectiveness of chaos methods and for interpretation of the outcomes. It is this 'minimum' data size that has motivated numerous studies, especially immediately after the development of many of the chaos methods in the 1980s, to address the data size issue. Such studies have resulted in the proposal of a number of guidelines on the minimum data size, especially linking the minimum data size requirement with the embedding dimension for the phase space reconstruction or the correlation dimension of the time series (e.g. Smith 1988; Nerenberg and Essex 1990). These guidelines have subsequently formed the basis for criticisms on the applications of chaos studies to real time series. In what follows, a brief account of the effects of data size on the estimation of invariants, guidelines for the minimum data size requirement, as well as the associated issues is presented.

7.3.1 Effects of Data Size

The size of data is an important issue in almost all time series analysis methods, not just in chaos analysis methods. Even the most fundamental statistical methods, let alone the so-called 'data-driven' methods, often require a 'long' time series to obtain reliable results. For instance, the estimate of the mean of a time series can be considered reliable only when the data size is reasonably long to represent the underlying system, and the estimate becomes more reliable when the data size gets longer. What is 'long' is, of course, often subjective, and needs to be considered in the context of the purpose at hand and the method to be used, as different purposes and methods may require different lengths of data. For instance, in many water resources engineering applications (e.g. design of storage structures), the minimum length of data required is generally considered to be 30 years (monthly scale). However, in many large-scale climate studies, the minimum length of data required is generally in the order of hundreds of years (annual scale).

It is fair to say that the size of data has some influence on the outcomes of all chaos identification and prediction methods; see, for example, Theiler (1986, 1990, 1991). However, the extent of influence of the data size is often different for different chaos methods. For instance, the effect of small data size is generally considered to be more serious on the outcomes of the correlation dimension estimation than on the predictions from the local approximation method. This can be explained as follows. The correlation exponent and, hence, the correlation dimension are computed from the slope of the scaling region in the Log $C(r)$ versus Log r plot or in the local slope versus Log r plot (see Chap. 6, Sect. 6.4 for details). It is always desirable to have a larger scaling region to determine the slope, since the determination of the slope for a smaller scaling region may be difficult and possibly result in errors. A longer data set results in a larger scaling region due to the inclusion of a large number of points (or vectors) on the reconstructed phase space. However, if the data set were smaller, there would be only a few points on the reconstructed phase space, which makes the slope determination difficult.

The recognition that the effects of data size may be different for different chaos identification and prediction methods has led researchers to attempt to estimate the minimum data requirement for different methods. Most of these attempts have been directed at the correlation dimension method (e.g. Smith 1988; Havstad and Ehlers 1989; Nerenberg and Essex 1990; Ramsey and Yuan 1990; Lorenz 1991; Tsonis et al. 1993; Sivakumar et al. 2002b; Sivakumar 2005a), but there have also been several notable studies on other methods as well (e.g. Wolf et al. 1985; Briggs

1990; Zeng et al. 1991; Rosenstein et al. 1993; Bonachela et al. 2008). As the issue of data size in the correlation dimension method has and continues to receive far more attention, it is discussed in more detail here.

7.3.2 *Minimum Data Size*

Since the proposal of the correlation dimension method by Grassberger and Procaccia (1983a, b), numerous studies have attempted to determine the minimum number of data points ($N_{\min}$) for a reliable estimation of the correlation dimension and also offered different guidelines. Many of these studies and guidelines relate the minimum data size to the embedding dimension (m) or correlation dimension (d), but several other approaches have also been adopted to determine the effect of data size. Some examples of such studies are as follows.

The study by Smith (1988) was the first study to address the minimum data size for correlation dimension estimation in terms of the embedding dimension. Smith (1988) concluded that the minimum data size was equal to 42^m, where m is the smallest integer above the dimension of the attractor. Nerenberg and Essex (1990) demonstrated that the procedure by Smith (1988) to obtain the 42^m estimate was flawed and that the data requirements might not be so extreme. They suggested that the minimum number of points required for the dimension estimate is $N_{\min} \sim 10^{2+0.4m}$. Other suggested guidelines include $N^{\min} \approx 10^m$ or $N_{\min} \approx [2(m + 1)]^m$ (Essex 1991).

Havstad and Ehlers (1989) used a variant of the nearest neighbor dimension algorithm to compute the dimension of the time series generated from the Mackey–Glass equation (Mackey and Glass 1977), whose actual dimension is 7.5. Using a data set of as small as 200 points, Havstad and Ehlers (1989) reported an underestimation of the dimension by about 11 %. Ramsey and Yuan (1990) concluded that for small sample sizes, dimension could be estimated with upward bias for chaotic systems and with downward bias for random noise as the embedding dimension is increased. They proved that, due to these bias effects, a correlation dimension estimate of 0.214 could imply an actual correlation dimension value of as high as 1.68.

Using data generated by a mathematical system whose dimensions can be evaluated by other means, Lorenz (1991) found that the Grassberger–Procaccia algorithm yielded systematic underestimates of the correlation dimension for sample sizes of ~4000. However, Lorenz (1991) also argued that different climatic variables yield different estimates of correlation dimension and that a suitably selected variable could yield a fairly accurate estimate of dimension even if the number of points were not large.

A number of studies have followed and/or supported one or more of the above-mentioned guidelines as to the potential underestimation of the correlation dimension when the data size is small (e.g. Tsonis et al. 1993; Wang and Gan 1998; Schertzer et al. 2002). However, several counter-arguments to these guidelines have also been made (e.g. Sivakumar et al. 2002a, b; Sivakumar 2005a); see below for some details. As of now, a clear-cut guideline on the minimum data size for the correlation dimension estimation continues to be elusive.

In the absence of clear-cut guidelines, a practical way to address the minimum data size requirement is by decreasing (or increasing) the length of the time series step-by-step and estimating the correlation dimension for each of the resulting time series. The length of data below which significant changes are observed can be taken as the minimum data size required. While this procedure may have some drawbacks when it comes to real time series, as the properties (e.g. correlation dimension) of such time series are not known a priori, it is nevertheless still useful, if sufficient caution is exercised in its implementation and interpretation of the outcomes. This approach has been adopted by some studies (e.g. Jayawardena and Lai 1994; Mikosch and Wang 1995; Wang and Gan 1998; Sivakumar et al. 1999a; Sivakumar 2005a). The study by Sivakumar (2005a) is discussed in more detail here, to illustrate the effectiveness of this procedure in determining the minimum data size.

Sivakumar (2005a) carried out the correlation dimension analysis for various data sizes from each of three types of time series: (a) stochastic series of maximum length of 5000 values—artificially generated using a random number generation technique (see Chap. 6, Sect. 6.2 for details); (b) chaotic series of maximum length of 5000 values—artificially generated using the Henon map equation (Henon 1976) (see Chap. 5, Sect. 5.11.2 and Chap. 6, Sect. 6.2 for details); and (c) real hydrologic series of maximum length of 1560 values—monthly streamflow data observed over a period of 130 years (January 1807—December 1936) in the Göta River basin in Sweden. Analysis and results for the stochastic and chaotic series are presented here; see Chap. 12 for a discussion on the Göta River basin.

Figure 7.1a, b shows the stochastic and chaotic time series, respectively. For each of these series, Sivakumar (2005a) considered 11 different data lengths, ranging from 100 to 5000 values; the lengths were selected at an irregular increasing order, such that the results could reasonably reflect the sensitivity of the

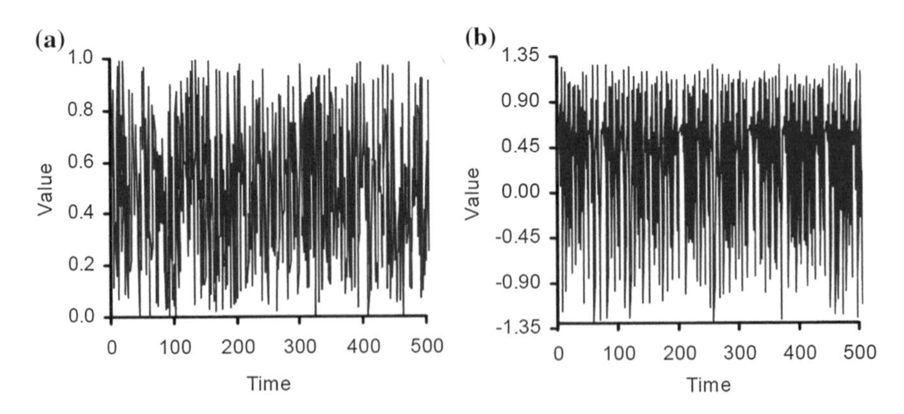

Fig. 7.1 Time series plot of **a** stochastic series (artificial random series); and **b** chaotic series (artificial Henon map) (*source* Sivakumar (2005a))

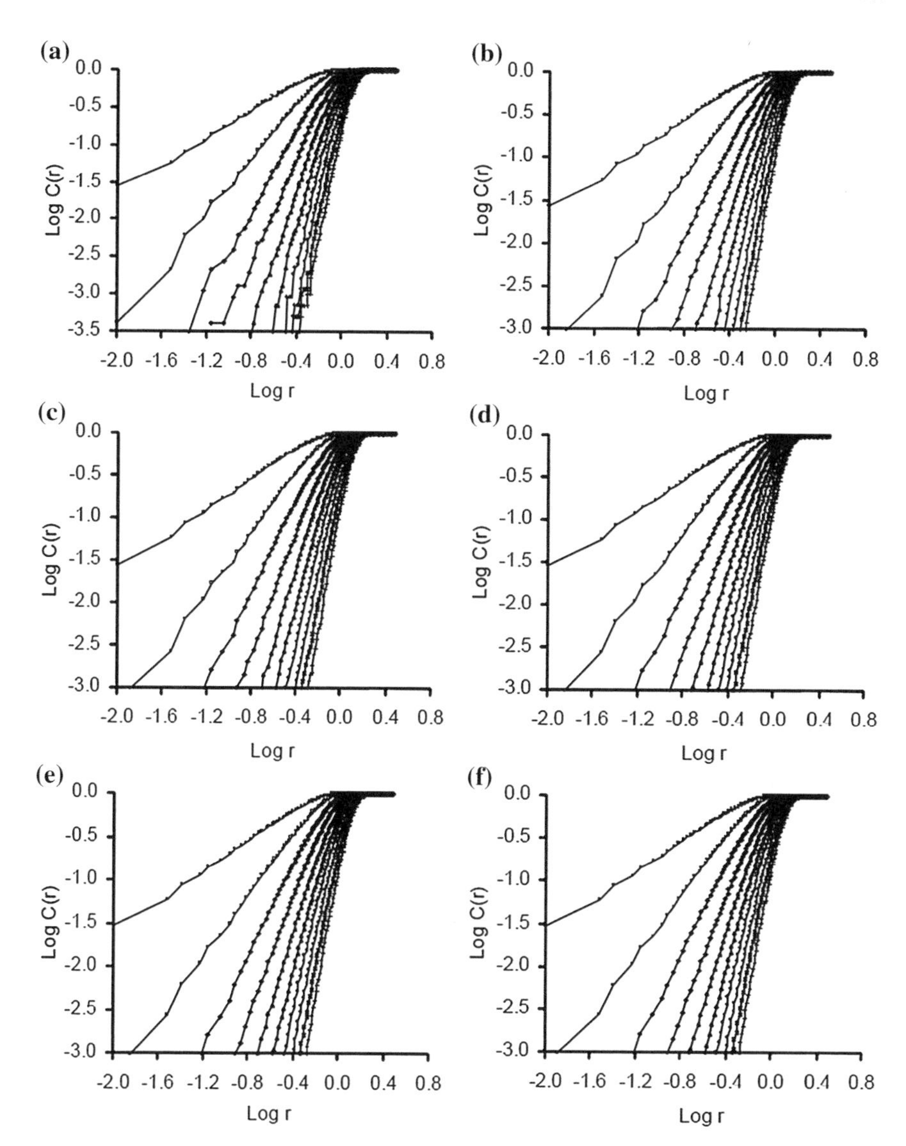

Fig. 7.2 Correlation dimension results for stochastic series with **a** 100 points; **b** 200 points; **c** 300 points; **d** 500 points; **e** 1000 points; and **f** 5000 points. Embedding dimension m = 1–10 (from *left* to *right*) (*source* Sivakumar (2005a))

dimension results to data length. The effect of data size was evaluated through a visual inspection of the scaling regimes in the correlation dimension plots (and even the entire plots). Figures 7.2 and 7.3 present the results for six of these different data lengths for each of the two series—stochastic series (Fig. 7.2) and chaotic series (Fig. 7.3). The results show that: (a) for the stochastic series, there is no change in

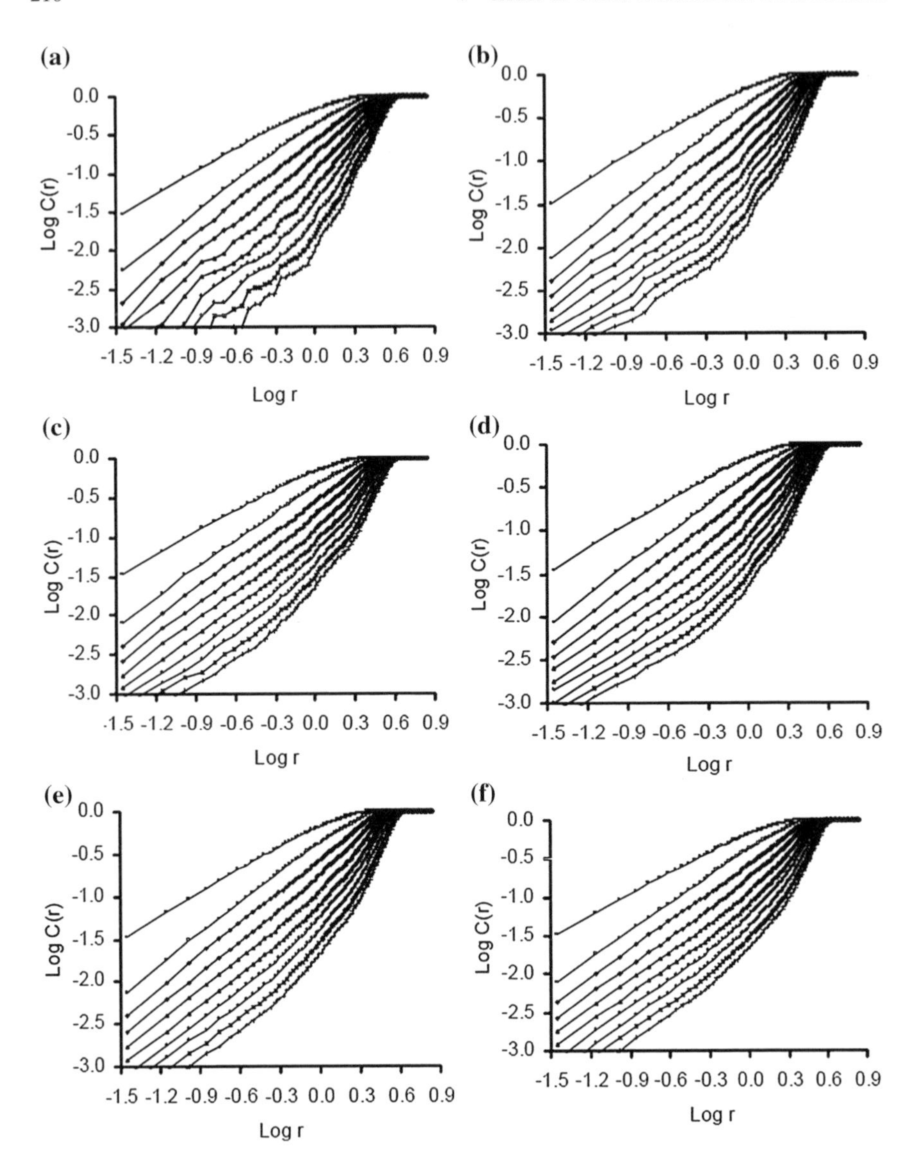

Fig. 7.3 Correlation dimension results for chaotic series with **a** 100 points; **b** 200 points; **c** 300 points; **d** 500 points; **e** 1000 points; and **f** 5000 points. Embedding dimension $m = 1$–10 (from *left* to *right*) (*source* Sivakumar (2005a))

the scaling regimes for lengths of 300 data points and above, irrespective of m; and (b) for the chaotic series, there is no change in the scaling regimes for lengths of 1000 data points and above, irrespective of m, and only a little change for lengths above 300 and below 1000 points.

Considering that the true dynamic properties of real time series are not known a priori, it would be more appropriate to make interpretations based on the results obtained for the artificial stochastic and chaotic time series. The results for these two series are indeed interesting and informative, since almost the same, and very small, size of data (just 300 values) is found to be sufficient for the estimation of the correlation dimension, despite the fact that the two time series are generated from totally different systems—random (high-dimensional) and chaotic (low-dimensional), respectively. This clearly helps eliminate the need to link the data size to the embedding dimension or the correlation dimension, i.e. $N_{\min}$ is *not* a function of m or d.

The above results suggest that the crucial question one must ask in regards to the data size issue in the correlation dimension estimation is whether the time series is long enough to sufficiently represent the changes the system undergoes over a period of time. Any concern based purely on the number of values in the time series, including relating $N_{\min}$ with m or d, is not only unhelpful but also inappropriate; see also, for example, the study by Sivakumar et al. (2002a) that points out the potential flaws in claiming such links. It is also important to note that even the availability of a very very long time series does not necessarily solve the minimum data size issue in the correlation dimension method. This is because other factors, such as the sampling frequency, may also play important roles. For instance, a streamflow time series with, say, 10^6 values, but collected at 1-s interval, is not going to be of much use in representing the changes in the flow dynamics of a large-scale river basin, as it does not cover a long enough period to contain enough information about the system changes. Therefore, despite the availability of a long time series and despite the time series meeting the criterion linking, for example, $N_{\min}$ and m, the correlation dimension for such a series may still be underestimated.

7.4 Data Noise

Another important assumption in the development of chaos identification and prediction methods is that the data are noise-free. However, all real data are, to some extent, contaminated by noise. This means, we usually observe a noisy time series X_i, $i = 1, 2, \ldots, N$, which is composed of a clean signal Y_i from a deterministic system $X_{i+1} = f(X_i)$ and some amount of noise η_i, according to:

$$X_i = Y_i + \eta_i \tag{7.3}$$

There are two types of noise: measurement noise and dynamical noise. Measurement noise refers to the corruption of observations by errors, which are external and independent of the dynamics, and may be caused by, for example, the measuring device. Dynamical noise, in contrast to measurement noise, is a feedback process wherein the system is perturbed by a small random amount at each timestep. This type of noise, internal to the system, arises from the propagation of minor random fluctuations in the settings of the main system parameters causing random-like fluctuations that are not specific to the system. It might also be caused by the influence of intrinsic system events taking place at random (Schouten et al. 1994). Dynamical noise directly influences the evolution of the system in time.

The presence of noise can affect the outcomes of the chaos identification and prediction methods (e.g. Schreiber and Kantz 1996; Sivakumar et al. 1999b; Kantz and Schreiber 2004), and the effects of measurement noise and dynamical noise may also be different. However, this does not mean that chaos analysis cannot be performed on real data. Rather, one just needs to exercise proper caution in applying the methods and interpreting the results. Indeed, the outcomes of chaos methods can themselves offer clues as to the presence/level of noise. For instance, the outcomes of the false nearest neighbor (FNN) algorithm (e.g. Kennel et al. 1992), the local approximation prediction method (e.g. Farmer and Sidorowich 1987; Casdagli 1989), and the deterministic versus stochastic (DVS) algorithm (e.g. Casdagli 1992) are useful to interpret whether the time series is generated from a purely random system or from a deterministic system that is contaminated by some level of noise and how the effects of noise amplify when higher-dimensional phase spaces are reconstructed. Nevertheless, an even better way to approach chaos-related studies is to first remove the noise and then apply the chaos methods.

A number of nonlinear noise reduction methods have been proposed in the literature (e.g. Kostelich and Yorke 1988; Hammel 1990; Schreiber and Grassberger 1991; Cawley and Hsu 1992; Sauer 1992; Kantz et al. 1993; Schreiber 1993b; Davies 1994; Luo et al. 2005; Sun et al. 2007; Chelidze 2014). As reliable information about the level of noise is required for proper implementation of noise reduction methods, attempts have also been made to determine the level of noise in a time series (e.g. Cawley and Hsu 1992; Schreiber 1993a; Schouten et al. 1994; Heald and Stark 2000; Nakamura and Small 2006). There is already an enormous amount of literature on the effects of noise on chaos methods as well as techniques for noise level determination and noise reduction. Some details are presented below.

7.4.1 *Effects of Noise*

Noise affects the performance of many techniques of identification, modeling, prediction, and control of deterministic systems. The severity of the influence of noise depends largely on the level and the nature of noise. For example, dynamical noise induces much greater problems in data processing than does measurement noise, since in the latter case a nearby clean trajectory of the underlying deterministic system exists. Furthermore, what one interprets to be dynamical noise may sometimes be a higher-dimensional deterministic part of the dynamics with small amplitude. Even if this is not the case, dynamical noise may be essential for the observed dynamics because transitions to qualitatively different behavior can be induced or delayed by dynamical noise.

In general, most dynamical measures of determinism are reasonably robust to small levels of noise, but as the noise level approaches a few percent, estimates can become quite unreliable (Schreiber and Kantz 1996; Kantz and Schreiber 2004). Therefore, the estimation of the level of noise in data is important to understand its possible effects. A brief account of the specific effects of noise on the correlation dimension and nonlinear prediction methods is presented here.

The presence of noise influences the estimation of the correlation dimension primarily from the identification of the scaling region. Noise may corrupt the scaling behavior at all length scales, but its effects are significant especially at smaller length scales. If the data are noisy, then below a length scale of a few multiples of the noise level, the data points are not confined to the fractal structure but smeared out over the whole available phase space. Thus, the local scaling exponents may increase. It has been observed that even small levels of noise significantly complicate estimates of dimension, a quantity that in principle should be straightforward to measure (e.g. Schreiber and Kantz 1996). This, however, may not always be the case, as it may depend strongly on the system. For instance, Sivakumar et al. (1999b), through analysis of the Henon time series (and also rainfall time series from Singapore), reported that the influence of noise on the correlation dimension estimate is not that significant.

Noise is one of the most prominent limiting factors for the predictability of deterministic systems. Noise limits the accuracy of predictions in three possible ways: (1) the prediction error cannot be smaller than the noise level, since the noise part of the future measurement cannot be predicted; (2) the values on which the predictions are based are themselves noisy, inducing an error proportional to and of the order of the noise level; and (3) in the generic case, where the dynamical evolution has to be estimated from the data, this estimate will be affected by noise (Schreiber and Kantz 1996). In the presence of the above three effects, the prediction error will increase faster than linearly with the noise level.

The sensitivity of the correlation dimension (or any other invariant) and the prediction accuracy to the presence of noise is the price one has to pay for using these to identify chaos. The definitions of these involve the limit of small length scales because it is only then that the quantity becomes independent of the details of the measurement technique, the data processing, and the phase space reconstruction method. The permissible noise level for a practical application of these methods depends, in a complicated way, on the details of the underlying system and the measurement.

These observations clearly indicate that noise present in data should be taken into account if the analysis is to remain realistic. The important first step is to be aware of the problem and to recognize its effects on the data analysis techniques by estimating the level and the nature of noise. If it is found that the level of noise is only moderate, and there are hints that there is a strong deterministic component in the signal, then one can attempt the second step of separating the deterministic signal from the noise. While this two-step procedure is the most appropriate approach to assess the effects of noise and their mitigation, it is often difficult to implement this for real data, since determination of the level of noise for such data is a very complicated process. Therefore, many studies have employed only the second step, i.e. noise reduction. However, some of those studies have also attempted to make useful interpretations on the level of noise based on the outcomes of noise reduction. In some cases, noise level determination and noise reduction may go hand-in-hand.

7.4.2 Noise Level Determination

Several methods have been proposed in the literature to determine the noise level in a time series in the specific context of nonlinear analysis (e.g. Cawley and Hsu 1992; Schreiber 1993a; Schouten et al. 1994; Diks 1999; Heald and Stark 2000; Siefert et al. 2003; Nakamura and Small 2006; Urbanowicz and Hołyst 2006; Jayawardena et al. 2008). The methods may also make different assumptions about the nature of noise, even if they use somewhat similar approaches. For instance, both the method proposed by Schreiber (1993a) and the method proposed by Schouten et al. (1994) are based on the correlation integral. However, the method by Schreiber (1993a) assumes that noise is Gaussian, while the method of Schouten et al. (1994) assumes that the noise is strictly bounded in magnitude. A brief account of the method proposed by Schouten et al. (1994) is presented here as an example to explain the procedure for noise level determination.

In the noise level determination method of Schouten et al. (1994), a simple analytical expression is derived for the rescaled correlation integral

$$C(r) = \frac{2}{N(N-1)} \sum_{\substack{i,j \\ (1 \le i < j \le N)}} H\left(r - \left\|\mathbf{Y}_i - \mathbf{Y}_j\right\|\right) \tag{7.4}$$

where H is the Heaviside step function; see Chap. 6, Sect. 6.4.2 for details. If the time series is characterized by an attractor, the correlation integral exhibits a power law with $C(r) \approx r^{\nu}$ (and hence r^d), as $r \to 0$ and $N \to \infty$. The exponent ν can be obtained from the Log $C(r)$ versus Log r plot, and the saturation value of ν (with an increasing embedding dimension) is the correlation dimension d. This scaling relationship holds well as long as the data is noise-free. However, the presence of noise may corrupt the scaling behavior at all length scales and, consequently, the Log $C(r)$ versus Log r plot may not show a linear part at all. This means that, the power law relationship, r^{ν} (and hence r^d), does not give a good representation of the inter-point correlations. The influence of noise on the correlation integral can be evaluated as follows.

Let us consider two points $\boldsymbol{Y}_i$ and $\boldsymbol{Y}_j$ that are located on the reconstructed attractor on different orbits. Since these points are not disturbed by noise, they may be considered as *true* points satisfying the exact (chaotic) dynamics of the system. The maximum norm distance between these points is given by:

$$\left\| \boldsymbol{Y}_i - \boldsymbol{Y}_j \right\| = \max_{0 \leq k \leq m-1} \left| X_{i+k} - X_{j+k} \right| \quad (7.5)$$

with

$$\boldsymbol{Y}_i = (X_i, X_{i+1}, \ldots, X_{i+m-1}) \quad (7.6a)$$

and

$$\boldsymbol{Y}_j = \left(X_j, X_{j+1}, \ldots, X_{j+m-1}\right) \quad (7.6b)$$

Let us now consider that each point in the time series is corrupted by noise that is bounded in magnitude with maximum possible amplitude of $\pm 1/2\delta X_{\max}$. Let us now also assume that there exists a trajectory satisfying the true dynamics of the chaotic system sufficiently close to the measured, noise-corrupted trajectory. In this case, the elements ($Z_{i,k}$ and $Z_{j,k}$) of the noise-corrupted vectors are assumed to be composed of a noise-free part ($X_{i,k}$ and $X_{j,k}$) and a noisy part ($\delta X_{i,k}$ and $\delta X_{j,k}$) according to:

$$Z_{i,k} = X_{i,k} + \delta X_{i,k} \quad (7.7a)$$

and

$$Z_{j,k} = X_{j,k} + \delta X_{j,k} \quad (7.7b)$$

with

$$-1/2\delta X_{\max} \leq \delta X_i \leq +1/2\delta X_{\max} \quad (7.8a)$$

and

$$-1/2\delta X_{\max} \leq \delta X_j \leq +1/2\delta X_{\max} \quad (7.8b)$$

When the number of vector elements or the embedded dimension is infinite, i.e. $m \rightarrow \infty$, the probability of finding two corresponding elements $Z_{i,k}$ and $Z_{j,k}$ that are maximally corrupted with $-1/2\delta X_{\max}$ and $+1/2\delta X_{\max}$, respectively, will be unity. It is also necessary that the maximally-corrupted pair to be the pair for which $|X_{i,k} - X_{j,k}|$ is maximal. If the embedding dimension is sufficiently large, while $X_{i,k}$ and $X_{j,k}$ depend smoothly on k, then this coincidence can be well approximated. The maximum norm distance between the noise-corrupted vectors is thus found from

$$\begin{aligned} r_z &= \lim_{m\to\infty} \max_{0\le k\le m-1} \left| Z_{i,k} - Z_{j,k} \right| \\ &= \lim_{m\to\infty} \max_{0\le k\le m-1} \left| (X_{i,k} + \delta X_{i,k}) - (X_{j,k} + \delta X_{j,k}) \right| \\ &= \lim_{m\to\infty} \max_{0\le k\le m-1} \left| (X_{i,k} - X_{i,k}) \right| + \delta X_{\max} \\ &= r_x + r_n \end{aligned} \tag{7.9}$$

where r_z is the corrupted distance, r_x is the noise-free distance, and $r_n = \delta X_{\max}$ is the maximum noise distance. Equation (7.9) illustrates that the probability of finding inter-point distances r_n below $r_z = \delta X_{\max}$ is zero. This means that $C(r_z \le r_n) = 0$ and $C(r_z > r_n) > 0$, implying that the maximum noise scale can be directly obtained from the correlation integral.

When the power law dependency holds for the noise-free distances r_x according to $C(r_x) \sim (r_n)^d$, then it can be written as:

$$C(r_z | r_z > r_n) \sim (r_z - r_n)^d \tag{7.10}$$

since $r_x = r_z - r_n$ [from Eq. (7.9)]. Also, with the requirements that $C(r_z = r_n) = 0$ and $C(r_z = r_0) = 1$, it can be written as:

$$C(r_z) = \left[\frac{r_z - r_n}{r_0 - r_n} \right]^d, \; r_n \le r_z \le r_0 \tag{7.11}$$

All distances are normalized with respect to the maximum scaling distance r_0, using $l = r_z / r_0$ and $l_n = r_n / r_0$, so that

$$C(l) = \left[\frac{l - l_n}{1 - l_n} \right]^d, \; l_n \le l \le 1 \tag{7.12}$$

This expression illustrates that the corrupted distances r_z have been effectively rescaled in order to let the correlation integral obey the power law function again. The parameters l_n and d can be estimated from a nonlinear least-squares fit of the above integral function to the experimentally determined correlation integral. For details of the selection of the various parameters involved in this method, see Schouten et al. (1994).

7.4.3 *Noise Reduction*

The classical statistical tool for distinguishing noise and signal is the power spectrum. Random noise has a flat, or at least a broad, spectrum, whereas periodic or quasi-periodic signals have sharp spectral lines. After both components have been identified in the spectrum, a Wiener filter or other band-pass filters can be used to separate the time series accordingly. The basic idea behind such filtering approach is that the noise can be modeled as a collection of high-frequency components and these can be subtracted from a power spectrum of the input data. The resulting transform can be inverted to yield a new time series with some of the high-frequency components removed.

Although the filtering approach works well for linear systems, such an approach fails for nonlinear, and especially deterministic chaotic, dynamic systems. This is because, the output of nonlinear systems usually leads to broad band spectra itself and, thus, possesses spectral properties generally attributed to random noise. Even if parts of the spectrum can be clearly associated with the signal, a separation into signal and noise fails for most parts of the frequency domain. Moreover, the suppression of certain frequencies can alter the dynamics of the filtered output signal (e.g. Badii et al. 1988; Chennaoui et al. 1990). For example, Badii et al. (1988) demonstrated that such an approach might introduce an extra Lyapunov exponent that depends on the cutoff frequency.

The difficulties in the use of filtering approaches to separate the noise and signal in chaotic time series led to the development of nonlinear noise reduction methods. Consequently, a number of nonlinear noise reduction methods for chaotic time series have been proposed in the literature (e.g. Kostelich and Yorke 1988, 1990; Hammel 1990; Farmer and Sidorowich 1991; Marteau and Abarbanel 1991; Schreiber and Grassberger 1991; Cawley and Hsu 1992; Davies 1992, 1994; Sauer 1992; Enge et al. 1993; Grassberger et al. 1993; Kantz et al. 1993; Schreiber 1993b; Luo et al. 2005; Sun et al. 2007; Chelidze 2014). The different noise reduction methods differ in the way the dynamics are approximated, how the trajectory is adjusted, and how the approximation and the adjustment steps are linked to each other; see Kostelich and Schreiber (1993) for a survey. It has been reported that most of these methods reduce noise by a similar amount and their performances do not differ much (Kantz and Schreiber 2004). However, noise reduction algorithms are generally chosen on the basis of their robustness, ease of use and implementation, and the computing resources needed.

To explain the general concept and procedure behind nonlinear noise reduction methods, an extremely simple but robust noise reduction method proposed by Schreiber (1993b) is considered here. This method has been found to be suitable for trajectories contaminated with high noise levels. Suppose we have a scalar time series X_i, $i = 1, 2,..., N$, where the X_i are composed of a clean signal Y_i with some noise η_i added so that $X_i = Y_i + \eta_i$. The main idea of the noise reduction method is to replace each measurement X_i by the average value of this coordinate over points in a suitably chosen neighborhood. The neighborhoods are defined in a $(k + 1 + l)$-dimensional phase space reconstructed by delay coordinates using information on k past coordinates and l future coordinates given by

$$Y_i = (X_{i-k}, X_{i-k+1}, \ldots, X_i, \ldots, X_{i+1-l}, X_{i+l}) \tag{7.13}$$

Further, choosing a radius r for the neighborhoods, for each value of X_i, a set Im_i^r of all neighbors X_j, for which $\|\boldsymbol{Y}_j - \boldsymbol{Y}_i\| \approx r$, is found. The present coordinate of X_i is then replaced by X_i^{corr} given by

$$X_i^{corr} = \frac{1}{|\mathfrak{I}_i^r|} \sum_{\mathfrak{I}_i^r} X_j \qquad (7.14)$$

According to this procedure, only the central coordinate in the delay window is corrected, since only this coordinate is optimally controlled from past and future. The X_i^{corr} values can then be used to reconstruct the phase space and the procedure can be repeated. The selection of the neighborhood, and other parameters, is important in this procedure, and some details are available in Schreiber (1993b).

Most of the aforementioned techniques are dynamical noise reduction techniques and remove noise from each and every point of a trajectory. In many cases, however, this may not be what we really want. Often, we may only be interested simply in cleaner statistical properties of the time series, such as power spectra and dimension. For this purpose, any cleaner time series with the same statistical properties as the 'true' time series is good enough. This is called statistical noise reduction. A straightforward and simple approach for statistical noise reduction is to find a global model to the noisy data and iterate it (starting from some initial condition) to obtain a new time series. As demonstrated in Eubank and Farmer (1990), this type of noise reduction could be quite effective if the underlying attractor is not very sensitive to the parameters involved. Since statistical noise reduction produces a new orbit with the same statistical properties as the true orbit for every initial condition, it could be used in a way similar to the way bootstrapping is used to artificially obtain many data points. Such an approach might be useful for approximating invariant measures like dimensions from an initially small amount of data; see Casdagli (1989).

Finally, one has to be careful in dealing with the dynamical noise. Since dynamical noise is an inherent part of the system, whether the dynamical noise should be removed is debatable. Consequently, in most cases, noise reduction is performed only to remove the measurement noise, and especially additive noise. Indeed, most of the noise reduction methods have been developed for this type of noise. There also exist methods to clearly identify the specific type of noise. For instance, Heald and Stark (2000) proposed a method whereby dynamical noise and measurement noise can be measured very precisely, but only if the dynamical equations are known. Siefert et al. (2003) presented an approach to clarify which kind of noise is present, even when the dynamics are unknown.

7.4.4 Coupled Noise Level Determination and Reduction: An Example

As mentioned above, the most appropriate approach to deal with noisy data in chaotic analysis is to first determine the noise level and then reduce that level of noise. However, this approach has not been widely followed. Many studies, especially in the early years of chaos analysis, have employed the noise reduction methods, without making any attempt to determine the noise level. In general, noise level determination and noise reduction have been done independently.

Sivakumar et al. (1999b) proposed a systematic approach to noise reduction in chaotic time series by combining a noise level determination method (Schouten et al. 1994) and a nosie reduction method (Schreiber 1993b) for estimation of a probable noise level. They used prediction accuracy as the main diagnostic tool to verify the success of the noise reduction, since prediction accuracy can be determined without any knowledge of the noise-free signal or the underlying system dynamics and it is also sensitive to under- or over-removal of noise. However, the procedure can be effectively implemented with any other invariant as a diagnostic tool as well, as was demonstrated by Sivakumar et al. (1999b) with correlation dimension. The combined noise level determination–noise reduction procedure of Sivakumar et al. (1999b) can be generalized as follows:

1. Estimate the noise level in the time series, and use it as the initial estimate of the noise level;
2. Apply the noise reduction method with different neighborhood sizes and number of iterations;
3. Determine the combinations of neighborhood sizes and number of iterations that remove exactly (or near accurately) the initial noise level estimated in Step 1;
4. Determine the prediction accuracy (or estimate any invariant) of the different sets of noise-reduced data with the combinations of neighborhood sizes and number of iterations obtained from Step 3;
5. Select the combination of neighborhood size and number of iterations that yields the best prediction accuracy (or the most reliable estimate of any invariant);
6. Repeat Steps 2–5 to remove other higher (or lower) noise levels than that estimated in Step 1, and check whether the prediction accuracy (or invariant estimate) improves. This is to take into account any underestimation or overestimation of noise level estimated by the noise determination method; for instance, the method by Schouten et al. (1994) has been found to generally underestimates the noise level; and
7. Select the noise level, from Step 6, that yields the best prediction accuracy (or the most reliable invariant estimate). This noise level can be considered as the probable noise level in the data.

When Step 6 is executed for the first time, noise levels selected are any value greater (or smaller) than the noise level initially resulted from the method, Step 1. However, to reduce the number of trial and error, noise levels recommended are multiples (or divisions) of the level estimated in Step 1. Further refinement, if necessary, on noise level can be done once the most probable noise level has been derived in Step 7.

Sivakumar et al. (1999b) implemented this procedure first on synthetically generated Henon time series (Henon 1976) and then on real rainfall data observed in Singapore. For Henon time series, they used additive, independent, and uniformly distributed noise, bounded in magnitude, with maximum noise bound δX_{max} equal to 0.05, 0.10, and 0.20, respectively. The noise level was defined as the ratio of the standard deviation of the noise generated to the standard deviation of the noise-free Henon time series. The standard deviation of the above three levels of noise added are about 0.03, 0.06, and 0.12, respectively, and the standard deviation of the Henon time series generated is about 0.76. Therefore, the above added noise levels correspond to 4, 8, and 16 %, respectively. For the purpose of demonstration

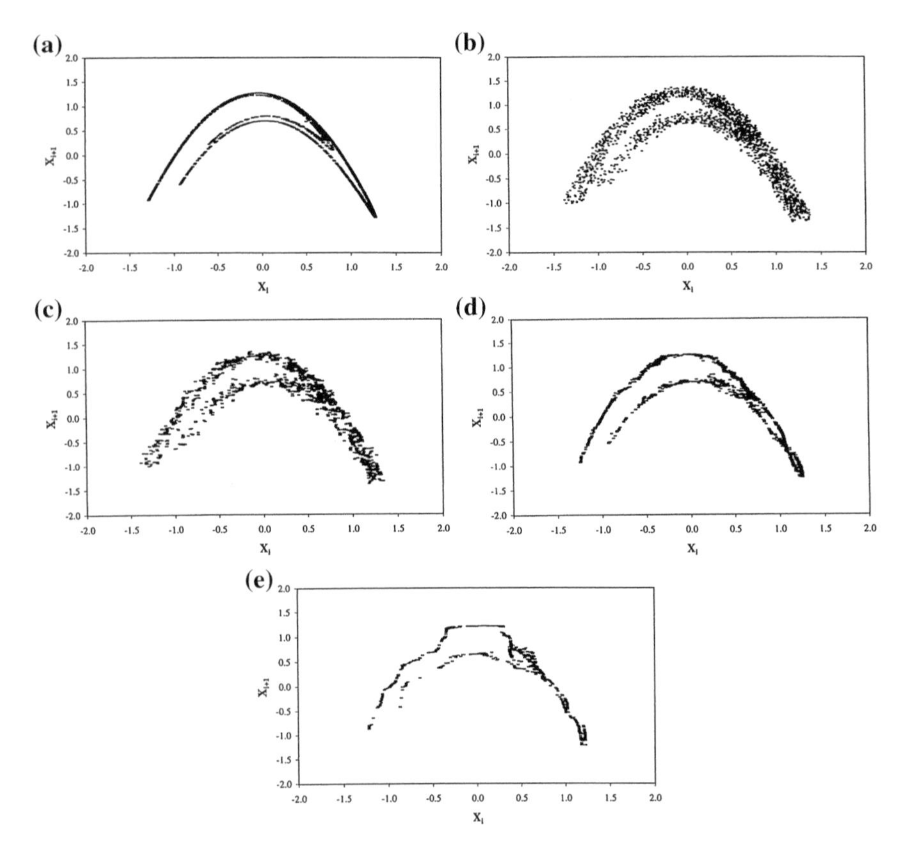

Fig. 7.4 Phase space plots for Henon data: **a** noise-free; **b** 8 % noisy; **c** 4.1 % noise-reduced; **d** 8.2 % noise-reduced; and **e** 12.3 % noise-reduced (*source* Sivakumar et al. (1999b))

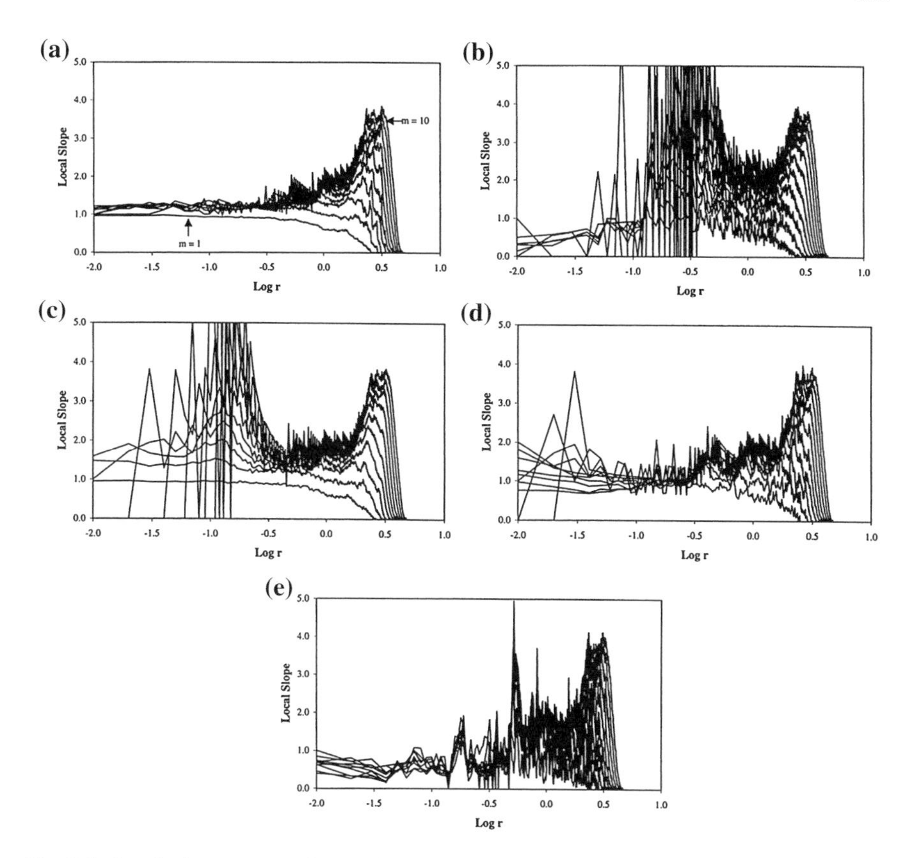

Fig. 7.5 Local slopes versus Log r for Henon data: **a** noise-free; **b** 8 % noisy; **c** 4.1 % noise-reduced; **d** 8.2 % noise-reduced; and **e** 12.3 % noise-reduced (*source* Sivakumar et al. (1999b))

of the effectiveness of the method, some results obtained for the case of 8 % noise level are presented here.

Application of the method of Schouten et al. (1994) to the 8 % noisy Henon time series yielded an initial estimate of the noise level of about 4.1 % (standard deviation = 0.0309), which is a significant underestimation. So, this level of noise was reduced from the time series, which was followed by 8.2 and 12.3 % noise reduction (i.e. multiples of 4.1 %). Figure 7.4 shows the phase space diagram for the noise-free, 8 % noisy, 4.1 % noise-reduced, 8.2 % noise-reduced, and 12.3 % noise-reduced series. Figure 7.5 shows the variations of the local slopes against Log r for the same five series, while Fig. 7.6 shows the relationship between the estimated correlation exponent and the embedding dimension. The results shown are the best among the different combinations of the neighborhood sizes and number of iterations selected in each level of noise reduction.

The figures indicate that the results obtained for the 8.2 % noise-reduced time series closely resemble those of the noise-free data, whereas those obtained for 4.1

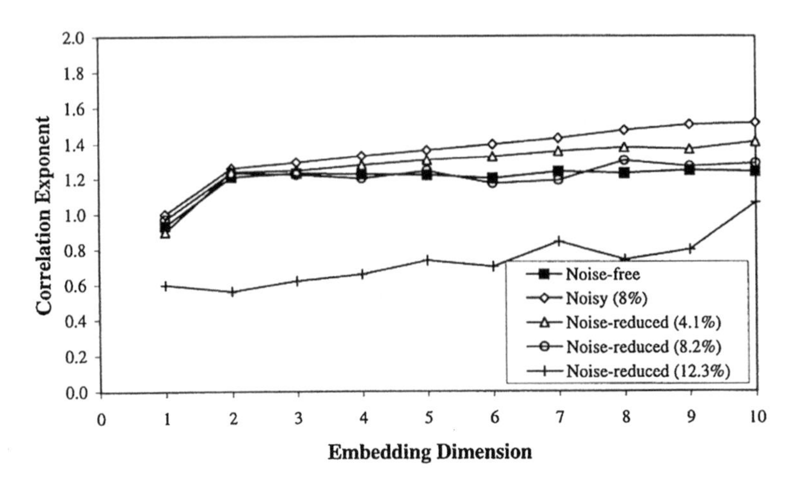

Fig. 7.6 Relationship between correlation exponent and embedding dimension for noise-free, 8 % noisy, and noise-reduced Henon data (*source* Sivakumar et al. (1999b))

and 12.3 % noise-reduced time series are significantly different. These results clearly indicate the effectiveness of a noise reduction method when coupled with a noise level determination method. Following these positive outcomes, Sivakumar et al. (1999b) tested the effectiveness of the method on real rainfall data from Singapore. The results indicated a noise level ranging from 9 and 11 % in the daily rainfall data. These noise levels are also in good agreement with the noise levels estimated for tipping-bucket-gaged rainfall data through other means (e.g. Sevruk 1996).

7.5 Zeros in Data

Most of the issues associated with chaos identification and prediction methods are common to almost all fields. For instance, delay time and embedding dimension for phase space reconstruction are inherent issues in chaos methods and, thus, need to be considered, regardless of the time series. Similarly, smaller data size and presence of noise are common problems in dealing with observed time series, regardless of the field of study. These are the reasons why these issues have received considerable attention in the analysis of chaos in real time series.

However, there are also other problems, which may be as serious as the above issues, that have not received the necessary attention in chaos analysis, because they are essentially encountered in just one or a few fields. For instance, the issue of the presence of a large number of zeros in time series is perhaps unique to the field of hydrology and water resources, as is the case with rainfall observations (and also streamflow observations, in many cases), especially at fine temporal resolutions.

One possible influence of the presence of a large number of zeros (or any other single value) is that the reconstructed hyper-surface in the phase space will tend to a point and may result in an underestimation of the correlation dimension (e.g. Tsonis et al. 1994; Sivakumar 2001).

It is not yet clear how to deal with the issue of a large number of zeros in a time series. Since zero values are also often indicative of, and equally important to understand, how the dynamics of the system evolve, it is questionable whether they should be eliminated from the time series in chaos analysis for more reliable results, although there can be exceptions depending upon the task at hand (e.g. Sivakumar et al. 2001). In fact, removal of the zeros can have serious effects and result in unrealistic outcomes, although the nature and extent of effects depend on the problem at hand. What is needed, therefore, is a careful understanding of the potential influences of the inclusion or exclusion of the zero values with respect to the particular task at hand. On the other hand, there may be alternative means to address the problem of zeros in chaos analysis, such as the verification of the results from one method with that of another; see, for example, Sivakumar et al. (1999a).

7.6 Other Issues

A number of other methodological and data issues have also been identified to influence the outcomes of chaos identification and prediction methods. Some of these issues are also related to one or more of the issues discussed above and among themselves as well. They include, for example, selection of optimal embedding dimension, selection of neighborhood, sampling frequency, temporal correlation, and scale of data, and even the potential inability of the methods to clearly distinguish between stochastic and chaotic systems. Identification of such issues has also resulted in a number of improvements to the methods. Extensive details of such issues, improvements to methods, and interpretations of the outcomes are available in the literature, and the interested reader is directed to the following studies, among others: Broomhead and King (1986), Theiler (1987), Čenys and Pyragas (1988), Grassberger (1988, 1990), Osborne and Provenzale (1989), Provenzale et al. (1991), Theiler et al. (1992), Tsonis et al. (1993, 1994), Zeng and Pielke (1993), Schreiber and Schmitz (1996), Cao (1997), Judd and Mees (1998), Sivakumar (2000, 2001, 2005b), Jayawardena et al. (2002, 2010), Small and Tse (2004), and Xu (2009).

7.7 Summary

Due to the basic assumptions that are difficult to meet for real data (e.g. infinite and noise-free time series) and lack of definitive guidelines in the implementation of analysis techniques (e.g. delay time, embedding dimension), applications of the ideas of chaos theory to real time series and their outcomes have often been subjected to skepticisms and criticisms. There is no question that many of these issues are important, since they can influence the reliability of the outcomes. This does not, however, mean that chaos methods cannot/should not be applied for real time series. What is actually needed is proper caution in the application of the methods and interpretation of the outcomes, not a literal translation of the assumptions and associated limitations. For instance, attempting to interpret the correlation dimension estimate of a time series in the context of the system and problem at hand is a useful endeavor, whereas looking for an 'infinite' and 'noise-free' time series to implement the correlation dimension algorithm is futile, as such a time series simply does not exist.

A majority of the studies that have employed chaos theory-based methods so far have indeed exercised proper caution in applying the methods to real data and interpreting the outcomes. Recognizing the many associated methodological and data issues, studies have not only offered numerous improvements to many of the earlier methods but also developed new methods that are more suited for real applications. In many cases, more than one method have been employed for the same time series and the results cross-verified to gain confidence in the reliability of the outcomes. This chapter has cited a number of important studies across various fields that present evidence to these. In the following chapters, the focus is on the applications of chaos theory in hydrology.

References

Aguirre LA (1995) A nonlinear correlation function for selecting the delay time in dynamical reconstructions. Phys Lett 203A(2,3):88–94

Albano AM, Passamante A, Farrell ME (1991) Using high-order correlations to define an embedding window. Physica D 54(1–2):85–97

Badii R, Broggi G, Derighetti B, Ravani M, Ciliberto S, Politi A, Rubio MA (1988) Dimension increase in filtered chaotic signals. Phys Rev Lett 60(11):979–982

Bonachela JA, Hinrichsen H, Muñoz M (2008) Entropy estimations of small data set. J Phys A: Math Theor 41:202001

Briggs K (1990) An improved method for estimating Lyapunov exponents of chaotic time series. Phys Lett A 151:27–32

Brock WA, Hsieh DA, Lebaron B (1991) Nonlinear dynamics, chaos, and instability: statistical theory and economic evidence. MIT Press, Cambridge

Brock WA, Dechert WA, Scheinkman JA, LeBaron B (1996) A test for independence based on the correlation dimension. Econ Rev 15(3):197–235

Broomhead DS, King GP (1986) Extracting qualitative dynamics from experimental data. Physica D 20:217–236

Cao L (1997) Practical method for determining the minimum embedding dimension of a scalar time series. Physica D 110:43–50

Cao L, Mees A, Judd K (1998) Dynamics from multivariate time series. Physica D 121:75–88

Casdagli M (1989) Nonlinear prediction of chaotic time series. Physica D 35:335–356

Casdagli M (1992) Chaos and deterministic versus stochastic nonlinear modeling. J Royal Stat Soc B 54(2):303–328

Casdagli M, Eubank S, Farmer JD, Gibson J (1991) State space reconstruction in the presence of noise. Physica D 51:52–98
Cawley R, Hsu GH (1992) Local-geometric-projection method for noise reduction in chaotic maps and flows. Phys Rev A 46(6):3057–3082
Čenys and Pyragas (1988) Estimation of the number of degrees of freedom from chaotic time series. Phys Lett A 129(4):227–230
Chelidze D (2014) Smooth local subspace projection for nonlinear noise reduction. Chaos 24:013121
Chennaoui A, Pawelzik K, Liebert W, Schuster H, Pfister G (1990) Attractor reconstruction from filtered chaotic time series. Phys Rev A 41(8):4151–4159
Davies M (1992) Noise reduction by gradient descent. Int J Bifurcation Chaos 3:113–118
Davies M (1994) Noise reduction schemes for chaotic time series. Physica D 79:174–192
Diks C (1999) Nonlinear time series analysis, methods and applications. World Scientific, Singapore
Enge N, Buzug Th, Pfister G (1993) Noise reduction on chaotic attractors. Phys Lett A 175:178–186
Essex C (1991) Correlation dimension and data sample size. In: Schertzer D, Lovejoy S (eds) Non-linear variability in geophysics, scaling and fractals. Kluwer, Dordrecht, The Netherlands, pp 93–98
Eubank SG, Farmer D (1990) An introduction to chaos and randomness. In: Jen E (ed) 1989 Lectures in complex systems. Santa Fe Institute Studies in the Sciences of Complexity, Addison-Wesley, Redwood City, CA, pp 75–190
Farmer DJ, Sidorowich JJ (1987) Predicting chaotic time series. Phys Rev Lett 59:845–848
Farmer JD, Sidorowich JJ (1991) Optimal shadowing and noise reduction. Physica D 47:373–392
Fraser AM, Swinney HL (1986) Independent coordinates for strange attractors from mutual information. Phys Rev A 33(2):1134–1140
Frison T (1994) Nonlinear data analysis techniques. In: Deboeck GJ (ed) Trading on the edge: neural, genetic, and fuzzy systems for chaotic financial markets. Wiley, New York, pp 280–296
Grassberger P (1988) Finite sample corrections to entropy and dimension estimates. Phys Lett A 128(6–7):369–373
Grassberger P (1990) An optimized box-assisted algorithm for fractal dimensions. Phys Lett A 148:63–68
Grassberger P, Procaccia I (1983a) Measuring the strangeness of strange attractors. Physica D 9:189–208
Grassberger P, Procaccia I (1983b) Characterisation of strange attractors. Phys Rev Lett 50 (5):346–349
Grassberger P, Procaccia I (1983c) Estimation of the Kolmogorov entropy from a chaotic signal. Phys Rev A 28:2591–2593
Grassberger P, Schreiber T, Schaffrath C (1991) Non-linear time sequence analysis. Int J Bifur Chaos 1(3):521–547
Grassberger P, Hegger R, Kantz H, Schaffrath C (1993) On noise reduction methods for chaotic data. Chaos 3(2):127–141
Hammel SM (1990) A noise reduction method for chaotic systems. Phys Lett A 148(8/9):421–428
Havstad JW, Ehlers CL (1989) Attractor dimension of nonstationary dynamical systems from small data sets. Phys Rev A 39(2):845–853
Heald JPM, Stark J (2000) Estimation of noise levels for models of chaotic dynamical systems. Phys Rev Lett 84(11):2366–2369
Henon M (1976) A two-dimensional mapping with a strange attractor. Commun Math Phys 50:69–77
Hirata Y, Suzuki H, Aihara K (2006) Reconstructing state phases from multivariate data using variable delays. Phys Rev E 47:026202
Holzfuss J, Mayer-Kress G (1986) An approach to error-estimation in the application of dimension algorithms. In: Mayer-Kress G (ed) Dimensions and entropies in chaotic systems. Springer, New York, pp 114–122

Jayawardena AW, Lai F (1994) Analysis and prediction of chaos in rainfall and stream flow time series. J Hydrol 153:23–52
Jayawardena AW, Li WK, Xu P (2002) Neighborhood selection for local modeling and prediction of hydrological time series. J Hydrol 258:40–57
Jayawardena AW, Xu PC, Li WK (2008) A method of estimating the noise level in a chaotic time series. Chaos 18(2):023115. doi:10.1063/1.2903757
Jayawardena AW, Xu PC, Li WK (2010) Modified correlation entropy estimation for a noisy chaotic time series. Chaos 20:023104
Judd K, Mees AI (1998) Embedding as a modeling problem. Physica D 120:273–286
Kantz H, Schreiber T (2004) Nonlinear time series analysis. Cambridge University Press, Cambridge
Kantz H, Schreiber T, Hoffmann I, Buzug T, Pfister G, Flepp LG, Simonet J, Badii R, Brun E (1993) Nonlinear noise reduction: a case study on experimental data. Phys Rev E 48(2):1529–1538
Kembe G, Fowler AC (1993) A correlation function for choosing time delays in phase portrait reconstructions. Phys Lett A 179(2):72–80
Kennel MB, Brown R, Abarbanel HDI (1992) Determining embedding dimension for phase space reconstruction using a geometric method. Phys Rev A 45:3403–3411
Kim HS, Eykholt R, Salas JD (1998) Delay time window and plateau onset of the correlation dimension for small data sets. Phys Rev E 58(5):5676–5682
Kim HS, Eykholt R, Salas JD (1999) Nonlinear dynamics, delay times, and embedding windows. Physica D 127(1–2):48–60
Kim HS, Lee KH, Kyoung MS, Sivakumar B, Lee ET (2009) Measuring nonlinear dependence in hydrologic time series. Stoch Environ Res Risk Assess 23:907–916
Kostelich EJ, Yorke JA (1988) Noise reduction in dynamical systems. Phys Rev A 38(3):1649–1652
Kostelich EJ, Yorke JA (1990) Noise reduction: finding the simplest dynamical system consistent with the data. Physica D 41:183–196
Kostelich EJ, Schreiber T (1993) Noise reduction in chaotic timeseries data: a survey of common methods. Phys Rev E 48(3):1752–1763
Kugiumtzis D (1996) State space reconstruction parameters in the analysis of chaotic time series—the role of the time window length. Physica D 95(1):13–28
Liebert W, Schuster HG (1989) Proper choice of the time delay for the analysis of chaotic time series. Phys Lett A 141:386–390
Lin J, Wang Y, Huang Z, Shen Z (1999) Selection of proper time-delay in phase space reconstruction of speech signals. Signal Process 15(2):220–225
Lorenz EN (1991) Dimension of weather and climate attractors. Nature 353:241–244
Luo XD, Zhang J, Small M (2005) Optimal phase space projection for noise reduction. Phys Rev E 72:046710
Mackey MC, Glass L (1977) Oscillation and chaos in physiological control systems. Science 197 (4300):287–289
Marteau PF, Abarbanel HDI (1991) Noise reduction in chaotic time series using scaled probabilistic method. J Nonlinear Sci 1:313–349
Mikosch T, Wang Q (1995) A Monte-Carlo method for estimating the correlation dimension. J Stat Phys 78:799–813
Martinerie JM, Albano AM, Mees AI, Rapp PE (1992) Mutual information, strange attractors, and the optimal estimation of dimension. Phys Rev A 45:7058–7064
Nakamura T, Small M (2006) Nonlinear dynamical system identification with dynamic noise and observational noise. Physica D 223:54–68
Nerenberg MAH, Essex C (1990) Correlation dimension and systematic geometric effects. Phys Rev A 42(12):7065–7074
Osborne AR, Provenzale A (1989) Finite correlation dimension for stochastic systems with power-law spectra. Physica D 35:357–381

Packard NH, Crutchfield JP, Farmer JD, Shaw RS (1980) Geometry from a time series. Phys Rev Lett 45(9):712–716

Palit SK, Mukherjee S, Bhattacharya DK (2013) A high dimensional delay selection for the reconstruction of proper phase space with cross auto-correlation. Neurocomputing 113:49–57

Pecora LM, Moniz L, Nichols J, Carroll TL (2007) A unified approach to attractor reconstruction. Chaos 17, 013110. doi:10.1063/1.2430294

Pfister G, Buzug Th (1992a) Optimal delay time and embedding dimension for delay-time coordinates by analysis of the global static and local dynamical behaviour of strange attractors. Phys Rev A 45(10):7073–7084

Pfister G, Buzug Th (1992b) Comparison of algorithms calculating optimal embedding parameters for delay time coordinates. Physica D 58(1–4):127–137

Porporato A, Ridolfi R (2001) Multivariate nonlinear prediction of river flows. J Hydrol 248(1–4):109–122

Provenzale A, Osborne AR, Soj R (1991) Convergence of the K2 entropy for random noises with power law spectra. Physica D 47:361–372

Ramsey JB, Yuan HJ (1990) The statistical properties of dimension calculations using small data sets. Nonlinearity 3:155–176

Rosenstein MT, Collins JJ, De Luca CJ (1993) A practical method for calculating largest Lyapunov exponents from small data sets. Physica D 65:117–134

Rosenstein MT, Colins JJ, De Luca CJ (1994) Reconstruction expansion as a geometry-based framework for choosing proper delay times. Physica D 73(1):82–98

Rössler OE (1976) An equation for continuous chaos. Phys Lett A 57:397–398

Sangoyomi TB, Lall U, Abarbanel HDI (1996) Nonlinear dynamics of the Great Salt Lake: dimension estimation. Water Resour Res 32(1):149–159

Sauer T (1992) A noise reduction method for signals from nonlinear systems. Physica D 58:193–201

Sauer T, Yorke JA, Casdagli M (1991) Embedology. J Stat Phys 65(3/4):579–616

Schertzer D, Tchiguirinskaia I, Lovejoy S, Hubert P, Bendjoudi H (2002) Which chaos in the rainfall-runoff process? A discussion on 'Evidence of chaos in the rainfall-runoff process' by Sivakumar et al. Hydrol Sci J 47(1):139–147

Schouten JC, Takens F, van den Bleek CM (1994) Estimation of the dimension of a noisy attractor. Phys Rev E 50(3):1851–1861

Schreiber T (1993a) Determination of the noise level of chaotic time series. Phys Rev E 48(1): R13–R16

Schreiber T (1993b) Extremely simple nonlinear noise reduction method. Phys Rev E 47(4):2401–2404

Schreiber T, Grassberger P (1991) A simple noise reduction method for real data. Phys Lett A 160:411–418

Schreiber T, Kantz H (1996) Observing and predicting chaotic signals: is 2% noise too much? In: Kadtke JB (ed) Kravtsov YuA. Predictability of complex dynamical systems, Springer Series in Synergetics, Springer, Berlin, pp 43–65

Schreiber T, Schmitz A (1996) Improved surrogate data for nonlinearity tests. Phys Rev Lett 77 (4):635–638

Schuster HG (1988) Deterministic chaos. VCH, Weinheim

Sevruk B (1996) Adjustment of tipping-bucket precipitation gage measurement. Atmos Res 42:237–246

Siefert M, Kittel A, Friedrich R, Peinke J (2003) On a quantitative method to analyze dynamical and measurement noise. Europhys Lett 61(4):466–472

Sivakumar B (2000) Chaos theory in hydrology: important issues and interpretations. J Hydrol 227 (1–4):1–20

Sivakumar B (2001) Rainfall dynamics at different temporal scales: a chaotic perspective. Hydrol Earth Syst Sci 5(4):645–651

Sivakumar B (2005a) Correlation dimension estimation of hydrologic series and data size requirement: myth and reality. Hydrol Sci J 50(4):591–604

Sivakumar B (2005b) Chaos in rainfall: variability, temporal scale and zeros. J Hydroinform 7 (3):175–184
Sivakumar B, Liong SY, Liaw CY, Phoon KK (1999a) Singapore rainfall behavior: chaotic? J Hydrol Engg 4(1):38–48
Sivakumar B, Phoon KK, Liong SY, Liaw CY (1999b) A systematic approach to noise reduction in chaotic hydrological time series. J Hydrol 219(3–4):103–135
Sivakumar B, Sorooshian S, Gupta HV, Gao X (2001) A chaotic approach to rainfall disaggregation. Water Resour Res 37(1):61–72
Sivakumar B, Berndtsson R, Olsson J, Jinno K (2002a) Reply to 'which chaos in the rainfall-runoff process?' by Schertzer et al. Hydrol Sci J 47(1):149–158
Sivakumar B, Persson M, Berndtsson R, Uvo CB (2002b) Is correlation dimension a reliable indicator of low-dimensional chaos in short hydrological time series? Water Resour Res 38(2). doi:10.1029/2001WR000333
Sivakumar B, Berndtsson R, Persson M, Uvo CB (2005) A multi-variable time series phase-space reconstruction approach to investigation of chaos in hydrological processes. Int J Civil Environ Engg 1(1):35–51
Small M, Tse CK (2004) Optimal embedding: a modelling paradigm. Physica D 194:283–296
Smith LA (1988) Intrinsic limits on dimension calculations. Phys Lett A 133(6):283–288
Sun J, Zhao Y, Zhang J, Luo X, Small M (2007) Reducing coloured noise for chaotic time series in the local phase space. Phys Rev E 76:026211
Takens F (1981) Detecting strange attractors in turbulence. In: Rand DA, Young LS (eds) Dynamical systems and turbulence, vol 898., Lecture Notes in Mathematics Springer-Verlag, Berlin, Germany, pp 366–381
Theiler J (1986) Spurious dimension from correlation algorithms applied to limited time series data. Phys Rev A 34:2427–2432
Theiler J (1987) Efficient algorithm for estimating the correlation dimension from a set of discrete points. Phys Rev A 36:4456–4462
Theiler J (1990) Statistical precision in dimension estimators. Phys Rev A 41:3038–3051
Theiler J (1991) Some comments on the correlation dimension of $1/f^{\alpha}$ noise. Phys Lett A 155:480–493
Theiler J, Eubank S, Longtin A, Galdrikian B, Farmer JD (1992) Testing for nonlinearity in time series: the method of surrogate data. Physica D 58:77–94
Tsonis AA (1992) Chaos: from theory to applications. Plenum Press, New York
Tsonis AA, Elsner JB (1988) The weather attractor over short timescales. Nature 333:545–547
Tsonis AA, Elsner JB, Georgakakos KP (1993) Estimating the dimension of weather and climate attractors: important issues about the procedure and interpretation. J Atmos Sci 50:2549–2555
Tsonis AA, Triantafyllou GN, Elsner JB, Holdzkom JJ II, Kirwan AD Jr (1994) An investigation on the ability of nonlinear methods to infer dynamics from observables. Bull Amer Meteor Soc 75:1623–1633
Urbanowicz K, Hołyst JA (2006) Noise estimation by use of neighboring distances in Takens space and its applications to stock market data. Int J Bifurcation Chaos 16(6):1865–1869
Uzal LC, Grinblat GL, Verdes PF (2011) Optimal reconstruction of dynamical systems: a noise amplification approach. Phys Rev E 83:016223
Wang Q, Gan TY (1998) Biases of correlation dimension estimates of streamflow data in the Canadian prairies. Water Resour Res 34(9):2329–2339
Wolf A, Swift JB, Swinney HL, Vastano A (1985) Determining Lyapunov exponents from a time series. Physica D 16:285–317
Xu P (2009) Differential phase space reconstructed from chaotic time series. Appl Math Model 33:999–1013
Zeng X, Pielke RA (1993) What does a low-dimensional weather attractor mean? Phys Lett A 175:299–304
Zeng X, Eykholt R, Pielke RA (1991) Estimating the Lyapunov-exponent spectrum from short time series of low precision. Phys Rev Lett 66:3229–3232

Part III
Applications of Chaos Theory in Hydrology

Chapter 8
Overview

Abstract Over the past three decades, the concepts of chaos theory have been extensively applied in hydrology. Applications of chaos theory in hydrology started with the basic identification of chaos in rainfall data and subsequently explored a wide range of problems in different types of hydrologic data. The problems studied include identification and prediction of chaos, scaling and disaggregation, missing data estimation, and catchment classification, among others. The data studied include rainfall, river flow, rainfall-runoff, lake volume and level, sediment transport, and groundwater, among others. Many studies have also addressed some of the important issues in the applications of chaos methods in hydrology, including delay time, data size, data noise, and presence of zeros in data. This chapter presents an overview of chaos studies in hydrology. The presentation is organized to reflect three stages of development: early stage (1980s–1990s), change of course (2000–2006), and studies on global-scale challenges (since 2007).

8.1 Introduction

Development of chaos identification methods in the early 1980s, such as phase space reconstruction (e.g. Packard et al. 1980; Takens 1981), correlation dimension method (e.g. Grassberger and Pracaccia 1983a, b), Kolmogorov entropy method (e.g. Grassberger and Pracaccia 1983c), and Lyapunov exponent method (e.g. Wolf et al. 1985), led to the initial applications of such approaches in hydrology in the latter part of that decade (e.g. Hense 1987; Rodriguez-Iturbe et al. 1989). With subsequent improvements to these methods as well as development of others for nonlinearity and chaos detection and prediction, such as nonlinear local approximation prediction method (e.g. Farmer and Sidorowich 1987; Casdagli 1989, 1992; Sugihara and May 1990), false nearest neighbor algorithm (e.g. Kennel et al. 1992), noise level determination and reduction methods (e.g. Schreiber and Grassberger 1991; Grassberger et al. 1993; Schreiber 1993a, b; Schouten et al. 1994), surrogate data method (e.g. Theiler et al. 1992; Schreiber and Schmitz 1996), close returns plot (e.g. Gilmore 1993), method of redundancy (e.g. Paluš 1995; Prichard and

B. Sivakumar, *Chaos in Hydrology*, DOI 10.1007/978-90-481-2552-4_8

Theiler 1995), nonlinear interpolation (e.g. Amritkar and Pradeep Kumar 1995), and multi-variate analysis technique (e.g. Cao et al. 1998), concepts and ideas gained from chaos theory have found numerous applications in hydrology since the 1990s. Table 8.1 presents a short list of such studies, especially those that have reported the dimensionality (obtained solely or primarily using the correlation dimension method) of the time series analyzed.

Further details on many of these studies are presented in the forthcoming chapters. Considering that investigations on rainfall and river flow time series have dominated chaos studies in hydrology, such are discussed in separate chapters, with Chap. 9 for rainfall and Chap. 10 for river flow. Chapter 11 discusses chaos studies on other hydrologic time series, including rainfall-runoff, lake volume and level, sediment transport, and groundwater, among others. Chapter 12 discusses the studies on hydrologic data-related issues in the applications of chaos theory. In the current chapter, only a brief account of the development of chaos theory applications in hydrology is presented, in three different stages.

8.2 Early Stage (1980s–1990s)

The very first study on chaos theory application in a hydrologic context was probably the one conducted by Hense (1987) on rainfall time series, although chaos in rainfall time series had already been investigated previously in the context of climate and weather (e.g. Fraedrich 1986, 1987). However, the first study that actually became known to many researchers in hydrology, both for the authors' field of expertise (i.e. hydrology) and for the popularity of the journal among water researchers (i.e. *Water Resources Research*), was the one entitled "Chaos in hydrology" by Rodriguez-Iturbe et al. (1989). The study investigated the presence of chaos in two different rainfall time series (data of a storm event in Boston and data of weekly rainfall in Genoa) using correlation dimension method and Lyapunov exponent method, and reported mixed results for the two series; see also Ghilardi and Rosso (1990) and Rodriguez-Iturbe et al. (1990) for related discussions, including the existence of the study by Hense (1987) and issues related to chaos identification methods and data limitations.

The study by Rodriguez-Iturbe et al. (1989) was a significant motivation to many other studies on chaos theory applications in hydrology in the ensuing years. Studies during the first few years of the 1990s focused mainly on the investigation and prediction of chaos in rainfall, river flow, and lake volume time series in a purely single-variable data reconstruction sense (e.g. Sharifi et al. 1990; Wilcox et al. 1991; Islam et al. 1993; Berndtsson et al. 1994; Jayawardena and Lai 1994; Waelbroeck et al. 1994; Georgakakos et al. 1995; Abarbanel and Lall 1996; Abarbanel et al. 1996; Puente and Obregon 1996; Sangoyomi et al. 1996; Porporato and Ridolfi 1996). In addition to the correlation dimension and Lyapunov exponent methods, these studies also employed other chaos identification and prediction methods, including Kolmogorov entropy method (e.g. Jayawardena and Lai 1994;

Table 8.1 A short list of chaos studies in hydrology

Data	Dimension	References
Rainfall		
1. Monthly rainfall in Nauru Island	2.5–4.5	Hense (1987)
2. 15-s rainfall intensity in Boston	3.78	Rodriguez-Iturbe et al. (1989)
3. Weekly rainfall in Genoa	No low	Rodriguez-Iturbe et al. (1989)
4. 15-s rain intensity in Boston (3 stations)	3.35, 3.75, 3.60	Sharifi et al. (1990)
5. 10-s rain intensity from cloud model	1.5	Islam et al. (1993)
6. Time between successive raingage signals each corresponding to 0.01 mm rain	2.4	Tsonis et al. (1993)
7. Daily rainfall in Hong Kong (3 stations)	0.95, 1.76, 1.65	Jayawardena and Lai (1994)
8. Monthly rainfall in Lund, Sweden		
Raw	No low	Berndtsson et al. (1994)
Noise-reduced	<4	Berndtsson et al. (1994)
9. Storm events in Iowa City (11 storms)		
High-intensity storms	2.8–7.9	Georgakakos et al. (1995)
Low-intensity storms	0.5–1.6	Georgakakos et al. (1995)
One storm event	No low	Georgakakos et al. (1995)
10. 15-min rainfall in Greece	No low	Koutsoyiannis and Pachakis (1996)
11. Daily rainfall in Singapore (6 stations)	1.01, 1.03, 1.06, 1.03, 1.02, 1.06	Sivakumar et al. (1998, 1999a, b)
12. Monthly rainfall in Göta River, Sweden	6.4	Sivakumar et al. (2000, 2001a)
13. Rainfall in Leaf River basin, USA (daily, 2-day, 4-day, 8-day)	4.82, 5.26, 6.42, 8.87	Sivakumar (2001a)
14. Monthly rainfall in Coaracy Nunes/Araguari River watershed	4.4	Sivakumar et al. (2005a)
15. Rainfall in California (Daily, Weekly, Biweekly, Monthly, Winter daily)	0.76, 1.81, 4.82, 7.13, 1.71	Sivakumar et al. (2006)
16. Monthly rainfall in Seoul		
Observed rainfall (1971–1999)	5.30	Kyoung et al. (2011)
GCM-generated rainfall (1951–1999)	4.50	Kyoung et al. (2011)
GCM-projected rainfall (2000–2099)	3.65	Kyoung et al. (2011)
17. Monthly rainfall in Western Australia (62 stations)	4.63–8.29	Sivakumar et al. (2014)
18. Daily rainfall in Koyna reservoir, India (average of nine stations)	<1.0	Jothiprakash and Fathima (2013)

(continued)

Table 8.1 (continued)

Data	Dimension	References
River flow		
1. Daily runoff in southwest Idaho, USA	No low	Wilcox et al. (1991)
2. Daily flow in Hong Kong (two stations)	0.455, 0.460	Jayawardena and Lai (1994)
3. Daily flow in Po River, Italy	<4	Porporato and Ridolfi (1996, 1997)
4. Daily flow in Canadian Prairies (6 rivers)	7–9	Wang and Gan (1998)
5. Flow from Uhlirska basin, Czech Republic (Daily and 30-min)	No low, 2.89	Stehlik (1999)
6. Daily flow in Scandinavian region	Low	Krasovskaia et al. (1999)
7. Daily flow in Western Run, MD, USA	No low	Pasternack (1999)
8. Daily discharge of Chao Phraya River, Thailand (at Nakhon Sawan)	2.9	Jayawardena and Gurung (2000)
9. Daily discharge of Mekong River in Thailand (at Nong Khai and Pakse)	1.69, 1.58	Jayawardena and Gurung (2000)
10. Daily discharge of spring Almyros, Greece	3–4	Lambrakis et al. (2000)
11. Monthly flow in Göta River, Sweden	5.5	Sivakumar et al. (2000, 2001a)
12. Daily flow in Adige River, Trento, Italy	2.8	Lisi and Villi (2001)
13. Monthly flow in Coaracy/Nunes, Brazil	3.62	Sivakumar et al. (2001c, 2002c)
14. Daily flow in Little River and Reed Creek, VA, USA	1.19, 1.07	Elshorbagy et al. (2001)
15. Daily flow in English River, Canada	2.4	Elshorbagy et al. (2002a, b)
16. Daily flow in Lindenborg, Denmark	3.76	Islam and Sivakumar (2002)
17. Daily flow in Tryggevaelde, Denmark	1.4	Phoon et al. (2002)
18. Daily flow in Altamaha River, USA	0.85	Phoon et al. (2002)
19. Daily flow in Mississippi River, MO, USA	2.32	Sivakumar and Jayawardena (2002)
20. Annual flood series in Huaihe River Basin in China	4.66	Zhou et al. (2002)
21. Kentucky River, Kentucky, USA (Daily, 5-day, Weekly)	4.22, 4.63, 4.87	Regonda et al. (2004)
22. Merced River, California, USA (Daily, 5-day, Weekly)	3.5, 4.7, 5.5	Regonda et al. (2004)

(continued)

Table 8.1 (continued)

Data	Dimension	References
23. Stillaguamish River, Washington State, USA (Daily, 5-day, Weekly)	No low, no low, no low	Regonda et al. (2004)
24. Daily flow from Mahanathi River, India (2 stations)	6–7	Dhanya and Nagesh Kumar (2011b)
25. Daily flow from Sogutluhan hydrometric station, Turkey (Stage, Discharge)	2.9, 2.4	Khatibi et al. (2012)
26. Monthly flow in the western United States (117 stations)	Ranging from low to high	Sivakumar and Singh (2012)
27. Daily flow in seven major sub-basins of Rhine River	Ranging from low to high	Tongal et al. (2013)
Rainfall-Runoff		
Monthly runoff coefficient in Göta River, Sweden	7.8	Sivakumar et al. (2000, 2001a)
Lake volume/level		
1. Bi-weekly volume in the Great Salt Lake, USA	3.4	Sangoyomi et al. (1996), Abarbanel and Lall (1996), Abarbanel et al. (1996)
2. Daily lake water level in Sweden (Vänern, Vättern, and Mälaren)	3.37, 3.97, 4.44	Tongal and Berndtsson (2014)
Sediment transport		
1. Daily discharge, suspended sediment conc and load in Mississippi River, USA	2.32, 2.55, 2.41	Sivakumar and Jayawardena (2002)
2. Daily suspended sediment concentration in Yellow River, Tongguan, China	6.6	Shang et al. (2009)
Groundwater		
1. Simulated annual solute transport in San Joaquin Valley system, California	2.12	Sivakumar et al. (2005b)
2. Arsenic contamination in Bangladesh (3085 shallow wells)		
All wells (Whole of Bangladesh)	10–11	Hossain and Sivakumar (2006)
Holocene deposits (Southwest region)	10–11	Hossain and Sivakumar (2006)
Pleistocene deposits (Northwest region)	8–9	Hossain and Sivakumar (2006)

Puente and Obregon 1996), nonlinear prediction method (e.g. Jayawardena and Lai 1994; Abarbanel and Lall 1996; Abarbanel et al. 1996; Porporato and Ridolfi 1996), and Poincaré maps (e.g. Porporato and Ridolfi 1996). A few of these studies and several others that followed in the latter years of that decade also addressed, in addition to chaos identification/prediction, important methodological and data issues, including minimum data size requirement for correlation dimension estimation (e.g. Tsonis et al. 1993; Sivakumar et al. 1998, 1999a), effects of data noise on chaos identification and prediction, including noise level determination and reduction (e.g. Berndtsson et al. 1994; Tsonis et al. 1994; Porporato and Ridolfi 1997; Sivakumar et al. 1999b, c), effects of the presence of zeros on chaos identification and prediction (e.g. Tsonis et al. 1994; Koutsoyiannis and Pachakis 1996; Wang and Gan 1998; Sivakumar et al. 1999a), delay time/delay window selection for phase space reconstruction (Sangoyomi et al. 1996; Sivakumar et al. 1998, 1999a; Pasternack 1999), stochastic processes possibly leading to chaos identification (e.g. Wang and Gan 1998; Pasternack 1999; Sivakumar et al. 1999a), and others (e.g. Liu et al. 1998; Krasovskaia et al. 1999; Stehlik 1999).

8.3 Change of Course (2000–2006)

At the very beginning of this century, Sivakumar (2000) published the first ever review of chaos theory applications in hydrology. In light of earlier and continuing criticisms on chaos studies in hydrology (because of limitations of methods and data) and skepticisms on the reported outcomes (positive evidence of chaos) (e.g. Ghilardi and Rosso 1990; Koutsoyiannis and Packakis 1996; Wang and Gan 1998), the review put particular emphasis on addressing the important issues in chaos theory applications in hydrology and also interpreting the reported results. The comprehensive and balanced nature of the review significantly helped allay many of the earlier fears and misgivings about chaos studies in hydrology and their outcomes and completely changed the course of chaos theory in hydrology for ever. Although there have and continue to be some issues related to chaos studies in hydrology (which are discussed in the forthcoming chapters), there is no question that the study by Sivakumar (2000) clearly put such issues in a proper perspective for constructive future discussions and deliberations, as is normally the case with any other scientific theory as well.

The study by Sivakumar (2000) led, either directly or indirectly, to rapid and exciting advances in chaos studies in hydrology in the years that immediately followed, both in theory and in application. During that period, chaos theory was also applied to still other hydrologic processes and associated problems. The hydrologic processes studied include: rainfall-runoff (e.g. Sivakumar et al. 2000, 2001a; Dodov and Foufoula-Georgiou 2005), sediment transport (e.g. Sivakumar 2002; Sivakumar and Jayawardena 2002, 2003; Sivakumar and Wallender 2004, 2005), soil nutrient cycles (e.g. Manzoni et al. 2004), and subsurface flow and solute transport (e.g. Faybishenko 2002; Sivakumar et al. 2005b), including arsenic contamination in groundwater (e.g. Hossain and Sivakumar 2006). The hydrologic problems studied include: scaling and data aggregation/disaggregation (e.g. Sivakumar 2001a, b; Sivakumar et al. 2001b, 2004b; Regonda et al. 2004; Sivakumar and Wallender 2004; Salas et al. 2005; Gaume et al. 2006), including the development of a new chaotic approach for disaggregation of hydrologic data

(Sivakumar et al. 2001b), missing data estimation (e.g. Elshorbagy et al. 2001, 2002a), reconstruction of system equations (e.g. Zhou et al. 2002), regional hydrology and river flow regimes (Sivakumar 2003), parameter estimation (Hossain et al. 2004), and model integration (e.g. Sivakumar 2004b). Further, several studies addressed the issues of data size, noise, zeros, selection of optimal parameters, and others (e.g. Jayawardena and Gurung 2000; Sivakumar 2001a, 2005a, 2005b; Islam and Sivakumar 2002; Elshorbagy et al. 2002b; Jayawardena et al. 2002; Phoon et al. 2002; Schertzer et al. 2002; Sivakumar et al. 2002a, c, 2006; Laio et al. 2004; Khan et al. 2005; Salas et al. 2005; She and Basketfield 2005; Koutsoyiannis 2006). Some studies also compared hydrologic predictions based on chaos methods with those based on other techniques, such as stochastic methods and artificial neural networks (e.g. Jayawardena and Gurung 2000; Lambrakis et al. 2000; Lisi and Villi 2001; Sivakumar et al. 2002b, c; Laio et al. 2003). Further, attempts were also made to perform chaos analysis in hydrologic time series based on multi-variable phase space reconstruction (e.g. Porporato and Ridolfi 2001; Jin et al. 2005; Sivakumar et al. 2005a), rather than reconstruction based on just a single variable. Some studies also used other data analysis methods for prediction of hydrologic time series that exhibit chaotic behavior (e.g. Karunasinghe and Liong 2006).

A comprehensive review of many of these studies (and earlier ones) was presented by Sivakumar (2004a), within the broader context of reviewing chaos theory applications in geophysics. Pointing out the fundamental ideas of chaos theory (nonlinear interdependence, determinism and order, and sensitivity to initial conditions) as well as their obvious relevance in hydrology, Sivakumar (2004a) also argued that chaos theory should not be viewed as a separate theory but rather as a theory that connects the two dominant theories in existence: deterministic and stochastic. Research into advancing this idea towards developing a balanced 'middle-ground' approach for hydrologic modeling, as a potential alternative to our existing dominant one-sided 'extreme-view' deterministic and stochastic approaches, has and continues to be pursued since then. The study by Sivakumar (2004a) and also the publication of a special issue (in the journal *Nonlinear Processes in Geophysics*), exclusively focusing on the status and challenges in the study of nonlinear deterministic dynamics in hydrologic systems (Sivakumar et al. 2004a), helped put chaos theory in hydrology at an even higher pedestal. Chaos theory studies became part of 'mainstream' hydrology.

8.4 Looking at Global-scale Challenges (2007–)

With significant inroads made in the application of chaos theory and related concepts in hydrology over two decades, there has been increasing realization in more recent years on the opportunities and possibilities for studying large-scale problems using chaos theory. This realization has certainly come at an important time, as we are currently (and will also most likely be in the future) facing some tremendously challenging global-scale issues in hydrology and water resources, including (1) assessment of the impacts of global climate change on water resources for devising appropriate adaptation and mitigation strategies; (2) development of a generic catchment classification framework for more effective and efficient hydrologic modeling and forecasting; (3) identification and evaluation of connections between hydrologic systems and other systems, including ecologic, human, and economic systems; and (4) study of issues related to transboundary waters (river basins as well as aquifers) for improving planning and management of such waters and eliminating/alleviating water crisis and conflicts, among others. Further details on these global-scale issues and the associated challenges can be found in, for example, McDonnell and Woods (2004), Paola et al. (2006), Wagener et al. (2010), and Sivakumar (2011a, b) and, therefore, are not reported herein; see also Sivakumar and Singh (2015) for a compilation of some of the grand challenges in hydrology.

In the last few years, some preliminary, yet important, attempts have been made to address at least two of these global-scale hydrologic problems using ideas from chaos theory. Sivakumar et al. (2007) explored the utility of phase space reconstruction for assessing the complexity of hydrologic systems and thus their classification, first by demonstrating their approach on two artificial (stochastic and chaotic) time series and then by testing it on several real river-related hydrologic series. Following up on this, Sivakumar and Singh (2012) and Sivakumar et al. (2015) present a more comprehensive scientific background and discussion for proposing system complexity as a basis for catchment classification framework and nonlinear dynamic concepts as a suitable methodology for assessing system complexity. Kyoung et al. (2011) investigated the dynamic characteristics of rainfall under conditions of climate change, through analysis of observed and global climate model (GCM)-simulated (present and future) monthly rainfall in the Korean Peninsula. They reported that the nature of rainfall dynamics falls more on the chaotic dynamic spectrum than on the linear stochastic spectrum and also that future (GCM)-simulated rainfall exhibits stronger nonlinearity and chaos compared to the present rainfall. These results emphasize the need for a chaotic dynamic-based framework for downscaling outputs from GCMs.

While applications of chaos theory to study these global-scale hydrologic challenges have been gaining momentum lately, other chaos studies in hydrology have been growing as well, including studies on sediment transport (e.g. Sivakumar and Chen 2007; Shang et al. 2009), arsenic contamination in groundwater (e.g. Hill et al. 2008), ensemble prediction of chaotic hydrologic time series (including use of wavelets and multivariate prediction with climate inputs) (e.g. Dhanya and Nagesh Kumar 2010, 2011a, b), river stage and discharge (Khatibi et al. 2012), and others (e.g. Sivakumar 2007; Kim et al. 2009; Khatibi et al. 2012; Sivakumar et al. 2014; Tongal and Berndtsson 2014). A review of some of the more recent chaos studies in hydrology, since Sivakumar (2004a), is presented in Sivakumar (2009), which reiterates the need for a middle-ground approach in hydrology and the role chaos theory can play in its formulation. Sivakumar (2011c) makes further philosophical and pragmatic arguments to this end.

8.5 Summary

Since the early development of chaos identification methods in the 1980s, chaos theory has found numerous applications in hydrology. In addition to chaos identification and prediction in hydrologic time series, a variety of other problems encountered in hydrology have also been studied. Many studies have also addressed some of the issues associated with the applications of chaos methods in hydrology, especially those related to data constraints. Recent and current efforts indicate that chaos theory is gaining momentum in addressing broader and global-scale issues in hydrology, including studies on catchment classification framework and global climate model outputs. This chapter has presented an overview of studies on chaos theory in hydrology. The next four chapters will review chaos studies on rainfall data (Chap. 9), river flow data (Chap. 10), other hydrologic data (Chap. 11), and on hydrologic data-related issues (Chap. 12).

References

Abarbanel HDI, Lall U (1996) Nonlinear dynamics of the Great Salt Lake: system identification and prediction. Climate Dyn 12:287–297

Abarbanel HDI, Lall U, Moon YI, Mann M, Sangoyomi T (1996) Nonlinear dynamics and the Great Salk Lake: a predictable indicator of regional climate. Energy 21(7/8):655–666

Amritkar RE, Pradeep Kumar P (1995) Interpolation of missing data using nonlinear and chaotic system analysis. J Geophys Res 100(D2):3149–3154

Berndtsson R, Jinno K, Kawamura A, Olsson J, Xu S (1994) Dynamical systems theory applied to long-term temperature and precipitation time series. Trends Hydrol 1:291–297

Cao L, Mees A, Judd K (1998) Dynamics from multivariate time series. Physica D 121:75–88

Casdagli M (1989) Nonlinear prediction of chaotic time series. Physica D 35:335–356

Casdagli M (1992) Chaos and deterministic versus stochastic nonlinear modeling. J Royal Stat Soc B 54(2):303–328

Dhanya CT, Nagesh Kumar D (2010) Nonlinear ensemble prediction of chaotic daily rainfall. Adv Water Resour 33:327–347

Dhanya CT, Nagesh Kumar D (2011a) Predictive uncertainty of chaotic daily streamflow using ensemble wavelet networks approach. Water Resour Res 47:W06507. doi:10.1029/2010WR010173

Dhanya CT, Nagesh Kumar D (2011b) Multivariate nonlinear ensemble prediction of daily chaotic rainfall with climate inputs. J Hydrol 403:292–306

Dodov B, Foufoula-Georgiou E (2005) Incorporating the spatio-temporal distribution of rainfall and basin geomorphology into nonlinear analysis of streamflow dynamics. Adv Water Resour 28(7):711–728

Elshorbagy A, Panu US, Simonovic SP (2001) Analysis of cross-correlated chaotic streamflows. Hydrol Sci J 46(5):781–794

Elshorbagy A, Simonovic SP, Panu US (2002a) Estimation of missing streamflow data using principles of chaos theory. JHydrol 255:123–133

Elshorbagy A, Simonovic SP, Panu US (2002b) Noise reduction in chaotic hydrologic time series: facts and doubts. J Hydrol 256:147–165

Farmer DJ, Sidorowich JJ (1987) Predicting chaotic time series. Phys Rev Lett 59:845–848

Faybishenko B (2002) Chaotic dynamics in flow through unsaturated fractured media. Adv Water Resour 25(7):793–816

Fraedrich K (1986) Estimating the dimensions of weather and climate attractors. J Atmos Sci 43:419–432

Fraedrich K (1987) Estimating weather and climate predictability on attractors. J Atmos Sci 44:722–728

Gaume E, Sivakumar B, Kolasinski M, Hazoumé L (2006) Identification of chaos in rainfall disaggregation: application to a 5-minute point series. J Hydrol 328(1–2):56–64

Georgakakos KP, Sharifi MB, Sturdevant PL (1995) Analysis of high-resolution rainfall data. In: Kundzewicz ZW (ed) New uncertainty concepts in hydrology and water resources. Cambridge University Press, New York, pp 114–120

Ghilardi P, Rosso R (1990) Comment on "Chaos in rainfall". Water Resour Res 26(8):1837–1839

Gilmore CG (1993) A new test for chaos. J Econ Behavior Organiz 22:209–237

Grassberger P, Procaccia I (1983a) Measuring the strangeness of strange attractors. Physica D 9:189–208

Grassberger P, Procaccia I (1983b) Characterisation of strange attractors. Phys Rev Lett 50(5):346–349

Grassberger P, Procaccia I (1983c) Estimation of the Kolmogorov entropy from a chaotic signal. Phys Rev A 28:2591–2593

Grassberger P, Hegger R, Kantz H, Schaffrath C (1993) On noise reduction methods for chaotic data. Chaos 3(2):127–141

Hense A (1987) On the possible existence of a strange attractor for the southern oscillation. Beitr Phys Atmos 60(1):34–47

Hill J, Hossain F, Sivakumar B (2008) Is correlation dimension a reliable proxy for the number of dominant influencing variables for modeling risk of arsenic contamination in groundwater? Stoch Environ Res Risk Assess 22(1):47–55

Hossain F, Sivakumar B (2006) Spatial pattern of arsenic contamination in shallow wells of Bangladesh: regional geology and nonlinear dynamics. Stoch Environ Res Risk Assess 20(1–2):66–76

Hossain F, Anagnostou EN, Lee KH (2004) A non-linear and stochastic response surface method for Bayesian estimation of uncertainty in soil moisture simulation from a land surface model. Nonlinear Process Geophys 11:427–440

Islam MN, Sivakumar B (2002) Characterization and prediction of runoff dynamics: A nonlinear dynamical view. Adv Water Resour 25(2):179–190

Islam S, Bras RL, Rodriguez-Iturbe I (1993) A possible explanation for low correlation dimension estimates for the atmosphere. J Appl Meteor 32:203–208

Jayawardena AW, Gurung AB (2000) Noise reduction and prediction of hydrometeorological time series: dynamical systems approach vs. stochastic approach. J Hydrol 228:242–264

Jayawardena AW, Lai F (1994) Analysis and prediction of chaos in rainfall and stream flow time series. J Hydrol 153:23–52

Jayawardena AW, Li WK, Xu P (2002) Neighborhood selection for local modeling and prediction of hydrological time series. J Hydrol 258:40–57

Jin YH, Kawamura A, Jinno K, Berndtsson R (2005) Nonlinear multivariate analysis of SOI and local precipitation and temperature. Nonlinear Process Geophys 12:67–74

Jothiprakash V, Fathima TA (2013) Chaotic analysis of daily rainfall series in Koyna reservoir catchment area, India. Stoch Environ Res Risk Assess 27:1371–1381

Karunasinghe DSK, Liong SY (2006) Chaotic time series prediction with a global model: Artificial neural network. J Hydrol 323:92–105

Khatibi R, Sivakumar B, Ghorbani MA, Kişi Ö, Kocak K, Zadeh DF (2012) Investigating chaos in river stage and discharge time series. J Hydrol 414–415:108–117

Kennel MB, Brown R, Abarbanel HDI (1992) Determining embedding dimension for phase space reconstruction using a geometric method. Phys Rev A 45:3403–3411

Khan S, Ganguly AR, Saigal S (2005) Detection and predictive modeling of chaos in finite hydrological time series. Nonlinear Process Geophys 12:41–53

Kim HS, Lee KH, Kyoung MS, Sivakumar B, Lee ET (2009) Measuring nonlinear dependence in hydrologic time series. Stoch Environ Res Risk Assess 23:907–916
Koutsoyiannis D (2006) On the quest for chaotic attractors in hydrological processes. Hydrol Sci J 51(6):1065–1091
Koutsoyiannis D, Pachakis D (1996) Deterministic chaos versus stochasticity in analysis and modeling of point rainfall series. J Geophys Res 101(D21):26441–26451
Krasovskaia I, Gottschalk L, Kundzewicz ZW (1999) Dimensionality of Scandinavian river flow regimes. Hydrol Sci J 44(5):705–723
Kyoung MS, Kim HS, Sivakumar B, Singh VP, Ahn KS (2011) Dynamic characteristics of monthly rainfall in the Korean peninsula under climate change. Stoch Environ Res Risk Assess 25(4):613–625
Laio F, Porporato A, Revelli R, Ridolfi L (2003) A comparison of nonlinear flood forecasting methods. Water Resour Res 39(5). 10.1029/2002WR001551
Laio F, Porporato A, Ridolfi L, Tamea S (2004) Detecting nonlinearity in time series driven by non-Gaussian noise: the case of river flows. Nonlinear Process Geophys 11:463–470
Lambrakis N, Andreou AS, Polydoropoulos P, Georgopoulos E, Bountis T (2000) Nonlinear analysis and forecasting of a brackish karstic spring. Water Resour Res 36(4):875–884
Lisi F, Villi V (2001) Chaotic forecasting of discharge time series: A case study. J Am Water Resour Assoc 37(2):271–279
Liu Q, Islam S, Rodriguez-Iturbe I, Le Y (1998) Phase-space analysis of daily streamflow: characterization and prediction. Adv Water Resour 21:463–475
Manzoni S, Porporato A, D'Odorico P, Laio F, Rodriguez-Iturbe I (2004) Soil nutrient cycles as a nonlinear dynamical system. Nonlinear Process Geophys 11:589–598
McDonnell JJ, Woods RA (2004) On the need for catchment classification. J Hydrol 299:2–3
Packard NH, Crutchfield JP, Farmer JD, Shaw RS (1980) Geometry from a time series. Phys Rev Lett 45(9):712–716
Paola C, Foufoula-Georgiou E, Dietrich WE, Hondzo M, Mohrig D, Parker G, Power ME, Rodriguez-Iturbe I, Voller V, Wilcock P (2006) Toward a unified science of the Earth's surface: opportunities for synthesis among hydrology, geomorphology, geochemistry, and ecology. Water Resour Res 42:W03S10. doi:10.1029/2005WR004336
Pasternack GB (1999) Does the river run wild? Assessing chaos in hydrological systems. Adv Water Resour 23(3):253–260
Paluš M (1995) Testing for nonlinearity using redundancies: quantitative and qualitative aspects. Physica D 80:186–205
Phoon KK, Islam MN, Liaw CY, Liong SY (2002) A practical inverse approach for forecasting of nonlinear time series analysis. ASCE J Hydrol Eng 7(2):116–128
Porporato A, Ridolfi L (1996) Clues to the existence of deterministic chaos in river flow. Int J Mod Phys B 10:1821–1862
Porporato A, Ridolfi R (1997) Nonlinear analysis of river flow time sequences. Water Resour Res 33(6):1353–1367
Porporato A, Ridolfi R (2001) Multivariate nonlinear prediction of river flows. J Hydrol 248(1–4):109–122
Prichard D, Theiler J (1995) Generalized redundancies for time series analysis. Physica D 84:476–493
Puente CE, Obregon N (1996) A deterministic geometric representation of temporal rainfall. Results for a storm in Boston. Water Resour Res 32(9):2825–2839
Regonda S, Sivakumar B, Jain A (2004) Temporal scaling in river flow: can it be chaotic? Hydrol Sci J 49(3):373–385
Rodriguez-Iturbe I, De Power FB, Sharifi MB, Georgakakos KP (1989) Chaos in rainfall. Water Resour Res 25(7):1667–1675
Rodriguez-Iturbe I, De Power FB, Sharifi MB, Georgakakos KP (1990) Reply. Water Resour Res 26(8):1841–1842

Salas JD, Kim HS, Eykholt R, Burlando P, Green TR (2005) Aggregation and sampling in deterministic chaos: implications for chaos identification in hydrological processes. Nonlinear Process Geophys 12:557–567
Sangoyomi TB, Lall U, Abarbanel HDI (1996) Nonlinear dynamics of the Great Salt Lake: dimension estimation. Water Resour Res 32(1):149–159
Schertzer D, Tchiguirinskaia I, Lovejoy S, Hubert P, Bendjoudi H (2002) Which chaos in the rainfall-runoff process? A discussion on 'Evidence of chaos in the rainfall-runoff process' by Sivakumar et al. Hydrol Sci J 47(1):139–147
Schouten JC, Takens F, van den Bleek CM (1994) Estimation of the dimension of a noisy attractor. Phys Rev E 50(3):1851–1861
Schreiber T (1993a) Determination of the noise level of chaotic time series. Phys Rev E 48(1): R13–R16
Schreiber T (1993b) Extremely simple nonlinear noise reduction method. Phys Rev E 47 (4):2401–2404
Schreiber T, Grassberger P (1991) A simple noise reduction method for real data. Phys Lett A 160:411–418
Schreiber T, Schmitz A (1996) Improved surrogate data for nonlinearity tests. Phys Rev Lett 77(4):635–638
Shang P, Na X, Kamae S (2009) Chaotic analysis of time series in the sediment transport phenomenon. Chaos Soliton Fract 41(1):368–379
Sharifi MB, Georgakakos KP, Rodriguez-Iturbe I (1990) Evidence of deterministic chaos in the pulse of storm rainfall. J Atmos Sci 47:888–893
She N, Basketfield D (2005) Streamflow dynamics at the Puget Sound, Washington: application of a surrogate data method. Nonlinear Process Geophys 12:461–469
Sivakumar B (2000) Chaos theory in hydrology: important issues and interpretations. J Hydrol 227(1–4):1–20
Sivakumar B (2001a) Rainfall dynamics at different temporal scales: A chaotic perspective. Hydrol Earth Syst Sci 5(4):645–651
Sivakumar B (2001b) Is a chaotic multi-fractal approach for rainfall possible? Hydrol Process 15(6):943–955
Sivakumar B (2002) A phase-space reconstruction approach to prediction of suspended sediment concentration in rivers. J Hydrol 258:149–162
Sivakumar B (2003) Forecasting monthly streamflow dynamics in the western United States: a nonlinear dynamical approach. Environ Model Softw 18(8–9):721–728
Sivakumar B (2004a) Chaos theory in geophysics: past, present and future. Chaos Soliton Fract 19(2):441–462
Sivakumar B (2004b) Dominant processes concept in hydrology: moving forward. Hydrol Process 18(12):2349–2353
Sivakumar B (2005a) Correlation dimension estimation of hydrologic series and data size requirement: myth and reality. Hydrol Sci J 50(4):591–604
Sivakumar B (2005b) Chaos in rainfall: variability, temporal scale and zeros. J Hydroinform 7(3):175–184
Sivakumar B (2007) Nonlinear determinism in river flow: prediction as a possible indicator. Earth Surf Process Landf 32(7):969–979
Sivakumar B (2009) Nonlinear dynamics and chaos in hydrologic systems: latest developments and a look forward. Stoch Environ Res Risk Assess 23:1027–1036
Sivakumar B (2011a) Global climate change and its impacts on water resources planning and management: assessment and challenges. Stoch Environ Res Risk Assess 25(4):583–600
Sivakumar B (2011b) Water crisis: from conflict to cooperation – an overview. Hydrol Sci J 56(4):531–552
Sivakumar B (2011c) Chaos theory for modeling environmental systems: Philosophy and pragmatism. In: Wang L, Garnier H (eds) System identification, environmental modelling, and control system design. Springer-Verlag, London Limited, pp 533–555

Sivakumar B, Chen J (2007) Suspended sediment load transport in the Mississippi River basin at St. Louis: temporal scaling and nonlinear determinism. Earth Surf Process Landf 32(2):269–280
Sivakumar B, Jayawardena AW (2002) An investigation of the presence of low-dimensional chaotic behavior in the sediment transport phenomenon. Hydrol Sci J 47(3):405–416
Sivakumar B, Jayawardena AW (2003) Sediment transport phenomenon in rivers: an alternative perspective. Environ Model Softw 18(8–9):831–838
Sivakumar B, Singh VP (2012) Hydrologic system complexity and nonlinear dynamic concepts for a catchment classification framework. Hydrol Earth Syst Sci 16:4119–4131
Sivakumar B, Singh VP (2015) Special issue: Grand challenges in hydrology. ASCE J Hydrol Eng 20(1)
Sivakumar B, Wallender WW (2004) Deriving high-resolution sediment load data using a nonlinear deterministic approach. Water Resour Res 40:W05403. doi:10.1029/2004WR003152
Sivakumar B, Wallender WW (2005) Predictability of river flow and sediment transport in the Mississippi River basin: a nonlinear deterministic approach. Earth Surf Process Landf 30:665–677
Sivakumar B, Liong SY, Liaw CY (1998) Evidence of chaotic behavior in Singapore rainfall. J Am Water Resour Assoc 34(2):301–310
Sivakumar B, Liong SY, Liaw CY, Phoon KK (1999a) Singapore rainfall behavior: chaotic? ASCE J Hydrol Eng 4(1):38–48
Sivakumar B, Phoon KK, Liong SY, Liaw CY (1999b) A systematic approach to noise reduction in chaotic hydrological time series. J Hydrol 219(3–4):103–135
Sivakumar B, Phoon KK, Liong SY, Liaw CY (1999c) Comment on "Nonlinear analysis of river flow time sequences" by Amilcare Porporato and Luca Ridolfi. Water Resour Res 35(3):895–897
Sivakumar B, Berndtsson R, Olsson J, Jinno K, Kawamura A (2000) Dynamics of monthly rainfall-runoff process at the Göta basin: A search for chaos. Hydrol Earth Syst Sci 4(3):407–417
Sivakumar B, Berndttson R, Olsson J, Jinno K (2001a) Evidence of chaos in the rainfall-runoff process. Hydrol Sci J 46(1):131–145
Sivakumar B, Sorooshian S, Gupta HV, Gao X (2001b) A chaotic approach to rainfall disaggregation. Water Resour Res 37(1):61–72
Sivakumar B, Berndtsson R, Persson M (2001c) Monthly runoff prediction using phase-space reconstruction. Hydrol Sci J 46(3):377–387
Sivakumar B, Berndtsson R, Olsson J, Jinno K (2002a) Reply to 'which chaos in the rainfall-runoff process?' by Schertzer et al. Hydrol Sci J 47(1):149–158
Sivakumar B, Jayawardena AW, Fernando TMGH (2002b) River flow forecasting: use of phase-space reconstruction and artificial neural networks approaches. J Hydrol 265 (1–4):225–245
Sivakumar B, Persson M, Berndtsson R, Uvo CB (2002c) Is correlation dimension a reliable indicator of low-dimensional chaos in short hydrological time series? Water Resour Res 38(2). doi:10.1029/2001WR000333
Sivakumar B, Berndtsson R, Lall U (2004a) Nonlinear deterministic dynamics in hydrologic systems: present activities and future challenges. Special Issue, Nonlinear Process Geophys
Sivakumar B, Wallender WW, Puente CE, Islam MN (2004b) Streamflow disaggregation: a nonlinear deterministic approach. Nonlinear Process Geophys 11:383–392
Sivakumar B, Berndtsson R, Persson M, Uvo CB (2005a) A multi-variable time series phase-space reconstruction approach to investigation of chaos in hydrological processes. Int J Civil Environ Eng 1(1):35–51
Sivakumar B, Harter T, Zhang H (2005b) Solute transport in a heterogeneous aquifer: a search for nonlinear deterministic dynamics. Nonlinear Process Geophys 12:211–218
Sivakumar B, Wallender WW, Horwath WR, Mitchell JP, Prentice SE, Joyce BA (2006) Nonlinear analysis of rainfall dynamics in California's Sacramento Valley. Hydrol Process 20(8):1723–1736

Sivakumar B, Jayawardena AW, Li WK (2007) Hydrologic complexity and classification: a simple data reconstruction approach. Hydrol Process 21(20):2713–2728
Sivakumar B, Woldemeskel FM, Puente CE (2014) Nonlinear analysis of rainfall variability in Australia. Stoch Environ Res Risk Assess 28(1):17–27
Sivakumar B, Singh V, Berndtsson R, Khan S (2015) Catchment classification framework in Hydrology: challenges and directions. J Hydrol Eng 20:A4014002
Stehlik J (1999) Deterministic chaos in runoff series. J Hydrol Hydromech 47(4):271–287
Sugihara G, May RM (1990) Nonlinear forecasting as a way of distinguishing chaos from measurement error in time series. Nature 344:734–741
Takens F (1981) Detecting strange attractors in turbulence. In: Rand DA, Young LS (eds) Dynamical systems and turbulence, vol 898., Lecture notes in mathematicsSpringer-Verlag, Berlin, Germany, pp 366–381
Theiler J, Eubank S, Longtin A, Galdrikian B, Farmer JD (1992) Testing for nonlinearity in time series: the method of surrogate data. Physica D 58:77–94
Tongal H, Berndtsson R (2014) Phase-space reconstruction and self-exciting threshold modeling approach to forecast lake water levels. Stoch Environ Res Risk Assess 28(4):955–971
Tongal H, Demirel MC, Booij MJ (2013) Seasonality of low flows and dominant processes in the Rhine River. Stoch Environ Res Risk Assess 27:489–503
Tsonis AA, Elsner JB, Georgakakos KP (1993) Estimating the dimension of weather and climate attractors: important issues about the procedure and interpretation. J Atmos Sci 50:2549–2555
Tsonis AA, Triantafyllou GN, Elsner JB, Holdzkom JJ II, Kirwan AD Jr (1994) An investigation on the ability of nonlinear methods to infer dynamics from observables. Bull Amer Meteor Soc 75:1623–1633
Waelbroeck H, Lopex-Pena R, Morales T, Zertuche F (1994) Prediction of tropical rainfall by local phase space reconstruction. J Atmos Sci 51(22):3360–3364
Wagener T, Sivapalan M, Troch PA, McGlynn BL, Harman CJ, Gupta HV, Kumar P, Rao PSC, Basu NB, Wilson JS (2010) The future of hydrology: an evolving science for a changing world. Water Resour Res 46:W05301. doi:10.1029/2009WR008906
Wang Q, Gan TY (1998) Biases of correlation dimension estimates of streamflow data in the Canadian prairies. Water Resour Res 34(9):2329–2339
Wilcox BP, Seyfried MS, Matison TM (1991) Searching for chaotic dynamics in snowmelt runoff. Water Resour Res 27(6):1005–1010
Wolf A, Swift JB, Swinney HL, Vastano A (1985) Determining Lyapunov exponents from a time series. Physica D 16:285–317
Zhou Y, Ma Z, Wang L (2002) Chaotic dynamics of the flood series in the Huaihe River Basin for the last 500 years. J Hydrol 258:100–110

Chapter 9
Applications to Rainfall Data

Abstract Initial applications of the ideas of chaos theory in hydrology were on rainfall data. Early studies essentially addressed the identification and prediction of chaotic behavior of rainfall data. Encouraging outcomes from these studies subsequently led to investigations on the chaotic nature of scaling relationships in rainfall and disaggregation of data, including development of a new chaotic approach for rainfall disaggregation. More recently, some studies have examined the spatial variability and classification of rainfall. In addition to these, a number of studies have also addressed the important methodological and data issues in the applications of chaos methods to rainfall data. This chapter presents a review of chaos studies on rainfall data. The presentation is organized into three parts to address three important problems associated with rainfall: identification and prediction of chaos, scaling and disaggregation, and spatial variability and classification. An example is presented for each of these to demonstrate the utility and effectiveness of chaos concepts and methods to study these different problems.

9.1 Introduction

Since the very early studies on the applications of chaos theory to identify the dynamic nature of rainfall in the late 1980s (e.g. Fraedrich 1986, 1987; Hense 1987; Rodriguez-Iturbe et al. 1989), numerous studies have investigated the utility, suitability, and effectiveness of nonlinear dynamic and chaos concepts for rainfall dynamics; see Sivakumar (2000, 2004, 2009) for some general reviews. Such studies have attempted, among others, identification and prediction (e.g. Sharifi et al. 1990; Islam et al. 1993; Tsonis et al. 1993; Berndtsson et al. 1994; Jayawardena and Lai 1994; Waelbroeck et al. 1994; Georgakakos et al. 1995; Koutsoyiannis and Pachakis 1996; Puente and Obregon 1996; Sivakumar et al. 1998, 1999a, b, 2000, 2001a, 2006; Jin et al. 2005; Tsonis and Georgakakos 2005; Koutsoyiannis 2006; Kim et al. 2009; Dhanya and Nagesh Kumar 2010, 2011; Kyoung et al. 2011; Jothiprakash and Fathima 2013); scaling and disaggregation (e.g. Sivakumar 2001a, b; Sivakumar et al. 2001b, 2006; Gaume et al. 2006; Jothiprakash and Fathima 2013); and spatial variability and classification

B. Sivakumar, *Chaos in Hydrology*, DOI 10.1007/978-90-481-2552-4_9

(e.g. Sivakumar et al. 2014). Several studies have also addressed the potential issues in the application of chaos theory to rainfall time series, such as parameter selection for phase space reconstruction and prediction, minimum data size, data noise and noise reduction, and presence of zeros (e.g. Tsonis et al. 1993, 1994; Berndtsson et al. 1994; Sivakumar 2001a, b, 2005; Sivakumar et al. 1999a, b, 2001b, 2006; Jayawardena and Gurung 2000; Jayawardena et al. 2002; Kim et al. 2009). This chapter presents a review of chaos studies in rainfall, with emphasis on some of the more important studies. The review is roughly organized to address the following: identification and prediction, scaling and disaggregation, spatial variability and classification, and issues in chaos theory application for rainfall. A few selected studies are discussed in far more detail, for different reasons, including for their novelty, significance, representativeness, and attention they have received.

9.2 Identification and Prediction

The first studies on the application of chaos theory to rainfall time series were carried out by Fraedrich (1986, 1987), but such were in the context of climate and weather. In the specific context of hydrology, however, the study by Hense (1987) was the first on chaos identification in rainfall time series. Hense (1987) investigated the dynamic nature of rainfall observed in Nauru Island. Applying the correlation dimension method (e.g. Grassberger and Procaccia 1983a, b) to a series of 1008 values of monthly rainfall and obtaining a low correlation dimension value (between 2.5 and 4.5), Hense (1987) reported the presence of chaos in the rainfall dynamics. Rodriguez-Iturbe et al. (1989) investigated the presence of chaos in two different rainfall time series: (1) a record of 1990 rainfall values, measured with a sampling frequency of 8 Hz and then aggregated at equally spaced intervals of 15 s, from a single storm event in Boston, USA; and (2) weekly rainfall data over a period of 148 years observed in Genoa, Italy. Employing the correlation dimension method, they reported the presence of chaos (correlation dimension 3.78) in the storm data from Boston and absence of chaos (no finite correlation dimension) in the weekly rainfall data from Genoa. They also supported their claim on the presence of chaos in the storm event through application of the Lyapunov exponent method (e.g. Wolf et al. 1985) and observation of a positive Lyapunov exponent (0.0002 bits/s). Ghilardi and Rosso (1990), however, commented on the study and results reported by Rodriguez-Iturbe et al. (1989), in the context of both the study by Hense (1987) and the issue of data size; see also Rodriguez-Iturbe et al. (1990).

Further evidence on the presence of chaos in storm rainfall was presented by Sharifi et al. (1990). Employing the correlation dimension method to fine-increment data from three storms (with 4000, 3991, and 3316 data points, respectively) observed in Boston, they reported low correlation dimensions of 3.35, 3.75, and 3.60, respectively. Islam et al. (1993) reported the presence of chaos in rainfall by analyzing rainfall intensity data set of 7200 values, generated at 10-s timesteps from a three-dimensional cloud model, and observing a correlation dimension of 1.5.

Tsonis et al. (1993) studied data representing the time between successive raingage signals each corresponding to a collection of 0.01 mm of rain. Applying the correlation dimension method and observing a correlation dimension of 2.4, they reported the presence of chaos in rainfall.

Berndtsson et al. (1994) investigated the presence of chaos in a 238-year monthly rainfall dataset recorded in Lund, Sweden. They performed the correlation dimension analysis on two different types of this dataset: raw rainfall series and noise-reduced rainfall series. They found no evidence of chaos in the raw rainfall series, but the noise-reduced series was found to exhibit chaos with a dimension less than 4. Incidentally, their study was the first ever to attempt noise reduction in hydrologic time series in the context of chaos theory application. Jayawardena and Lai (1994) attempted identification and prediction of chaos in the daily rainfall from Hong Kong. Employing the correlation dimension method, Lyapunov exponent method, Kolmogorov entropy method (e.g. Grassberger and Procaccia 1983c), and local approximation prediction method (e.g. Farmer and Sidorowich 1987; Casdagli 1989) to three rainfall data sets (4015 points), they reported the presence of chaos. This was also probably the first ever study to attempt prediction of rainfall from a chaotic perspective. Further, comparing the results from the local approximation prediction method with those obtained from the traditional linear autoregressive moving average (ARMA) method, they also reported the superiority of the former. Waelbroeck et al. (1994) attempted prediction of daily tropical rainfall using local approximation method. They observed that the prediction skill for daily rainfall dropped off quickly within a timescale of two days but that the prediction was much better for the 10-day rainfall accumulations.

Georgakakos et al. (1995) investigated the presence of chaos in 11 storm events observed in Iowa City, USA. Applying the correlation dimension method, they reported the possible presence of chaos in all of these storm events, except one. They found that the correlation dimensions ranged from 2.8 to 7.9 in the high-intensity scaling region and from 0.5 to 1.6 in the low-intensity scaling region. Puente and Obregon (1996) conducted a more detailed investigation on the presence of chaos in the Boston storm event, which had been studied previously by Rodriguez-Iturbe et al. (1989). Applying the correlation dimension method, Kolmogorov entropy method, false nearest neighbor method (e.g. Kennel et al. 1992), and Lyapunov exponent method, they reported the presence of chaos in the storm data. Through presentation of a deterministic fractal-multifractal (FM) approach for modeling the storm event, they also hinted that a stochastic framework for rainfall modeling might not be necessary. However, Koutsoyiannis and Pachakis (1996) defended the use of stochastic models for rainfall. Applying chaos concepts to incremental rainfall depths measured every 15 min, they concluded that a synthetic continuous rainfall series generated by a well-structured stochastic model might be practically indistinguishable from a historic rainfall series even if one used the tools of chaotic dynamic theory for characterization.

Sivakumar et al. (1998) investigated the presence of chaos in the rainfall data in Singapore. Applying the correlation dimension method to daily rainfall time series observed at six stations across the country, they reported the presence of low correlation dimensions and possible presence of chaos in rainfall dynamics. They also addressed the issue of minimum data size requirement for correlation dimension estimation by analyzing rainfall records at different lengths (1–30 years). Sivakumar et al. (1999a) extended the above investigation through employing the local approximation prediction method, including the deterministic versus stochastic (DVS) approach (Casdagli 1992), and presented further evidence to the presence of chaos. To put the above results in a more solid footing, they also employed, for the first time in rainfall studies (and hydrologic studies at large), the surrogate data method (e.g. Theiler et al. 1992) to detect the absence of linearity. The outcomes from these studies, in particular the presence of chaos and low prediction accuracy, led Sivakumar et al. (1999b) to study the influence of presence of noise (measurement error) on the correlation dimension and prediction accuracy estimates. For this purpose, coupling a noise level determination method (Schouten et al. 1994) and a noise reduction method (Schreiber 1993), they proposed a systematic approach for noise reduction in rainfall (or any other) time series; see also Sivakumar et al. (1999c). The results provided further support to the presence of a deterministic component in the rainfall phenomenon in Singapore and also possible reasons for the low prediction accuracy estimates in the earlier studies.

Sivakumar et al. (2000, 2001a) analyzed monthly rainfall data observed over a period of 131 years (January 1807–December 1937) in the Göta River basin in Sweden, as part of their search for chaos in the rainfall-runoff process. Employing the correlation dimension method and the local approximation prediction method and observing a correlation dimension value of 6.4 and good prediction results, they reported the presence of chaos in this rainfall series. However, the outcomes of these studies, especially the correlation dimension value reported by Sivakumar et al. (2001a), have been subjected to some criticisms (Schertzer et al. 2002); see also Sivakumar et al. (2002) for additional details. Therefore, this rainfall time series is considered here as a representative series to illustrate the analysis to identify the presence of chaos in rainfall, and details are presented in Sect. 9.2.1. Sivakumar (2001a, b) and Sivakumar et al. (2001b) investigated the presence of chaos in the rainfall data in the Leaf River basin, Mississippi, USA, focusing on scaling and disaggregation. Similar scaling-based studies in the context of chaos theory were also performed by Gaume et al. (2006) and Sivakumar et al. (2006). Details of these scaling-based studies are discussed in Sect. 9.3.

Tsonis and Georgakakos (2005) addressed the problem of rainfall estimation from satellite imagery while investigating the possibility of deriving useful insights about the variability of the system from only a part of the complete state vector. Postulating a low-order observable vector and a system response as linear functions of portions of the state vector, they first conducted a numerical study on a toy model representing a low-dimensional dynamic system (Lorenz map; see Chap. 5) and then applied the approach to satellite images (spatial resolution: 4 km × 4 km; temporal resolution: 3 h) over the Des Moines River basin in Iowa, USA. They reported that, while reducing the number of observables reduces the correlation between actual and inferred rainfall amounts, good estimates for extremes are still recoverable. As part of the investigation on the potential problems in the application of chaos identification methods to real hydrologic series, Koutsoyiannis (2006) analyzed the daily rainfall series from Vakari in western Greece (and also daily streamflow series from Pinios River in Greece), and reported absence of chaos (in both). Koutsoyiannis (2006) also showed, through theoretical analyses, that specific peculiarities of hydrologic processes on fine timescales (e.g. asymmetric, J-shaped distribution functions, intermittency, and high autocorrelations) are synergistic factors that could lead to misleading conclusions regarding the presence of low-dimensional deterministic chaos.

Kim et al. (2009) investigated the dynamic characteristics of rainfall observed in Seoul, South Korea, as part of their assessment on the suitability and effectiveness of the C–C method (Kim et al. 1999) for real hydrologic time series. In particular, they focused on the estimation of the general dependence of the time series. Applying the C–C method to daily rainfall data observed over a period of 10 years (1987–1996), they reported that the rainfall dynamics were dominated by linear stochastic behavior. Kyoung et al. (2011) investigated the dynamic characteristics of rainfall under conditions of climate change, through analysis of observed and global climate model (GCM)-simulated monthly rainfall in the Korean Peninsula. They studied both 'present rainfall' (observed rainfall for the period 1971–1999 and GCM-simulated rainfall for the period 1951–1999) and 'future rainfall' (GCM-simulated rainfall for the period 2000–2009) from Seoul. Applying four different methods, namely autocorrelation function, phase space reconstruction, correlation dimension, and close returns plot (e.g. Gilmore 1993), they reported that the nature of rainfall dynamics falls more on the chaotic dynamic spectrum than on the linear stochastic spectrum. Their study also revealed that the future GCM-simulated rainfall exhibits stronger nonlinearity and chaos compared to the present rainfall.

Dhanya and Nagesh Kumar (2010) studied the chaotic dynamic behavior and prediction of daily rainfall from three different regions in India (Malaprabha, Mahanadi, and All-India). For identification of chaos, they used the correlation dimension method, false nearest neighbor method, Lyapunov exponent method, nonlinear prediction method, and surrogate data method. The results from these methods indicated the presence of chaos in the three rainfall series. However, different methods resulted in slightly different embedding dimensions, and also delay times and neighborhood size, for each of the three rainfall series. To take these differences into account, they also used an appropriate range of embedding dimension, delay time, and neighborhood size and generated an ensemble of predictions of rainfall. Subsequently, Dhanya and Nagesh Kumar (2011) investigated the limit to rainfall predictability due to sensitivity to initial conditions and ineffectiveness of the model. They presented a multivariate nonlinear ensemble prediction approach for quantifying the uncertainties involved. To this end, they used a climate data set of 16 variables to study the predictability of daily rainfall in the Malaprabha basin. They reported that the ensembles generated from multivariate predictions were better than those from univariate predictions and that the uncertainty in predictions decreased (or predictability increased) when multivariate nonlinear ensemble prediction was adopted.

Jothiprakash and Fathima (2013) investigated the presence of chaos in the daily rainfall dynamics in the Koyna reservoir catchment in Maharashtra, India. They applied the correlation dimension method to rainfall data observed (average of nine stations in the catchment) over a period of 49 years (1961–2009), and reported the presence of chaotic behavior. They also addressed the effect of the radius (r) and the scaling region in the Log $C(r)$ versus Log r on the correlation exponent estimation, as well as the effect of longer length of zeros. They also studied data at different scales.

9.2.1 Chaos Analysis of Rainfall: An Example—Göta River Basin

To illustrate the analysis for identification and prediction of chaos in rainfall, monthly rainfall time series observed over a period of 131 years (January 1807–December 1937) in the Göta River basin in Sweden is considered here. Figure 9.1a shows the variation of this rainfall time series. A visual inspection of this series indicates significant peaks every few years, but the seemingly irregular and random behavior does not indicate anything regarding the presence (or absence) of chaotic behavior. Figure 9.1b presents an example of a higher-dimensional phase-space reconstruction of this rainfall time series, according to Takens' delay embedding theorem (Takens 1981); see Chap. 6, Eq. (6.3). More specifically, the figure presents the reconstruction of the series in a two-dimensional phase space ($m = 2$), i.e. the projection of the attractor on the plane $\{X_i, X_{i+1}\}$. The projection yields a reasonably good structure (with trajectories lying within a specific region of the phase space), but it is neither as well-defined as the one that is normally observed for a very low-dimensional system nor as scattered as the one that is normally observed for a very high-dimensional system.

After the reconstruction of the rainfall series in higher-dimensional phase space, the correlation functions and, hence, the correlation exponents are computed, according to the Grassberger–Procaccia correlation dimension algorithm, described in Chap. 6 (Sect. 6.4.2). Figure 9.1c shows the relationship between the correlation integral, $C(r)$, and the radius, r, for embedding dimensions, m, from 1 to 20. The Log $C(r)$ versus Log r plots exhibit large and clear scaling regions, allowing fairly reliable estimation of the correlation exponents. Figure 9.1d presents the relationship between the correlation exponent values and the embedding dimension values. As may be seen, the correlation exponent value increases with the embedding dimension up to a certain point and saturates beyond that point. Such a saturation of the correlation exponent is an indication of the presence of deterministic dynamics. The saturation value of the correlation exponent (or correlation dimension) for the rainfall series is about 6.4. The finite, low, and non-integer correlation dimension obtained may be an indication that rainfall dynamics exhibit low-dimensional chaotic behavior. As the nearest integer above the correlation dimension value generally provides the number of variables dominantly governing the dynamics of the underlying system, the correlation dimension obtained for the rainfall series

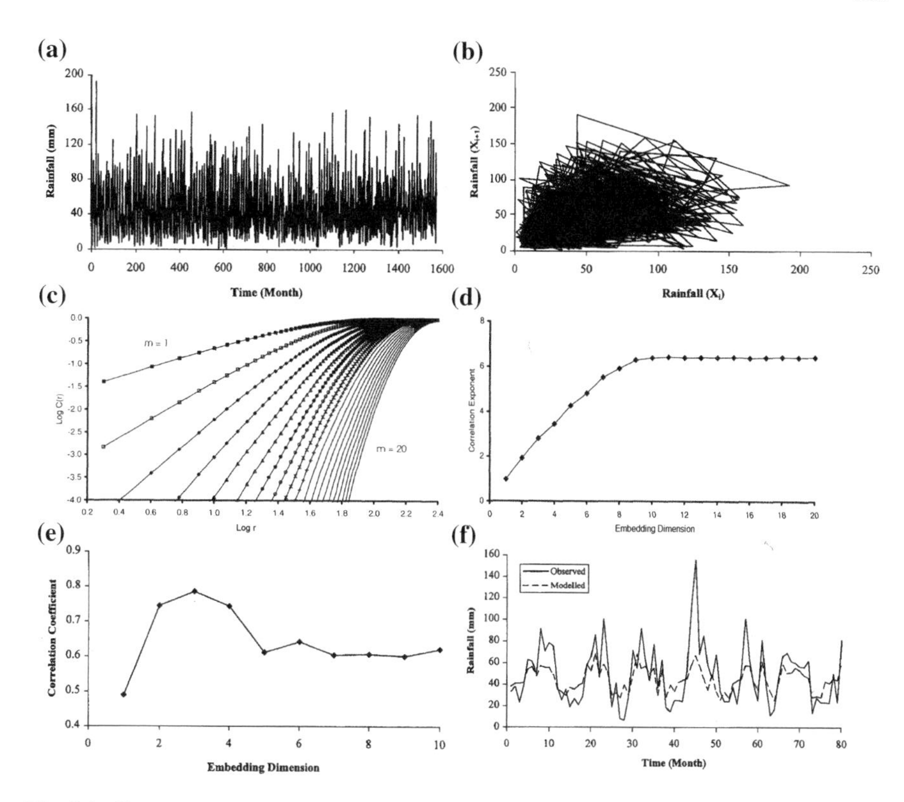

Fig. 9.1 Chaos analysis of monthly rainfall from Göta River basin: **a** time series; **b** phase space; **c** Log C(r) versus Log r; **d** relationship between correlation exponent and embedding dimension; **e** relationship between correlation coefficient and embedding dimension; and **f** comparison between time series plot of predicted and observed values (*source* Sivakumar et al. (2000, 2001a))

indicates that at least seven variables are essential for a reliable representation of the monthly rainfall dynamics in the Göta River basin.

With this encouraging result, the local approximation method is employed for prediction of rainfall; see Chap. 6 (Sect. 6.11) for details. The first 1440 data are used for phase space reconstruction (i.e. training or learning set) to predict the subsequent 80 values. One timestep ahead predictions are made, and the local maps are learned in the form of local polynomials (Abarbanel 1996). Figure 9.1e presents the variation of the correlation coefficient against the embedding dimension for the rainfall series. The correlation coefficient increases with the embedding dimension up to $m = 3$ and then decreases when the dimension is increased further. The presence of an optimal embedding dimension value, $m_{opt} = 3$, indicates the possible presence of chaos in the rainfall series. Figure 9.1f compares, using time series plots, the observed and the predicted values of rainfall; the results are those obtained for the optimal embedding dimension (i.e. $m = 3$). Although the predicted values are not in very good agreement with the observed values, the trends (rises and falls) in the values seem to be fairly well captured. The reasonably good prediction results obtained using the local approximation method seem to indicate the suitability of the method to model and predict the dynamics of the rainfall process.

A brief discussion about the presence of chaos and the prediction results is now in order. According to the concept of chaos theory, for a chaotic time series with an attractor dimension, d, (a) accurate short-term predictions can be achieved when it is embedded in a sufficient phase space, m_{opt} or higher; and (b) the prediction accuracy will remain constant for any embedding dimension higher than m_{opt}. However, the results in Fig. 9.1e, f indicate that: (a) the rainfall predictions are far from accurate; and (b) the prediction accuracy does not remain constant beyond the optimal embedding dimension, rather decreases when the embedding dimension is increased further. A possible explanation for these observations is the presence of noise in the time series, as noise is one of the most prominent limiting factors for the predictability of deterministic chaotic systems (e.g. Schreiber and Kantz 1996); see Chap. 7 (Sect. 7.4) for general details about the influence of noise on chaos identification and prediction.

9.3 Scaling and Disaggregation/Downscaling

9.3.1 Chaotic Scale-invariance: An Example—Leaf River Basin

While many early studies applying the concepts of chaos theory to rainfall data (e.g. Hense 1987; Rodriguez-Iturbe et al. 1989; Tsonis et al. 1993; Berndtsson et al. 1994; Jayawardena and Lai 1994; Georgakakos et al. 1995; Sivakumar et al. 1998, 1999a, 2000) had reported the presence of chaos in rainfall at different temporal scales, they could not provide any useful information as to the presence of chaos in rainfall across different temporal scales. This is because, those studies had investigated rainfall data at different temporal scales from different locations: for instance, monthly rainfall from Nauru (Hense 1987) and Sweden (Berndtsson et al. 1994; Sivakumar et al. 2000), weekly rainfall from Genoa (Rodriguez-Iturbe et al. 1989), daily rainfall from Hong Kong (Jayawardena and Lai 1994) and Singapore (Sivakumar et al. 1998, 1999a), and 15-s rainfall from storms in Boston (Rodriguez-Iturbe et al. 1989) and storm events in Iowa (Georgakakos et al. 1995). To address the issue of scaling or fractal, rainfall data at different temporal scales from the same location need to be studied.

The study by Sivakumar (2001a) was the first to investigate the presence of chaos in rainfall at different temporal scales from the same location. Sivakumar (2001a) analyzed rainfall data observed at four different temporal scales (daily, 2-, 4-, and 8-day) in the Leaf River basin in Mississippi, USA. The underlying assumption in studying rainfall data at these different scales is that the behavior of

Table 9.1 Statistics and correlation dimension results for rainfall data at different temporal scales in the Leaf River basin, Mississippi, USA (*source* Sivakumar (2001a))

Statistic	Daily	2-day	4-day	8-day
Number of data	8192	4096	2048	1024
Mean (mm)	4.03	8.06	16.12	32.24
Standard deviation (mm)	10.47	15.61	22.08	31.90
Variance (mm^2)	109.46	243.56	487.62	1017.62
Coefficient of variation	2.60	1.94	1.37	0.99
Maximum value (mm)	221.52	221.52	221.52	234.03
Minimum value (mm)	0.00	0.00	0.00	0.00
Number of zeros	4467 (54.53 %)	1633 (39.87 %)	412 (20.12 %)	62 (6.05 %)
Correlation dimension	4.82	5.26	6.42	8.87
Number of variables	5	6	7	9

the dynamics of rainfall process at the individual scales provides important information about the dynamics of the overall rainfall transformation between the scales. With the available daily rainfall data, the 2-, 4-, and 8-day rainfall data were obtained by simply adding the rainfall values corresponding to the number of days. Table 9.1 presents some of the important statistics of the rainfall data at the four scales. The correlation dimension method was employed to investigate the presence of chaos in these rainfall time series.

Figure 9.2a shows, for instance, the variation of the daily rainfall series observed over a period of 25 years (January 1963–December 1987) in the Leaf River basin. For this series, Fig. 9.2b shows the Log $C(r)$ versus Log r plot for embedding dimensions, m, from 1 to 20, and Fig. 9.2c shows the relationship between the correlation exponent and the embedding dimension. As can be seen, the correlation exponent value increases with the embedding dimension up to a certain point and saturates beyond that point. The correlation dimension of the rainfall series is 4.82, suggesting the presence of chaos in the rainfall process at the daily scale. The correlation dimension results for the 2-, 4-, and 8-day rainfall series (see Fig. 9.2d) also suggest the presence of chaos, with dimension values of 5.26, 6.42, and 8.87, respectively (see Table 9.1). All these results also suggest the presence of chaos in the scaling relationship in rainfall between the four scales.

While the presence of a chaotic scale-invariant behavior in rainfall is encouraging, the study by Sivakumar (2001a) raised other questions. For instance, a comparison of the correlation dimension values and the coefficient of variation values (see Table 9.1) reveals an inverse relationship between the two, i.e. higher dimension for lower coefficient of variation and vice versa. This inverse relationship is contrary to the concepts of correlation dimension and coefficient of variation, as they both are representation of the degree of variability of rainfall. The reason for this inverse relationship is not clear—whether correlation dimension (a nonlinear dynamic measure) or coefficient of variation (a linear statistical measure). However,

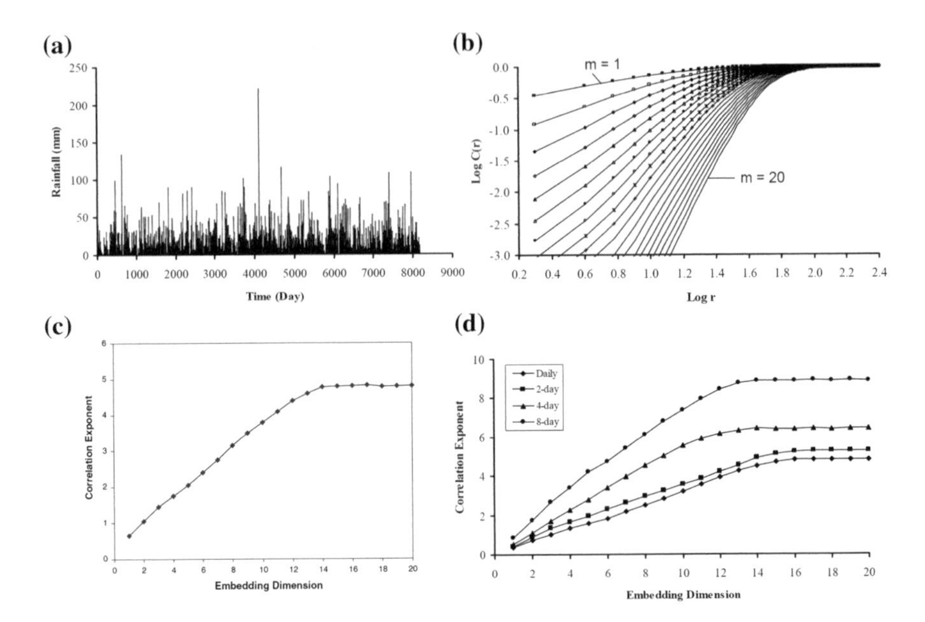

Fig. 9.2 Correlation dimension analysis of rainfall data at different temporal scales from the Leaf River basin: **a** time series of daily data; **b** Log C(r) versus Log r plot for daily data; **c** correlation exponent versus embedding dimension for daily data; and **d** correlation dimension versus embedding dimension for daily, 2-, 4-, and 8-day data (*source* Sivakumar (2001a))

Sivakumar (2001a) pointed out the potential limitations of the correlation dimension method, especially in the context of the presence of zeros and noise in the rainfall data. For instance, the presence of a large number of zeros in the finer-resolution series (see Table 9.1) could result in an underestimation of the dimension, and the presence of a higher level of noise in the coarser-resolution time series could result in an overestimation of the dimension. The correlation dimension results obtained for the four rainfall time series, with different data lengths (see Table 9.1), also led Sivakumar (2001a) to interpret that the issue of (minimum) data size requirement was not as severe as it was believed to be; for instance, rainfall series with shorter lengths have higher correlation dimensions when compared to those with longer lengths.

As fractal theory-based methods have been widely used to study the scaling behavior of rainfall (and other hydrologic) data over the past several decades, application of such methods, along with chaos-based methods, could shed further light regarding the chaotic scale-invariant behavior of rainfall. To this end, Sivakumar (2001b) presented both the fractal and the chaos analyses of the daily rainfall series from the Leaf River basin, studied earlier by Sivakumar (2001a). Autocorrelation function, power spectrum, empirical probability distribution function, and statistical moment scaling function were used as indicators of fractal behavior, and correlation dimension was used as an indicator of chaos. The results indicated the existence of both fractal and chaotic behaviors in the rainfall series.

Based on this, Sivakumar (2001b) suggested the possibility of a chaotic multi-fractal approach for modeling rainfall process, rather than just a stochastic (i.e. random cascade) fractal approach that has been prevalent in hydrology; see also Puente and Obregon (1996) for some relevant details.

9.3.2 *Chaotic Disaggregation: An Example—Leaf River Basin*

The studies by Sivakumar (2001a, b) offered important information as to the presence of chaotic scaling nature in rainfall dynamics. However, they considered only rainfall data at different individual scales, but not the actual transformation of rainfall between the scales. Therefore, the question whether the actual rainfall transformation process would also exhibit chaotic behavior still remained. This question needed to be addressed in any attempt to use the concepts of chaos theory for rainfall disaggregation, and related, purposes. At the core of this question is the nature of the distribution of rainfall between two (or more) scales; in other words, the weights of disaggregation/aggregation between scales. Sivakumar et al. (2001b) extensively addressed this problem, and also proposed a new chaotic approach for rainfall disaggregation. Some details of their study are presented here.

Let us assume that we have a rainfall time series X_i, $i = 1, 2, \ldots, N$, at a certain resolution T_1, and the task at hand is to obtain the (disaggregated) rainfall values $(Z_i)_k$, $k = 1, 2, \ldots, p$, at a higher (finer) resolution T_2, where $p = T_1/T_2$. Let us also assume that the values of X_i are distributed into $(Z_i)_k$ according to $(Z_i)_k = (W_i)_k * X_i$, where $(W_i)_k$ are the distributions of weights of X_i to $(Z_i)_k$ and $\sum_{k=1}^{p} (W_i)_k = 1$. If, for example, only rainfall data at successively doubled temporal resolutions are considered for disaggregation purposes, the parameter p will then be given by $p = T_1/T_2 = 2$. A schematic diagram depicting such a disaggregation situation is presented in Fig. 9.3.

It is important to emphasize, at this point, that theoretically the transformation of rainfall data from one resolution to another is possible only if the distributions of weights between the two resolutions are available. This is why the determination of

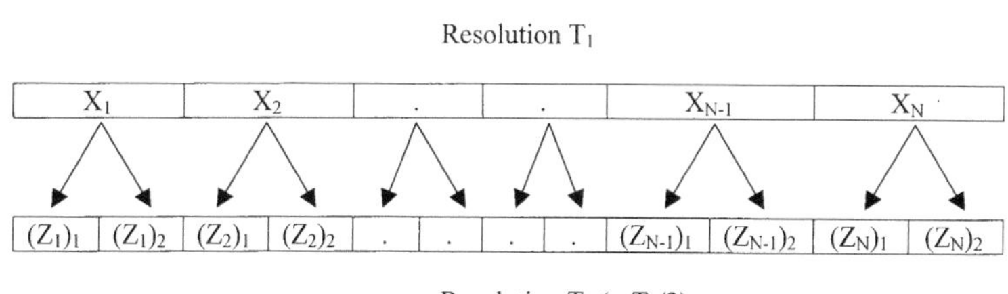

Fig. 9.3 Schematic representation of distributions of weights of rainfall transformation between two different temporal scales (*source* Sivakumar et al. (2001b))

the behavior (chaotic or stochastic) of distributions of weights is essential in understanding the behavior of transformation of data from one resolution to another. In reality, however, the distributions of weights are not known a priori (deriving which is, in fact, the task at hand) and, therefore, have to be computed from the values of X_i and $(Z_i)_k$.

Sivakumar et al. (2001b) studied the rainfall data observed over a period of 25 years (January 1963–December 1987) from the Leaf River basin to investigate the presence of chaos in the actual rainfall transformation process. They analyzed the distributions of weights $(W_i)_k$ between rainfall data at six successively doubled temporal resolutions: 6-, 12-, 24-, 48-, 92-, and 192-h. The weights considered for the analysis were those obtained from the transformation of only the non-zero rainfall values at a particular resolution (e.g. 12-h) to a successively doubled higher resolution (e.g. 6-h). The zero values were eliminated for at least two reasons: (1) the presence of a large number of zeros could significantly influence the correlation dimension results; and (2) the presence of a zero value in rainfall time series of a particular resolution does not contribute anything to its disaggregation to another (higher) resolution, as the disaggregation values of a zero rainfall value are also zeros (see Sivakumar et al. 2001b for further details).

Table 9.2 presents some important statistics of rainfall weights between the above successively doubled resolutions from the Leaf River basin. A comparison of the statistics of the actual rainfall data and of the weights reveals that the number of zeros in the weights is significantly reduced by the exclusion of zero values in the computation of the weights. For example, the percentage of zeros in the weights between 12- and 6-h resolutions is about 21 % when the zeros are excluded in the

Table 9.2 Characteristics and correlation dimension results for rainfall weights between different resolutions in the Leaf River basin, Mississippi, USA (*source* Sivakumar et al. (2001b))

Statistic	192- to 96-h	96- to 48-h	48- to 24-h	24- to 12-h	12- to 6-h
Number of data	1924	3272	4926	7450	10,254
Mean	0.50	0.50	0.50	0.50	0.50
Standard deviation	0.3846	0.4315	0.4394	0.4427	0.4395
Variance	0.1479	0.1862	0.1931	0.1960	0.1932
Maximum value	1.00	1.00	1.00	1.00	1.00
Minimum value	0.00	0.00	0.00	0.00	0.00
Number of zeros	288 (14.97 %)	809 (24.72 %)	981 (19.91 %)	1726 (23.17 %)	2165 (21.12 %)
Correlation dimension	3.46	2.61	2.23	1.65	1.86
Number of variables	4	3	3	2	2

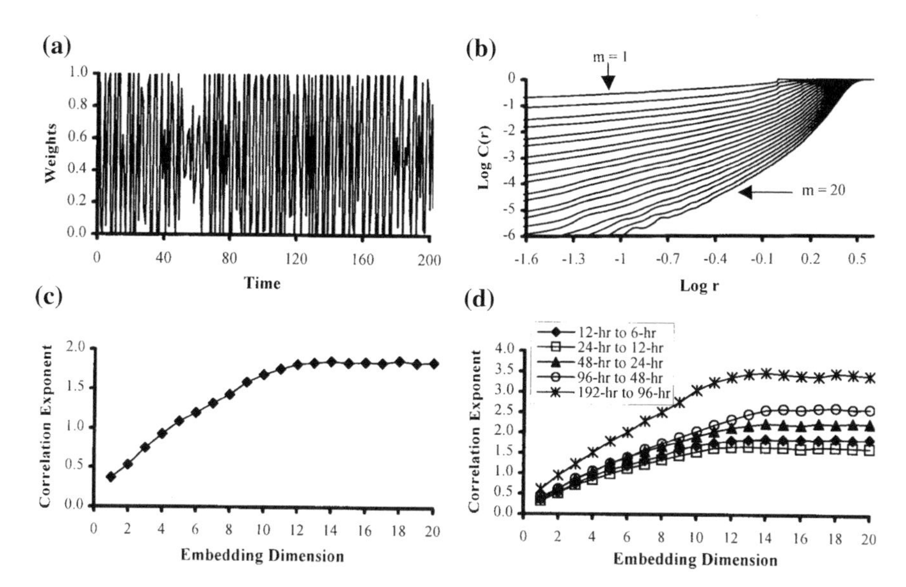

Fig. 9.4 Correlation dimension analysis of distributions of weights of rainfall data in the Leaf River basin: **a** time series of weights between 12- and 6-h resolutions; **b** Log C(r) versus Log r for **a**; **c** correlation exponent versus embedding dimension for **a**; and **d** correlation exponent versus embedding dimension for distributions of weights of rainfall data between different successively doubled resolutions (*source* Sivakumar et al. (2001b))

computation of the weights, whereas the corresponding number would be about 75 % if zeros are also included. While it may be argued that the weights obtained in the former case could still lead to an underestimation of the dimension because of the presence of zeros, the underestimation would be significantly less compared to the latter.

Figure 9.4a, for example, shows the variation of the distributions of weights of rainfall data between 12- and 6-h resolutions. For this series, Fig. 9.4b shows the Log $C(r)$ versus Log r plot for embedding dimensions, m, from 1 to 20, and Fig. 9.4c presents the relationship between the correlation exponent and the embedding dimension. As can be seen, the correlation exponent value increases with the embedding dimension up to a certain point and saturates beyond that point. The correlation dimension value is 1.86, suggesting the presence of chaos in the distribution of weights in rainfall between 12- and 6-h resolutions. The correlation dimension results obtained for the weights for the other resolutions are also low: 1.65 for 24- to 12-h, 2.23 for 48- to 24-h, 2.61 for 92- to 48-h, and 3.46 for 192- to 96-h (see Fig. 9.4d). The existence of low-dimensional chaotic behavior in the transformation of data between successively doubled resolutions suggests the suitability of a chaotic approach for rainfall disaggregation.

The chaotic disaggregation approach proposed by Sivakumar et al. (2001b) is somewhat similar to the approach adopted in chaos prediction (e.g. Farmer and Sidorowich 1987) (see Chap. 6, Sect. 6.11), and is described here. As the purpose is rainfall disaggregation (rather than prediction), the procedure is simplified by working with only the available rainfall series (rather than predicting/generating the future rainfall values and disaggregating them). Let us now assume that information is available about the history of distributions of weights $(W_i)_k$ (or X_i and $(Z_i)_k$), $i = 1, 2, \ldots, n$, where $n < N$, and the task at hand is to obtain the distributions of weights $(W_i)_k$ and, hence, the rainfall values $(Z_i)_k$ at a finer resolution, where $i = n + 1, n + 2, \ldots, N$ and $k = 1, 2, \ldots, p$. In other words, rainfall values X_i, $i = 1, 2, \ldots, n$, are used as the "training set" for the model to learn the dynamics of disaggregation (or transformation), while rainfall values X_i, $i = n + 1, n + 2, \ldots, N$ are used as the "testing test" to assess the model performance. Based on these information, the chaotic disaggregation approach is developed as follows.

Let us now consider determining how the rainfall data X_{n+1} (i.e. the value at time $n + 1$) at resolution T_1 is disaggregated into values at resolution T_2, i.e. determining the distributions of weights $(W_{n+1})_k$. The phase space $\boldsymbol{Y}_j$ for this case can be reconstructed using the series X_i, $i = 1, 2, \ldots, n + 1$, according to Chap. 6, Eq. (6.3), where $j = 1, 2, \ldots, (n + 1) - (m - 1)\tau/\Delta t$. Then, the disaggregation of X_{n+1} is made based on $\boldsymbol{Y}_j$, $j = (n + 1) - (m - 1)\tau/\Delta t$, and its neighbors $\boldsymbol{Y}'_j$ for all $j' < j$. The neighbors of $\boldsymbol{Y}_j$ are found on the basis of the minimum values of $\left\|\boldsymbol{Y}_j - \boldsymbol{Y}'_j\right\|$. If only one neighbor is considered, then the distributions of weights $(W_{n+1})_k$ of X_{n+1} would be the distributions of weights of the corresponding element X_j in the nearest vector $\boldsymbol{Y}'_j$. This is called the zeroth-order approximation. An improvement to this is the first-order approximation, which considers k' number of neighbors, and the distributions of weights $(W_{n+1})_k$ of X_{n+1} is taken as an average of the k' values' distributions of weights of the corresponding elements X_j in the nearest vectors. The optimal value of k' (i.e., k'_{opt}) is determined by trial and error. Having determined the weights, the disaggregation of the rainfall value X_{n+1} observed at the resolution T_1 to rainfall values $(Z_{n+1})_k$ at resolution T_2 is obtained according to $(Z_{n+1})_k = (W_{n+1})_k * X_{n+1}$. The above procedure is repeated to obtain the distributions of weights of rainfall values $X_{n+2}, X_{n+3}, \ldots, X_N$, i.e. $(W_{n+2})_k, (W_{n+3})_k, \ldots, (W_N)_k$, and hence the rainfall values at the resolution T_2, i.e. $(Z_{n+2})_k, (Z_{n+3})_k \ldots, (Z_N)_k$. The accuracy of disaggregation can be evaluated by comparing the actual and the modeled disaggregated values using any of the standard statistical measures.

Sivakumar et al. (2001b) applied the above approach to disaggregate data between successively doubled resolutions in the Leaf River basin. Only non-zero rainfall values were considered. The number of neighbors considered in the disaggregation procedure ranged from 1 to 200, and the procedure was carried out for embedding dimensions from 1 to 10. In each case (i.e. disaggregation between different successively doubled resolutions) the coarser-resolution time series was split into two parts, with the last 100 points, disaggregated to yield 200 finer-resolution values, forming the test set to evaluate the disaggregation accuracy.

Figure 9.5a, for example, presents the relationship between the disaggregation accuracy (correlation coefficient) and the number of neighbors for the case of rainfall disaggregation from 12-h resolution to 6-h resolution, and Fig. 9.5b shows

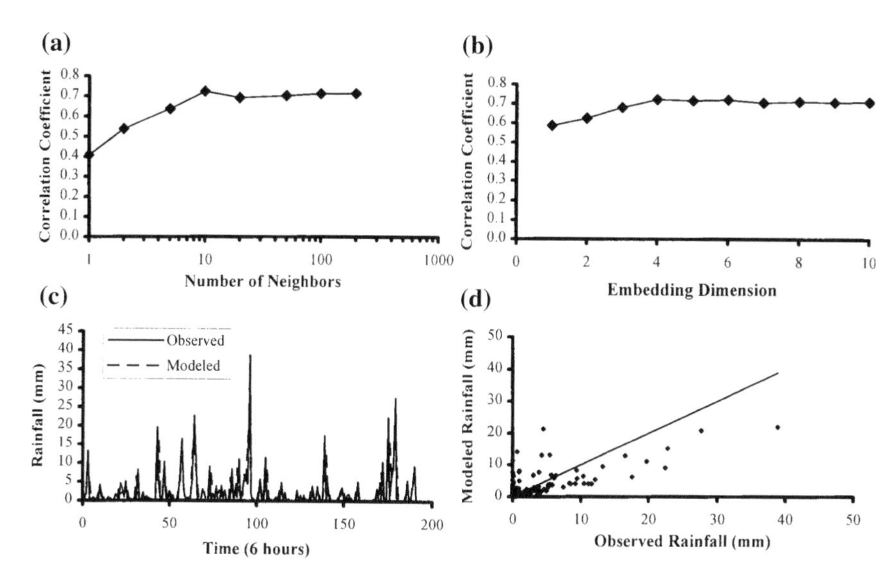

Fig. 9.5 Chaotic disaggregation of rainfall data from 12-h resolution to 6-h resolution in the Leaf River basin: **a** relationship between correlation coefficient and number of neighbors ($m_{opt} = 4$); **b** relationship between correlation coefficient and embedding dimension ($k'_{opt} = 10$); **c** comparison between time series of modeled and observed rainfall values ($m_{opt} = 4$; $k'_{opt} = 10$); and **d** scatterplot of modeled and observed rainfall values ($m_{opt} = 4$; $k'_{opt} = 10$) (*source* Sivakumar et al. (2001b))

the disaggregation accuracy against embedding dimension for the same case. As can be seen, the optimal number of neighbors (k'_{opt}) is 10, and the optimal embedding dimension (m_{opt}) is 4. The existence of an optimal and low embedding dimension suggests the presence of chaos, which is also supported by the observation that the best disaggregation is achieved for a small number of neighbors. The disaggregation results are reasonably good, with a correlation coefficient value of 0.723, root mean square error value of 3.42, and coefficient of efficiency value of 0.567. For this case, Fig. 9.5c presents a time series comparison between the modeled and observed values, and Fig. 9.5d shows the scatterplot. The plots show fairly accurate matching between the modeled and the observed values. Table 9.3 presents the rainfall disaggregation results for all the successively doubled resolutions studied. The results indicate reasonably good disaggregation. All these results suggest the suitability and effectiveness of a chaotic approach for rainfall disaggregation.

The rainfall disaggregation results presented in Fig. 9.5d and Table 9.3 also offer some other important findings. The correlation dimension results (Fig. 9.4d) indicate that the lowest correlation dimension is obtained for the case of distribution of weights of rainfall from 24- to 12-h. This means that the rainfall transformation dynamics between 24- and 12-h have a greater degree of determinism when compared to the other resolutions studied, including that between 12- and 6-h. This,

Table 9.3 Results of rainfall disaggregation in the Leaf River basin, Mississippi, USA (*source* Sivakumar et al. (2001b))

Statistic	192- to 96-h	96- to 48-h	48- to 24-h	24- to 12-h	12- to 6-h
Correlation dimension	3.46	2.61	2.23	1.65	1.86
Correlation coefficient	0.7148	0.6772	0.7165	0.7530	0.7234
Root mean square error	15.62	14.41	9.91	4.89	3.42
Coefficient of efficiency	0.401	0.458	0.507	0.588	0.567
Optimal dimension	6	6	4	4	4
Optimal neighbors	10	10	10	10	10

in turn, suggests that rainfall disaggregation will be more effective between 12- and 6-h scales than other scales studied. This observation has broad implications for rainfall scaling and disaggregation studies, since it may be widely surmised, rightly so in a purely scale-invariance perspective, that disaggregation of rainfall may be possible across all scales, from extremely coarse (e.g. annual) to extremely fine (e.g. minute). The rainfall disaggregation results (Table 9.3) also support the above observations, as the best disaggregation is achieved for the case of 24- to 12-h among all the cases. The consistency between the correlation dimension and the disaggregation results clearly indicates that the disaggregation model performs sufficiently well, and according to theoretical expectations.

9.3.3 Others

The studies by Sivakumar (2001a, b) and Sivakumar et al. (2001b) led several others to connect chaos and scaling in rainfall. Gaume et al. (2006) studied rainfall weights to investigate the presence of chaos, in a somewhat similar manner to the study by Sivakumar et al. (2001b). Employing the correlation dimension method to an 8-year rainfall weight series obtained by disaggregating a 10-min series to a 5-min series, they neither observed low-dimensional chaotic behavior nor that the data were composed of independent and identically distributed random variables. The results also suggested that the correlation dimension method could be an effective tool for exploring data also in other contexts in addition to chaos analysis. Sivakumar et al. (2006) investigated the dynamic nature of rainfall in California's Sacramento Valley. They studied rainfall data observed at four different temporal scales between daily and monthly scales (i.e. daily, weekly, biweekly, and monthly). Employing the correlation dimension method, they reported that the rainfall dynamics were dominated by a large number of variables at all these scales but also that dynamics at coarser resolutions were more irregular than at finer resolutions. Comparison of all-year and winter rainfall, in an attempt to investigate the effects of zeros, showed that winter rainfall had a higher variability. Jothiprakash and Fathima (2013) investigated the presence of chaos in daily (full-year and monsoon), weekly, 10-day, monthly, and seasonal rainfall data observed in the Koyna reservoir catchment in India. Applying the correlation dimension method for chaos identification, they reported that the daily full-year, weekly, and 10-day rainfall showed chaotic behavior, while daily monsoon, monthly, and seasonal rainfall exhibited stochastic behavior.

9.4 Spatial Variability and Classification

While a large number of studies have investigated the chaotic dynamic behavior of rainfall, they have essentially focused on rainfall dynamics at a single location or only at a very few locations without any attempt to specifically examine the dynamics of rainfall in space, the only exception being Sivakumar et al. (2014). However, understanding the spatial rainfall dynamic variability is important for various purposes, including interpolation/extrapolation of rainfall and classification of catchments.

Sivakumar et al. (2014) examined the utility of chaos concepts to study the spatial rainfall variability in Western Australia. They employed the correlation dimension method to monthly rainfall data observed over a period of 67 years (January 1937–December 2003) at 62 raingage stations across the state. These 62 stations and the rainfall observed have a wide range of characteristics. As for station characteristics, the elevation ranges from as low as 4 m to as high as 670 m. In terms of rainfall statistics, the mean ranges from 16.90 to 75.53 mm, standard deviation from 24.32 to 100.86 mm, maximum from 131.80 to 764.30 mm, and no-rainfall months from 0 to 40.92 % (i.e. 329 months). These statistics clearly reveal the significant spatial variability of rainfall in Western Australia, including indicating a four-to-six fold difference in the range (maximum and minimum) values in mean, standard deviation, and maximum rainfall observed among all the 62 stations, and an even greater difference in terms of number of zeros. Figure 9.6a–d presents a graphical representation of these statistics for the 62 stations, for better visualization. There are also significant differences in the 'relationships' among these 62 stations and rainfall observed, as the correlations between stations, distance between stations, and correlation versus distance indicate (figures not shown here; see Sivakumar et al. 2014).

Application of the correlation dimension method to rainfall data from these 62 stations provides correlation dimension values ranging from 4.63 to 8.29. Figure 9.6e presents the correlation dimension values for these 62 stations. The figure also shows grouping (or classification) of these 62 stations according to five different ranges of dimension values: 4.00–4.99, 5.00–5.99, 6.00–6.99, 7.00–7.99, and 8.00–8.99, roughly to indicate the number of variables dominantly governing the rainfall dynamics.

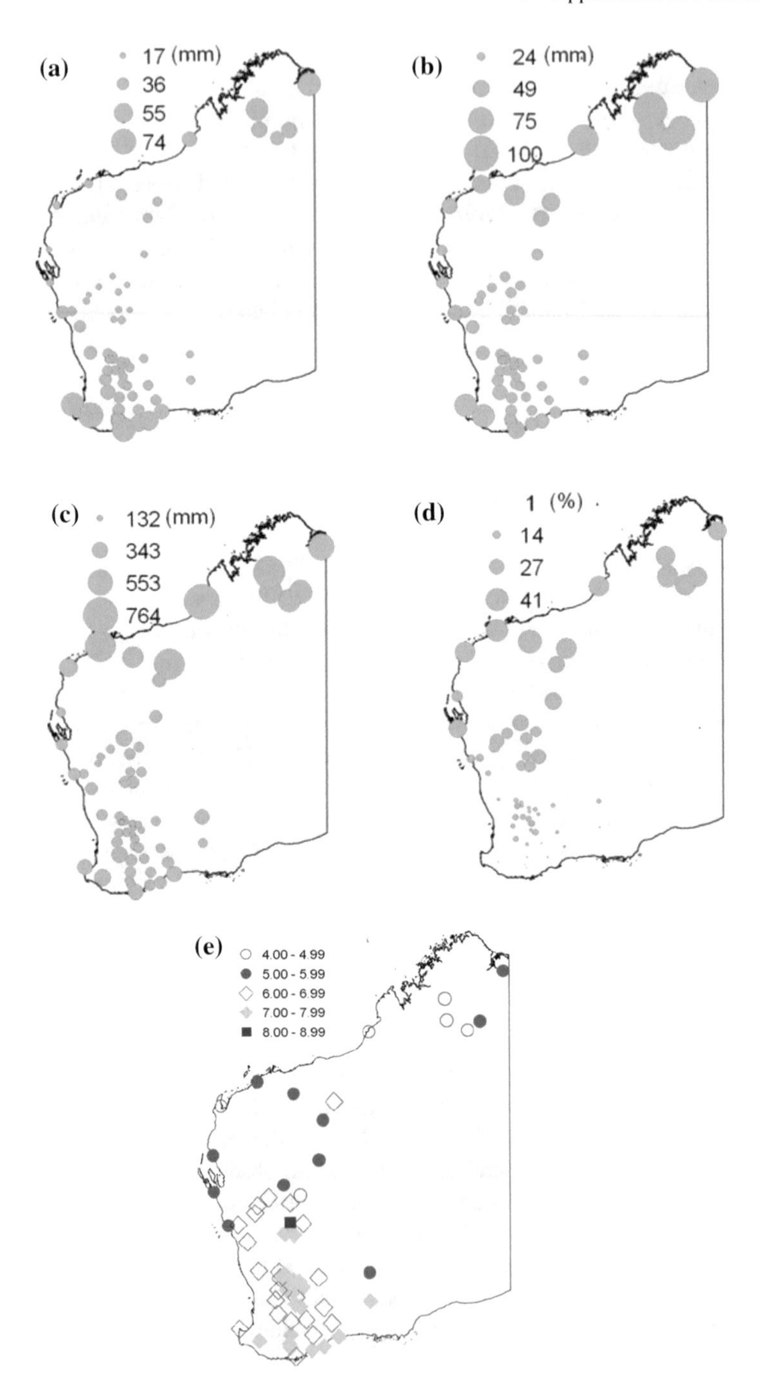

Fig. 9.6 Analysis of variability of monthly rainfall in Western Australia: **a** mean; **b** standard deviation; **c** maximum; **d** percentage number of zeros; and **e** correlation dimension (*source* Sivakumar et al. (2014))

The correlation dimension values clearly indicate that the rainfall dynamics are far more complex in the southwest part of Western Australia ($d > 6$) compared to the other parts of the state; even within the southwest, the far south exhibits greater level of complexity in rainfall dynamics ($d > 7$). The mid-western region seems to show slightly more complex rainfall dynamics ($5 < d < 6$) than that in the northwest ($d < 5$). This grouping of stations/regions is certainly useful to identify/develop appropriate complexity of the models and interpolation/extrapolation of rainfall data, among others.

9.5 Others (Data Size, Noise Reduction, Zeros)

Although the primary purpose of chaos studies on rainfall has and continues to be on addressing the major hydrologic problems (e.g. system identification, prediction, scaling/disaggregation), many studies have, directly or indirectly, also addressed the issues associated with the application of chaos theory-based methods to rainfall data. Among the issues addressed are delay time, data size, data noise, zeros, and others. The issue of delay time in rainfall analysis has been addressed by Rodriguez-Iturbe et al. (1989), Tsonis (1992), Sivakumar et al. (1999a), Koutsoyiannis (2006), and Kim et al. (2009), among others. Studies that have addressed the data size issue include Tsonis et al. (1993), Jayawardena and Lai (1994), Sivakumar et al. (1998, 1999a, 2002), Sivakumar (2001a), and Schertzer et al. (2002). Rainfall data noise and noise reduction issues have been addressed by Berndtsson et al. (1994), Sivakumar et al. (1999b), and Jayawardena and Gurung (2000), among others. Tsonis et al. (1994), Koutsoyiannis and Pachakis (1996), Sivakumar et al. (1999a, 2001b, 2006), Sivakumar (2001a, 2005), and Koutsoyiannis (2006) have addressed the issue of the presence of zeros in rainfall data. Studies that have addressed other issues, such as selection of scaling region in the correlation dimension estimation and neighborhood selection in nonlinear prediction/disaggregation, include Koutsoyiannis and Pachakis (1996), Sivakumar et al. (2001b), and Jothiprakash and Fathima (2013). Some of these studies will be discussed in more detail in Chap. 12.

9.6 Summary

Applications of the concepts of chaos theory in hydrology started with the analysis of rainfall data in the late 1980s. Since then, numerous studies have applied the chaos concepts to rainfall data for different purposes. This chapter has presented a review of chaos studies on rainfall, with an example for identification and prediction of chaotic behavior, investigation of scaling and disaggregation, and analysis of spatial variability and classification. The outcomes of such studies are very encouraging, especially in the context of the broader scope and potential chaos theory can have in studying rainfall dynamics. In addition to these studies, many attempts have also been made to address the issues in the applications of chaos methods to rainfall data, including data size, data noise, and presence of zeros. Chapter 12 will discuss the details of some of these studies.

References

Abarbanel HDI (1996) Analysis of observed chaotic data. Springer, New York

Berndtsson R, Jinno K, Kawamura A, Olsson J, Xu S (1994) Dynamical systems theory applied to long-term temperature and precipitation time series. Trends Hydrol 1:291–297

Casdagli M (1989) Nonlinear prediction of chaotic time series. Physica D 35:335–356

Casdagli M (1992) Chaos and deterministic versus stochastic nonlinear modeling. J Royal Stat Soc B 54(2):303–328

Dhanya CT, Nagesh Kumar D (2010) Nonlinear ensemble prediction of chaotic daily rainfall. Adv Water Resour 33:327–347

Dhanya CT, Nagesh Kumar D (2011) Multivariate nonlinear ensemble prediction of daily chaotic rainfall with climate inputs. J Hydrol 403:292–306

Farmer DJ, Sidorowich JJ (1987) Predicting chaotic time series. Phys Rev Lett 59:845–848

Fraedrich K (1986) Estimating the dimensions of weather and climate attractors. J Atmos Sci 43:419–432

Fraedrich K (1987) Estimating weather and climate predictability on attractors. J Atmos Sci 44:722–728

Gaume E, Sivakumar B, Kolasinski M, Hazoumé L (2006) Identification of chaos in rainfall disaggregation: application to a 5-minute point series. J Hydrol 328(1–2):56–64

Georgakakos KP, Sharifi MB, Sturdevant PL (1995) Analysis of high-resolution rainfall data. In: Kundzewicz ZW (ed) New uncertainty concepts in hydrology and water resources. Cambridge University Press, New York, pp 114–120

Ghilardi P, Rosso R (1990) Comment on "Chaos in rainfall". Water Resour Res 26(8):1837–1839

Gilmore CG (1993) A new test for chaos. J Econ Behav Organ 22:209–237

Grassberger P, Procaccia I (1983a) Measuring the strangeness of strange attractors. Physica D 9:189–208

Grassberger P, Procaccia I (1983b) Characterisation of strange attractors. Phys Rev Lett 50(5):346–349

Grassberger P, Procaccia I (1983c) Estimation of the Kolmogorov entropy from a chaotic signal. Phys Rev A 28:2591–2593

Hense A (1987) On the possible existence of a strange attractor for the southern oscillation. Beitr Phys Atmos 60(1):34–47

Islam S, Bras RL, Rodriguez-Iturbe I (1993) A possible explanation for low correlation dimension estimates for the atmosphere. J Appl Meteor 32:203–208

Jayawardena AW, Gurung AB (2000) Noise reduction and prediction of hydrometeorological time series: dynamical systems approachvs. stochastic approach. J Hydrol 228:242–264

Jayawardena AW, Lai F (1994) Analysis and prediction of chaos in rainfall and stream flow time series. J Hydrol 153:23–52

Jayawardena AW, Li WK, Xu P (2002) Neighborhood selection for local modeling and prediction of hydrological time series. J Hydrol 258:40–57

Jin YH, Kawamura A, Jinno K, Berndtsson R (2005) Nonlinear multivariate analysis of SOI and local precipitation and temperature. Nonlinear Processes Geophys 12:67–74

Jothiprakash V, Fathima TA (2013) Chaotic analysis of daily rainfall series in Koyna reservoir catchment area, India. Stoch Environ Res Risk Assess 27:1371–1381

Kennel MB, Brown R, Abarbanel HDI (1992) Determining embedding dimension for phase space reconstruction using a geometric method. Phys Rev A 45:3403–3411

Kim HS, Eykholt R, Salas JD (1999) Nonlinear dynamics, delay times, and embedding windows. Physica D 127(1–2):48–60

Kim HS, Lee KH, Kyoung MS, Sivakumar B, Lee ET (2009) Measuring nonlinear dependence in hydrologic time series. Stoch Environ Res Risk Assess 23:907–916

Koutsoyiannis D (2006) On the quest for chaotic attractors in hydrological processes. Hydrol Sci J 51(6):1065–1091

Koutsoyiannis D, Pachakis D (1996) Deterministic chaos versus stochasticity in analysis and modeling of point rainfall series. J Geophys Res 101(D21):26441–26451

Kyoung MS, Kim HS, Sivakumar B, Singh VP, Ahn KS (2011) Dynamic characteristics of monthly rainfall in the Korean peninsula under climate change. Stoch Environ Res Risk Assess 25(4):613–625

Puente CE, Obregon N (1996) A deterministic geometric representation of temporal rainfall. Results for a storm in Boston. Water Resour Res 32(9):2825–2839

Rodriguez-Iturbe I, De Power FB, Sharifi MB, Georgakakos KP (1989) Chaos in rainfall. Water Resour Res 25(7):1667–1675

Rodriguez-Iturbe I, De Power FB, Sharifi MB, Georgakakos KP (1990) Reply. Water Resour Res 26(8):1841–1842

Schertzer D, Tchiguirinskaia I, Lovejoy S, Hubert P, Bendjoudi H (2002) Which chaos in the rainfall-runoff process? A discussion on 'Evidence of chaos in the rainfall-runoff process' by Sivakumar et al. Hydrol Sci J 47(1):139–147

Schouten JC, Takens F, van den Bleek CM (1994) Estimation of the dimension of a noisy attractor. Phys Rev E 50(3):1851–1861

Schreiber T (1993) Extremely simple nonlinear noise reduction method. Phys Rev E 47 (4):2401–2404

Schreiber T, Kantz H (1996) Observing and predicting chaotic signals: is 2 % noise too much? In: Kravtsov YuA, Kadtke JB (eds) Predictability of complex dynamical systems. Springer Series in Synergetics. Springer, Berlin, pp 43–65

Sharifi MB, Georgakakos KP, Rodriguez-Iturbe I (1990) Evidence of deterministic chaos in the pulse of storm rainfall. J Atmos Sci 47:888–893

Sivakumar B (2000) Chaos theory in hydrology: important issues and interpretations. J Hydrol 227(1–4):1–20

Sivakumar B (2001a) Rainfall dynamics at different temporal scales: a chaotic perspective. Hydrol Earth Syst Sci 5(4):645–651

Sivakumar B (2001b) Is a chaotic multi-fractal approach for rainfall possible? Hydrol Process 15(6):943–955

Sivakumar B (2004) Chaos theory in geophysics: past, present and future. Chaos Soliton Fract 19(2):441–462

Sivakumar B (2005) Chaos in rainfall: variability, temporal scale and zeros. J Hydroinform 7(3):175–184

Sivakumar B (2009) Nonlinear dynamics and chaos in hydrologic systems: latest developments and a look forward. Stoch Environ Res Risk Assess 23:1027–1036

Sivakumar B, Liong SY, Liaw CY (1998) Evidence of chaotic behavior in Singapore rainfall. J Am Water Resour Assoc 34(2):301–310

Sivakumar B, Liong SY, Liaw CY, Phoon KK (1999a) Singapore rainfall behavior: chaotic? ASCE J Hydrol Eng 4(1):38–48

Sivakumar B, Phoon KK, Liong SY, Liaw CY (1999b) A systematic approach to noise reduction in chaotic hydrological time series. J Hydrol 219(3–4):103–135

Sivakumar B, Phoon KK, Liong SY, Liaw CY (1999c) Comment on "Nonlinear analysis of river flow time sequences" by Amilcare Porporato and Luca Ridolfi. Water Resour Res 35 (3):895–897

Sivakumar B, Berndtsson R, Olsson J, Jinno K, Kawamura A (2000) Dynamics of monthly rainfall-runoff process at the Göta basin: A search for chaos. Hydrol Earth Syst Sci 4 (3):407–417

Sivakumar B, Berndttson R, Olsson J, Jinno K (2001a) Evidence of chaos in the rainfall-runoff process. Hydrol Sci J 46(1):131–145

Sivakumar B, Sorooshian S, Gupta HV, Gao X (2001b) A chaotic approach to rainfall disaggregation. Water Resour Res 37(1):61–72

Sivakumar B, Berndtsson R, Olsson J, Jinno K (2002) Reply to 'which chaos in the rainfall-runoff process?' by Schertzer et al. Hydrol Sci J 47(1):149–158

Sivakumar B, Wallender WW, Horwath WR, Mitchell JP, Prentice SE, Joyce BA (2006) Nonlinear analysis of rainfall dynamics in California's Sacramento Valley. Hydrol Process 20(8):1723–1736

Sivakumar B, Woldemeskel FM, Puente CE (2014) Nonlinear analysis of rainfall variability in Australia. Stoch Environ Res Risk Assess 28:17–27

Takens F (1981) Detecting strange attractors in turbulence. In: Rand DA, Young LS (eds) Dynamical systems and turbulence, Lecture notes in mathematics 898. Springer, Berlin, pp 366–381

Theiler J, Eubank S, Longtin A, Galdrikian B, Farmer JD (1992) Testing for nonlinearity in time series: the method of surrogate data. Physica D 58:77–94

Tsonis AA (1992) Chaos: from theory to applications. Plenum Press, New York

Tsonis AA, Georgakakos KP (2005) Observing extreme events in incomplete state spaces with application to rainfall estimation from satellite images. Nonlinear Processes Geophys 12:195–200

Tsonis AA, Elsner JB, Georgakakos KP (1993) Estimating the dimension of weather and climate attractors: important issues about the procedure and interpretation. J Atmos Sci 50:2549–2555

Tsonis AA, Triantafyllou GN, Elsner JB, Holdzkom JJ II, Kirwan AD Jr (1994) An investigation on the ability of nonlinear methods to infer dynamics from observables. Bull Amer Meteor Soc 75:1623–1633

Waelbroeck H, Lopex-Pena R, Morales T, Zertuche F (1994) Prediction of tropical rainfall by local phase space reconstruction. J Atmos Sci 51(22):3360–3364

Wolf A, Swift JB, Swinney HL, Vastano A (1985) Determining Lyapunov exponents from a time series. Physica D 16:285–317

Chapter 10
Applications to River Flow Data

Abstract Chaos theory has found widespread applications in studies on river flow data. Indeed, river flow is the most studied data in the context of chaos studies in hydrology. Early applications mainly focused on identification and prediction of chaotic behavior in river flow dynamics. Later years witnessed studies on a wide range of problems associated with river flow, including scaling and disaggregation, missing data estimation, reconstruction of system equations, multivariable analysis, and spatial variability and classification. Many studies have also addressed the important issues in the applications of chaos theory to river flow data, including data size, data noise, and selection of parameters involved in chaos identification and prediction methods. This chapter presents a review of applications of chaos theory to river flow data. The studies are roughly grouped into three categories to represent the following aspects: identification and prediction, scaling and disaggregation, and spatial variability and classification. These applications are also illustrated through examples.

10.1 Introduction

Chaos studies in river flow have probably received, and continue to receive, far more attention than those on any other hydrologic phenomenon. Since the initial studies by Wilcox et al. (1990, 1991), chaos studies on river flow have skyrocketed. They have addressed chaos identification and prediction (e.g. Jayawardena and Lai 1994; Porporato and Ridolfi 1996, 1997; Liu et al. 1998; Sivakumar et al. 2001b; Islam and Sivakumar 2002), scaling and disaggregation (e.g. Regonda et al. 2004; Sivakumar et al. 2004), missing data estimation (e.g. Elshorbagy et al. 2001, 2002a), reconstruction of system equations (e.g. Zhou et al. 2002), and spatial variability and classification (e.g. Sivakumar and Singh 2012; Vignesh et al. 2015). Several studies have also addressed the potential issues in the application of chaos theory to river flow time series, such as parameter selection for phase space reconstruction, data size, and data noise and noise reduction (e.g. Porporato and Ridolfi 1997; Phoon et al. 2002; Sivakumar et al. 2002c; Sivakumar 2005; Kim

B. Sivakumar, *Chaos in Hydrology*, DOI 10.1007/978-90-481-2552-4_10

et al. 2009; Dhanya and Nagesh Kumar 2011). Some studies have also attempted multi-variable analysis of river flow time series (e.g. Porporato and Ridolfi 2001; Sivakumar et al. 2005). Sivakumar (2000, 2004a, 2009) discuss many of these studies. This chapter reviews the chaos studies in river flow. The presentation is divided into the following topics: identification and prediction, scaling and disaggregation, spatial variability and classification, and others. For the purpose of demonstration, details of a few selected studies are presented.

10.2 Identification and Prediction

To my knowledge, the first studies on chaos in river flow were conducted by Wilcox et al. (1990, 1991). These studies investigated the daily snowmelt runoff measured from the Reynolds Mountain catchment in the Owyhee Mountains of southwestern Idaho, USA. Applying the correlation dimension method (e.g. Grassberger and Procaccia 1983a, b) to this series and observing no low dimension, they concluded that the random-appearing behavior of snowmelt runoff was generated from the complex interactions of many factors, rather than low dimensional chaotic dynamics. Kember and Flower (1993) were the first ones to attempt prediction of river flow using concepts of chaos theory. They employed the nearest neighbor method to predict daily river flows at Spruce Falls in Northern Ontario, Canada. They found that the best results were obtained for a low embedding dimension ($m = 6$), which suggested the presence of low dimensional dynamics. Comparison of these prediction results with those from an autoregressive integrated moving average (ARIMA) model indicated that the nearest neighbor method provided improved forecasts.

Jayawardena and Lai (1994) attempted identification and prediction of chaos in the daily streamflow in Hong Kong. They employed the correlation dimension method, Lyapunov exponent method (e.g. Wolf et al. 1985), Kolmogorov entropy method (e.g. Grassberger and Procaccia 1983c), and local approximation prediction method (e.g. Farmer and Sidorowich 1987; Casdagli 1989) to two flow series (7300 and 6205 points). Their study provided convincing evidence of the existence of chaos in these time series. Although the prediction accuracy estimates from the local approximation method were found to be low, they were still found superior to those achieved from the linear autoregressive moving average (ARMA) method. Porporato and Ridolfi (1996) provided clues to the presence of chaos in the flow dynamics observed at Dora Baltea, a tributary of the river Po in Italy. Application of the correlation dimension and the local approximation prediction (and also phase space and Poincaré section) methods to the daily flow series consisting of 14,246 points indicated the existence of a strong deterministic component. The study also led to a more detailed analysis of the flow phenomenon (Porporato and Ridolfi 1997), including noise reduction, interpolation, and nonlinear prediction, which provided important confirmations regarding its nonlinear deterministic behavior.

Liu et al. (1998) applied the local approximation prediction method to the daily streamflow series observed at 28 selected stations in the continental United States. They reported that streamflow signals spanned a wide dynamic range between deterministic chaos and periodic signal contaminated with additive noise. Wang and Gan (1998) employed the correlation dimension to investigate the presence of chaos in the unregulated streamflow series of six rivers in the Canadian prairies. The results provided possible signs of deterministic chaotic behavior, as correlation dimensions of about 3.0 for these series were obtained. However, observing a consistent underestimation of the correlation dimension for the randomly re-sampled data by an amount of 4–6, they interpreted that the actual dimensions of these streamflow series should be between 7 and 9, and that the streamflow process might be stochastic. Krasovskaia et al. (1999) investigated the dimensionality (in terms of fractal and intrinsic dimensions) of the Scandinavian river flow regimes in their effort to identify the possible presence of chaos. For this purpose, they analyzed, using a variety of methods, river flow series observed in a number of stations in the region. The results revealed a variety of fractal and intrinsic dimensions for the different series that were well in agreement with the stability character of the investigated regime types, i.e. the less stable the regime, the higher the fractal and intrinsic dimensions and the number of variables required for its description.

Pasternack (1999) reported the absence of chaos in the river flow process, based on the analysis of daily flow series observed in Western Run in upper Baltimore county, Maryland in the Gunpowder River basin in USA. Citing the pitfalls in the implementation of the correlation dimension algorithm, and based on analysis using the surrogate data method (e.g. Theiler et al. 1992), Pasternack (1999) criticized the earlier studies that had reported presence of chaos in river flow process in particular, and hydrologic processes in general. However, Liaw et al. (2001) questioned the study by Pasternack (1999). They commented that the embedding parameters (mainly the delay time) for phase space reconstruction had not been properly selected by Pasternack (1999). They argued that the attractor reconstructed in the correlation dimension analysis and surrogate data analysis could not actually represent the dynamic behavior of the underlying dynamics of the system; see also Pasternack (2001) for a response.

Jayawardena and Gurung (2000) made a systematic effort to address the issue of the presence/absence of chaos in river flow (and other hydrologic) dynamics, by employing both nonlinear dynamic and linear stochastic approaches to flow series from the Chao Phraya River (at Nakhon Sawan) and the Mekong River (at Nong Khai and Pakse) in Thailand. The nonlinear dynamic analysis included correlation dimension estimation, noise reduction, and local approximation prediction, whereas the linear stochastic analysis included model identification, formulation, diagnostic tests, and prediction. Based on the predictions and determinism test results, they concluded that the flow phenomena were indeed nonlinear deterministic. Lambrakis et al. (2000) presented evidence to nonlinearity and chaotic nature of the discharge process of the spring of Almyros, Iraklion, Crete, Greece, through application of the correlation dimension and surrogate data methods. Based on this, they also attempted short-term forecasting of the discharge series using local approximation prediction method and artificial neural networks. Both techniques were found to yield a very satisfactory predictive ability, with neural networks performing slightly better.

Sivakumar et al. (2000, 2001a) provided convincing evidence as to the presence of chaos in the monthly flow series in the Göta River basin in Sweden, as they observed a low correlation dimension of 5.5 and also extremely good predictions. Analysis of this flow series was part of their study on the rainfall-runoff process; see also Chaps. 9 (Sect. 9.2.1) and 11 (Sect. 11.2) for relevant details. Lisi and Villi (2001) checked for evidence of chaotic behavior in the daily river flow process of the Adige River in Italy, as part of their study on forecasting the time series. Using correlation dimension and Lyapunov exponent as indicators, they observed a nonlinear deterministic chaotic type dynamics in the flow process. Employing the local approximation method and comparing the predictions with those achieved using a stochastic (ARIMA) model, they reported superior performance of the former by a margin of 12–18 % for up to 3-day ahead predictions. Porporato and Ridolfi (2001) extended the phase space reconstruction and local approximation prediction concepts to a multi-variate form to include information from other time series in addition to that of river flow. The effectiveness of their proposed multi-variate prediction approach was tested on the daily river flow phenomenon at Dora Baltea river in Italy, for which signs of low dimensional determinism had been found earlier (Porporato and Ridolfi 1996, 1997). Using rainfall and temperature as additional variables in the multi-variate approach, they reported that such an approach was much better than the univariate approach for river flow predictions, especially in the prediction of highest flood peaks. This nonlinear multi-variate approach was also found to be superior than a stochastic ARMAX (Autoregressive Moving Average with Extraneous inputs) model. The flexibility of the multi-variate model to adapt to the different sources of information was also observed. A similar multi-variable approach for streamflow, based on only rainfall and streamflow, was also presented by Sivakumar et al. (2005).

Sivakumar et al. (2001b) employed the local approximation method to the monthly flow series in Coaracy Nunes/Araguari River basin in northern Brazil: (1) to predict the runoff dynamics; and (2) to detect possible presence of chaos in runoff dynamics using the prediction results. The local approximation method yielded extremely good predictions for the flow series. Based on observations of a low optimal embedding dimension and a clear decrease in prediction accuracy with increasing lead time, they concluded that the flow process exhibited low dimensional chaotic behavior. These results were subsequently verified and confirmed by Sivakumar et al. (2002c) through the very good predictions achieved using neural networks and low correlation dimension of 3.62 obtained for the flow series. These results also facilitated addressing the issue of data size requirement for a reliable estimation of correlation dimension in river flow (and other hydrologic) time series. Due to the extensive nature of the analysis and results available, this river flow series is considered here as a representative series to illustrate the analysis to identify the presence of chaos in river flow dynamics, and details are presented in Sect. 10.2.1.

Islam and Sivakumar (2002) performed an extensive analysis towards characterizing and predicting the river flow dynamics at the Lindenborg catchment in Denmark. Through a host of both standard statistical techniques and specific nonlinear and chaotic dynamic techniques, they reported presence of chaos in this river flow dynamics, with a low correlation dimension of 3.76. The study was the first one to employ the false nearest neighbor algorithm (e.g. Kennel et al. 1992) for chaos identification in river flow. Applying a local approximation prediction method, the study also reported near-accurate prediction results for the flow series and provided further support as to the utility and suitability of a chaotic approach for modeling and predicting river flow dynamics. Zhou et al. (2002) investigated the presence of chaos in the last 500-year annual flood series in the Huaihe River Basin in eastern China. Employing power spectrum and correlation dimension techniques, they reported presence of chaotic behavior in the flood dynamics with an attractor dimension of 4.66. Using the concepts of chaos theory and the inverted theorem of differential equations, they also attempted reconstruction of the flood series in three dimensions and second power. Sivakumar and Jayawardena (2002) analyzed the daily flow data observed in the Mississippi River basin (at St. Louis, Missouri), USA, As part of their investigation of chaos in the sediment transport phenomenon. Observing a low correlation dimension value of 2.32, they suggested possible presence of chaos in the river flow dynamics. Sivakumar and Wallender (2005) advanced the chaos analysis on this time series by attempting nonlinear prediction. They reported: (1) extremely good one-day ahead predictions; (2) optimal embedding dimension; and (3) decreasing prediction accuracy with increasing lead time. All these offered further evidence regarding the presence of chaos in the river flow series. However, the decrease in prediction accuracy with increasing lead time for the flow series was not that significant, especially when compared against the ones observed for suspended sediment concentration and suspended sediment load series.

Sivakumar et al. (2002b) employed the nonlinear local approximation approach for forecasting the flow dynamics of the Chao Phraya River (Nakhon Sawan station) in Thailand. The analysis was performed in tandem with the forecasting of the flow series using artificial neural networks, in order to compare the performance of the two approaches. Attempting 1-day and 7-day ahead forecasts, they reported reasonably good prediction performance of both the approaches. However, the local approximation approach was found to be superior to neural networks; this result was attributed to the representation of the flow dynamics in the phase space step by step in local neighborhoods in the former approach, rather than a global approximation as done in the latter. In a similar vein, but in a multi-variable sense, Laio et al. (2003) employed local approximation method and artificial neural networks for forecasting flood (water stages) of river Tanaro in Alba, Italy, and reported slightly better performance of the former at short forecast times and the reverse situation for longer times.

10.2.1 Chaos Analysis in River Flow: An Example—Coaracy Nunes/Araguari River Basin

For illustration of chaos identification and prediction of river flow dynamics, monthly river flow data from the Coaracy Nunes/Araguari River basin in northern Brazil is considered here. This basin has an area of approximately 24,200 km^2. The data analyzed was observed over a period of 48 years (January 1945–December 1992). Figure 10.1a shows the variation of this runoff time series, and Fig. 10.1b presents its two-dimensional phase-space reconstruction, according to Takens' delay embedding theorem; see Chap. 6, Eq. (6.3). The projection shown in Fig. 10.1b corresponds to a delay time value $\tau = 1$. The reconstruction yields a well-defined attractor, providing initial clues to the presence of low dimensional deterministic dynamics. Indeed, a well-defined attractor seems to be present when $\tau \leq 3$, which also happens to be the lag time at which the autocorrelation function first crosses the zero line, a guideline (e.g. Holzfuss and Mayer-Kress 1986) widely used for delay time selection in chaos analysis. The attractor seems to become less and less clear when τ is increased further, but becomes clear again when $\tau = 12$. This supports the observation, from the time series plot (Fig. 10.1a), of the presence of an annual cycle in the runoff dynamics.

With this reconstruction, the correlation functions and exponents are computed, following the Grassberger-Procaccia algorithm; see Chap. 6, Sect. 6.4.2. Figure 10.1c shows the relationship between the correlation integral, $C(r)$, and the radius, r,

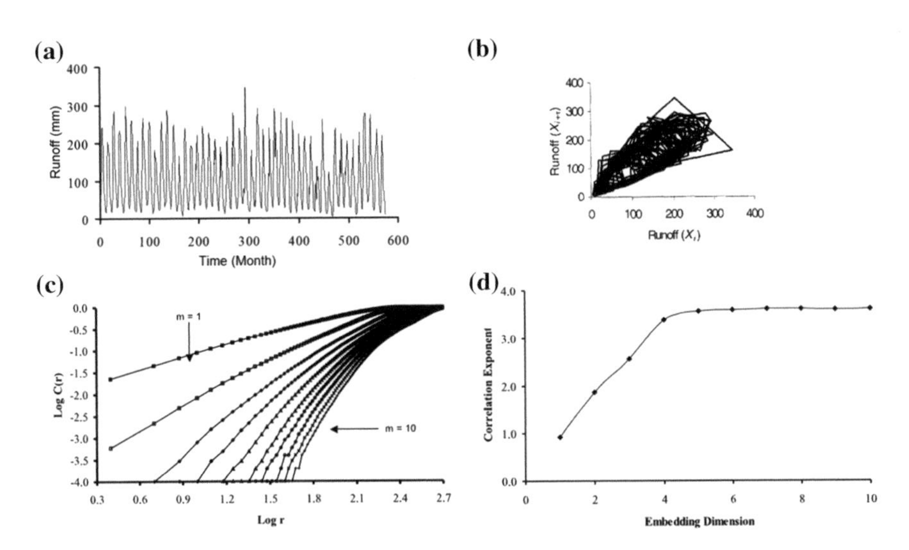

Fig. 10.1 Correlation dimension analysis of monthly runoff from Coaracy Nunes/Araguari River basin in Brazil: **a** time series; **b** phase space; **c** Log C(r) versus Log r; and **d** relationship between correlation exponent and embedding dimension (*source* Sivakumar et al. (2001b, 2005))

for embedding dimensions, m, from 1 to 10. The Log $C(r)$ versus Log r curves exhibit large and clear scaling regions and allow reliable estimation of the correlation exponents. Figure 10.1d presents the relationship between the correlation exponent and embedding dimension values. The correlation exponent increases with the embedding dimension up to a certain point and saturates beyond that. The saturation value of the correlation exponent (i.e. correlation dimension) is 3.76. The finite, low, and non-integer correlation dimension indicates the presence of low dimensional chaotic behavior in the monthly flow dynamics at the Coaracy Nunes/Araguari River basin. The correlation dimension value 3.76 also suggests that the flow dynamics are dominantly governed by about four variables.

After identifying the presence of chaos, prediction of this river flow series is attempted using the local approximation method. The first 480 values in the time series are used for phase-space reconstruction (i.e. training or learning set) and to predict the subsequent 70 values. Embedding dimensions, m, from 1 to 10 are considered for phase space reconstruction, and the predictions are made for lead times (T) from 1 to 10. The local maps are learned in the form of local polynomials (Abarbanel 1996). The prediction accuracy is evaluated using three statistical measures: correlation coefficient (CC), root mean square error ($RMSE$), and coefficient of efficiency (E^2). Time series plots and scatter plots are also used to choose the best prediction results, among a large combination of results obtained for different embedding dimensions and lead times.

Figure 10.2 presents, for lead time $T = 1$, the variation of the above three statistical evaluation measures against the embedding dimension for the flow series. As can be seen, the prediction accuracy increases with the embedding dimension up to a certain value ($m = 3$) and then saturates (or even slightly decreases) beyond that value. The presence of an optimal embedding dimension value, $m_{opt} = 3$, indicates the possible presence of low dimensional chaos in the flow series. Figure 10.3a compares, using time series plots, the observed and the predicted runoff values. The plots shown correspond to the results obtained with embedding dimensions 2, 3, 4, 5, 7, and 10. Figure 10.3b shows the corresponding scatter plots. As can be seen, the prediction results are in good agreement with the observed ones for all the embedding dimensions. Even very high and very low runoff values are reasonably well predicted, as are the trends. The good agreement in the time series plots and scatter plots and also the high E^2 values ($E^2 > 0.91$) (see Table 10.1) indicate the suitability and effectiveness of the nonlinear local approximation method for predicting the runoff dynamics.

The results presented in Fig. 10.3 and Table 10.1 also reveal that the best predictions are achieved only when $m = 3$ (i.e. m_{opt}), and the results are almost the same or slightly worse for $m > 3$ and noticeably worse for $m < 3$. The slight decrease in the prediction accuracy when higher phase spaces ($m > 3$) are used for reconstruction could be due to the presence of noise (e.g. measurement error) in the runoff series, as the influence of noise at higher embedding dimension could be

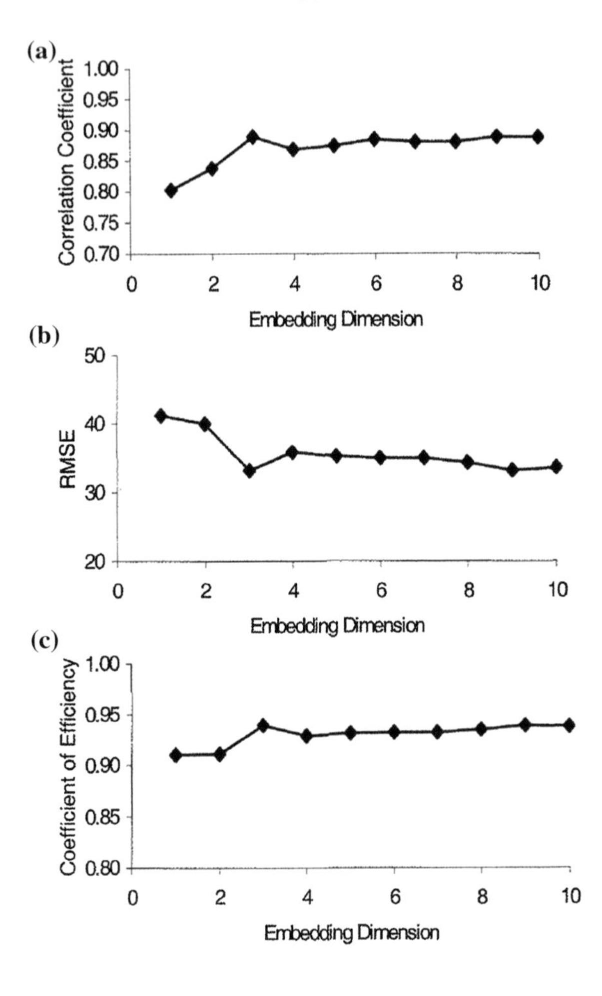

Fig. 10.2 Nonlinear local approximation prediction analysis of monthly runoff from Coaracy Nunes/Araguari River basin in Brazil: **a** correlation coefficient; **b** RMSE; and **c** coefficient of efficiency (*source* Sivakumar et al. (2001b))

greater; see Sivakumar et al. (1999). The noticeably worse results obtained for m < 3 indicate that the reconstruction at these dimensions is an incomplete representation of the system dynamics.

Figure 10.4 shows the relationship between the prediction accuracy and lead time; see also Table 10.1. The results presented are those obtained for the optimal embedding dimension, i.e. $m = 3$. As can be seen, a rapid decrease in the prediction accuracy is observed when the lead time is increased (i.e. when predictions are made further into the future), which is a typical characteristic of chaotic systems (e.g. Sugihara and May 1990). This provides additional support to the observation made earlier regarding the possible presence of low dimensional chaos in the runoff dynamics.

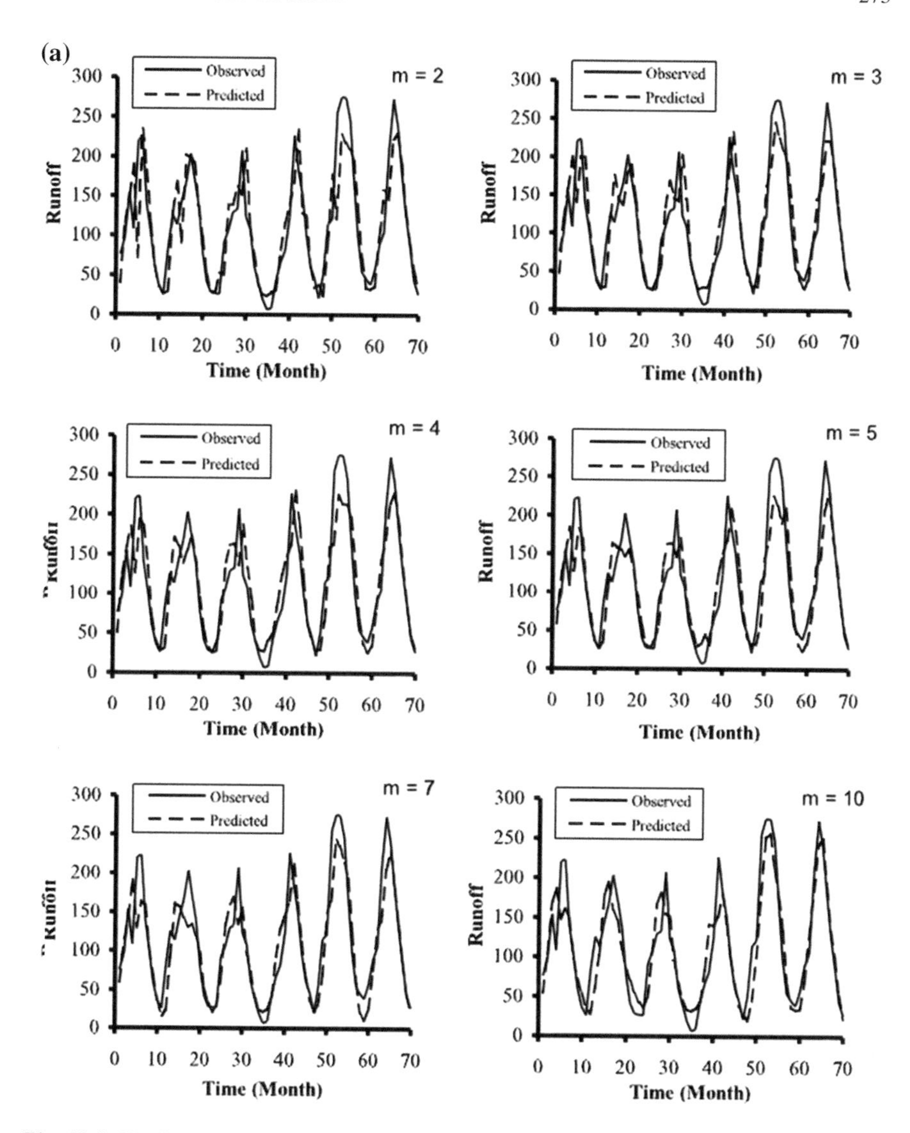

Fig. 10.3 Nonlinear local approximation prediction analysis of monthly runoff from Coaracy Nunes/Araguari River basin in Brazil: **a** time series comparison (*source* Sivakumar et al. (2002c)); and **b** scatter plot (*source* Sivakumar et al. (2001b))

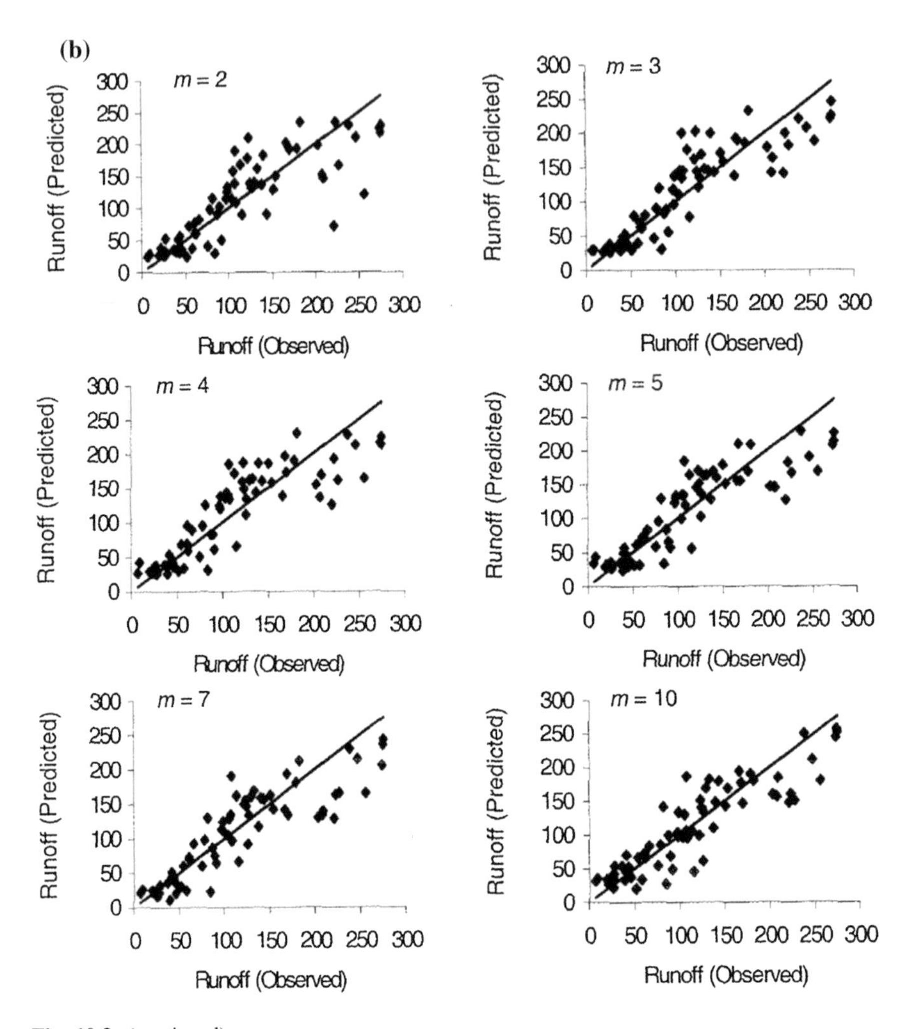

Fig. 10.3 (continued)

10.3 Scaling and Disaggregation

10.3.1 Chaos and Scaling

While most of the early studies investigating the presence of chaos in river flow data had reported the presence of chaos in flow dynamics at different temporal scales, none of them, except the one by Stehlik (1999), had investigated flow data at different temporal scales from the same location. As a result, they could not provide any useful information as to the presence of chaos in runoff process across different temporal scales. Stehlik (1999) studied the runoff process at two different temporal

Table 10.1 Nonlinear local approximation prediction results for monthly runoff data from the Coaracy Nunes/Araguari River in Brazil (*source* Sivakumar et al. (2001b))

$T = 1$ m	CC	RMSE (mm)	E^2	$m = 3$ T	CC	RMSE (mm)	E^2
1	0.8023	41.216	0.9103	1	0.8895	33.138	0.9388
2	0.8375	39.991	0.9109	2	0.7892	45.011	0.8692
3	0.8895	33.138	0.9388	3	0.7612	47.804	0.8518
4	0.8687	35.805	0.9285	4	0.7447	49.393	0.8318
5	0.8744	35.214	0.9309	5	0.7747	47.566	0.8553
6	0.8845	34.928	0.9320	6	0.7945	45.972	0.8326
7	0.8804	34.935	0.9319	7	0.7979	45.989	0.8335
8	0.8805	34.261	0.9346	8	0.7769	48.645	0.8484
9	0.8889	33.122	0.9380	9	0.7381	52.233	0.8382
10	0.8879	33.523	0.9377	10	0.7234	52.853	0.8435

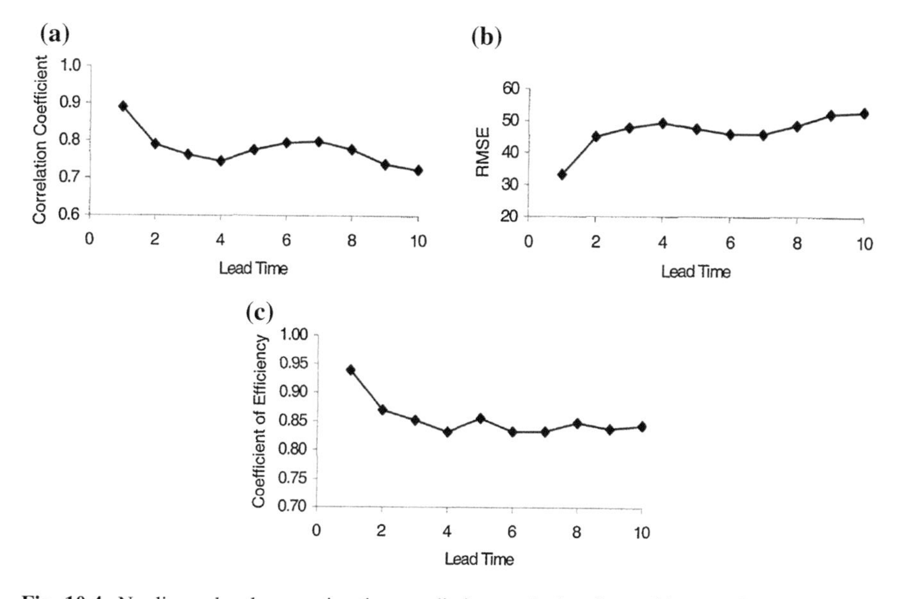

Fig. 10.4 Nonlinear local approximation prediction analysis of monthly runoff from Coaracy Nunes/Araguari River basin in Brazil for different lead times: **a** correlation coefficient; **b** RMSE; and **c** coefficient of efficiency (*source* Sivakumar et al. (2001b))

scales (daily and 30-min) observed at the experimental basin Uhlirska in the Jizera Mountains in Czech Republic. Applying the correlation dimension method, he reported the presence of chaos in the 30-min series with a correlation dimension of 2.89 but absence of chaos in the daily series. Despite these results, Stehlik (1999) did not make any serious attempt to discuss the scaling or scale-invariant behavior of river flow from a chaotic perspective.

The study by Regonda et al. (2004) was the first ever to specifically address the scaling behavior in river flow from a chaotic perspective. Regonda et al. (2004)

investigated the type of scaling behavior (stochastic or chaotic) in the temporal dynamics of river flow. For this purpose, they considered flow dynamics at three different temporal scales (from the same location): daily, 5-day, and 7-day. To put the analysis on a more solid footing, they analyzed flow data from three river basins in the United States: Kentucky River in Kentucky (period of data: January 1960–December 1989), Merced River in California (period of data: January 1960–December 1989), and Stillaguamish River in Washington state (period of data January 1929–December 2000). They applied the correlation dimension method to identify the presence of chaos.

Figure 10.5 presents the correlation dimension results for the flow series at three different temporal scales from these rivers. The results indicate mixed/contrasting scenarios regarding the dynamic behavior of individual flow series and, hence, of the scaling relationship between them. The results for the Kentucky River (Fig. 10.5a) indicate a convincingly clear chaotic behavior in the flow dynamics at each of the three scales, with low correlation dimensions of 4.22, 4.63, and 4.87 for daily, 5-day, and 7-day series, respectively. Chaotic behavior is present also in the

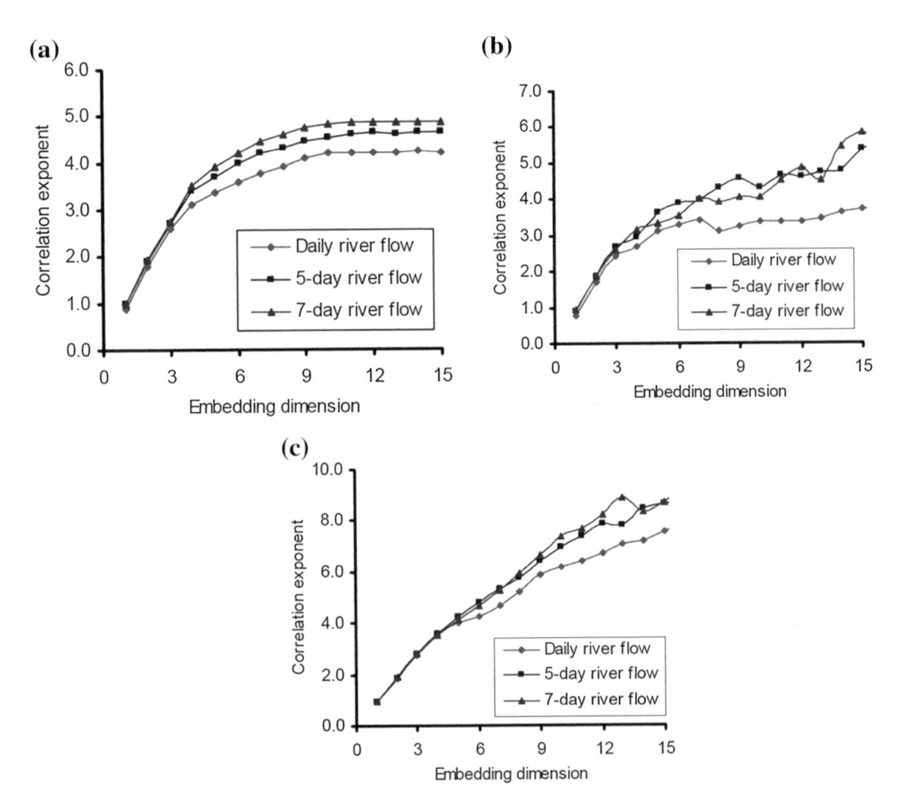

Fig. 10.5 Correlation dimension of streamflow at different temporal scales: **a** Kentucky River, Kentucky, USA; **b** Merced River, California, USA; and **c** Stillaguamish River, Washington State, USA (*source* Regonda et al. (2004))

flow dynamics of different scales at the Merced River, with correlation dimension values of 3.5, 4.7, and 5.5 for daily, 5-day, and 7-day series, respectively. However, the results for the flow series from the Stillaguamish River indicate stochastic behavior in all of the three scales studied. Another important finding from these results, regardless of the type of dynamic behavior (chaotic or stochastic), is the observation of an increase in the dimensionality (or complexity) of the flow dynamics with the scale of aggregation. In other words, dynamics changing from a less complex (more deterministic) behavior to a more complex (more stochastic) behavior with aggregation in time. Some subsequent studies investigating chaos in river flow also report the same (e.g. Salas et al. 2005; Sivakumar et al. 2007). This finding is interesting, since it is widely believed that temporal aggregation generally results in some kind of 'smoothing' and, hence, less complex (and more predictable) behavior.

Salas et al. (2005) investigated the effects of aggregation and sampling of streamflow data on the identification of the dynamics underlying the system. They analyzed daily streamflows from two rivers in Florida, USA: the St. Marys River near MacClenny and the Ocklawaha River near Conner. Based on estimates of delay time, delay window, and correlation integral (see also Kim et al. (1999)), they reported that the Ocklawaha River, which has a stronger basin storage contribution, departs significantly from the behavior of a chaotic system, while the departure was less significant for the St. Marys River, which has a smaller basin storage contribution. The study also suggested, through analysis of Lorenz map (see Chap. 5, Sect. 5.11.3), that increase in aggregation and sampling scales deteriorates chaotic behavior and eventually ceases to show evidence of low dimensional determinism. Sivakumar et al. (2007) used the phase space reconstruction concept to analyze river flow at four different scales (daily, 2-day, 4-day, and 8-day) from the Mississippi River basin (at St. Louis, Missouri), USA. They observed, for each scale, a clear attractor in a well-defined region in the phase space, suggesting possibly low dimensional determinism. However, the region of attraction increases in order with increasing scale of data aggregation. Based on these results, Sivakumar et al. (2007) interpreted that the flow dynamics between daily and weekly scales are simple, with the dynamics approaching intermediate complexity at the weekly scale. This is also supported by the results reported by Sivakumar et al. (2004), but from a streamflow disaggregation perspective, which is discussed next.

10.3.2 Chaotic Disaggregation of River Flow

Streamflow disaggregation has and continues to be a challenging problem in hydrology. The past few decades have witnessed numerous studies on streamflow disaggregation and, consequently, a large number of mathematical models; see, for example, Salas et al. (1995) for details. Traditionally, streamflow disaggregation approaches have involved some variant of a linear model of the form

$$\boldsymbol{X}_t = \boldsymbol{A}Z_t + \boldsymbol{B}\boldsymbol{V}_t \tag{10.1}$$

where $\boldsymbol{X}_t$ is the vector of disaggregate variables at time t, Z_t is the aggregate variable, $\boldsymbol{V}_t$ is a vector of independent random innovations (usually drawn from a Gaussian distribution), and **A** and **B** are parameter matrices. As Eq. (10.1) involves linear combinations of random variables, it is compatible mainly with Gaussian distributions (with only a few exceptions). Therefore, if the marginal distribution of the streamflow variables involved is not Gaussian, normalizing transformations are required for each streamflow component, in which Eq. (10.1) would be applied to the normalized flow variables. It is often difficult to find a general normalizing transformation and retain statistical properties of the streamflow process in the untransformed multi-variable space. The linear nature of Eq. (10.1) limits it from representing any strong nonlinearity in the dependence structure between variables, except through the normalizing transformation used. In view of the limitations with the parametric approach, Tarboton et al. (1998) developed a nonparametric approach for streamflow disaggregation, following up on the studies by Lall and Sharma (1996) and Sharma et al. (1997) for streamflow simulation. While the approach by Tarboton et al. (1998) addressess the nonlinearity of flow behavior and disaggregation phenomenon, it does not specifically address the nonlinear deterministic nature. To this end, Sivakumar et al. (2004) introduced a chaotic approach for streamflow disaggregation.

The approach introduced by Sivakumar et al. (2004) for streamflow disaggregation followed the approach of Sivakumar et al. (2001c) for rainfall, with appropriate modifications suitable for streamflow. The concept and the procedure for the chaotic disaggregation approach have already been discussed in Chap. 9 (Sect. 9.3.2), and the reader is directed to such for details. Sivakumar et al. (2004) employed the chaotic approach to disaggregate streamflow observed at the St. Louis gaging station in the Mississippi River basin, USA. It is relevant to note that the (daily) streamflow dynamics at the St. Louis gaging station had been identified to exhibit chaotic behavior (e.g. Sivakumar and Jayawardena 2002); see also Sivakumar and Wallender (2005) and Sivakumar et al. (2007) for subsequent studies.

To evaluate the effectiveness of the disaggregation approach, Sivakumar et al. (2004) used an aggregation-disaggregation scheme (aggregation followed by disaggregation). First, the available daily flow values were aggregated (by simple addition) to obtain flow data at four successively doubled coarser resolutions (i.e. 2-day, 4-day, 8-day, and 16-day). Then, the chaotic approach was employed to disaggregate these aggregated data series to obtain flow data at the successfully doubled finer resolutions (i.e. from 16-day to 8-day, from 8-day to 4-day, from 4-day to 2-day, and from 2-day to daily). Table 10.2 presents some of the important characteristics of these five flow series. As the minimum values indicate, there are no zero values in the flow series. This eliminates the problems faced by Sivakumar et al. (2001c) in their study of disaggregation of rainfall series observed in the Leaf

Table 10.2 Statistics of streamflow data at different temporal resolutions in the Mississippi River basin at St. Louis, Missouri, USA (unit = $m^3\ s^{-1}\ d_s$, where d_s is the scale of observations in days) (*source* Sivakumar et al. (2004))

Statistic	Daily	2-day	4-day	8-day	16-day
Number of data	8192	4096	2048	1024	512
Mean	5513.9	11027.7	22055.4	44110.8	88221.6
Standard deviation	3462.6	6908.1	13713.4	26995.2	52251.5
Maximum value	24100	48100	94300	183300	338500
Minimum value	980	1990	4030	8280	17430
Coefficient of variation	0.6280	0.6264	0.6218	0.6120	0.5923
Skew	1.4779	1.4771	1.4704	1.4559	1.4122
Kurtosis	2.5031	2.5081	2.5078	2.5066	2.3898

River basin (see Chap. 9, Sect. 9.3.2), even though this cannot be generalized for every streamflow time series.

Each of the five series is used as follows in the implementation of the disaggregation procedure. The entire series is divided into two halves. The first half of the series is used for phase space reconstruction to represent the dynamics of the disaggregation process, i.e. training or learning set. The disaggregation is then made only for one-fourth of the second half of the series (that immediately follows the first half), i.e. testing set. Therefore, the training and testing sets are selected in such a way that disaggregation is made for the same period, irrespective of the disaggregation resolution. This is done to allow useful and consistent comparisons between the disaggregation results obtained for the four disaggregation cases. For each of the four disaggregation cases, the flow series is reconstructed in embedding dimensions, m, from 1 to 10 to represent the transformation dynamics, and several different number of neighbors (k') are considered: 1, 2, 5, 10, 20, 50, 100, 150, and 200.

Figure 10.6, for example, presents the accuracy of disaggregation, in terms of correlation coefficient (CC) and root mean square error (RMSE), against the number of neighbors, for different embedding dimensions, when the 2-day flow series is disaggregated into daily flow series. Overall, for any embedding dimension, the disaggregation accuracy increases with increasing number of neighbors up to a certain point and then saturates (or even decreases) beyond that point. Table 10.3a presents the disaggregation results in a slightly different form, including the optimal number of neighbors, k'_{opt}. As can be seen, different k'_{opt} values are obtained for different embedding dimensions. Again, the disaggregation results show a trend of increase in accuracy with increasing embedding dimension up to a certain point and then saturation (or even decrease) in accuracy beyond that point. The smallest embedding dimension corresponding to such a saturation point is the optimal embedding dimension, m_{opt}.

Figure 10.6 and Table 10.3a indicate that, even though almost all of the ten combinations of m and nine combinations of k' yield very good results, the best disaggregation results (with CC = 0.9991 and RMSE = 183.801) are achieved

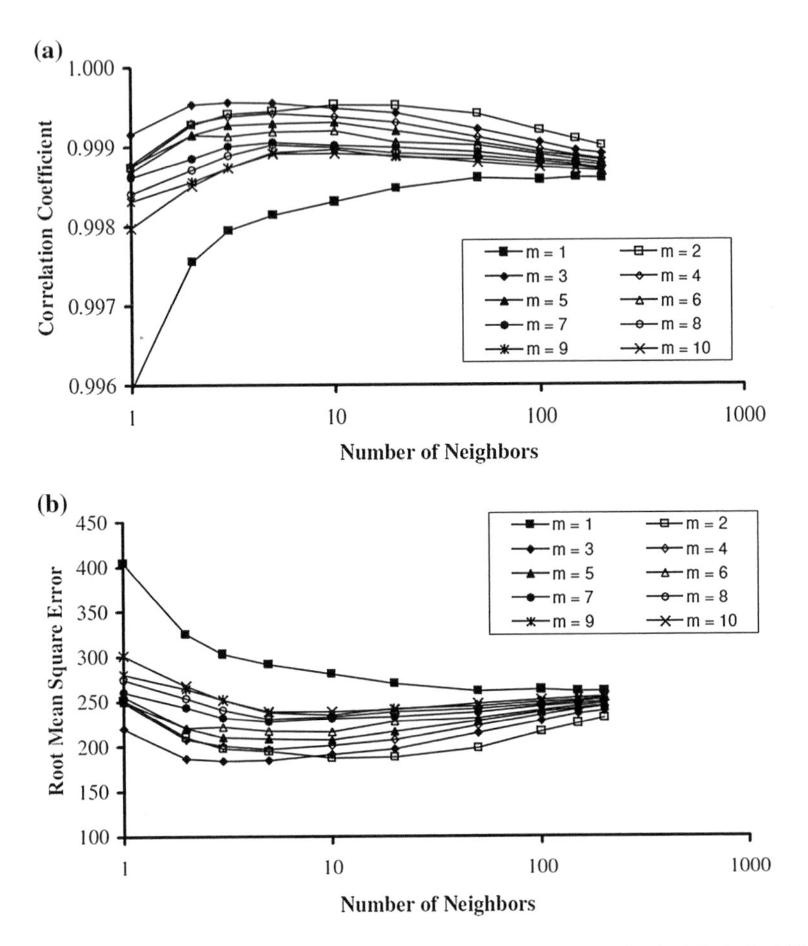

Fig. 10.6 Chaotic disaggregation of streamflow from 2-day to daily scale in the Mississippi River basin at St. Louis, Missouri, USA—effect of number of neighbors: **a** relationship between correlation coefficient and number of neighbors; and **b** relationship between root mean square error and number of neighbors (*source* Sivakumar et al. (2004))

when $m = 3$ and $k' = 3$. For this case, Fig. 10.7a compares, using scatter plot, the actual daily flow series and the daily flow series disaggregated from the 2-day series. As can be seen, the disaggregated values are in excellent agreement with the actual flow series.

The fact that the best disaggregation results are achieved for $m = 3$ could be an indication that there are at least three variables dominantly influencing the dynamics of the flow transformation process between 2-day and daily scales. This suggests that the disaggregation dynamics can be understood and modeled using a low dimensional approach. The near-accurate disaggregation results achieved using such an approach seem to provide further support. The observations of low m_{opt} (=3) and small k'_{opt} (=3) values also seem to present clues to the presence of low dimensional deterministic behavior in the underlying transformation dynamics. The decrease in disaggregation accuracy beyond m_{opt} and k'_{opt} may be attributed to the presence of noise in the data.

Tables 10.3b, 10.3c, and 10.3d show the results obtained when the flow is disaggregated from 4-day to 2-day, 8-day to 4-day, and 16-day to 8-day, respectively. The results indicate that the best disaggregation results are generally obtained at low embedding dimensions: $m_{opt} = 3$ for 4-day to 2-day, $m_{opt} = 2$ for 8-day to 4-day, and

Table 10.3a 2-day to daily streamflow disaggregation results in the Mississippi River basin at St. Louis, Missouri, USA (*source* Sivakumar et al. (2004))

Embedding dimension (m)	Correlation coefficient (CC)	Root mean square error (RMSE)	Optimal number of neighbors (k'_{opt})
1	0.9981	260.867	150
2	0.9990	187.025	10
3	**0.9991**	**183.801**	**3**
4	0.9989	196.865	5
5	0.9988	207.081	10
6	0.9987	216.099	10
7	0.9986	227.645	5
8	0.9985	230.183	5
9	0.9985	234.772	10
10	0.9984	238.474	10

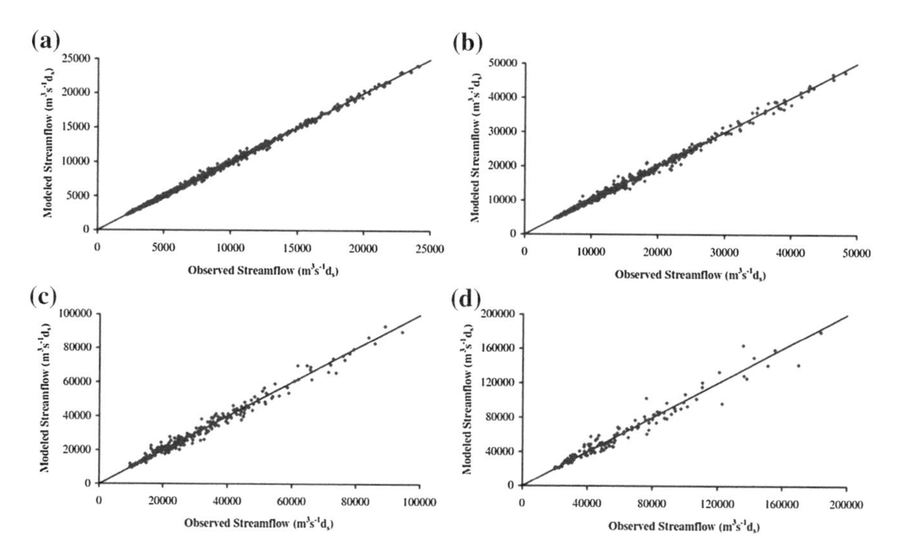

Fig. 10.7 Chaotic disaggregation of streamflow between successively doubled resolutions in the Mississippi River basin at St. Louis, Missouri, USA: **a** 2-day to daily (m = 3, k′ = 3); **b** 4-day to 2-day (m = 3; k′ = 5); **c** 8-day to 4-day (m = 2, k′ = 50); and **d** 16-day to 8-day (m = 3; k′ = 20) (*source* Sivakumar et al. (2004))

Table 10.3b 4-day to 2-day streamflow disaggregation results in the Mississippi River basin at St. Louis, Missouri, USA (*source* Sivakumar et al. (2004))

Embedding dimension (m)	Correlation coefficient (CC)	Root mean square error (RMSE)	Optimal number of neighbors (k'_{opt})
1	0.9941	920.248	200
2	0.9961	745.735	20
3	**0.9966**	**702.532**	**5**
4	0.9958	770.948	5
5	0.9951	833.635	10
6	0.9948	860.788	10
7	0.9947	871.089	20
8	0.9945	882.668	20
9	0.9945	885.667	20
10	0.9945	882.886	20

Table 10.3c 8-day to 4-day streamflow disaggregation results in the Mississippi River basin at St. Louis, Missouri, USA (*source* Sivakumar et al. (2004))

Embedding dimension (m)	Correlation coefficient (CC)	Root mean square error (RMSE)	Optimal number of neighbors (k'_{opt})
1	0.9892	2470.02	200
2	**0.9902**	**2350.76**	**50**
3	0.9899	2381.36	50
4	0.9898	2401.80	100
5	0.9897	2411.06	100
6	0.9896	2425.62	100
7	0.9896	2418.28	150
8	0.9896	2425.59	200
9	0.9894	2440.80	200
10	0.9894	2445.44	200

Table 10.3d 16-day to 8-day streamflow disaggregation results in the Mississippi River basin at St. Louis, Missouri, USA (*source* Sivakumar et al. (2004))

Embedding dimension (m)	Correlation coefficient (CC)	Root mean square error (RMSE)	Optimal number of neighbors (k'_{opt})
1	0.9747	7441.11	100
2	0.9750	7398.83	150
3	**0.9754**	**7342.64**	**20**
4	0.9753	7358.45	20
5	0.9744	7478.98	200
6	0.9751	7388.94	10
7	0.9755	7325.99	10
8	**0.9759**	**7258.19**	**10**
9	0.9743	7493.35	200
10	**0.9756**	**7315.44**	**5**

$m_{opt} = 3$ for 16-day to 8-day (although $m = 8$ and $m = 10$ also provide similar or slightly better results). This suggests that the flow transformation dynamics between these successively-doubled resolutions exhibit low dimensional chaotic behavior. Similarly, the best disaggregation results are obtained when the number of neighbors is very small: $k'_{opt} = 5$ for 4-day to 2-day, $k'_{opt} = 50$ for 8-day to 4-day, and $k'_{opt} = 20$ for 16-day to 8-day (also $k'_{opt} = 10$ and $k'_{opt} = 5$ for the other best results). These results offer further support to the presence of low dimensional deterministic behavior in the streamflow transformation process. Figure 10.7b–d show the comparisons of actual and modeled values for flow disaggregation from 4-day to 2-day, 8-day to 4-day, and 16-day to 8-day, respectively. The plots correspond to the optimum cases identified in Tables 10.3b, 10.3c, and 10.3d. As can be seen, the disaggregated values are in excellent agreement with the actual flow series, for all the cases. The results also allow interpretations similar to the ones made above for the case of disaggregation of flow between 2-day and daily scales.

10.4 Spatial Variability and Classification

While some earlier studies performed chaos analysis of streamflow data at many different stations (e.g. Liu et al. 1998; Krasovskaia et al. 1999; Sivakumar 2003), they made no attempt to specifically address the spatial variability and classification. For instance, Liu et al. (1998) applied the local approximation prediction method to streamflow data from 28 stations in the continental United States. They reported that streamflow signals spanned a wide dynamic range between deterministic chaos and periodic signal contaminated with additive noise, but made no attempt to discuss the spatial variability. They also reported that there was no direct relationship between the nature of the underlying streamflow characteristics and basin area, but did not discuss any further about classification. Krasovskaia et al. (1999) studied the dimensionality of the Scandinavian river flow regimes in their effort to identify the possible presence of chaos. Although they considered specific regions to identify flow regimes, the number of streamflow stations was too small for an extensive analysis of spatial variability and classification. Sivakumar (2003) studied the dynamic behavior of monthly streamflow in the western United States, through application of the local approximation method to data observed at 79 stations across 11 states. The analysis was carried out by grouping the 79 stations under three categories on the basis of the magnitude of mean streamflow as: (1) low-flow stations; (2) high-flow stations; and (3) medium-flow stations. Sivakumar (2003) reported good predictions of streamflow dynamics irrespective of the flow regime and also that predictions for the low-flow stations were relatively better than those for the medium-flow and high-flow stations. Despite the above grouping based on mean streamflows, Sivakumar (2003) made no attempt to discuss the results in the context of spatial variability and catchment classification.

In the specific context of catchment classification, Sivakumar et al. (2007) explored the utility of phase space reconstruction approach to classify catchments, following up on the suggestion by Sivakumar (2004b) in addressing the 'dominant processes concept' (DPC). They used the 'region of attraction of trajectories' in the phase space to identify the streamflow data as exhibiting 'simple' or 'intermediate' or 'complex' behavior and, correspondingly, classify the catchment as potentially low-, medium- or high dimensional. The idea was first demonstrated on artificial time series and then tested on streamflow (and sediment) data representing different geographic regions, climatic conditions, basin sizes, processes, and scales. However, the number of streamflow series studied was still too few for any reliable conclusion regarding classification and spatial variability. Nevertheless, the study led to a more advanced and extensive investigation by Sivakumar and Singh (2012), some details of which are presented here.

To examine the utility of the concepts of chaos theory for catchment classification purposes, Sivakumar and Singh (2012) employed the correlation dimension method, in addition to phase space construction, to streamflow data from a large network of 117 gaging stations across 11 states in the western United States. The analysis was performed on monthly streamflow data observed over a period of 52 years (1951–2002). Figure 10.8 shows the locations of these 117 stations, and Table 10.4 presents some basic catchment characteristics and streamflow statistics.

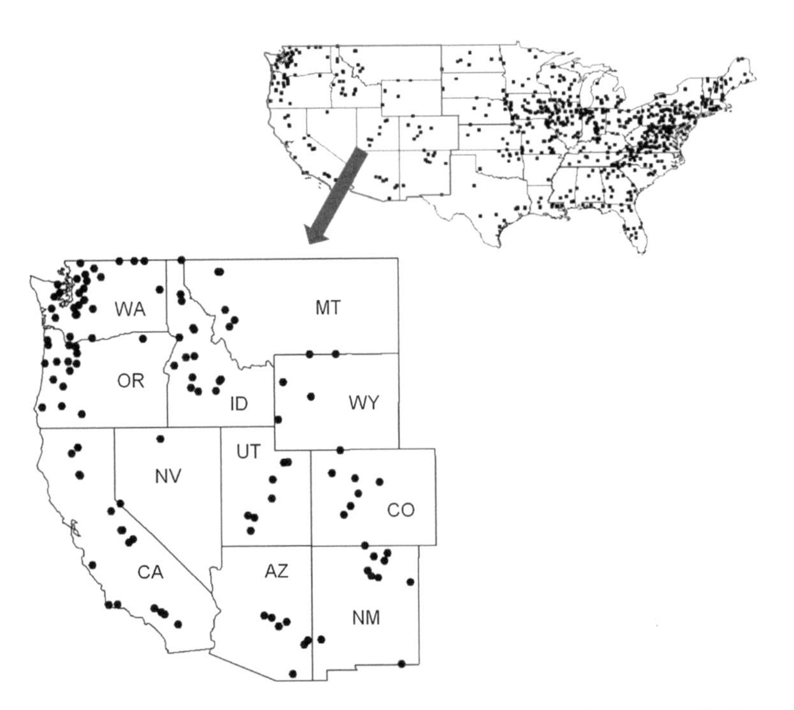

Fig. 10.8 Western United States and locations of 117 streamflow stations (*source* Sivakumar and Singh (2012))

Table 10.4 Overall statistics of streamflow data at 117 stations in the western United States

Statistic	Minimum	Maximum	Station
Station			
Drainage area (km^2)	22.79	35094	11058500 (CA), 13317000 (ID)
Flow			
Mean (m^3/s)	0.06	322	11063500 (CA), 13317000 (ID)
Std. dev. (m^3/s)	0.11	373.5	11063500 (CA), 13317000 (ID)
CV	0.295	4.324	11367500 (CA), 10258500 (CA)

As can be seen from Table 10.4, the gaging stations and streamflow data possess a wide range of variability.

Based on careful examination of phase space diagrams and correlation dimension results of all 117 streamflow series, Sivakumar and Singh (2012) were able to identify four reasonably distinct groups. The identification was made based on the correlation dimension (d) value as the primary criterion, but the consistency between the dimensionality and the attractor shape in the phase space diagram (for each group) was also given emphasis for a more reliable grouping. The four groups and the associated dimensionalities are: (1) low dimensional, with $d \leq 3.0$; (2) medium dimensional, with $3.0 < d \leq 6.0$; (3) high dimensional, with $d > 6.0$; and (4) unidentifiable. The selection of the number of groups and the range of dimension values for each group was somewhat arbitrary. Nevertheless, this grouping according to correlation dimensions is also reasonable in the context of process/model complexity, since the influence of more than six dominant governing variables (i.e. $d > 6.0$) often leads to high complexity in dynamics (requiring 'complex' models), whereas that of 3 or less variables can confidently be considered to lead to simpler dynamics (requiring 'simple' models), with the other in between (medium-complexity dynamics, requiring medium-complexity models). For the purpose of discussion here, results for two time series from each of the above four groups are presented. (1) low dimensional—Station #10032000 (WY) and Station #13317000 (ID); (2) medium dimensional—Station #11315000 (CA) and Station #11381500 (CA); (3) high dimensional—Station #12093500 (WA) and Station #14185000 (OR); and (4) unidentifiable – Station #8408500 (NM) and Station #11124500 (CA).

Figure 10.9a–h shows the two-dimensional phase space diagrams for streamflow series from the above eight stations. The following general observations may be made: (1) the plots on the first row exhibit reasonably well-structured attractors, suggesting that the systems are likely less complex and low dimensional; (2) the second row plots indicate slightly wider scattering of the attractor, suggesting systems of medium complexity and medium dimension; (3) the plots on the third row exhibit much wider scattering (especially with one or a few outliers), suggesting highly complex and high dimensional systems; and (4) the last two plots do not show any identifiable patterns, thus making it hard to include them in any of the above three groups.

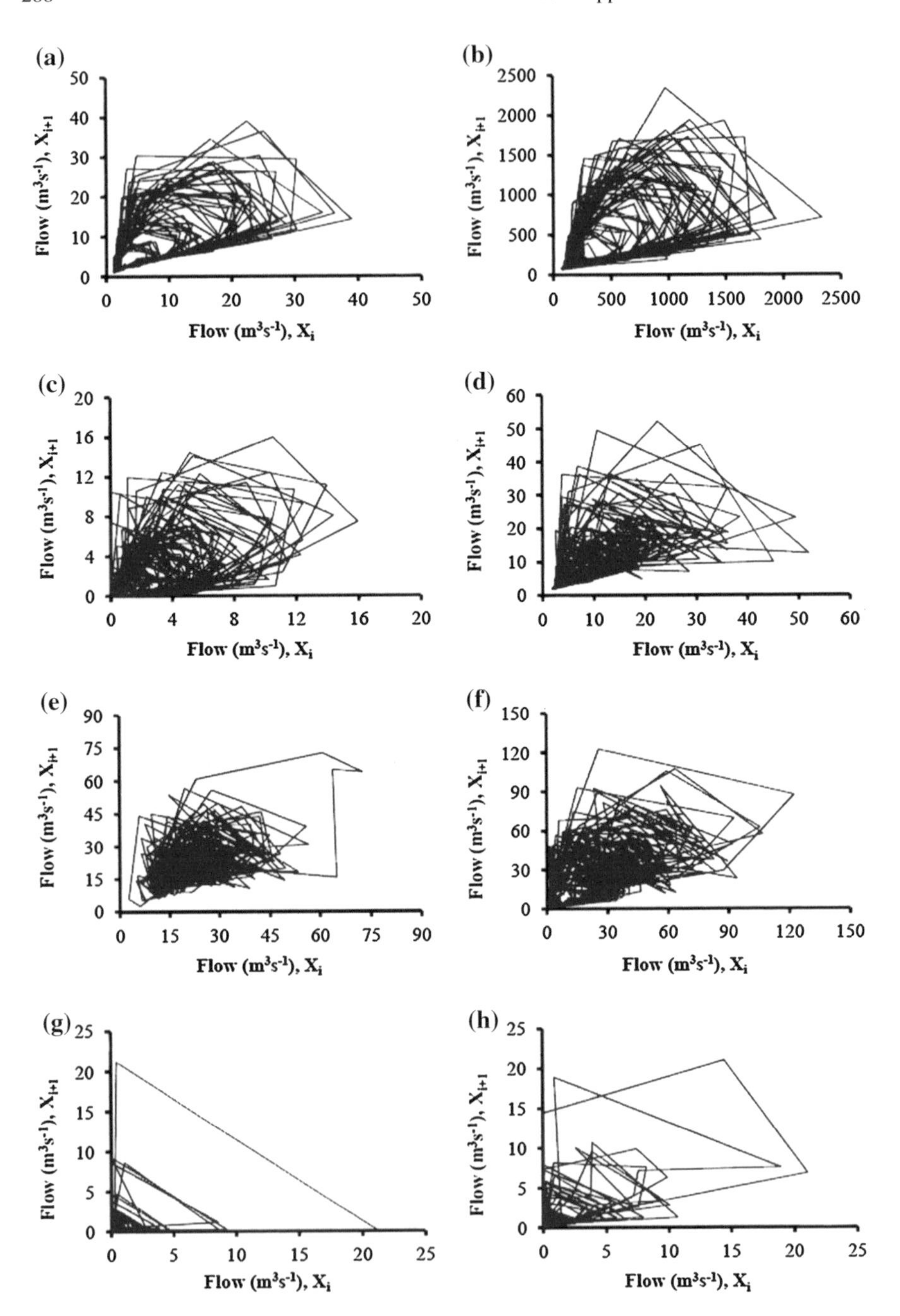

Fig. 10.9 Phase space diagram: **a** Station #10032000; **b** Station #13317000; **c** Station 11315000; **d** Station #11381500; **e** Station #12093500; **f** Station #14185000; **g** Station #8408500; and **h** Station #11124500 (*source* Sivakumar and Singh (2012))

Figure 10.10a–h presents the correlation dimension results for the corresponding eight streamflow series. The plots show the correlation exponent (v) versus Log r, for embedding dimensions, m, from 1 to 20 (bottom to top curves). The results suggest the following: (1) the top row plots reveal saturation of correlation exponent at a value less than 3 (shown using a thick horizontal line), suggesting low dimensional and less complex systems; (2) the second row plots yield slightly higher dimensions (but less than 6), suggesting medium dimensional and slightly more complex systems; (3) the plots on the third row do not indicate any saturation of correlation exponent, suggesting high dimensional and highly complex systems; and (4) the last two plots do not show any clear indication regarding the dimension value or group (as they show neither saturation of correlation exponent nor high dimensionality) and, therefore, are considered 'unidentifiable.'

Figure 10.11 presents the grouping of the above 117 streamflow series, based on correlation dimension value (and also phase space). The grouping shows some kind of "homogeneity" in the dimensionality and complexity of streamflow dynamics within certain regions. For instance: (1) streamflow dynamics in the far northwest (i.e. western parts of WA and OR) are generally high dimensional; (2) the dimensionality of streamflows in the far south and southwest (southern CA, southern AZ, southern NM) is generally unidentifiable; (3) the complexity of streamflow dynamics in the west (northern CA and NV) is generally medium dimensional; and (4) low-dimensional complexity is generally observed for streamflows in Wyoming. Despite this, it is also important to note that this "homogeneity" is not true for every region, and there are indeed strong exceptions. For example: (1) both low-dimensional and medium-dimensional complexity of streamflow dynamics are observed in some other regions, especially in the east and north (including CO, ID, MT, and some parts of WA); and (2) streamflow dynamic complexity in some regions is rather very mixed, ranging from low dimensional to medium dimensional to unidentifiable (UT and, to some extent, northern NM).

The above classification of streamflow based on complexity and nonlinear dynamic concepts, with dimensionality as a criterion, is both useful and interesting. In particular, the dimension estimates and the grouping of streamflow time series clearly show that: (1) the dimensionality concept captures the complexity of streamflow dynamics at individual stations independently and then allows classification regardless of the proximity of catchments, without resorting to a 'nearest neighbor' approach that is widely used in hydrology; and (2) a nearest neighbor approach, even for monthly streamflows, is not necessarily the right way to classification, despite the close proximity of some catchments. This observation has important implications for interpolation/extrapolation of streamflow data, including in the context of predictions in ungauged basins (PUBs).

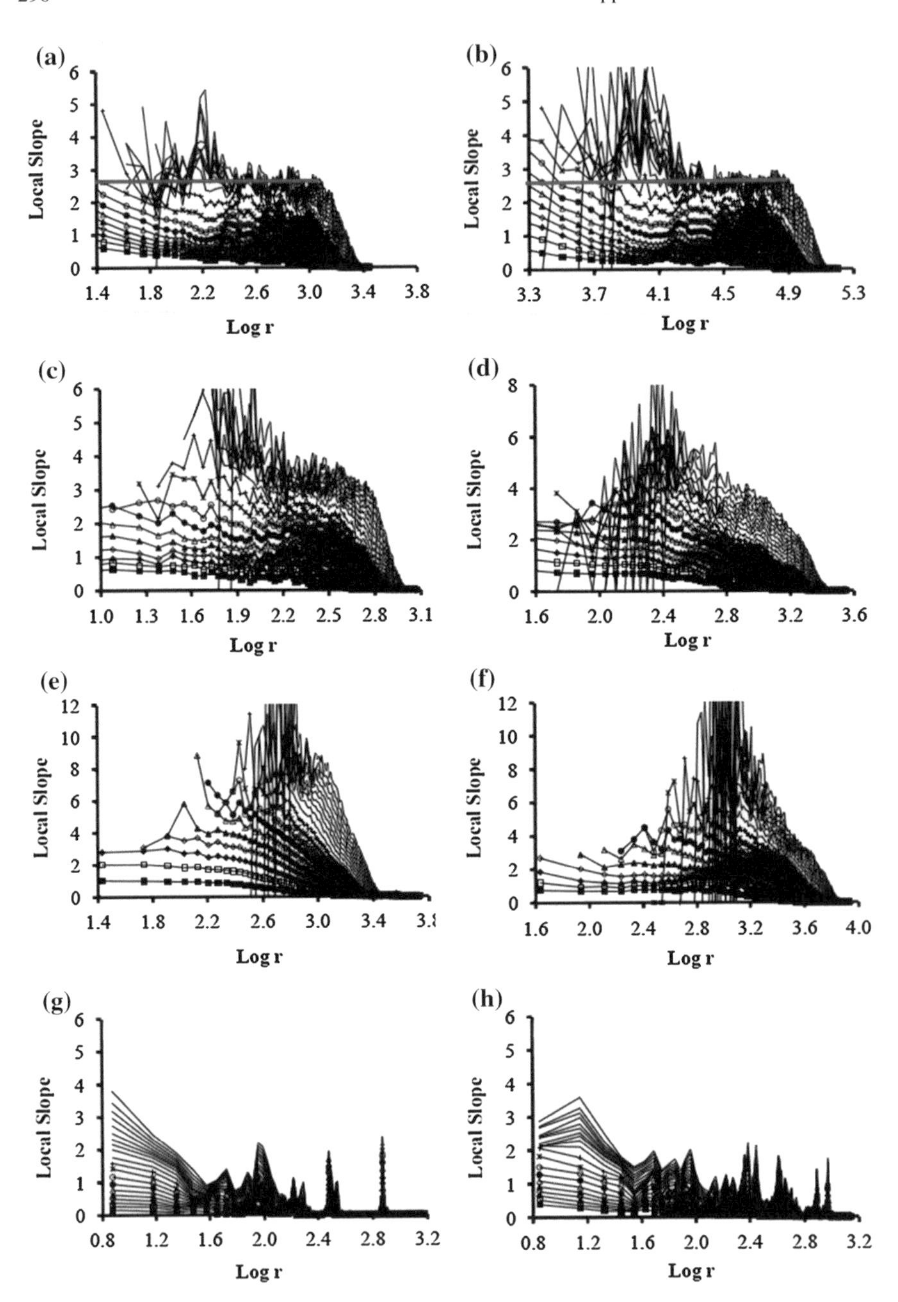

Fig. 10.10 Correlation dimension: **a** Station #10032000; **b** Station #13317000; **c** Station 11315000; **d** Station #11381500; **e** Station #12093500; **f** Station #14185000; **g** Station #8408500; and **h** Station #11124500 (*source* Sivakumar and Singh (2012))

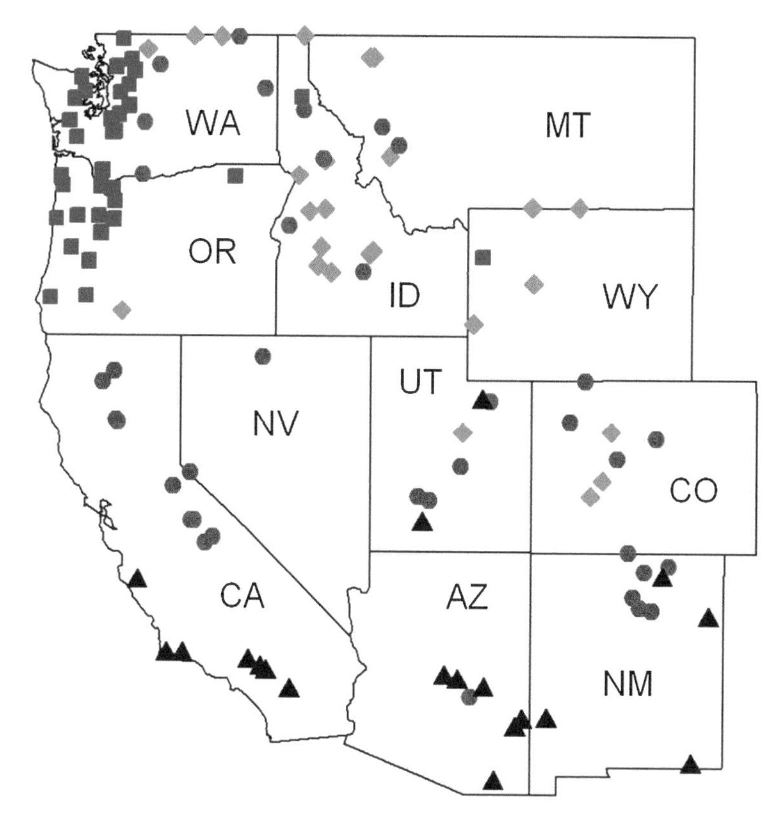

Fig. 10.11 Grouping of streamflow stations according to correlation dimension (d) estimates: low dimensional ($d \leq 3$); medium dimensional ($3 < d \leq 6$); high dimensional ($d > 6.0$); and unidentifiable (d not identifiable) (*source* Sivakumar and Singh (2012))

10.5 Others

Elshorbagy et al. (2001, 2002a) employed the principles of chaos theory for estimation of missing streamflow data. In the study by Elshorbagy et al. (2001), missing data in one streamflow series were estimated from another complete and cross-correlated flow series, whereas Elshorbagy et al. (2002a) used only a single streamflow series for estimating consecutive missing values in the series. Two general steps were followed: (1) identification of chaos in streamflow series; and (2) estimation of missing data. Streamflow series observed from the Little River and the Reed Creek in Virginia, USA were considered in Elshorbagy et al. (2001), wherein the correlation dimension method was used for chaos identification and the missing data were estimated by fitting one global model and multiple local models; the superior performance of the local models was attributed to the chaotic behavior of the two series. In the study by Elshorbagy et al. (2002a), flow series from the English River in Ontario, Canada, was studied; chaos identification was done through four different methods and missing data were estimated using artificial neural networks (ANNs) and K-nearest neighbors (K-nn).

In addition, several other studies have attempted identification and prediction of chaos in river flow data and also addressed a number of associated methodological/data issues. Such studies include those by Elshorbagy et al. (2002b), Jayawardena et al. (2002), Phoon et al. (2002), Sivakumar et al. (2002a), Laio et al. (2004), Khan et al. (2005), She and Basketfield (2005), Sivakumar (2007), Kim et al. (2009), Dhanya and Nagesh Kumar (2011), Khabiti et al. (2012), Tongal et al. (2013), and Vignesh et al. (2015). Some of these studies will be discussed in Chaps. 12 and 14.

10.6 Summary

River flow has, arguably, been the most studied data in the context of chaos studies in hydrology. Since the initial studies in the early 1990s, numerous studies have employed the concepts of chaos theory to river flow data. Such studies have largely focused on the identification and prediction of chaotic behavior, investigation of scaling and disaggregation, and study of spatial variability and classification. This chapter has presented a review of chaos studies on river flow, with an example for each of the above. In addition to these, some studies have also attempted missing data estimation, reconstruction of system equations, and multi-variable analysis. Still some other studies have also investigated the influence of data size, data noise, and other data-related issues in the applications of chaos theory to river flow data. Details of some of these studies will be presented in Chaps. 12 and 14.

References

Abarbanel HDI (1996) Analysis of observed chaotic data. Springer, New York, USA

Casdagli M (1989) Nonlinear prediction of chaotic time series. Physica D 35:335–356

Dhanya CT, Nagesh Kumar D (2011) Predictive uncertainty of chaotic daily streamflow using ensemble wavelet networks approach. Water Resour Res 47:W06507. doi:10.1029/2010WR010173

Elshorbagy A, Panu US, Simonovic SP (2001) Analysis of cross-correlated chaotic streamflows. Hydrol Sci J 46(5):781–794

Elshorbagy A, Simonovic SP, Panu US (2002a) Estimation of missing streamflow data using principles of chaos theory. J Hydrol 255:123–133

Elshorbagy A, Simonovic SP, Panu US (2002b) Noise reduction in chaotic hydrologic time series: facts and doubts. J Hydrol 256:147–165

Farmer DJ, Sidorowich JJ (1987) Predicting chaotic time series. Phys Rev Lett 59:845–848

Grassberger P, Procaccia I (1983a) Measuring the strangeness of strange attractors. Physica D 9:189–208

Grassberger P, Procaccia I (1983b) Characterisation of strange attractors. Phys Rev Lett 50 (5):346–349

Grassberger P, Procaccia I (1983c) Estimation of the Kolmogorov entropy from a chaotic signal. Phys Rev A 28:2591–2593

Holzfuss J, Mayer-Kress G (1986) An approach to error-estimation in the application of dimension algorithms. In: Mayer-Kress G (ed) Dimensions and entropies in chaotic systems. Springer, New York, pp 114–122

Islam MN, Sivakumar B (2002) Characterization and prediction of runoff dynamics: a nonlinear dynamical view. Adv Water Resour 25(2):179–190

Jayawardena AW, Gurung AB (2000) Noise reduction and prediction of hydrometeorological time series: dynamical systems approach vs. stochastic approach. J Hydrol 228:242–264

Jayawardena AW, Lai F (1994) Analysis and prediction of chaos in rainfall and stream flow time series. J Hydrol 153:23–52

Jayawardena AW, Li WK, Xu P (2002) Neighborhood selection for local modeling and prediction of hydrological time series. J Hydrol 258:40–57

Kember G, Flower AC (1993) Forecasting river flow using nonlinear dynamics. Stoch Hydrol Hydraul 7:205–212

Kennel MB, Brown R, Abarbanel HDI (1992) Determining embedding dimension for phase space reconstruction using a geometric method. Phys Rev A 45:3403–3411

Khan S, Ganguly AR, Saigal S (2005) Detection and predictive modeling of chaos in finite hydrological time series. Nonlinear Processes Geophys 12:41–53

Khatibi R, Sivakumar B, Ghorbani MA, Kisi Ö, Kocak K, Zadeh DF (2012) Investigating chaos in river stage and discharge time series. J Hydrol 414–415:108–117

Kim HS, Eykholt R, Salas JD (1999) Nonlinear dynamics, delay times, and embedding windows. Physica D 127(1–2):48–60

Kim HS, Lee KH, Kyoung MS, Sivakumar B, Lee ET (2009) Measuring nonlinear dependence in hydrologic time series. Stoch Environ Res Risk Assess 23:907–916

Krasovskaia I, Gottschalk L, Kundzewicz ZW (1999) Dimensionality of Scandinavian river flow regimes. Hydrol Sci J 44(5):705–723

Laio F, Porporato A, Revelli R, Ridolfi L (2003) A comparison of nonlinear flood forecasting methods. Water Resour Res 39(5). doi:10.1029/2002WR001551

Laio F, Porporato A, Ridolfi L, Tamea S (2004) Detecting nonlinearity in time series driven by non-Gaussian noise: the case of river flows. Nonlinear Processes Geophys 11:463–470

Lall U, Sharma A (1996) A nearest neighbor bootstrap for time series resampling. Water Resour Res 32(3):679–693

Lambrakis N, Andreou AS, Polydoropoulos P, Georgopoulos E, Bountis T (2000) Nonlinear analysis and forecasting of a brackish karstic spring. Water Resour Res 36(4):875–884

Liaw CY, Islam MN, Phoon KK, Liong SY (2001) Comment on "Does the river run wild? Assessing chaos in hydrological systems". Adv Water Resour 24(5):575–578

Lisi F, Villi V (2001) Chaotic forecasting of discharge time series: A case study. J Am Water Resour Assoc 37(2):271–279

Liu Q, Islam S, Rodriguez-Iturbe I, Le Y (1998) Phase-space analysis of daily streamflow: characterization and prediction. Adv Water Resour 21:463–475

Pasternack GB (1999) Does the river run wild? Assessing chaos in hydrological systems. Adv Water Resour 23(3):253–260

Pasternack GB (2001) Reply to "Comment on 'Does the river run wild? Assessing chaos in hydrological systems'" by Pasternack. Adv Water Resour 24(5):578–580

Phoon KK, Islam MN, Liaw CY, Liong SY (2002) A practical inverse approach for forecasting of nonlinear time series analysis. ASCE J Hydrol Eng 7(2):116–128

Porporato A, Ridolfi L (1996) Clues to the existence of deterministic chaos in river flow. Int J Mod Phys B 10:1821–1862

Porporato A, Ridolfi R (1997) Nonlinear analysis of river flow time sequences. Water Resour Res 33(6):1353–1367

Porporato A, Ridolfi R (2001) Multivariate nonlinear prediction of river flows. J Hydrol 248(1–4):109–122

Regonda S, Sivakumar B, Jain A (2004) Temporal scaling in river flow: can it be chaotic? Hydrol Sci J 49(3):373–385
Salas JD, Delleur JW, Yevjevich V, Lane WL (1995) Applied modeling of hydrologic time series. Water Resources Publications, Littleton, Colorado
Salas JD, Kim HS, Eykholt R, Burlando P, Green TR (2005) Aggregation and sampling in deterministic chaos: implications for chaos identification in hydrological processes. Nonlinear Processes Geophys 12:557–567
Sharma A, Tarboton DG, Lall U (1997) Streamflow simulation: a nonparametric approach. Water Resour Res 33(2):291–308
She N, Basketfield D (2005) Streamflow dynamics at the Puget Sound, Washington: application of a surrogate data method. Nonlinear Processes Geophys 12:461–469
Sivakumar B (2000) Chaos theory in hydrology: important issues and interpretations. J Hydrol 227 (1–4):1–20
Sivakumar B (2003) Forecasting monthly streamflow dynamics in the western United States: a nonlinear dynamical approach. Environ Model Softw 18(8–9):721–728
Sivakumar B (2004a) Chaos theory in geophysics: past, present andfuture. Chaos Soliton Fract 19 (2):441–462
Sivakumar B (2004b) Dominant processes concept in hydrology: moving forward. Hydrol Processes 18(12):2349–2353
Sivakumar B (2005) Correlation dimension estimation of hydrologic series and data size requirement: myth and reality. Hydrol Sci J 50(4):591–604
Sivakumar B (2007) Nonlinear determinism in river flow: prediction as a possible indicator. Earth Surf Process Landf 32(7):969–979
Sivakumar B (2009) Nonlinear dynamics and chaos in hydrologic systems: latest developments and a look forward. Stoch Environ Res Risk Assess 23:1027–1036
Sivakumar B, Jayawardena AW (2002) An investigation of the presence of low-dimensional chaotic behavior in the sediment transport phenomenon. Hydrol Sci J 47(3):405–416
Sivakumar B, Wallender WW (2005) Predictability of river flow and sediment transport in the Mississippi River basin: a nonlinear deterministic approach. Earth Surf Process Landf 30:665–677
Sivakumar B, Singh VP (2012) Hydrologic system complexity and nonlinear dynamic concepts for a catchment classification framework. Hydrol Earth Syst Sci 16:4119–4131
Sivakumar B, Phoon KK, Liong SY, Liaw CY (1999) A systematic approach to noise reduction in chaotic hydrological time series. J Hydrol 219(3–4):103–135
Sivakumar B, Berndtsson R, Olsson J, Jinno K, Kawamura A (2000) Dynamics of monthly rainfall-runoff process at the Göta basin: a search for chaos. Hydrol Earth Syst Sci 4(3):407–417
Sivakumar B, Berndttson R, Olsson J, Jinno K (2001a) Evidence of chaos in the rainfall-runoff process. Hydrol Sci J 46(1):131–145
Sivakumar B, Berndtsson R, Persson M (2001b) Monthly runoff prediction using phase-space reconstruction. Hydrol Sci J 46(3):377–387
Sivakumar B, Sorooshian S, Gupta HV, Gao X (2001c) A chaotic approach to rainfall disaggregation. Water Resour Res 37(1):61–72
Sivakumar B, Berndtsson R, Olsson J, Jinno K (2002a) Reply to 'which chaos in the rainfall-runoff process?' by Schertzer et al. Hydrol Sci J 47(1):149–158
Sivakumar B, Jayawardena AW, Fernando TMGH (2002b) River flow forecasting: use of phase-space reconstruction and artificial neural networks approaches. J Hydrol 265(1–4):225–245
Sivakumar B, Persson M, Berndtsson R, Uvo CB (2002c) Is correlation dimension a reliable indicator of low-dimensional chaos in short hydrological time series? Water Resour Res 38(2). doi:10.1029/2001WR000333
Sivakumar B, Wallender WW, Puente CE, Islam MN (2004) Streamflow disaggregation: a nonlinear deterministic approach. Nonlinear Processes Geophys 11:383–392

Sivakumar B, Berndtsson R, Persson M, Uvo CB (2005) A multi-variable time series phase-space reconstruction approach to investigation of chaos in hydrological processes. Int J Civil Environ Engg 1(1):35–51

Sivakumar B, Jayawardena AW, Li WK (2007) Hydrologic complexity and classification: a simple data reconstruction approach. Hydrol Process 21(20):2713–2728

Stehlik J (1999) Deterministic chaos in runoff series. J Hydrol Hydromech 47(4):271–287

Sugihara G, May RM (1990) Nonlinear forecasting as a way of distinguishing chaos from measurement error in time series. Nature 344:734–741

Tarboton DG, Sharma A, Lall U (1998) Disaggregation procedures for stochastic hydrology based on nonparametric density estimation. Water Resour Res 34(1):107–119

Theiler J, Eubank S, Longtin A, Galdrikian B, Farmer JD (1992) Testing for nonlinearity in time series: the method of surrogate data. Physica D 58:77–94

Tongal H, Demirel MC, Booij MJ (2013) Seasonality of low flows and dominant processes in the Rhine River. Stoch Environ Res Risk Assess 27:489–503

Vignesh R, Jothiprakash V, Sivakumar B (2015) Streamflow variability and classification using false nearest neighbor method. J Hydrol 531:706–715

Wang Q, Gan TY (1998) Biases of correlation dimension estimates of streamflow data in the Canadian prairies. Water Resour Res 34(9):2329–2339

Wilcox BP, Seyfried MS, Blackburn WH, Matison TH (1990) Chaotic characteristics of snowmelt runoff: a preliminary study. In: Symposium on watershed management. American Society of Civil Engineers, Durango, CO

Wilcox BP, Seyfried MS, Matison TM (1991) Searching for chaotic dynamics in snowmelt runoff. Water Resour Res 27(6):1005–1010

Wolf A, Swift JB, Swinney HL, Vastano A (1985) Determining Lyapunov exponents from a time series. Physica D 16:285–317

Zhou Y, Ma Z, Wang L (2002) Chaotic dynamics of the flood series in the Huaihe River Basin for the last 500 years. J Hydrol 258:100–110

Chapter 11
Applications to Other Hydrologic Data

Abstract Following the early chaos studies mainly on rainfall and river flow, the concepts of chaos theory started to find applications in studies on other hydrologic data as well. Although such applications have been noticeably less when compared to those on rainfall and river flow, they have studied various types of hydrologic data. The data studied include rainfall-runoff, lake volume and level, sediment transport, groundwater, and soil moisture, among others. Further, while most of these studies have mainly focused on identification and prediction of chaos and, to some extent, investigation of scaling relationships, several other problems associated with the data have also been addressed. This chapter presents a review of the above studies, with particular focus on rainfall-runoff, lake volume and level, sediment transport, and groundwater. Examples are also provided to illustrate the applications to rainfall-runoff (i.e. runoff coefficient), sediment transport (i.e. flow discharge, suspended sediment concentration, and suspended sediment load), and groundwater (solute transport and arsenic contamination).

11.1 Introduction

Although a significant majority of studies on chaos applications in hydrology have mainly focused on rainfall and river flow data (discussed in Chaps. 9 and 10), there have been a number of applications to many other hydrologic data as well. These include: rainfall-runoff (e.g. Sivakumar et al. 2000, 2001a; Dodov and Foufoula-Georgiou 2005), lake volume and water level (e.g. Sangoyomi et al. 1996; Abarbanel and Lall 1996; Abarbanel et al. 1996; Tongal and Berndtsson 2014), sediment transport (e.g. Sivakumar 2002; Sivakumar and Jayawardena 2002; Sivakumar and Wallender 2004, 2005; Sivakumar and Chen 2007; Shang et al. 2009), and groundwater flow and solute transport, including arsenic contamination (e.g. Sivakumar et al. 2005; Hossain and Sivakumar 2006; Hill et al. 2008), among many others; see also Sivakumar (2009) for a more recent review. These studies have helped expand our knowledge on the relevance, suitability, and effectiveness of chaos studies in hydrology. This chapter presents a brief review of these studies.

B. Sivakumar, *Chaos in Hydrology*, DOI 10.1007/978-90-481-2552-4_11

11.2 Rainfall-runoff

Despite the large number of applications to rainfall process and river flow process separately, chaos theory has, surprisingly, not found many applications in the study of rainfall-runoff as a whole. The only studies, to my knowledge, that have employed the concept of chaos theory to specifically study the rainfall-runoff process were those conducted by Sivakumar et al. (2000, 2001a), Laio et al. (2004), and Dodov and Foufoula-Georgiou (2005). However, Porporato and Ridolfi (2001) and Laio et al. (2003) addressed the rainfall-runoff process in a slightly different way, i.e. forecasting river flow/river stage in a multi-variable sense; see Chap. 14 (Sect. 14.6) for some additional details.

Sivakumar et al. (2000, 2001a) investigated the presence of chaos in the rainfall-runoff process at the Göta River basin in Sweden. To identify chaos in rainfall-runoff, they analyzed the rainfall series and the runoff series first separately and then jointly. The runoff coefficient, defined as the ratio of runoff to rainfall, was considered as a connector of rainfall and runoff, and a concentration time of 6 months was used for its calculation. The reason behind analyzing rainfall and runoff series separately was that their individual behaviors (input and output, respectively) could provide important information about the behavior of the joint rainfall-runoff process (input-output relationship), whereas the runoff coefficient was considered as a better representative of the rainfall-runoff process as a whole. They employed the correlation dimension method and the nonlinear local approximation prediction method to rainfall, runoff, and runoff coefficient data observed over a period of 131 years (January 1807–December 1937). As some key results for the rainfall series and the runoff series are discussed in Chaps. 7 (Sect. 7.3.2) and 12 (Sect. 12.3), respectively, they are not presented here. In what follows, some important results only for the runoff coefficient series are presented.

Figure 11.1a shows the variation of the runoff coefficient series from the Göta River basin. As can be seen, the runoff coefficient series is highly variable. It is also found to be more variable than the rainfall series and the runoff series. Figure 11.1b presents its two-dimensional phase space reconstruction, according to Takens' delay embedding theorem; see Chap. 6, Eq. (6.3). The reconstruction indicates neither a well-defined structure in the phase space nor a significant scattering of trajectories all over the phase space. The attractor is also more scattered than that observed for the rainfall series and the runoff series, indicating that the dynamics underlying the runoff coefficient series are more complex. With this reconstruction, the correlation functions and exponents were computed, following the Grassberger-Procaccia algorithm (Grassberger and Procaccia 1983a, b) (see Chap. 6, Sect. 6.4.2). Figure 11.1c presents the relationship between the correlation integral, $C(r)$, and the radius, r, for embedding dimensions, m, from 1 to 20, and Fig. 11.1d presents the relationship between the correlation exponent and embedding dimension values. The correlation exponent increases with the embedding dimension up to a certain point and saturates beyond that. The correlation dimension value is 7.8 (it is 6.4 for rainfall and 5.5 for runoff), suggesting the presence of

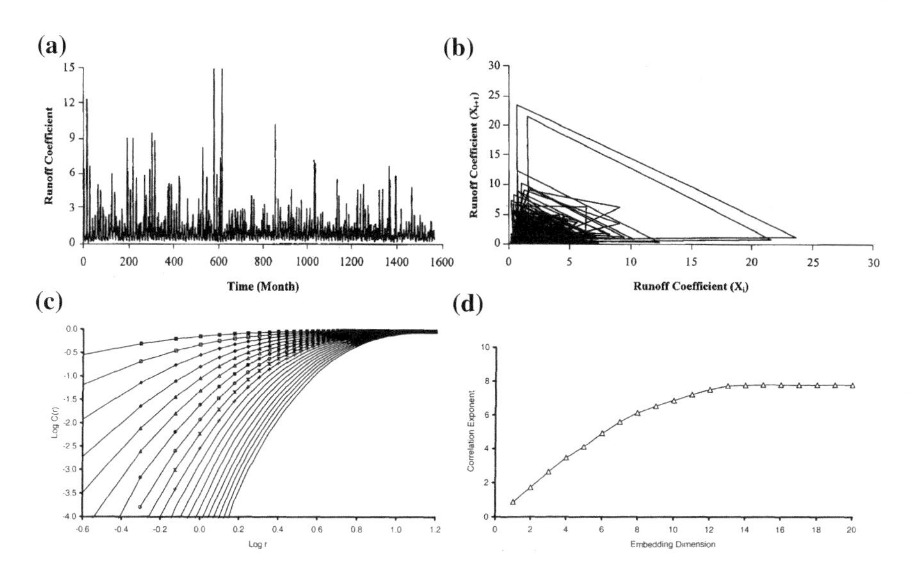

Fig. 11.1 Correlation dimension analysis of monthly runoff coefficient from Göta River basin: **a** time series; **b** phase space; **c** Log C(r) versus Log r; and **d** relationship between correlation exponent and embedding dimension (*source* Sivakumar et al. (2000, 2001a))

chaotic behavior in the monthly runoff coefficient series and, hence, in the rainfall-runoff dynamics at the Göta River basin.

In the local approximation prediction method (see Chap. 6, Sect. 6.11), the first 1440 runoff coefficient values were used for phase space reconstruction (i.e. training or learning set) to predict the subsequent 80 values. Embedding dimensions, m, from 1 to 10 were considered for phase space reconstruction. One time-step ahead predictions were made, and the local maps were learned in the form of local polynomials (Abarbanel 1996). The prediction accuracy was evaluated using three statistical measures: correlation coefficient (CC), root mean square error ($RMSE$), and coefficient of efficiency (E^2). Time series plots and scatter plots were also used to choose the best prediction results, among a large combination of results obtained for different embedding dimensions. Figure 11.2a, for instance, presents the variation of the correlation coefficient against the embedding dimension for the runoff coefficient series. The correlation coefficient increases with the embedding dimension up to $m = 7$ and then attains some kind of saturation when the dimension is increased further. The presence of an optimal embedding dimension value, $m_{opt} = 7$, indicates the possible presence of chaos in the runoff coefficient series. Consideration of the $RMSE$ and E^2 values and the time series and scatter plots indicates that the best results are obtained when $m = 6$, with $CC = 0.567$, $RMSE =$ 0.682 mm, and $CE = 0.220$ (which is significantly small). Figure 11.2b presents the time series comparisons of the observed and the predicted runoff coefficient values for this case. Although the predicted values are not in very good agreement with the observed ones, the trends (rises and falls) in the time series seem to be fairly well captured.

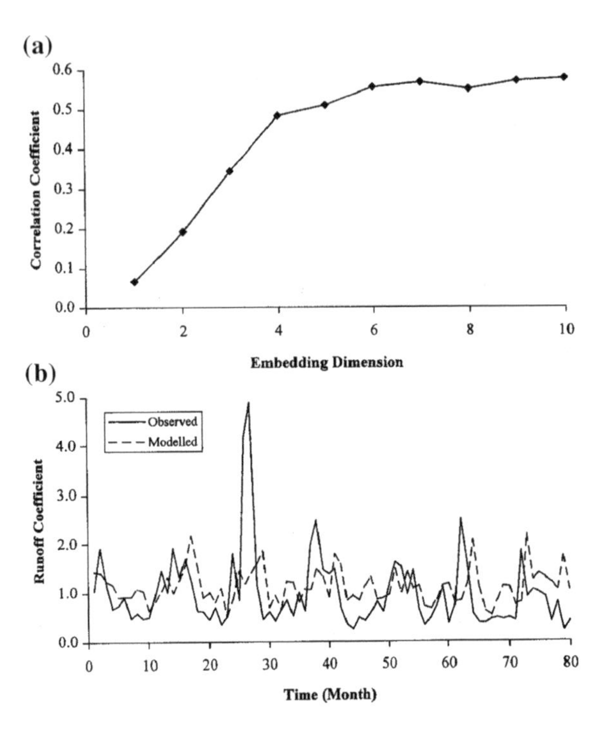

Fig. 11.2 Nonlinear local approximation prediction analysis of monthly runoff coefficient from Göta River basin: **a** relationship between correlation coefficient and embedding dimension; and **b** comparison between time series plots of predicted and observed values (*source* Sivakumar et al. (2000))

Despite these results, in view of the assumptions made in the use of rainfall, runoff, and runoff coefficient series for characterizing the rainfall-runoff process, Sivakumar et al. (2000, 2001a) cautioned on the interpretations of the results and recommended further verifications and confirmations. In particular, the usefulness of runoff coefficient as a parameter connecting rainfall and runoff was questioned. For instance, as Fig. 11.1a shows, several values of the runoff coefficient are greater than 1.0 as is the long-term mean value of 1.12, which are contrary to the acceptable definition of runoff coefficient (which should always be less than 1.0). The reason(s) for such values are unclear, although the concentration time may play a role in this. Nevertheless, the median, which describes the data better than the mean, is less than 1.0 (about 0.74).

Laio et al. (2004) employed the deterministic versus stochastic (DVS) method (e.g. Casdagli 1992) to daily river discharge from three Italian rivers in their investigation of nonlinearity in rainfall-runoff transformation. Dodov and Foufoula-Georgiou (2005) studied the nonlinear dependencies of rainfall and runoff and the effects of spatio-temporal distribution of rainfall on the dynamics of streamflow at flood timescales in two basins in Central North America. They proposed a framework based on 'hydrologically-relevant' rainfall-runoff phase space reconstruction, but with specific acknowledgment that rainfall-runoff is a stochastic spatially extended system rather than a deterministic multi-variate one. Three specific tasks were attempted: (1) quantification of the nonlinear dependencies between streamflow dynamics and the spatio-temporal dynamics of precipitation; (2) study of how streamflow predictability is affected by the trade-offs between the level of detail necessary to explain the spatial variability of rainfall and the reduction of complexity due to the smoothing effect of the basin; and (3) exploration of the possibility of incorporating process-specific information, in terms of catchment geomorphology and an a priori chosen uncertainty model, into nonlinear prediction. The results indicated the potential of using this framework for streamflow predictability and limits to prediction, as a function of the complexity of spatio-temporal forcing related to basin geomorphology, via nonlinear analysis of observations alone and without resorting to any particular rainfall-runoff model.

11.3 Lake Volume and Level

Studies that have employed the concept of chaos theory for understanding and predicting the dynamic changes in lakes have mainly focused on the Great Salt Lake (GSL) in the United States. Sangoyomi et al. (1996) investigated the presence of chaos in the GSL biweekly volume series. They employed the correlation dimension method, the nearest neighbor dimension method, and the false nearest neighbor dimension method (e.g. Kennel et al. 1992) (see Chap. 6 for details) to biweekly volume data observed over the period 1847–1992. Observing a correlation dimension value of 3.4, they suggested that the dynamics of the GSL biweekly volume series exhibit low-dimensional chaotic behavior, dominantly governed by four variables. Incidentally, the study was the first ever to use the concept of mutual information function for phase space reconstruction and the concept of false nearest neighbors for chaos identification in hydrologic time series. The presence of chaos in the GSL volume time series was further verified by Abarbanel and Lall (1996). They applied, in addition to the method used by Sangoyomi et al. (1996), the Lyapunov exponent method (e.g. Wolf et al. 1985) and determined the average predictability of the series as a few hundred days. They also attempted forecasting of the GSL series using local approximation method and constructing local polynomial maps (see Chap. 6). They tested the forecast skill for a variety of GSL conditions, such as lake average volume, near the beginning of a drought, near the end of a drought, and prior to a period of rapid lake rise. The results indicated excellent short-term predictions for the GSL series, but also revealed degrading predictions for longer time horizons. Abarbanel et al. (1996), subsequently, extended the above predictability study by also attempting multi-variate adaptive regression splines (MARS) and comparing the prediction results with those obtained using the local polynomials. Further details on the early applications of chaos theory to the GSL volume series can be found in Abarbanel (1996).

Regonda et al. (2005) studied the GSL biweekly volume time series to test the effectiveness of a nonparametric approach based on local polynomial regression for ensemble forecast of time series. They selected a suite of combinations of the four parameters involved in the nonparametric approach (i.e. embedding dimension, delay time, number of neighbors, and polynomial order) based on an objective criterion, called the Generalized Cross Validation (GCV). The ensemble approach (also providing the forecast uncertainty) yielded improved performance over the traditional method of providing a single mean forecast, and its superior performance was particularly realized for short noisy data. For the GSL, they presented blind predictions (i.e. no data outside the fitting subset used for prediction) for two cases: (i) the fall of the lake volume (during 1925–1930); and (ii) the dramatic rise and fall (during 1983–1987).

Tongal and Berndtsson (2014) investigated the presence of chaos in the daily water level dynamics of three lakes in Sweden: Vänern, Vättern, and Mälaren. Applying the correlation dimension method and observing dimensions of 3.37, 3.97, and 4.44, respectively, they reported the presence of chaos in the daily water level dynamics in these lakes. Identifying the (optimum) embedding dimensions for phase space reconstruction based on the correlation dimension values, they subsequently employed the *k*-nearest neighbor (*k*-NN) method for prediction of the daily water levels and obtained very good prediction results. Comparing the results from the *k*-NN approach with those from the self-exciting threshold autoregressive model (SETAR), they reported that both methods performed generally very well and that the phase space reconstruction-based *k*-NN method was superior in terms of the different efficiency criteria considered.

11.4 Sediment Transport

Initial studies on the applications of chaos theory for sediment transport phenomenon were driven by, among others, the recognition of key problems in the commonly used rating curve-based approaches and the need for an alternative approach for establishing relationships among water discharge, sediment concentration, and sediment load, and their predictions. For instance, Sivakumar and Jayawardena (2002) attempted to address this issue by studying the above three components independently, with the assumption that the dynamic behavior of these individual components could offer important information on the dynamic behavior of the relationships among them and, hence, the dynamic behavior of the overall sediment transport phenomenon. Sivakumar (2002) employed the local approximation method for prediction of suspended sediment concentration. Subsequent studies have addressed different aspects, including predictability, scaling, and disaggregation of one or more of the three components (e.g. Sivakumar and Jayawardena 2003; Sivakumar and Wallender 2004, 2005; Sivakumar and Chen 2007; Sivakumar et al. 2007; Shang et al. 2009). A majority of these studies have been conducted on the sediment transport phenomenon in the Mississippi River basin at St. Louis, Missouri, USA. Some details of these studies are presented here.

Sivakumar and Jayawardena (2002) employed the correlation dimension method to daily water discharge, suspended sediment concentration, and suspended sediment load in the Mississippi River basin to identify their dynamic behavior; see also

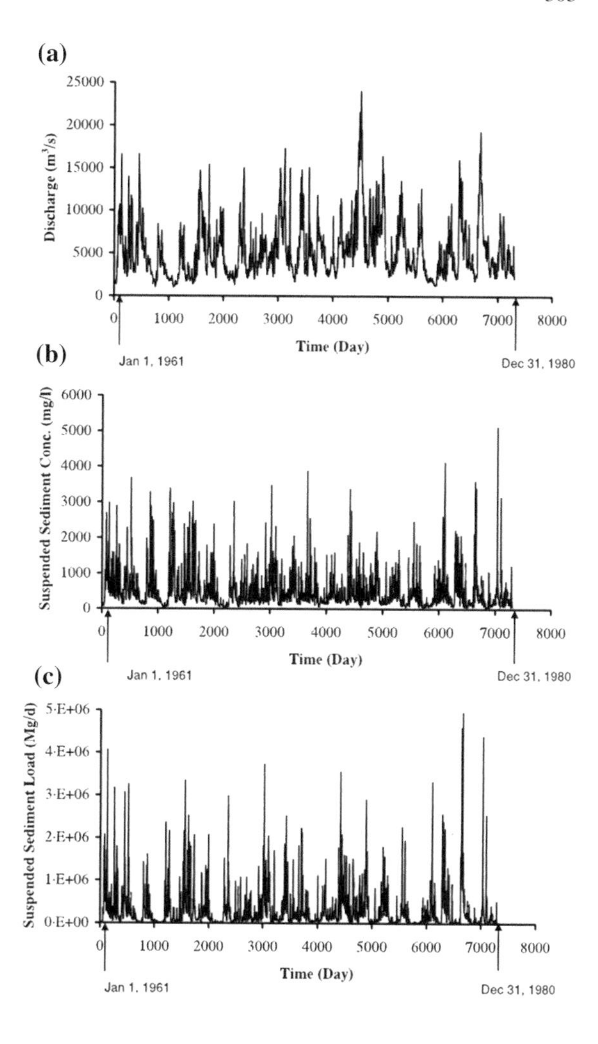

Fig. 11.3 Time series plots for sediment data from the Mississippi River basin at St. Louis, Missouri, USA: **a** discharge; **b** suspended sediment concentration; and **c** suspended sediment load (*source* Sivakumar and Wallender (2005))

Sivakumar and Wallender (2005) for additional analysis. The data analyzed were observed over a period of 20 years (January 1961–December 1980). Figure 11.3a–c presents the variations of the above three time series, respectively, and Table 11.1 presents some of their important statistics. Figure 11.4a–c presents the phase space diagram of the three time series. As can be seen, the phase space reconstruction yields reasonably well-defined structure or attractor for all the three series, with the one for the discharge series showing the most deterministic structure among the three. Figure 11.5 shows the correlation dimension results for the three time series, for embedding dimensions, m, from 1 to 20. As can be seen from Fig. 11.5b, d, f, for all the three series, the correlation exponent value increases with the embedding dimension up to a certain value and remains constant at higher dimensions, indicating deterministic dynamics. The correlation dimension value for the discharge, suspended sediment concentration, and suspended sediment load is 2.32, 2.55, and

Table 11.1 Statistics of daily discharge, suspended sediment concentration, and suspended sediment load data in the Mississippi River basin at St. Louis, Missouri, USA (*source* Sivakumar and Wallender (2005))

Statistic	Discharge (m^3 s^{-1})	Suspended sediment concentration (mg l^{-1})	Suspended sediment load (t day^{-1})
Mean	5309.97	468.28	283205
Standard deviation	3333.14	456.72	422233
Maximum value	24100.0	5140.0	4960000
Minimum value	980.0	12.0	2540
CV	0.6277	0.9753	1.491

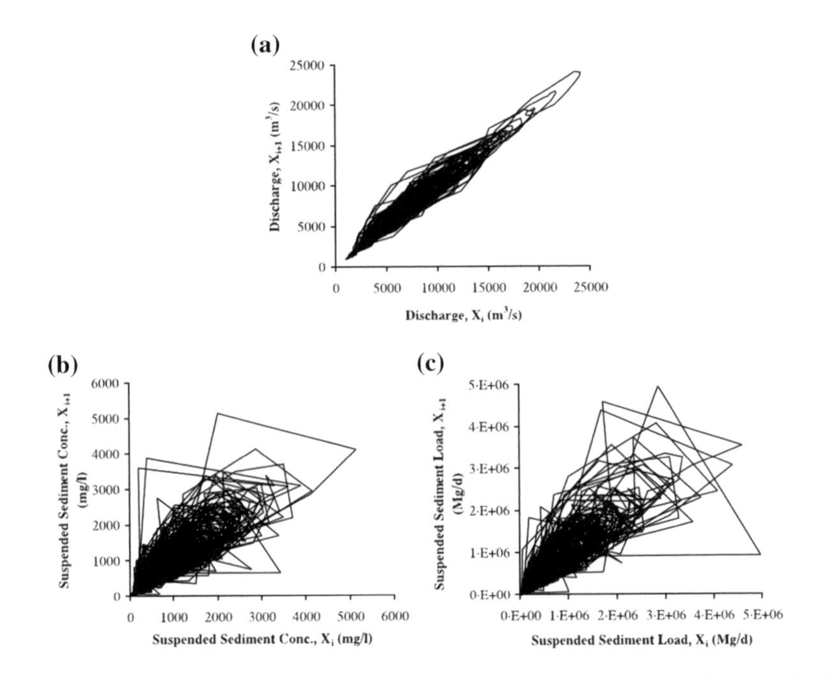

Fig. 11.4 Phasespace plots for sediment data from the Mississippi River basin at St. Louis, Missouri, USA: **a** discharge; **b** suspended sediment concentration; and **c** suspended sediment load (*source* Sivakumar and Wallender (2005))

2.41, respectively. These finite and low correlation dimensions suggest the possible presence of low-dimensional chaotic behavior in the dynamics of each of these components, with discharge being the most deterministic. The dimension values also suggest that each of the three components is dominantly governed by three variables. These results may imply that the entire sediment transport phenomenon exhibits chaotic dynamic behavior dominantly governed by only a few variables, although one has to be cautious in offering such an interpretation.

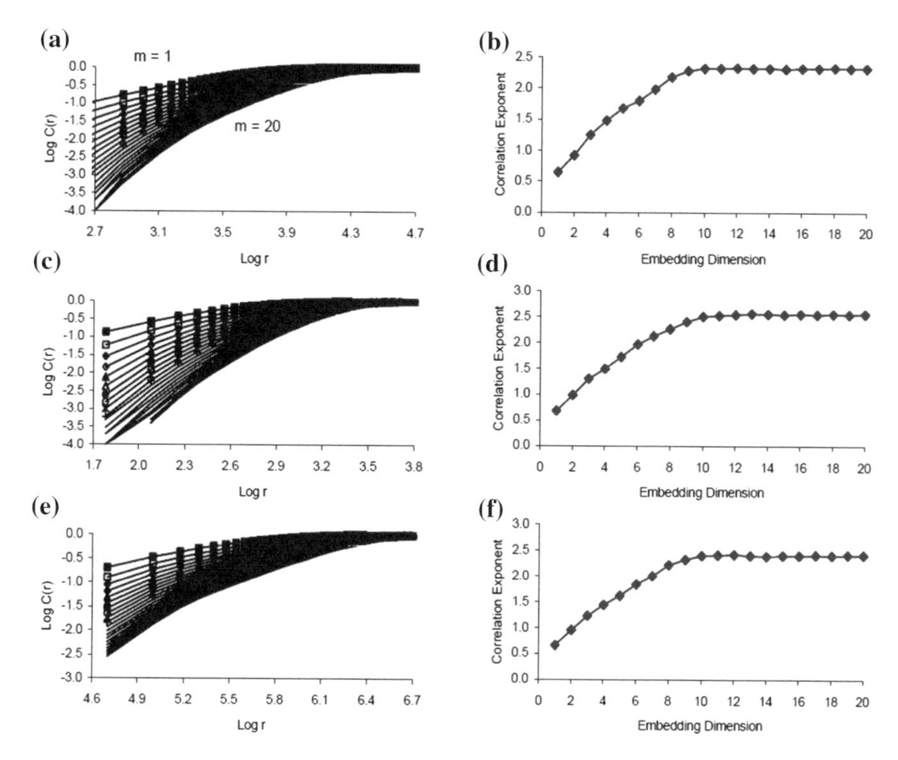

Fig. 11.5 Correlation dimension results for sediment data from the Mississippi River basin at St. Louis, Missouri, USA: **a**, **b** discharge; **c**, **d** suspended sediment concentration; and **e**, **f** suspended sediment load (*source* Sivakumar and Jayawardena (2002))

Sivakumar and Wallender (2005) addressed the predictability of discharge, suspended sediment concentration, and suspended sediment load dynamics in the Mississippi River basin. This study was an extension of the study by Sivakumar (2002), which attempted one-day ahead (i.e. lead time $T = 1$) prediction of suspended sediment concentration, in two specific ways: (1) predictions of river flow and suspended sediment load series were also made, in addition to the suspended sediment concentration series; and (2) predictions were made not just for one-day ahead but up to ten days ahead. They employed the nonlinear local approximation prediction method, involving local polynomials, to each of these series. Considering data over a period of 21 years (January 1961–December 1981), they used the first 20 years of data (the same data used by Sivakumar and Jayawardena 2002) for phase space reconstruction (i.e. training) for prediction of the subsequent 1-year of data (i.e. testing). The prediction accuracy was evaluated using three statistical evaluators, correlation coefficient (*CC*), root mean square error (*RMSE*), and coefficient of efficiency (R^2). The time series and scatter diagrams were also used to choose the best prediction results among a large combination of results

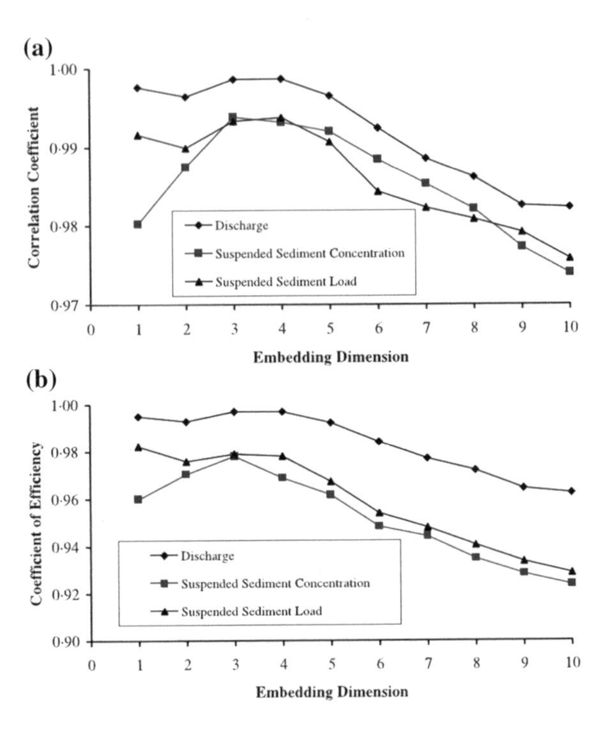

Fig. 11.6 One-day ahead prediction accuracy versus embedding dimension for daily discharge, suspended sediment concentration, and suspended sediment load in the Mississippi River basin at St. Louis, Missouri, USA: **a** correlation coefficient; and **b** coefficient of efficiency (*source* Sivakumar and Wallender (2005))

achieved with different embedding dimensions and number of neighbors (for a given lead time).

Figure 11.6 and Table 11.2 present a summary of the one-day ahead predictions achieved for the daily discharge, suspended sediment concentration, and suspended sediment load series (the RMSE plots are not presented in the figure, due to enormous differences in the values for the three cases). For each of the three series, overall, good predictions are achieved for any phase space dimension (with $CC > 0.98$ and $R^2 > 0.96$ for discharge; $CC > 0.97$ and $R^2 > 0.92$ for suspended sediment concentration; and $CC > 0.97$ and $R^2 > 0.92$ for suspended sediment load), but the best predictions are achieved when the series is reconstructed in a three-dimensional phase space (indicated in bold in Table 11.2). The time series and scatter plot comparisons also reveal that the best agreement between the observed and the predicted values is indeed achieved when the series is reconstructed in a three-dimensional phase space. For the identified best prediction case (i.e. for $m = 3$), Fig. 11.7 presents the time series and scatter plot comparisons between the observed and predicted values for each of the three series ((a) and (b) for discharge; (c) and (d) for suspended sediment concentration; and (e) and (f) for suspended sediment load). In general, for all the three series, the predicted values are in reasonably good agreement with the observed values. A closer look at these plots reveals that both the major trends (including the extreme values) as well as the minor fluctuations in the series are well captured.

Table 11.2 One-day ahead (T = 1) predictions for discharge, suspended sediment concentration, and suspended sediment load from the Mississippi River basin at St. Louis, Missouri, USA (*source* Sivakumar and Wallender (2005))

m	Discharge			Concentration			Load		
	CC	RMSE	R^2	CC	RMSE	R^2	CC	RMSE	R^2
1	0.998	198.16	0.995	0.98	85.79	0.96	0.992	56957	0.982
2	0.996	234.64	0.993	0.988	73.64	0.97	0.99	66484	0.976
3	**0.999**	**152.82**	**0.997**	**0.994**	**63.52**	**0.978**	**0.993**	**62010**	**0.979**
4	0.999	155.53	0.997	0.993	75.57	0.969	0.994	63341	0.978
5	0.997	246.61	0.992	0.992	83.83	0.962	0.991	77289	0.967
6	0.992	353.58	0.984	0.988	97.34	0.948	0.984	91645	0.954
7	0.989	425.22	0.977	0.985	101.15	0.944	0.982	97518	0.948
8	0.986	469.64	0.972	0.982	109.31	0.935	0.981	104222	0.941
9	0.982	529.58	0.964	0.977	114.56	0.928	0.979	110122	0.934
10	0.982	544.02	0.962	0.974	118.23	0.924	0.976	114084	0.929

An insight into Table 11.2 and Figs. 11.6 and 11.7 clearly indicates that discharge is more accurately predicted when compared to suspended sediment concentration and suspended sediment load. This observation is consistent with the one made earlier with reference to the phase space diagrams (Fig. 11.4) and correlation dimension results (Fig. 11.5) of the three series, that discharge is the most deterministic among the three series and the most predictable.

With very good one-day ahead predictions, it would be interesting to see the change in predictability when the lead time is increased, especially considering the enormous size of the basin (251,230 km^2) and flow. Figure 11.8 and Table 11.3 present a summary of the one-day to ten-day ahead predictions. The results presented are the ones achieved when each of the series is reconstructed in a three-dimensional phase space, i.e. the embedding dimension that yielded the best one-day ahead predictions. The results indicate a general trend of a decrease (i.e. fall-off) in prediction accuracy with an increase in predictability horizon (i.e. lead time), irrespective of the series. However, the fall-off is relatively slower up to five days when compared to that from six to ten days (the time series and scatter plots also support this). With the reasonably high *CC* and R^2 values (Fig. 11.8 and Table 11.3) and reasonably good agreement between the predicted and the observed values (figure not presented, see Fig. 7 of Sivakumar and Wallender 2005 for comparison of 1-day and 5-day ahead predictions), it may be said that predictions up to five days are reasonably good, particularly for discharge.

Following up on the earlier studies (e.g. Sivakumar and Jayawardena 2002; Sivakumar 2002), Sivakumar and Wallender (2004) introduced a chaotic approach for disaggregation of suspended sediment load data. This approach was based on the chaotic disaggregation approach proposed by Sivakumar et al. (2001b) for rainfall data (see Chap. 9, Sect. 9.3.2), with appropriate modifications to suit the suspended sediment load data (see also Chap. 10, Sect. 10.3.2 for the case of river

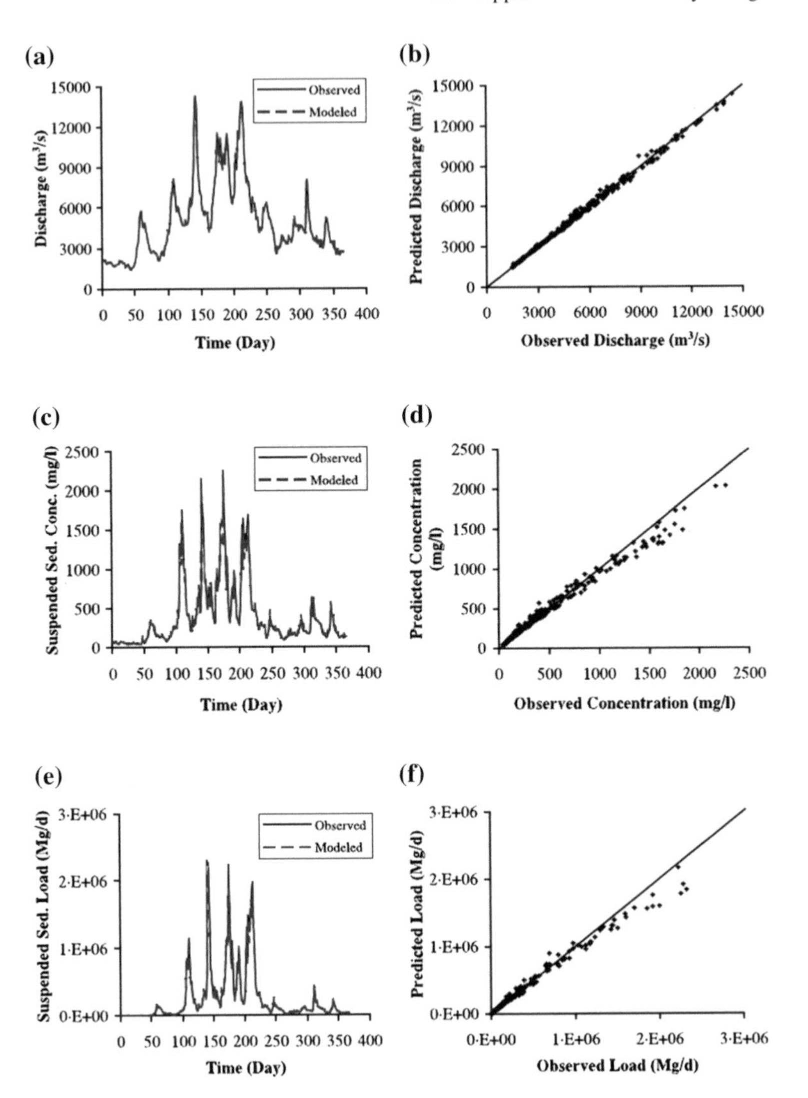

Fig. 11.7 Time series and scatter plot comparisons of one-day ahead predictions for daily sediment data in the Mississippi River basin at St. Louis, Missouri, USA: **a**, **b** discharge; **c**, **d** suspended sediment concentration; and **e**, **f** suspended sediment load (*source* Sivakumar and Wallender (2005))

flow). The approach was tested on the suspended sediment load data observed at the St. Louis gaging station, with disaggregation attempted for four successively doubled resolutions: 2-day to daily, 4-day to 2-day, 8-day to 4-day, and 16-day to 8-day. The study revealed the possible nonlinear deterministic nature of the sediment load transformation process at these scales, as the best results were achieved for low phase space dimension (<4) and relatively small number of neighbors

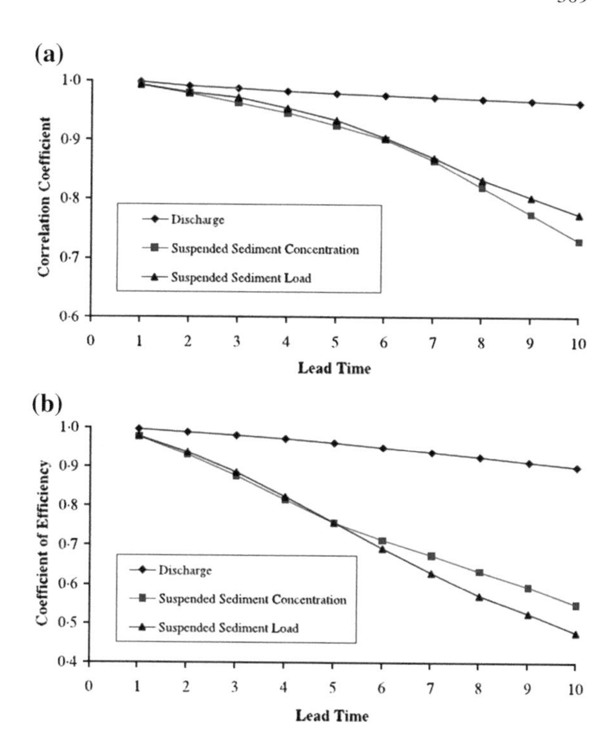

Fig. 11.8 Prediction accuracy versus lead time for daily discharge, suspended sediment concentration, and suspended sediment load in the Mississippi River basin at St. Louis, Missouri, USA: **a** correlation coefficient; and **b** coefficient of efficiency. Embedding dimension (m) = 3 (*source* Sivakumar and Wallender (2005))

Table 11.3 Predictions for different lead times for discharge, suspended sediment concentration, and suspended sediment load from the Mississippi River at St. Louis, Missouri, USA (m = 3 in all cases) (*source* Sivakumar and Wallender (2005))

T	Discharge			Concentration			Load		
	CC	RMSE	R^2	CC	RMSE	R^2	CC	RMSE	R^2
1	0.999	152.82	0.997	0.994	63.52	0.978	0.993	62010	0.979
2	0.992	290.94	0.989	0.979	111.95	0.932	0.981	106210	0.938
3	0.988	385.16	0.981	0.963	149.72	0.878	0.972	143968	0.887
4	0.983	463.97	0.973	0.946	183.31	0.817	0.954	179182	0.824
5	0.979	544.89	0.962	0.925	210.61	0.758	0.934	209892	0.759
6	0.976	627.52	0.95	0.902	228.71	0.715	0.905	236802	0.693
7	0.973	698.77	0.938	0.866	243.67	0.676	0.871	259655	0.631
8	0.97	763.83	0.926	0.82	258.77	0.635	0.834	279564	0.573
9	0.967	827.77	0.913	0.775	272.42	0.595	0.803	294108	0.527
10	0.964	884.22	0.901	0.73	287.08	0.55	0.775	308666	0.479

(<100). Sivakumar and Chen (2007) investigated, through the application of the correlation dimension method, the dynamic behavior of suspended sediment load transport at different temporal scales at the St. Louis gaging station. They analyzed data corresponding to five different temporal scales: daily, two-day, four-day, eight-day and 16-day. Observing correlation dimension values of 2.41, 2.54, 2.74, 3.15, and 3.62 for the five series, respectively, they reported the presence of low-dimensional determinism in the suspended sediment transport series at each of these five scales. These results also hinted the possible scale-invariance in the suspended sediment load dynamics, which was explored earlier using disaggregation by Sivakumar and Wallender (2004) and through fractal analysis by Sivakumar (2006). Additional details on chaos theory-based analysis of sediment transport phenomenon are also available in Sivakumar and Jayawardena (2003) and Sivakumar et al. (2007) (in the context of data classification).

Shang et al. (2009) attempted identification and prediction of chaotic dynamics in the sediment transport phenomenon in the Yellow River basin at Tongguan in Shanxi, China. They analyzed the daily suspended sediment concentration data observed over a period of 23 years. They employed a host of methods for identification and prediction purposes, including correlation dimension method, false nearest neighbor method, Lyapunov exponent method, phase space embedding-based weight predictor algorithm (PSEWPA). Observing a correlation dimension value of 6.6 and a positive value for the largest Lyapunov exponent (0.065), they reported the presence of chaotic behavior in the suspended sediment concentration dynamics. They also reported good prediction results from the PSEWPA method.

11.5 Groundwater

While surface water hydrology had witnessed a large number of chaos theory applications in the 1990s, subsurface hydrology had eluded the attention of chaos studies until earlier in this century. To my knowledge, Faybishenko (2002) was the first to introduce the concept of chaos in subsurface hydrology through one of his studies on complex flow processes in heterogeneous fractured media. He analyzed the time series of pressure fluctuations from two water-air flow experiments in replicas of rough-walled rock fractures under controlled laboratory conditions (Persoff and Pruess 1995), using a host of methods, including correlation dimension, global and local embedding dimensions, and Lyapunov exponents. The results were then also compared with the chaotic analysis of laboratory dripping-water experiments in fracture models and field-infiltration experiments in fractured basalt. Based on this comparison, it was conjectured that: (1) the intrinsic fracture flow and dripping, as well as extrinsic water dripping (from a fracture) subjected to a capillary-barrier effect, are deterministic-chaotic processes with a certain random component; and (2) the unsaturated fractured rock is a dynamic system that exhibits chaotic behavior because the flow processes are nonlinear, dissipative, and sensitive to initial conditions, with chaotic fluctuations generated by intrinsic properties of the system, not random external factors.

11.5.1 Solute Transport

Sivakumar et al. (2005) investigated the potential use of chaos theory to understand the dynamic nature of solute transport process in subsurface formations. They employed the correlation dimension method to time series of solute particle transport in a heterogeneous aquifer medium to identify the presence of chaos. Considering the western San Joaquin Valley aquifer system in California, USA as a reference system, the solute transport time series was simulated using an integrated transition probability/Markov chain (TP/MC) model, groundwater flow model, and particle transport model. To examine the sensitivity of the solute transport dynamics, four hydrostratigraphic parameters involved in the TP/MC model were also considered: (1) number of facies—two (sand and clay) versus three (sand, clay, and loam); (2) volume proportions of facies—30 combinations of proportions of two facies (sand from 15 to 60 % and clay forming the remainder) and one combination of proportions in three facies (i.e. sand 21.26 %, clay 53.28 %, and loam 25.46 %); (3) mean length and, thereby, anisotropy ratio of mean length—three sets of mean length ratios (ratios of dip to strike and dip to vertical facies mean length are 2:1 and 300:1, 5:1 and 300:1, and 2:1 and 50:1); and (4) juxtapositional tendencies (i.e. degree of entropy) among the facies—three combinations (maximum entropy or random juxtaposition of facies, intermediate entropy, and low entropy or highly structured order of facies). Some details of the analysis and results are presented here.

Figure 11.9a shows the time series of solulte transport or particle arrival (i.e. breakthrough curve) in two facies (sand 20 %, clay 80 %), simulated over a period

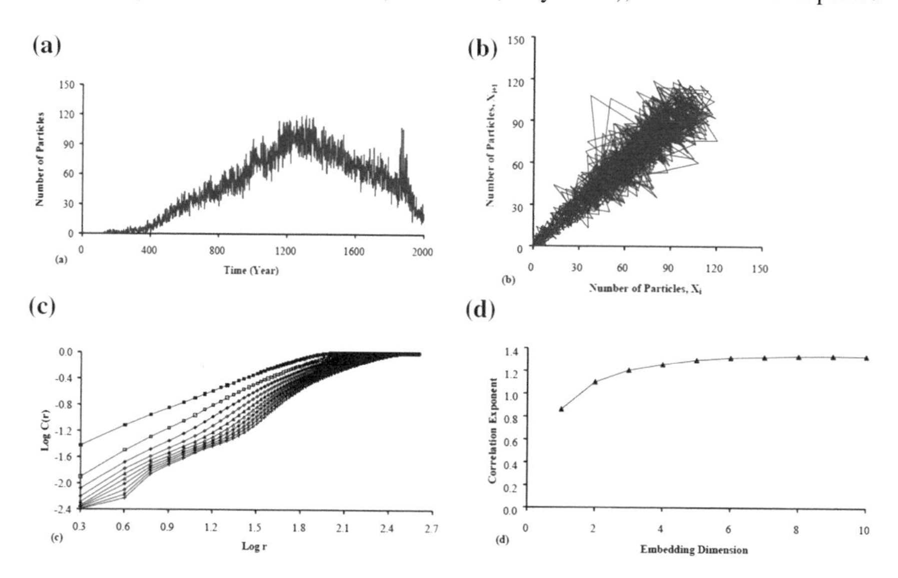

Fig. 11.9 Correlation dimension analysis of solute transport process in two facies medium (sand 20 %, clay 80 %) with anisotropy condition 2:1 and 50:1: **a** time series plot; **b** phase space diagram; **c** Log C(r) versus Log r plot; and **d** correlation exponent versus embedding dimension (*source* Sivakumar et al. (2005))

of 2000 years. The time series corresponds to mean length anisotropy ratios 2:1 (dip:strike) and 50:1 (dip:vertical). Figure 11.9b presents its reconstruction in two dimensions, with a delay time $\tau = 1$, which is a typical sampling interval for ambient groundwater monitoring. The reconstruction yields a well-defined attractor in the phase space, suggesting the possibility of deterministic dynamics. Figure 11.9c, d shows the results of the Grassberger-Procaccia correlation dimension analysis, for embedding dimensions, m, from 1 to 10. The correlation exponent increases with the embedding dimension up to a certain point and saturates beyond that. The saturation value of the correlation exponent is as low as 1.33, which suggests the presence of chaotic behavior in the solute transport dynamics, with two variables dominantly governing the system dynamics.

Figure 11.10 presents the time series, phase space, and correlation dimension results of the solute transport data in three facies medium. The data correspond to simulations with anisotropy condition 2:1 and 50:1 and field entropy. The phase space reconstruction of the time series yields a well-defined attractor, and the correlation exponent attains saturation at a value of 2.12. The low correlation dimension value is an indication of the presence of chaotic dynamics in solute transport in three facies, with three dominant governing variables.

To illustrate the effect of volume proportions on the dynamic behavior of solute transport, Fig. 11.11 presents the correlation dimension results for particle arrival time series simulated with four different sand proportions: 15, 25, 35, and 60 %. As can be seen, saturation of correlation exponent is observed for each of the four cases. The correlation dimension values for the four series are as low as 0.35, 0.62, 0.81, and 0.44, indicating that the solute transport dynamics exhibit

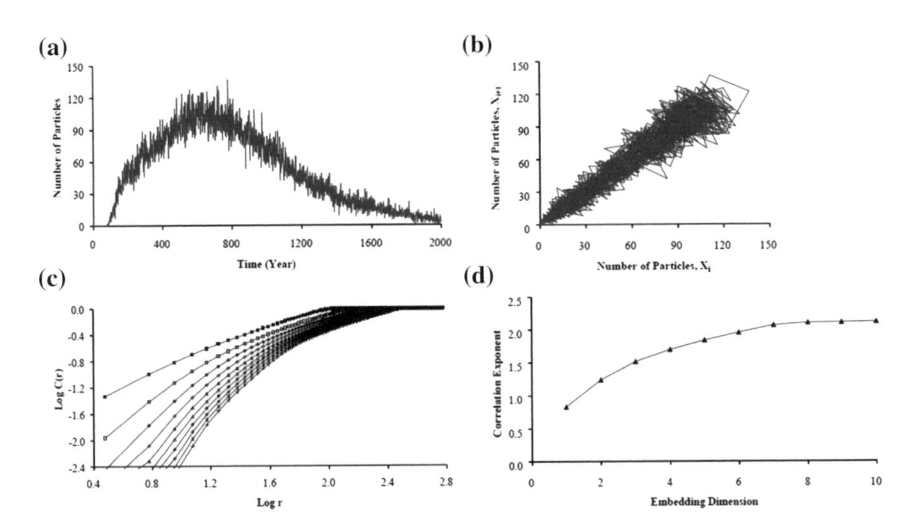

Fig. 11.10 Correlation dimension analysis of solute transport process in three facies medium (sand 21.26 %, clay 53.28 %, loam 25.46 %) with anisotropy condition 2:1 and 50:1 and field entropy: **a** time series plot; **b** phase space diagram; **c** Log C(r) versus Log r plot; and **d** correlation exponent versus embedding dimension (*source* Sivakumar et al. (2005))

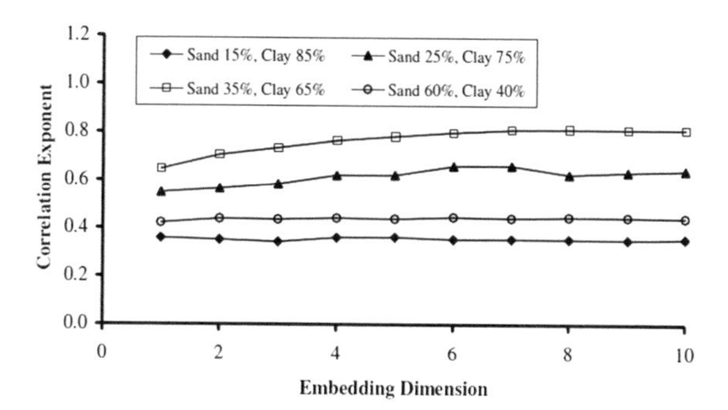

Fig. 11.11 Effect of volume proportions on solute transport behavior in two facies medium (with anisotropy condition 2:1 and 300:1): sand 15 %, clay 85 %; sand 25 %, clay 75 %; sand 35 %, clay 65 %; and sand 60 %, clay 40 % (*source* Sivakumar et al. (2005))

low-dimensional chaotic behavior. The results also indicate that the complexity of the process slightly increases with an increase in sand proportion up to a certain point and then decreases with further increase in sand (similar results are observed also for the other anisotropy conditions, i.e. 5:1 and 300:1, and 2:1 and 50:1). This is understandable, since only certain mechanisms may have dominant influence on the solute transport process in the presence of either very small or very large sand proportions, even though the connectivities may have opposite characteristics. The transport process becomes more heterogeneous when sand and clay proportions approach each other, as there may be additional mechanisms due to (significant) influence of both sand and clay. This is consistent with the fact that, at a given hydraulic conductivity contrast, the highest variance of the binary aquifer system is obtained when the two facies are present in equal proportions. It appears that there is a correlation between the dimension parameter and system variance.

Figure 11.12 presents the results from the correlation dimension analysis for the three anisotropy conditions studied: 2:1 and 300:1, 5:1 and 300:1, and 2:1 and 50:1. The results correspond to the two facies medium with sand proportion equal to 20 % (clay 80 %). The correlation dimension values are found to be 0.46, 0.64, and 1.33 for the three cases, respectively, suggesting the presence of low-dimensional chaotic behavior in the solute transport dynamics, regardless of the anisotropy condition. The extent of difference in the dimension results between the three anisotropy conditions indicates that a change in ratio of dip to vertical mean length

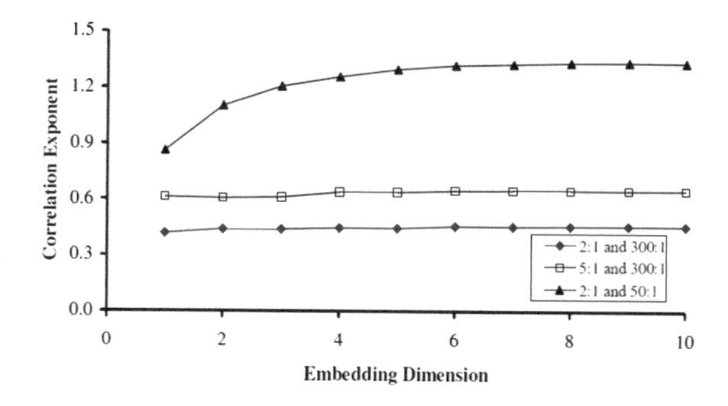

Fig. 11.12 Sensitivity of dynamic behavior of solute transport process to anisotropy conditions in two facies medium: sand 20 %, clay 80 % (*source* Sivakumar et al. (2005))

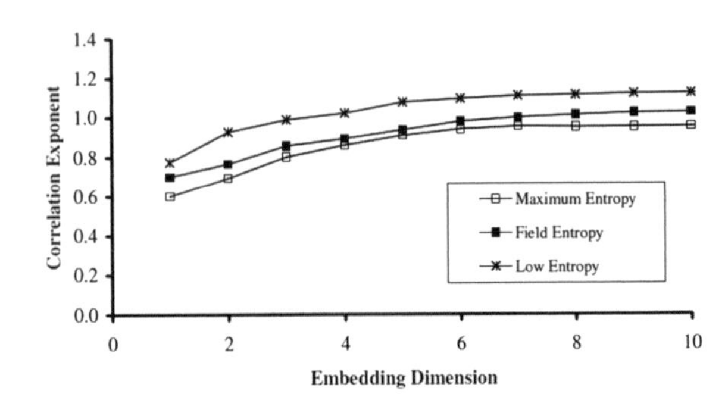

Fig. 11.13 Effect of entropy on solute transport behavior in three facies medium (sand 21.26 %, clay 53.28 %, loam 25.46 %) with anisotropy condition 5:1 and 300:1 (*source* Sivakumar et al. (2005))

(from 2:1 and 300:1 to 2:1 and 50:1) has much more influence on solute transport than a change in ratio of dip to strike mean length (from 2:1 and 300:1 to 5:1 and 300:1). This is essentially due to the vertical mean flow in the system. Similar results were observed also for other sand proportions in the two facies medium and for different entropy conditions in the three facies medium. However, the dynamics tend to become much more complex with anisotropy condition 2:1 and 50:1 for certain proportions and entropies. This is consistent with the fact the velocity field is more complex in the less stratified system (lower anisotropy). The significance lies in the fact that the dimension parameter appears sensitive to these changes in the velocity field complexity within the aquifer without being directly informed about the velocity.

Finally, Fig. 11.13 presents the results from the correlation dimension analysis of solute transport time series for the three different entropy conditions, to assess the influence of entropy on system dynamics. The results correspond to anisotropy condition 5:1 and 300:1. The correlation dimension values for the maximum, field, and low entropy conditions are 0.95, 1.02, and 1.11, respectively. These low correlation dimension values indicate the presence of low-dimensional chaotic dynamics in the solute transport process. The results also indicate increasing complexity of the solute transport process with an increasing order of facies in the aquifer medium (similar results are also observed for solute transport process simulated with the other two anisotropy conditions, i.e. 2:1 and 300:1 and 2:1 and 50:1, figures not presented). The effect is, again, related to the complexity of the velocity field, which can be (and in this example is) larger in media with a higher ordered structure than in media with a purely random conductivity distribution.

In summary, the results from the correlation dimension analysis of groundwater solute transport process generally indicate the nonlinear deterministic nature of solute transport dynamics, dominantly governed by only a very few variables, on the order of 3. However, more complex behavior can also be found under certain extreme hydrostratigraphic conditions.

11.5.2 Arsenic Contamination

Hossain and Sivakumar (2006) investigated the presence chaotic dynamic behavior in the spatial patterns of arsenic contamination in the shallow wells (<150 m) of Bangladesh. They employed the correlation dimension method to data observed at 3,085 shallow wells, with approximately one well per 37 km^2 ($\sim$6 km $\times$ 6 km). In the reconstruction of the phase space, they used delay distance (instead of delay time) to be some suitable multiple of the average intra-well distances Δs (instead of sampling time Δs). Giving particular emphasis to the role of regional geology on the spatial dynamics of arsenic contamination, they classified the arsenic contamination database into three categories: (A) Whole of Bangladesh (no distinction made in geology)—3,085 wells; (B) Holocene deposits of Southwest Bangladesh (geologic distinction—those regions usually high in arsenic)—848 wells; and (C) Pleistocene deposits of Northwest Bangladesh (geologic distinction—those regions usually low in arsenic)—872 wells.

Figure 11.14a–c, for instance, presents the correlation dimension results obtained for a selected well in the above three categories: focal well A-1, focal well B-1, and focal well C-1. The dimension values are about 9.8 for focal well A-1, 10.4 for focal well B-1, and 8.1 for focal well C-1. Considering the results from all the wells, the correlation dimensions values were found to range anywhere from 8 to 11, suggesting that the arsenic contamination in space exhibits a medium- to high-dimensional dynamics. In the context of regional geology, the correlation dimension values for Region A and Region B were found to be in the range 10–11, while those for Region C were found to be the range 8–9. These results indicate that the spatial dynamics of arsenic contamination may be moderately sensitive to geology, with Pleistocene aquifers requiring a minimum of about two less dominant processes or variables for its description when compared to that required by the Holocene aquifers.

Hill et al. (2008) attempted to further verify the correlation dimension results reported by Hossain and Sivakumar (2006). Using a logistic regression approach, they explored possible physical connections between the correlation dimension values and the mathematical modeling of risk of arsenic contamination in groundwater. Based on the correlation dimension values of 8 to 11 reported by Hossain and Sivakumar (2006), Hill et al. (2008) considered 11 variables as indicators of the aquifer's geochemical regime with potential influence on arsenic contamination, and a total of 2,048 possible combinations of these variables were included as candidate logistic regression models to delineate the impact of the number of variables on the prediction accuracy of the model. They found that the uncertainty associated with prediction of wells as 'safe' and 'unsafe' by the logistic regression model declined systematically as the total number of influencing variables increased from 7 to 11. The sensitivity of the mean predictive performance also increased noticeably for this range. They concluded that the consistent reduction in predictive uncertainty coupled with the increased sensitivity of the

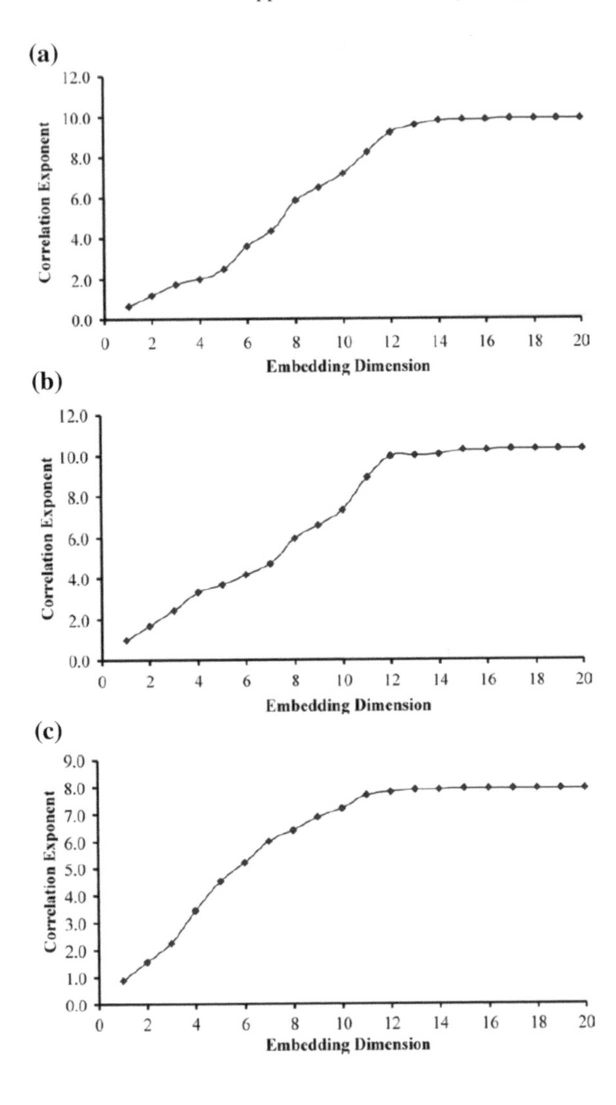

Fig. 11.14 Relationship between correlation exponent and embedding dimension for three different regions: **a** region A (whole of Bangladesh) focal well A-1; **b** region B (Holocene) focal well B-1; and **c** region C (Pleistocene) focal well C-1 (*source* Hossain and Sivakumar (2006))

mean predictive behavior within the universal sample space exemplify the ability of the correlation dimension method to function as a proxy for the number of dominant influencing variables.

11.6 Others

Rodriguez-Iturbe et al. (1992) studied the soil moisture balance equation over large spatial territories at seasonal timescales, with explicit consideration of local recycling of precipitation and dynamic effects of soil moisture in the generation and modification of mesoscale circulation patterns for parameterization that also incorporates time delays. They showed that the dynamics of the equations are quite complex, being capable of exhibiting fixed point, limit cycle, and chaotic type behavior. Manzoni et al. (2004) studied the soil carbon and nitrogen cycles from a dynamic system perspective, wherein the system nonlinearities and feedbacks were analyzed by considering the steady-state solution under deterministic hydro-climatic conditions.

Nordstrom et al. (2005) proposed the construction of a dynamic area fraction model (DAFM) that contains coupled parameterizations for all the major components of the hydrologic cycle involving liquid, solid, and vapor phases. Using this model, which shares some of the characteristics of an Earth system model of intermediate complexity, they investigated the nature of feedback processes in regulating Earth's climate as a highly nonlinear coupled system. Jin et al. (2005) studied the nonlinear relationships between southern oscillation index (SOI) and local precipitation and temperature (in Fukuoka, Japan), by representing this joint hydro-climatic system using a nonlinear multi-variate phase space reconstruction technique.

Wu et al. (2009) attempted prediction of dissolved oxygen time series by employing a 'global' chaos prediction method and its improved version, called reduced-dimension chaos prediction. Analyzing a very short and stable time series of weekly dissolved oxygen, they reported better performance of these two methods when compared to that of an autoregressive (AR) model.

Some other selected studies of interest are those that have focused on geomorphology (e.g. Phillips and Walls 2004; Phillips 2006a, b) and water level/stage in rivers/lagoons (e.g. Zaldivar et al. 2000; Khokhlov et al. 2008; Khatibi et al. 2012), and others (e.g. Ng et al. 2007).

11.7 Summary

The expansion of chaos theory-based studies in hydrology from the earlier ones that had mainly focused on rainfall (Chap. 9) and river flow (Chap. 10) to other processes has been tremendous. This chapter has reviewed the chaos studies on such data, including rainfall-runoff, lake volume and level, sediment transport, and groundwater flow and transport. The outcomes of these studies clearly highlight the scope and potential of chaos theory in almost all areas of hydrology. While identification, prediction, and scaling in these data have been the main problems studied thus far, there have also been attempts on a few other problems that are gaining considerable interest in hydrology at the current time, as highlighted in Chaps. 13 and 14. In addition, studies addressing some important data-related issues in the application of chaos methods in hydrology have also continued, as discussed in Chap. 12.

References

Abarbanel HDI (1996) Analysis of observed chaotic data. Springer-Verlag, New York, USA

Abarbanel HDI, Lall U (1996) Nonlinear dynamics of the Great Salt Lake: system identification and prediction. Climate Dyn 12:287–297

Abarbanel HDI, Lall U, Moon YI, Mann M, Sangoyomi T (1996) Nonlinear dynamics and the Great Salk Lake: A predictable indicator of regional climate. Energy 21(7/8):655–666

Casdagli M (1992) Chaos and deterministic versus stochastic nonlinear modeling. J Royal Stat Soc B 54(2):303–328

Dodov B, Foufoula-Georgiou E (2005) Incorporating the spatio-temporal distribution of rainfall and basin geomorphology into nonlinear analysis of streamflow dynamics. Adv Water Resour 28(7):711–728

Faybishenko B (2002) Chaotic dynamics in flow through unsaturated fractured media. Adv Water Resour 25(7):793–816

Grassberger P, Procaccia I (1983a) Measuring the strangeness of strange attractors. Physica D 9:189–208

Grassberger P, Procaccia I (1983b) Characterisation of strange attractors. Phys Rev Lett 50 (5):346–349

Hill J, Hossain F, Sivakumar B (2008) Is correlation dimension a reliable proxy for the number of dominant influencing variables for modeling risk of arsenic contamination in groundwater? Stoch Environ Res Risk Assess 22(1):47–55

Hossain F, Sivakumar B (2006) Spatial pattern of arsenic contamination in shallow wells of Bangladesh: regional geology and nonlinear dynamics. Stoch Environ Res Risk Assess 20(1–2):66–76

Jin YH, Kawamura A, Jinno K, Berndtsson R (2005) Nonlinear multivariate analysis of SOI and local precipitation and temperature. Nonlinear Processes Geophys 12:67–74

Kennel MB, Brown R, Abarbanel HDI (1992) Determining embedding dimension for phase space reconstruction using a geometric method. Phys Rev A 45:3403–3411

Khatibi R, Sivakumar B, Ghorbani MA, Kişi Ö, Kocak K, Zadeh DF (2012) Investigating chaos in river stage and discharge time series. J Hydrol 414:108–117

Khokhlov V, Glushkov A, Loboda N, Serbov N, Zhurbenko K (2008) Signatures of low-dimensional chaos in hourly water level measurements at coastal site of Mariupol, Ukraine. Stoch Environ Res Risk Assess 22:777–787

Laio F, Porporato A, Revelli A, Ridolfi L (2003) A comparison of nonlinear flood forecasting methods. Water Resour Res 39(5):1129. doi:10.1029/2002WR001551

Laio F, Porporato A, Ridolfi L, Tamea S (2004) Detecting nonlinearity in time series driven by non-Gaussian noise: the case of river flows. Nonlinear Processes Geophys 11:463–470

Manzoni S, Porporato A, D'Odorico P, Laio F, Rodriguez-Iturbe I (2004) Soil nutrient cycles as a nonlinear dynamical system. Nonlinear Processes Geophys 11:589–598

Ng WW, Panu US, Lennox WC (2007) Chaos based analytical techniques for daily extreme hydrological observations. J Hydrol 342:17–41

Nordstrom KM, Gupta VK, Chase TN (2005) Role of the hydrological cycle in regulating the planetary climate system of a simple nonlinear dynamical model. Nonlinear Processes Geophys 12:741–753

Persoff P, Pruess K (1995) Two-phase flow visualization and relative permeability measurement in natural rough-walled rock fractures. Water Resour Res 31(5):1175–1186

Phillips JD (2006a) Evolutionary geomorphology: thresholds and nonlinearity in landform response to environmental change. Hydrol Earth Syst Sci 10:731–742

Phillips JD (2006b) Deterministic chaos and historical geomorphology: a review and look forward. Geomorphology 76:109–121

Phillips JD, Walls MD (2004) Flow partitioning and unstable divergence in fluviokarst evolution in central Kentucky. Nonlinear Processes Geophys 11:371–381

Porporato A, Ridolfi L (2001) Multivariate nonlinear prediction of river flows. J Hydrol 248(1–4):109–122
Regonda S, Rajagopalan B, Lall U, Clark M, Moon YI (2005) Local polynomial method for ensemble forecast of time series. Nonlinear Processes Geophys 12:397–406
Rodriguez-Iturbe I, Entekhabi D, Lee J-S, Bras RL (1992) Nonlinear dynamics of soil moisture at climate scales 2. Chaotic analysis. Water Resour Res 27(8):1907–1915
Sangoyomi TB, Lall U, Abarbanel HDI (1996) Nonlinear dynamics of the Great Salt Lake: dimension estimation. Water Resour Res 32(1):149–159
Shang P, Na X, Kamae S (2009) Chaotic analysis of time series in the sediment transport phenomenon. Chaos Soliton Fract 41(1):368–379
Sivakumar B (2002) A phase-space reconstruction approach to prediction of suspended sediment concentration in rivers. J Hydrol 258:149–162
Sivakumar B (2006) Suspended sediment load estimation and the problem of inadequate data sampling: a fractal view. Earth Surf Process Landf 31:414–427
Sivakumar B (2009) Nonlinear dynamics and chaos in hydrologic systems: latest developments and a look forward. Stoch Environ Res Risk Assess 23:1027–1036
Sivakumar B, Chen J (2007) Suspended sediment load transport in the Mississippi River basin at St. Louis: temporal scaling and nonlinear determinism. Earth Surf Process Landf 32(2):269–280
Sivakumar B, Jayawardena AW (2002) An investigation of the presence of low-dimensional chaotic behavior in the sediment transport phenomenon. Hydrol Sci J 47(3):405–416
Sivakumar B, Jayawardena AW (2003) Sediment transport phenomenon in rivers: an alternative perspective. Environ Model Softw 18(8–9):831–838
Sivakumar B, Wallender WW (2004) Deriving high-resolution sediment load data using a nonlinear deterministic approach. Water Resour Res 40:W05403. doi:10.1029/2004WR003152
Sivakumar B, Wallender WW (2005) Predictability of river flow and sediment transport in the Mississippi River basin: a nonlinear deterministic approach. Earth Surf Process Landf 30:665–677
Sivakumar B, Berndtsson R, Olsson J, Jinno K, Kawamura A (2000) Dynamics of monthly rainfall-runoff process at the Göta basin: A search for chaos. Hydrol Earth Syst Sci 4(3): 407–417
Sivakumar B, Berndttson R, Olsson J, Jinno K (2001a) Evidence of chaos in the rainfall-runoff process. Hydrol Sci J 46(1):131–145
Sivakumar B, Sorooshian S, Gupta HV, Gao X (2001b) A chaotic approach to rainfall disaggregation. Water Resour Res 37(1):61–72
Sivakumar B, Harter T, Zhang H (2005) Solute transport in a heterogeneous aquifer: a search for nonlinear deterministic dynamics. Nonlinear Processes Geophys 12:211–218
Sivakumar B, Jayawardena AW, Li WK (2007) Hydrologic complexity and classification: a simple data reconstruction approach. Hydrol Process 21(20):2713–2728
Tongal H, Berndtsson R (2014) Phase-space reconstruction and self-exciting threshold modeling approach to forecast lake water levels. Stoch Environ Res Risk Assess 28(4):955–971
Wolf A, Swift JB, Swinney HL, Vastano A (1985) Determining Lyapunov exponents from a time series. Physica D 16:285–317
Wu J, Lu J, Wang J (2009) Application of chaos and fractal models to water quality time series prediction. Environ Modell Softw 24:632–636
Zaldivar JM, Gitiêrrez E, Galvân IM, Strozzi F, Tomasin A (2000) Forecasting high waters at Venice Lagoon using chaotic time series analysis and nonlinear neural networks. J Hydroinformatics 2:61–84

Chapter 12
Studies on Hydrologic Data Issues

Abstract Despite the tremendous growth in the applications of chaos theory in hydrology, there have been lingering criticisms. These criticisms have been based on the fundamental assumptions involved in the development of methods for identification and prediction of chaos (e.g. infinite and noise-free time series, lack of clear-cut guidelines on the selection of parameters involved) and/or the limitations of hydrologic data (e.g. short and noisy data, presence of zeros). A number of issues have been raised in this regard, but some have attracted far more attention than the others. This chapter presents a review of studies that have addressed such issues in chaos studies in hydrology. The review mainly focuses on four major issues: selection of an optimum delay time for phase space reconstruction, minimum data size for correlation dimension estimation, effects of data noise, and influence of the presence of zeros in data. Examples are also provided to illustrate how these issues have been addressed to gain more confidence in the applications of the methods and in the interpretation of the outcomes.

12.1 Introduction

As discussed in Chap. 7, there are several important issues in the applications of the concepts and methods of chaos theory to data from real systems. Such issues are concerned with the selection of parameters involved in the methods (e.g. delay time, embedding dimension, number of neighbors) and the quantity/quality/type of observed data (e.g. data size, data noise, presence of zeros), among others. For instance: (1) use of an inappropriate delay time for phase space reconstruction may result in an overestimation or underestimation of correlation dimension; (2) small data size may result in an underestimation of correlation dimension; and (3) presence of noise in the data may overestimate the correlation dimension. Many methods and guidelines have also been proposed to address these issues. These have already been extensively discussed in the literature; see, for instance, Fraser and Swinney (1986), Holzfuss and Mayer-Kress (1986), Theiler (1986), Havstad and Ehlers (1989), Osborne and Provenzale (1989), Nerenberg and Essex (1990), Tsonis et al. (1993, 1994), and Schreiber and Kantz (1996)

B. Sivakumar, *Chaos in Hydrology*, DOI 10.1007/978-90-481-2552-4_12

for some earlier studies. These issues naturally give rise to concerns on the suitability of chaos methods for real systems and the reliability of the outcomes reported by chaos studies on real data.

It is important to recognize that the above issues are highly relevant in the applications of chaos theory in hydrology. For instance: (1) information on the optimum parameters for phase space reconstruction of hydrologic systems is not available a priori—this may result in an inaccurate implementation of chaos methods; and (2) hydrologic data are often short and are always contaminated with noise —this may influence the outcomes of chaos methods. Therefore, studying these issues is important to obtain reliable outcomes in the applications of chaos theory in hydrology. A number of studies have addressed these issues in hydrology in various ways (e.g. Berndtsson et al. 1994; Jayawardena and Lai 1994; Koutsoyiannis and Pachakis 1996; Sangoyomi et al. 1996; Porporato and Ridolfi 1997; Wang and Gan 1998; Sivakumar et al. 1999a, c, 2001a, c, 2002a, c; Jayawardena and Gurung 2000; Sivakumar 2000, 2001, 2005a; Elshorbagy et al. 2002b; Jayawardena et al. 2002, 2010; Koutsoyiannis 2006; Dhanya and Nagesh Kumar 2011; Tongal and Berndtsson 2014; Vignesh et al. 2015). This chapter attempts to review such studies. Only a brief review is presented here, especially since some of these issues and studies have already been discussed in earlier chapters; see Chap. 7 for issues and Chap. 9 for the associated hydrologic studies. Further, as the issues of delay time, data size, data noise, and presence of zeros have received far more interest than the others in hydrologic studies, such are given particular attention.

12.2 Delay Time

As highlighted in Chap. 7, several methods and guidelines have been proposed in the literature for the selection of an appropriate delay time (τ) for phase space reconstruction and any subsequent chaos analysis. Among these, the autocorrelation function method and the mutual information method have been widely used, especially in hydrology. Most of the chaos studies in hydrology have adopted either the autocorrelation function method (and generally taking τ equal to the lag time at which the autocorrelation function first crosses the zero line) (e.g. Holzfuss and Mayer-Kress 1986) or the mutual information method (and generally taking τ equal to the lag time at which the first minimum of the mutual information occurs) (e.g. Fraser and Swinney). Some studies have used both these methods and cross-verified the outcomes. Furthermore, some studies have examined the effect of τ by considering different τ values in a trial-and-error manner, either considering values around that obtained using the autocorrelation function method and/or the mutual information method or considering arbitrary values. A brief account of these studies is presented here.

Sangoyomi et al. (1996) addressed the issue of delay time selection in their investigation of chaos in the biweekly volume time series of the Great Salt Lake, USA. They used both the autocorrelation function method and the mutual information method, and found no significant difference between the τ values obtained. The autocorrelation function method yielded a τ value of 13, whereas a value of τ between 9 and 13 was obtained using the mutual information method. Therefore, they reported the dimension and scaling results for $\tau = 9$. However, they also examined the influence of τ by using different τ values for dimension estimates ($6 < \tau < 24$) and for scaling methods ($8 < \tau < 20$) and found that the results were similar to those obtained for $\tau = 9$. The appropriateness of $\tau = 9$ for chaos analysis of the GSL biweekly volume time series was further supported by Lall et al. (1996), who examined the influence of τ on the forecasting of the time series and again found that $\tau = 9$ offered the best results. Koutsoyiannis and Pachakis (1996) employed both the autocorrelation function method and the mutual information method in their analysis of rainfall data from the Ortona Lock 2 station in Florida, USA. For four different resolutions of rainfall data considered (15-min, 1-h, 6-h, and 24-h), they observed τ value varying from 1 to 6 days from the autocorrelation function method, but about 12 days from the mutual information method, especially for the finest and the coarsest resolutions. In view of these, they considered $\tau = 12$ in the subsequent analysis.

Sivakumar et al. (1998, 1999a) adopted a trial-and-error approach to investigate the effect of delay time in the analysis of chaos in rainfall data at six stations in Singapore. They, however, used the τ value obtained from the autocorrelation function method as a basis to choose the other delay times for consideration. For instance, obtaining a τ value of 10 from the autocorrelation function method for one of the stations, they used τ values of 1, 2, 8, 12, 20, and 50, in addition to 10. They observed a slight underestimation/overestimation when τ was slightly smaller/larger than that obtained from the autocorrelation function method, but significant underestimation/overestimation when τ was considerably smaller/larger. Islam and Sivakumar (2002) considered both the autocorrelation function method and the mutual information method in the analysis of river flow dynamics in the Lindenborg catchment in Denmark. The autocorrelation function yielded a τ value of 200, while the mutual information method yielded $\tau = 7$. Comparing the phase space plot for $\tau = 1$ against those for $\tau = 200$ and $\tau = 7$, they chose $\tau = 7$ in the subsequent dimension and prediction methods, as it offered a compromise.

Kim et al. (2009) used the delay-window approach (Kim et al. 1998) and the C–C method (Kim et al. 1999) to study the general dependence and, hence, assess the nonlinear characteristics of three hydrologic time series: daily streamflow series from the St. Johns River near Cocoa, Florida, USA; biweekly volume time series from the Great Salt Lake, Utah, USA; and daily rainfall series from Seoul, South Korea. They also compared the results with those obtained using the autocorrelation function method. For the flow series from the St. Johns River and also for the GSL volume series, they found different τ values from the two methods: 51 (days) and 14 (weeks) from the autocorrelation function method, whereas 89 (days) and 11 (weeks) from the C–C method. However, for the daily rainfall series from Seoul, both methods yielded the same value of $\tau = 3$ (days). Based on these results, Kim et al. (2009) suggested that: (1) both the autocorrelation function method and the C–C method perform equally well when the time series is the outcome of a linear stochastic process (with small autocorrelations, fluctuating about zero); and (2) the C–C method performs better than the autocorrelation function method when the time series is the outcome of a nonlinear (stochastic or deterministic) process. Tongal and Berndtsson (2014) used both the autocorrelation function method and mutual information function method for the selection of τ in their analysis of water level time series in the three largest lakes in Sweden: Vänern, Vättern, and Mälaren. They found that both the methods yielded similar τ values.

Vignesh et al. (2015) studied the effect of delay time in the analysis of monthly streamflow data from 639 stations in the United States towards classification of catchments using the false nearest neighbor (FNN) algorithm (Kennel et al. 1992). They considered five different τ values: three of these values ($\tau = 1$, $\tau = 3$, and $\tau = 12$) were chosen to represent the monthly, seasonal, and annual separation of elements in the phase space reconstruction vector, and the remaining two values (each may be different for different stations) were obtained using the autocorrelation function method and the mutual information method. Figure 12.1, for example, shows the relationship between the percentage of false nearest neighbors and the embedding dimension for these five different τ values for the streamflow series from the Quinebaug River at Jewett City (USGS Station #1127000) in Connecticut. The results do not indicate significant differences in the FNN dimensions, as dimension values of 4, 5, 4, 4, and 4 are observed for these τ values. Similar observations were made for most of the remaining 638 flow series as well, except for when $\tau = 1$ and perhaps $\tau = 3$, which resulted in slightly higher FNN dimensions compared to those obtained using the other three τ values; see Fig. 12.2.

Other studies that have addressed the issue of delay time in the chaos analysis of hydrologic time series include those by Rodriguez-Iturbe et al. (1989), Wang and Gan (1998), Pasternack (1999), Sivakumar (2000), Liaw et al. (2001), Sivakumar et al. (2001b), Phoon et al. (2002), Regonda et al. (2004), Koutsoyiannis (2006), Dhanya and Nagesh Kumar (2011), and Tongal et al. (2013), among others.

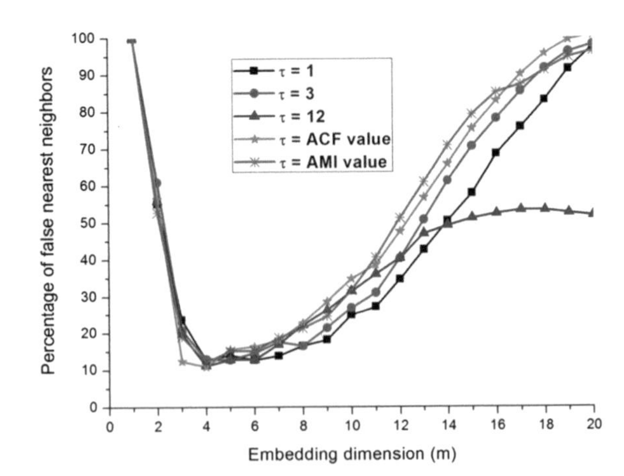

Fig. 12.1 Relationship between percentage of false nearest neighbors and embedding dimension for monthly streamflow time series from the Quinebaug River at Jewett City (USGS Station #1127000) in Connecticut, USA: effect of delay time

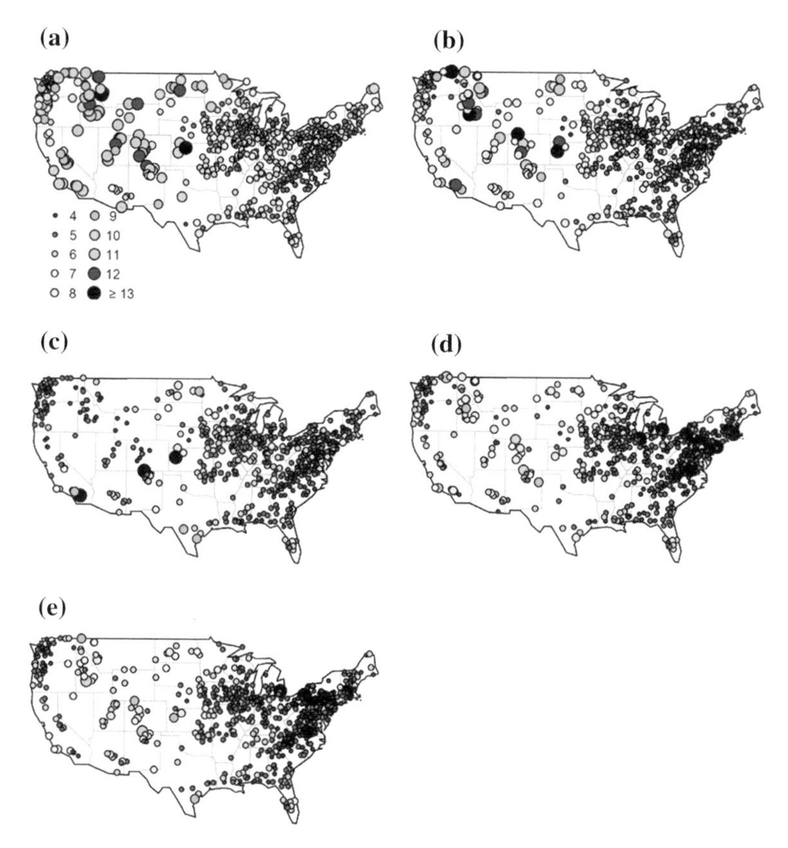

Fig. 12.2 False nearest neighbor dimensions for monthly streamflow time series from 639 stations in the United States: **a** delay time τ = 1; **b** delay time τ = 3; **c** delay time τ = 12; **d** delay time τ = ACF value; and **e** delay time τ = AMI value (*source* Vignesh et al. (2015))

12.3 Data Size

The issue of data size has received considerable attention in chaos studies in hydrology. The major reason for this are the proposed guidelines on 'minimum data size' (N_{min}) for correlation dimension estimation (e.g. Smith 1988; Nerenberg and Essex 1990; Essex 1991) as a function of embedding dimension (m) or attractor dimension (d), as such guidelines generally recommend very long time series, while hydrologic time series are often short.

Ghilardi and Rosso (1990) were the first to address the issue of data size in chaos studies in hydrology. They questioned the study by Rodriguez-Iturbe et al. (1989) on the report of low correlation dimension for rainfall (with only 1990 values), essentially based on the minimum data size requirement, $N_{min} \sim 42^m$ (Smith 1988). Tsonis et al. (1993) examined the data size issue in the chaos analysis of data representing the time between successive raingage signals each corresponding to a collection of 0.01 mm of rain. Applying the correlation dimension method and observing a dimension value of 2.4, they reported the presence of chaos. They also analyzed the Henon time series (Henon 1976) to examine the influence of embedding dimension on correlation exponent estimation, since most guidelines link the minimum data size to embedding dimension. Considering the guideline of $N_{min} \propto 10^{(2+0.4m)}$ proposed by Nerenberg and Essex (1990), they suggested that many of the studies reporting on low-dimensional attractors in weather and climate (including rainfall) come very close to satisfying the data size requirement.

In their study on the investigation of chaos in rainfall and streamflow in Hong Kong, Jayawardena and Lai (1994) adopted a trial-and-error approach to examine the data size issue, by implementing the correlation dimension method on different lengths of synthetic time series to check the reliability of the method before applying it to real rainfall and streamflow data. The results for the synthetic time series indicated that about 1200 data were sufficient for a reliable estimation of correlation dimension and gave confidence to implement the method to rainfall and streamflow data, which had 4015, 6205, and 7300 data. Sivakumar et al. (1998, 1999a) examined the effect of data size on chaos analysis of daily rainfall in six stations in Singapore. They considered eight different lengths of time series for each station (30, 20, 10, 5, 4, 3, 2, and 1 years). Their results indicated that noticeable differences in the dimensions occurred when the rainfall record length was less than 4 years (equivalent to 1461 points). Based on this, they suggested that the minimum number of data necessary to reasonably represent the dynamics of the daily rainfall process in Singapore might be about 1500. Wang and Gan (1998) investigated the effect of data size on the correlation dimension estimation of streamflow series of six rivers in the Canadian prairies. Analyzing different lengths of data and observing no significant difference in correlation dimensions between such lengths for a given embedding dimension, they suggested that sample size alone only marginally affects the correlation dimension estimation at low embedding dimensions and that the problem becomes more serious when the embedding dimension increases.

Sivakumar (2001) addressed the issue of data size in the study of rainfall dynamic behavior at different temporal scales in the Leaf River basin in Mississippi, USA. Employing the correlation dimension method to rainfall observed at daily, 2-day, 4-day, and 8-day resolutions yielded correlation dimensions of 4.82, 5.26, 6.42, and 8.87, respectively. Comparing the correlation dimension values and the coefficient of variation (CV) values of the four time series and observing an inverse relationship between the two, Sivakumar (2001) suggested that the presence of a large number of zeros at the finer-resolution time series (and the possible presence of a higher level of noise in the coarser-resolution series) might account for such a relationship. The effect of data size on the correlation dimension estimation was not evident.

Schertzer et al. (2002) raised the issue of data size in regards to the low correlation dimension results reported by Sivakumar et al. (2001a) for rainfall, river flow, and runoff coefficient time series in their study of rainfall-runoff process in the Göta River basin in Sweden. They essentially argued that the low correlation dimensions were a result of the small data sets used (1572 values) and that the rainfall-runoff process studied was indeed stochastic rather than low-dimensional chaotic. They attempted to support their claim by employing the Grassberger–Procaccia algorithm to a synthetic time series (4096 points) generated by a stochastic process and reporting a low correlation dimension (2.7–0.3). Sivakumar et al. (2002a) responded to the criticism of Schertzer et al. (2002) regarding data size in a systematic manner. Through explanation of the concept of phase space reconstruction and presentation of the correlation dimension results for a hypothetical stochastic series and an artificial chaotic series (Henon map), they assessed the reliability of the dimension results reported in Sivakumar et al. (2001a) and Schertzer et al. (2002). Based on such, they pointed out that the results reported by Schertzer et al. (2002) were significantly underestimated and that such an underestimation was not due to the small data size.

Sivakumar et al. (2002c) adopted an 'inverse approach' to address the issue of data size in the correlation dimension estimation of monthly runoff series observed at the Coaracy Nunes/Araguari River watershed in northern Brazil. According to this approach, predictions were first made using phase space reconstruction-based local approximation method and also artificial neural networks. The correlation dimension was then estimated independently and was compared with the prediction results; see also Chap. 10 (Sect. 10.2.1) for details on the correlation dimension results for this runoff series. With a runoff time series as short as only 576 values (48 years), the estimated correlation dimension of 3.62 was found to be in good agreement with the optimum embedding dimension (m_{opt}) in the phase space reconstruction prediction method (i.e. $m = 3$) and the optimum number of inputs in the artificial neural networks (3 inputs). Based on these results, Sivakumar et al. (2002c) suggested that the accuracy of the correlation dimension depends primarily on whether the time series is long enough to sufficiently represent the changes that the system undergoes over a period of time, rather than the data size in terms of the sheer number of values.

Sivakumar (2005a) presented an even more practical approach to investigate the effects of data size on the correlation dimension estimation of monthly runoff time series observed over a period of 130 years (January 1807–December 1936) in the Göta River basin in Sweden. This time series, shown in Fig. 12.3, had been studied earlier by Sivakumar et al. (2000, 2001a) to investigate the possible presence of chaos in the rainfall-runoff process. The approach adopted by Sivakumar (2005a) involved implementation of the dimension algorithm for different lengths of a given time series and inspection of the 'scaling regime' in the correlation dimension plots

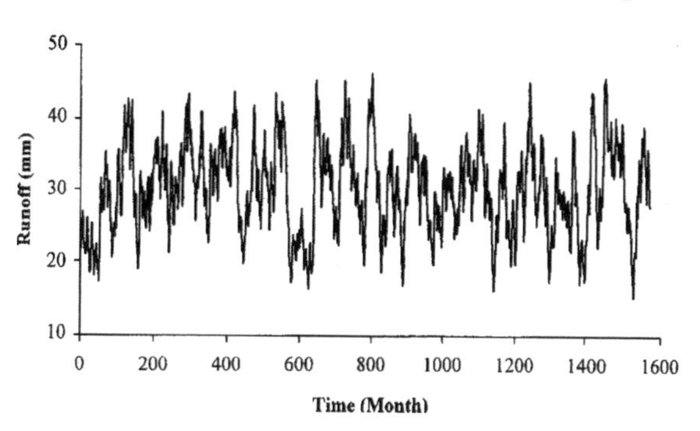

Fig. 12.3 Monthly runoff time series from Göta River basin, Sweden (*source* Sivakumar (2005a))

(and even the entire plots), somewhat similar to the approach adopted in Sivakumar et al. (1998, 1999a) for Singapore rainfall data. The procedure was first demonstrated on two artificial time series (stochastic and chaotic) (see Sect. 7.4.2 for details), and the results for these two series were then also used to interpret the results for the runoff series.

With 1560 values in the monthly runoff time series from the Göta River basin, Sivakumar (2005a) considered 12 different lengths, at a regular increasing order of 120 values, i.e. from 120 to 1560. Figure 12.4 presents the results for six of these

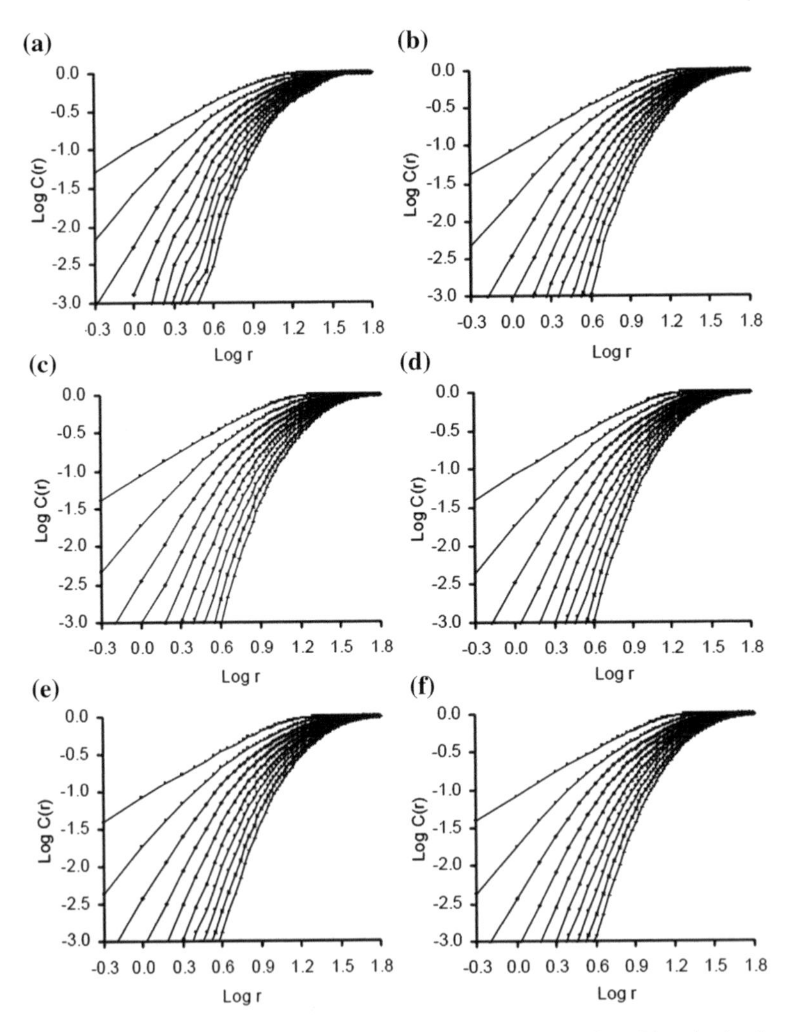

Fig. 12.4 Correlation dimension results for runoff time series from Göta River basin, Sweden: **a** 120 points; **b** 240 points; **c** 360 points; **d** 600 points; **e** 1080 points; and **f** 1560 points. Embedding dimension m = 1 to 10 (from *left* to *right*) (*source* Sivakumar (2005a))

different data lengths: 120, 240, 360, 600, 1080, and 1560. The results show that there is essentially no change in the scaling regimes for lengths of 360 points and above, irrespective of *m*, and also no noticeable difference for 240 points. Based on such results, Sivakumar (2005a) suggested that reliable estimation of the correlation dimension is possible even with a few hundred values (about 300) and further supported the study by Sivakumar et al. (2002c) regarding the (lack of) relationship between minimum data size and embedding dimension.

Several other studies have addressed the issue of data size on the implementation and outcomes of chaos methods in hydrologic time series, in one way or another. Such studies include Lorenz (1991), Tsonis (1992), Islam et al. (1993), Sangoyomi et al. (1996), Sivakumar (2000), Koutsoyiannis (2006), Hill et al. (2008), and Sivakumar et al. (2014), among others.

12.4 Data Noise

All hydrologic data are contaminated with some amount of noise, such as measurement error. Since the presence noise in the data may influence, sometimes significantly, the outcomes of chaos identification and prediction methods (e.g. Schreiber and Kantz 1996), it is important to examine such in chaos studies in hydrology. A number of studies have addressed this issue, either by noise level estimation or by noise reduction or both.

The study by Berndtsson et al. (1994) was the first to attempt nonlinear noise reduction of hydrologic time series in the context of chaos theory. Berndtsson et al. (1994) attempted noise reduction in the monthly rainfall observed over a period of 238 years in Lund, Sweden. They reported the presence of chaos in this noise-reduced rainfall series based on the observation of a correlation dimension of less than 4, while the raw rainfall data showed no evidence of chaos. Porporato and Ridolfi (1997) employed the noise reduction method of Schreiber and Grassberger (1991) to reduce the noise in the flow series of the river Dora Baltea in Italy. They applied a local averaging procedure iteratively until the mean absolute corrections between successive iterations became insignificant. The procedure was stopped after 200 iterations, since beyond that an unjustifiable calculation time was necessary to produce significant corrections. They reported improvements in the estimates of correlation dimension and prediction accuracy for the noise-reduced river flow series. Sivakumar et al. (1999b) identified four potential problems in the noise reduction procedure implemented by Porporato and Ridolfi (1997) for the river flow series from Dora Baltea. They essentially argued that the procedure adopted by Porporato and Ridolfi (1997) might have resulted in an overcorrection of the river flow data due to an inappropriate stopping criterion for the iteration procedure and, consequently, might have removed even some of the deterministic components in the data.

Sivakumar et al. (1999c), in their investigation of the influence of noise on the analysis of chaos in Singapore rainfall, proposed a systematic approach to noise reduction. Their approach combined a noise level determination method (e.g. Schouten et al. 1994) and a noise reduction method (Schreiber 1993b) for estimation of a probable noise level; see Chap. 7 (Sect. 7.4.3) for details of this approach. After demonstrating this approach on the synthetic Henon time series, they tested it on daily rainfall data observed at each of six stations in Singapore. Some results for data from one of the stations (Station 05) are presented here.

For the daily rainfall series from Station 05, application of the method of Schouten et al. (1994) yielded a noise level estimate of 4.6 %, which was considered as an initial estimate (see below). With this initial estimate, the noise reduction method of Schreiber (1993b) was employed, with several different neighborhood sizes, *r*. Figure 12.5a shows the relationship between the amount of noise removed (in terms of standard deviation) and the number of iterations. The results indicate that very small (e.g. $r = 2.50$) and very large (e.g. $r = 4.0$) neighborhood sizes are not desirable for noise reduction and, therefore, suggest the selection of a range of only intermediate neighborhood sizes (between $r = 2.60$ and $r = 3.50$). Figure 12.5b shows the prediction accuracy against the embedding dimension for the (noisy) original and the different 4.6 % noise-reduced rainfall data using different neighborhood sizes and associated number of iterations. The results show that the prediction results are improved for data resulting from 4.6 % noise reduction, and those corresponding to $r = 2.80$ are generally the best.

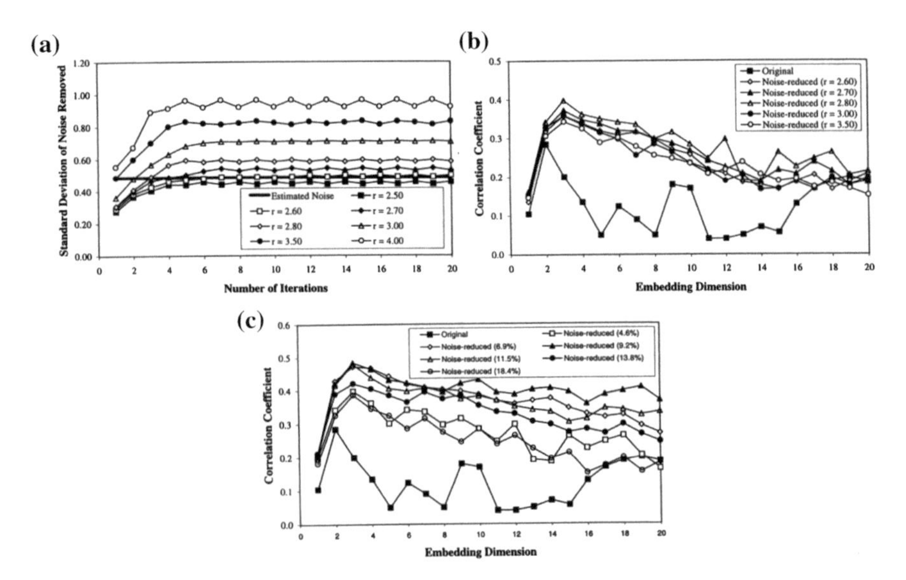

Fig. 12.5 Noise reduction results for daily rainfall data from Station 05: **a** relationship between noise level removed and number of iterations for different neighborhood sizes (*r*) with 4.6 % noise level (estimated) to be removed; and **b** relationship between correlation coefficient and embedding dimension for original and 4.6 % noise-reduced rainfall data with different neighborhood sizes; and **c** relationship between correlation coefficient and embedding dimension for original and different levels of noise-reduced rainfall data (*source* Sivakumar et al. (1999c))

Since the method of Schouten et al. (1994) was found to typically underestimate the noise level (based on the analysis of synthetic noisy Henon series), Sivakumar et al. (1999c) employed the noise reduction procedure with several higher noise levels as well: 6.9, 9.2, 11.5, 13.8, and 18.4 %. Figure 12.5c presents the prediction accuracy against the embedding dimension for the (noisy) original and these different levels of noise-reduced data. The results shown are the best results obtained for the different neighborhood sizes considered for each noise level. The best results are achieved when the noise reduction is 9.2 %, though the prediction results achieved for 6.9 and 11.5 % noise reduction closely follow this. However, a decrease in the prediction accuracy is observed when the noise level reduction is 13.8 %, and the decrease is even more clearly visible when the noise level reduction is further increased to 18.4 %, indicating the possible removal of some deterministic components at these levels of noise reduction. These results suggest that the range of the most probable noise level in the rainfall data may be between 6.9 and 11.5 %, but the most probable value seems to be around 9.2 %.

Sivakumar et al. (1999c) observed improvements in predictions of noise-reduced data for the remaining five stations as well. Considering all the six stations, the noise levels were found to range from 9 to 11 % in the daily rainfall data. This range is in good agreement with the noise levels estimated by some other means, especially for the tipping-bucket type rainfall gages (e.g. Sevruk 1996).

Jayawardena and Gurung (2000) attempted noise reduction of daily flow data from the Chao Phraya River at Nakhon Sawan in Thailand and from the Mekong River at Nong Khai in Thailand and at Pakse in Lao (in addition the Southern Oscillation Index time series). They employed three nonlinear noise reduction methods to reduce noise in these time series: Schreiber and Grassberger (1991), Grassberger et al. (1993), and Schreiber (1993b). They also used the method of Schreiber (1993a) to obtain an initial estimate of the noise level. They implemented these methods first on four synthetically generated time series: Henon series with 10 % noise and 20 % noise as well as Lorenz series (Lorenz 1963) with 10 and 20 % noise. Their results indicated that all the three noise reduction methods performed equally well but the method of Grassberger et al. (1993) was slightly superior. The prediction results obtained for the noise-reduced flow series were also found to be slightly better than those obtained for the raw flow series. Elshorbagy et al. (2002b) employed the noise reduction method of Schreiber (1993b) to the daily streamflow data observed in the English River (at Umferville, Ontario, Canada), which had earlier been identified to exhibit low-dimensional chaotic behavior (Elshorbagy et al. 2002a). They used nonlinear prediction, correlation dimension, Kolmogorov entropy, and Lyapunov exponent methods (see Chap. 6 for details of these methods) to compare the improvements after noise reduction. They found that the results regarding the presence of chaos were better for the noise-reduced flow data. However, they also cautioned that the commonly used algorithms for noise reduction in hydrologic data might also remove a significant part of the original signal and introduce an artificial chaoticity to the data.

Jayawardena et al. (2008) proposed a method to estimate the noise level in a chaotic time series and applied it to three hydrometeorologic time series.

The method, which uses a linear least-squares approach, was based on the correlation integral form obtained by Diks (1999) coupled with the special property of Kummer's confluent hypergeometric function. Jayawardena et al. (2008) implemented this method first on synthetically generated chaotic time series from the Henon map, the Lorenz equation, the Duffing equation, the Rössler equation, and the Chua's circuit, and reported better performance in identifying the noise level when compared to the performance of the method of Schreiber (1993a). They then applied the method to the Southern Oscillation Index time series, eastern equatorial Pacific sea surface temperature anomaly index, and the normalized Darwin-Tahiti mean sea level pressure differences. Jayawardena et al. (2010) proposed a method that is robust to noise to estimate the Kolmogorov-Sinai (KS) entropy, referred to as the modified correlation entropy (MCE), of a chaotic time series. The method employs the correlation integral equation obtained by Diks (1999) and Oltmans and Verheijen (1997) with Gaussian noise. The method was first applied to two synthetic time series (Lorenz and Rössler series) and then applied to two real-world river flow time series: daily flow of Mekong River at Nong Khai in Lao and daily flow of Chao Phraya at Nakhon Sawan in Thailand. Dhanya and Nagesh Kumar (2011) applied the noise reduction method of Schreiber (1993b) to streamflow data observed at two stations (Seorinarayan and Basantpur) in the Mahanadi River in India. They found no significant difference in the fraction of false nearest neighbors between the original data and the noise-reduced data, except some slight differences at low embedding dimensions. The FNN dimension was found to be the same (7) for both original and noise-reduced flow data.

Other studies that have addressed the issue of noise and/or applied noise reduction in hydrologic/hydrometeorologic time series in the context of chaos include those conducted by Jinno et al. (1995), Kawamura et al. (1998), Sivakumar (2000, 2001, 2005b), Laio et al. (2004), and Khan et al. (2005), among others.

12.5 Zeros in Data

The presence of a large number of zeros (or any other single value) in a time series may bias the outcomes of chaos methods, since the reconstructed hyper-surface in phase space will tend to a point (Tsonis et al. 1994). This issue often has enormous significance in hydrology, since it is common to observe zero values in hydrologic data. For instance, zero values are very common in rainfall, sometimes for very long stretches of periods. They are also common even during rainy periods, especially when finer resolutions (e.g. hourly) are considered. Therefore, a number of studies have addressed this issue on chaos analysis of hydrologic data.

Koutsoyiannis and Pachakis (1996) examined the influence of the dominance of zeros (dry periods) on the outcomes of chaos analysis in a 6-year record of rainfall observed at the Ortona Lock 2 station in Florida, USA. To put the analysis on a more solid footing, they also studied a 'corresponding' stochastic time series generated using a stochastic model that preserves the important properties of rainfall, such as intermittency, seasonality, and scaling behavior. Applying the correlation dimension method to 15-min, 1-, 6-, and 24-h resolutions of both the historic and the synthetic rainfall records, they reported that the applicability of such a method was limited due to the domination of voids (dry periods) in rainfall at a fine time resolution. In addition to delay time embedding, they also used Cantorian dust analogue method to estimate the dimension, and reported no substantial difference in behavior between the synthetic and the historic records. They found no evidence of low-dimensional determinism in these rainfall data.

Sivakumar et al. (1999a) adopted the surrogate data approach to examine the influence of the presence of zeros on the outcomes of chaos analysis in the daily rainfall observed at six stations in Singapore. They generated surrogate linear stochastic data that had approximately the same mean, standard deviation, number of zeros/non-zeros, and cumulative distribution as the original rainfall data. Application of the correlation dimension method to both the observed and the surrogate rainfall data sets indicated clear differences in the correlation dimensions between the two data sets and, hence, the presence of nonlinearity in the observed rainfall.

Sivakumar (2001) observed the influence of the presence of zeros on the outcomes of the correlation dimension method in the study of the dynamic behavior of rainfall at different temporal scales in the Leaf River basin in Mississippi. He studied rainfall at daily, 2-day, 4-day, and 8-day resolutions, and observed correlation dimensions of 4.82, 5.26, 6.42, and 8.87, respectively; see Fig. 12.6a. Comparison of the correlation dimension values and coefficient of variation (CV) values of the four time series revealed an inverse relationship between the two. The presence of a large number of zeros in the finer-resolution time series (and the presence of higher level of noise in the coarser-resolution time series) was considered as one of the reasons for this inverse relationship, as it could result in an underestimation of the dimension. The significance of this result was duly considered in the study by Sivakumar et al. (2001c) in the examination of the dynamic nature of transformation of rainfall process between different scales in the Leaf River basin and in the subsequent development of a chaotic disaggregation approach for rainfall; see Sect. 9.3 for further details. The study by Sivakumar (2005b) further investigated the issue of the presence of zeros on the outcomes of chaos analysis of rainfall in the Leaf River basin. For daily, 2-day, 4-day, and 8-day resolutions of rainfall, the study compared the results from the correlation dimension method and the nonlinear prediction methods. The study also carried out the correlation dimension analysis for non-zero rainfall data at the four resolutions, and obtained values of 5.92, 6.62, 8.16, and 9.46; see Fig. 12.6b. Comparison of the correlation dimension values and coefficient of variation values for the non-zero rainfall data at the four resolutions revealed an inverse relationship between the two, similar to the one observed by Sivakumar (2001) for the original data (including zeros). However, the correlation dimension values for the non-zero rainfall for the four scales (5.92, 6.62, 8.16, and 9.46) were found to be higher than the ones for the original rainfall data (4.82, 5.26, 6.42, and 8.87). These results indicate that the presence of zeros has some influence on the outcomes of the correlation dimension method (and other methods), although it is not significant.

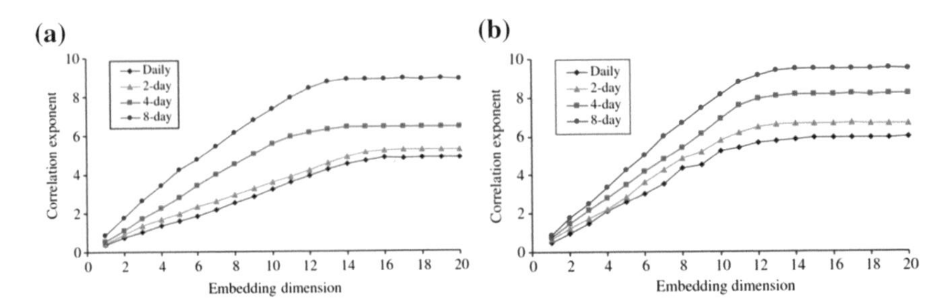

Fig. 12.6 Correlation dimension results for rainfall data of different resolutions in the Leaf River basin, Mississippi, USA: **a** data with zeros; and **b** data without zeros (*source* Sivakumar (2005b))

Sivakumar et al. (2006) observed the influence of zeros in their study of the dynamics of rainfall in California's Sacramento Valley. Applying the correlation dimension method to rainfall data at four different temporal scales (daily, weekly, biweekly, and monthly), they found no evidence of chaotic behavior in rainfall at any of these scales. However, the correlation dimension results also revealed that rainfall at coarser resolutions exhibited a higher degree of variability than rainfall at finer resolutions. Sivakumar et al. (2006) suggested that this could be due to the underestimation of the dimension for the finer-resolution series resulting from the significantly larger number of zeros when compared to the situation for coarser-resolution rainfall. To examine this further, they also studied the daily rainfall observed only during the winter (i.e. rainy period). The winter daily rainfall was found to have a higher dimension than the all-year daily rainfall, thus indicating a higher variability in the former and perhaps also indicating the influence of zeros on the dimension results.

Koutsoyiannis (2006) studied the influence of zeros on the outcomes of the correlation dimension analysis of daily rainfall observed in Vakari, western Greece. They found that the local slopes (i.e. correlation exponents) for all embedding dimensions became zero for small scales, and attributed this to the presence of a large number of zeros (60 % zero values in a total of 11,476 values) in the data. Jothiprakash and Fathima (2013) addressed the issue of zeros in their study of rainfall observed in the Koyna reservoir catchment in India. They studied daily, weekly, 10-day, monthly, and seasonal rainfall data. Their results were consistent with those reported by Sivakumar (2001) and Sivakumar et al. (2006), i.e. rainfall at coarser resolutions had a higher correlation dimension. They also compared the results for the monsoon daily (rainy season) rainfall and the all-year daily rainfall, and reported a higher correlation dimension for the monsoon rainfall, which further supported the earlier conclusion by Sivakumar et al. (2006).

Still other studies that have addressed the issue of zeros in the applications of chaos methods in hydrology include Gaume al. (2006), Sivakumar et al. (2014), and Fathima and Jothiprakash (2016), among others.

12.6 Others

While delay time, data size, data noise, and presence of zeros are some of the very important and well-known issues in the applications of chaos theory in hydrology, several other issues also exist. For instance, identification of optimal embedding dimension for phase space reconstruction and optimal neighborhood/number of neighbors in the dimension estimation/local approximation prediction plays an important role in achieving reliable outcomes of chaos analysis. Furthermore, there also exist connections between two or more of the above issues, and there may be better ways to address them. For instance, both delay time and embedding dimension are unknown parameters in the phase space reconstruction procedure. However, they can be addressed through the delay window embedding approach (e.g. Martinerie et al. 1992; Kim et al. 1998, 1999), as mentioned earlier. In view of these, several studies have addressed such issues, either independently or in a collective manner.

Phoon et al. (2002) proposed a practical inverse approach for optimal selection of the parameters involved in phase space reconstruction, chaos identification, and prediction (i.e. delay time, embedding dimension, and number of neighbors), in their study of daily river flow data from the Tryggevaelde catchment in Denmark and the Altamaha River in Georgia, USA. Considering optimum prediction accuracy as a single definite criterion, they first demonstrated the effectiveness of their approach on a theoretical chaotic time series (the Mackey-Glass series) (Mackey and Glass 1977) and then tested it on the above two river flow series. The approach was found to perform better than the standard approach (wherein one or more parameters are kept constant) both for prediction and system characterization.

Jayawardena et al. (2002) presented a new criterion, based on the generalized degrees of freedom, for optimum neighborhood selection for local modeling and prediction of river flow (and other hydrologic) series. It was first demonstrated on three artificial chaotic series (Lorenz map, Henon map, and Logistic map) and then tested on three daily river flow series: Chao Phraya River in Thailand and Mekong River in Thailand and Laos. Noise reduction was also performed on the flow series before employing the neighborhood selection scheme. The approach was found to be superior to the traditional one that arbitrarily fixes the number of neighbors.

Khan et al. (2005) examined the influence of both noise and seasonality in the chaos analysis of monthly river flow from the Arkansas River and daily river flow from the Colorado River in the United States. In their analysis for detection of chaos and predictability, they removed both noise and seasonality and compared the results obtained for such a series against those obtained for the original data. They first demonstrated the approach on synthetic Lorenz time series, by also adding white noise and seasonality. The results indicated that noise and seasonality have some influence on both correlation dimension estimation and predictability. They attributed this to the presence of thresholds, expressed in terms of noise to chaotic-signal and seasonality to chaotic-signal ratios. They suggested that the ability to detect chaos from hydrologic observations depends on whether the chaotic component is dominant enough to satisfy the thresholds.

Other studies that have addressed one or more of the above-mentioned issues include those by Lambrakis et al. (2000), Sivakumar et al. (1999a, 2002b), Hill et al. (2008), Dhanya and Nagesh Kumar (2011), and Tongal and Berndtsson (2014), among others.

12.7 Summary

Some fundamental assumptions in the development of chaos methods (infinite and noise-free time series) and absence of clear-cut guidelines for the selection of parameters involved in chaos methods (e.g. delay time, embedding dimension, number of neighbors) have led to concerns on the applications of chaos theory in hydrology. This is essentially because, real hydrologic data are often short and always contaminated with noise and the dynamic properties of the underlying systems are not known a priori. This chapter has reviewed studies that have addressed the issues in the applications of chaos methods in hydrology, with particular focus on four major issues: delay time selection, data length, data noise, and presence of zeros in data. The review reveals that many chaos studies in hydrology have exercised sufficient caution in the implementation of the methods and interpretation of the outcomes. The review also highlights that the limitations associated with chaos methods and hydrologic data are not as serious as they are generally perceived to be and that most of the outcomes reported by chaos studies in hydrology are indeed reliable.

References

Berndtsson R, Jinno K, Kawamura A, Olsson J, Xu S (1994) Dynamical systems theory applied to long-term temperature and precipitation time series. Trends Hydrol 1:291–297

Dhanya CT, Nagesh Kumar D (2011) Predictive uncertainty of chaotic daily streamflow using ensemble wavelet networks approach. Water Resour Res 47:W06507. doi:10.1029/2010WR010173

Diks C (1999) Nonlinear time series analysis, methods and applications. World Scientific, Singapore

Elshorbagy A, Simonovic SP, Panu US (2002a) Estimation of missing streamflow data using principles of chaos theory. J Hydrol 255:123–133

Elshorbagy A, Simonovic SP, Panu US (2002b) Noise reduction in chaotic hydrologic time series: facts and doubts. J Hydrol 256:147–165

Essex C (1991) Correlation dimension and data sample size. In: Schertzer D, Lovejoy S (eds) Non-linear variability in geophysics, scaling and fractals. Kluwer, Dordrecht, The Netherlands, pp 93–98

Fathima TA, Jothiprakash V (2016) Consequences of continuous zero values and constant values in time series modeling: understanding through chaotic approach. ASCE, J Hydrol Eng 05016012
Fraser AM, Swinney HL (1986) Independent coordinates for strange attractors from mutual information. Phys Rev A 33(2):1134–1140
Gaume E, Sivakumar B, Kolasinski M, Hazoumé L (2006) Identification of chaos in rainfall disaggregation: application to a 5-minute point series. J Hydrol 328(1–2):56–64
Ghilardi P, Rosso R (1990) Comment on "Chaos in rainfall". Water Resour Res 26(8):1837–1839
Grassberger P, Hegger R, Kantz H, Schaffrath C (1993) On noise reduction methods for chaotic data. Chaos 3(2):127–141
Havstad JW, Ehlers CL (1989) Attractor dimension of nonstationary dynamical systems from small data sets. Phys Rev A 39(2):845–853
Henon M (1976) A two-dimensional mapping with a strange attractor. Commun Math Phys 50:69–77
Hill J, Hossain F, Sivakumar B (2008) Is correlation dimension a reliable proxy for the number of dominant influencing variables for modeling risk of arsenic contamination in groundwater? Stoch Environ Res Risk Assess 22(1):47–55
Holzfuss J, Mayer-Kress G (1986) An approach to error-estimation in the application of dimension algorithms. In: Mayer-Kress G (ed) Dimensions and entropies in chaotic systems. Springer, New York, pp 114–122
Islam MN, Sivakumar B (2002) Characterization and prediction of runoff dynamics: A nonlinear dynamical view. Adv Water Resour 25(2):179–190
Islam S, Bras RL, Rodriguez-Iturbe I (1993) A possible explanation for low correlation dimension estimates for the atmosphere. J Appl Meteor 32:203–208
Jayawardena AW, Gurung AB (2000) Noise reduction and prediction of hydrometeorological time series: dynamical systems approach vs. stochastic approach. J Hydrol 228:242–264
Jayawardena AW, Lai F (1994) Analysis and prediction of chaos in rainfall and stream flow time series. J Hydrol 153:23–52
Jayawardena AW, Li WK, Xu P (2002) Neighborhood selection for local modeling and prediction of hydrological time series. J Hydrol 258:40–57
Jayawardena AW, Xu PC, Li WK (2008) A method of estimating the noise level in a chaotic time series. Chaos 18(2):023115. doi:10.1063/1.2903757
Jayawardena AW, Xu PC, Li WK (2010) Modified correlation entropy estimation for a noisy chaotic time series. Chaos 20:023104
Jinno K, Shiguo Xu, Berndtsson R, Kawamura A, Matsumoto M (1995) Prediction of sunspots using reconstructed chaotic system equations. J Geophys Res 100:14773–14781
Jothiprakash V, Fathima TA (2013) Chaotic analysis of daily rainfall series in Koyna reservoir catchment area, India. Stoch Environ Res Risk Assess 27:1371–1381
Kawamura A, McKerchar AI, Spigel RH, Jinno K (1998) Chaotic characteristics of the Southern Oscillation Index time series. J Hydrol 204:168–181
Kennel MB, Brown R, Abarbanel HDI (1992) Determining embedding dimension for phase space reconstruction using a geometric method. Phys Rev A 45:3403–3411
Khan S, Ganguly AR, Saigal S (2005) Detection and predictive modeling of chaos in finite hydrological time series. Nonlinear Processes Geophys 12:41–53
Kim HS, Eykholt R, Salas JD (1998) Delay time window and plateau onset of the correlation dimension for small data sets. Phys Rev E 58(5):5676–5682
Kim HS, Eykholt R, Salas JD (1999) Nonlinear dynamics, delay times, and embedding windows. Physica D 127(1–2):48–60
Kim HS, Lee KH, Kyoung MS, Sivakumar B, Lee ET (2009) Measuring nonlinear dependence in hydrologic time series. Stoch Environ Res Risk Assess 23:907–916
Koutsoyiannis D (2006) On the quest for chaotic attractors in hydrological processes. Hydrol Sci J 51(6):1065–1091
Koutsoyiannis D, Pachakis D (1996) Deterministic chaos versus stochasticity in analysis and modeling of point rainfall series. J Geophys Res 101(D21):26441–26451

Laio F, Porporato A, Ridolfi L, Tamea S (2004) Detecting nonlinearity in time series driven by non-Gaussian noise: the case of river flows. Nonlinear Processes Geophys 11:463–470

Lall U, Sangoyomi T, Abarbanel HDI (1996) Nonlinear dynamics of the Great Salt Lake: Nonparametric short-term forecasting. Water Resour Res 32(4):975–985

Lambrakis N, Andreou AS, Polydoropoulos P, Georgopoulos E, Bountis T (2000) Nonlinear analysis and forecasting of a brackish karstic spring. Water Resour Res 36(4):875–884

Lorenz EN (1963) Deterministic nonperiodic flow. J Atmos Sci 20:130–141

Lorenz EN (1991) Dimension of weather and climate attractors. Nature 353:241–244

Liaw CY, Islam MN, Phoon KK, Liong SY (2001) Comment on "Does the river run wild? Assessing chaos in hydrologic systems". Adv Water Resour 24(5):575–578

Mackey MC, Glass L (1977) Oscillation and chaos in physiological control systems. Science 197 (4300):287–289

Martinerie JM, Albano AM, Mees AI, Rapp PE (1992) Mutual information, strange attractors, and the optimal estimation of dimension. Phys Rev A 45:7058–7064

Nerenberg MAH, Essex C (1990) Correlation dimension and systematic geometric effects. Phys Rev A 42(12):7065–7074

Oltmans H, Verheijen PJT (1997) Influence of noise on power-law scaling functions and an algorithm for dimension estimations. Phys Rev E 56(1):1160–1170

Osborne AR, Provenzale A (1989) Finite correlation dimension for stochastic systems with power-law spectra. Physica D 35:357–381

Pasternack GB (1999) Does the river run wild? Assessing chaos in hydrological systems. Adv Water Resour 23(3):253–260

Phoon KK, Islam MN, Liaw CY, Liong SY (2002) A practical inverse approach for forecasting of nonlinear time series analysis. ASCE J Hydrol Eng 7(2):116–128

Porporato A, Ridolfi R (1997) Nonlinear analysis of river flow time sequences. Water Resour Res 33(6):1353–1367

Regonda S, Sivakumar B, Jain A (2004) Temporal scaling in river flow: can it be chaotic? Hydrol Sci J 49(3):373–385

Rodriguez-Iturbe I, De Power FB, Sharifi MB, Georgakakos KP (1989) Chaos in rainfall. Water Resour Res 25(7):1667–1675

Sangoyomi TB, Lall U, Abarbanel HDI (1996) Nonlinear dynamics of the Great Salt Lake: Dimension estimation. Water Resour Res 32(1):149–159

Schertzer D, Tchiguirinskaia I, Lovejoy S, Hubert P, Bendjoudi H (2002) Which chaos in the rainfall-runoff process? A discussion on 'Evidence of chaos in the rainfall-runoff process' by Sivakumar et al. Hydrol Sci J 47(1):139–147

Schouten JC, Takens F, van den Bleek CM (1994) Estimation of the dimension of a noisy attractor. Phys Rev E 50(3):1851–1861

Schreiber T (1993a) Determination of the noise level of chaotic time series. Phys Rev E 48(1): R13–R16

Schreiber T (1993b) Extremely simple nonlinear noise reduction method. Phys Rev E 47(4):2401–2404

Schreiber T, Grassberger P (1991) A simple noise reduction method for real data. Phys Lett A 160:411–418

Schreiber T, Kantz H (1996) Observing and predicting chaotic signals: is 2%noise too much? In: Kravtsov YuA, Kadtke JB (eds), Predictability of complex dynamical systems, Springer Series in Synergetics, Springer, Berlin, pp 43–65

Sevruk B (1996) Adjustment of tipping-bucket precipitation gage measurement. Atmos Res 42:237–246

Sivakumar B (2000) Chaos theory in hydrology: important issues and interpretations. J Hydrol 227 (1–4):1–20

Sivakumar B (2001) Rainfall dynamics at different temporal scales: A chaotic perspective. Hydrol Earth Syst Sci 5(4):645–651

Sivakumar B (2005a) Correlation dimension estimation of hydrologic series and data size requirement: myth and reality. Hydrol Sci J 50(4):591–604

Sivakumar B (2005b) Chaos in rainfall: variability, temporal scale and zeros. J Hydroinform 7 (3):175–184
Sivakumar B, Liong SY, Liaw CY (1998) Evidence of chaotic behavior in Singapore rainfall. J Am Water Resour Assoc 34(2):301–310
Sivakumar B, Liong SY, Liaw CY, Phoon KK (1999a) Singapore rainfall behavior: chaotic? ASCE J Hydrol Eng 4(1):38–48
Sivakumar B, Phoon KK, Liong SY, Liaw CY (1999b) Comment on "Nonlinear analysis of river flow time sequences" by Amilcare Porporato and Luca Ridolfi. Water Resour Res 35 (3):895–897
Sivakumar B, Phoon KK, Liong SY, Liaw CY (1999c) A systematic approach to noise reduction in chaotic hydrological time series. J Hydrol 219(3–4):103–135
Sivakumar B, Berndtsson R, Olsson J, Jinno K, Kawamura A (2000) Dynamics of monthly rainfall-runoff process at the Göta basin: A search for chaos. Hydrol Earth Syst Sci 4(3): 407–417
Sivakumar B, Berndttson R, Olsson J, Jinno K (2001a) Evidence of chaos in the rainfall-runoff process. Hydrol Sci J 46(1):131–145
Sivakumar B, Berndtsson R, Persson M (2001b) Monthly runoff prediction using phase-space reconstruction. Hydrol Sci J 46(3):377–388
Sivakumar B, Sorooshian S, Gupta HV, Gao X (2001c) A chaotic approach to rainfall disaggregation. Water Resour Res 37(1):61–72
Sivakumar B, Berndtsson R, Olsson J, Jinno K (2002a) Reply to 'which chaos in the rainfall-runoff process?' by Schertzer et al. Hydrol Sci J 47(1):149–158
Sivakumar B, Jayawardena AW, Fernando TMGH (2002b) River flow forecasting: use of phase-space reconstruction and artificial neural networks approaches. J Hydrol 265(1–4): 225–245
Sivakumar B, Persson M, Berndtsson R, Uvo CB (2002c) Is correlation dimension a reliable indicator of low-dimensional chaos in short hydrological time series? Water Resour Res 38(2), doi:10.1029/2001WR000333
Sivakumar B, Wallender WW, Horwath WR, Mitchell JP, Prentice SE, Joyce BA (2006) Nonlinear analysis of rainfall dynamics in California's Sacramento Valley. Hydrol Process 20 (8):1723–1736
Sivakumar B, Woldemeskel FM, Puente CE (2014) Nonlinear analysis of rainfall variability in Australia. Stoch Environ Res Risk Assess 28(1):17–27
Smith LA (1988) Intrinsic limits on dimension calculations. Phys Lett A 133(6):283–288
Theiler J (1986) Spurious dimension from correlation algorithms applied to limited time series data. Phys Rev A 34:2427–2432
Tongal H, Berndtsson R (2014) Phase-space reconstruction and self-exciting threshold modeling approach to forecast lake water levels. Stoch Environ Res Risk Assess 28(4):955–971
Tongal H, Demirel MC, Booij MJ (2013) Seasonality of low flows and dominant processes in the Rhine River. Stoch Environ Res Risk Assess 27:489–503
Tsonis AA (1992) Chaos: from theory to applications. Plenum Press, New York
Tsonis AA, Elsner JB, Georgakakos KP (1993) Estimating the dimension of weather and climate attractors: important issues about the procedure and interpretation. J Atmos Sci 50:2549–2555
Tsonis AA, Triantafyllou GN, Elsner JB, Holdzkom JJ II, Kirwan AD Jr (1994) An investigation on the ability of nonlinear methods to infer dynamics from observables. Bull Amer Meteor Soc 75:1623–1633
Vignesh R, Jothiprakash V, Sivakumar B (2015) Streamflow variability and classification using false nearest neighbor method. J Hydrol 531:706–715
Wang Q, Gan TY (1998) Biases of correlation dimension estimates of streamflow data in the Canadian prairies. Water Resour Res 34(9):2329–2339

Part IV
A Look Ahead

Chapter 13
Current Status

Abstract A review of chaos studies in hydrology over the past three decades indicates that we have explored a broad spectrum of hydrologic processes, problems, and data issues. The review also reveals that we now possess an ample level of understanding of the concepts and methods and are far more confident in applying the methods and interpreting the outcomes. This chapter discusses the current status of chaos theory in hydrology. In particular, five different aspects are considered for discussion: our ability to reliably identify the presence of chaos in hydrologic data; our ability to obtain better predictions of hydrologic data using chaos methods, especially when compared against other approaches; our success in extending the applications of chaos theory to several problems beyond simple identification and prediction; our knowledge of the limitations and concerns associated with chaos methods; and the discussions and debates that have and continue to improve our understanding of chaos theory.

13.1 Introduction

The preceding chapters (in Part C) make it abundantly clear that there has been a noticeable progress in the applications of chaos theory in hydrology, despite the fact that the theory is still in a fairly exploratory stage when compared to the far more established deterministic and stochastic theories. The inroads we have made, especially in recent years, including in the areas of scaling, parameter estimation, catchment classification, climate change, as well as data issues, are particularly significant, since these are arguably among the most important topics in hydrologic studies at the current time.

The preceding chapters also bring to light some important merits of chaos theory in the study of hydrologic systems. First, in the absence of knowledge of system equations (deterministic theories require system equations), chaos theory offers a more simplified view of hydrologic phenomena when compared to the view offered by stochastic theories. Second, chaos theory has been found to provide results that are often better than the results obtained using other theories in hydrologic

B. Sivakumar, *Chaos in Hydrology*, DOI 10.1007/978-90-481-2552-4_13

predictions, especially in the short term, although this cannot be generalized. Third, with its fundamental concepts of nonlinear interdependence, hidden order and determinism, and sensitivity to initial conditions, chaos theory can connect the deterministic and stochastic theories and serve as a more reasonable middle-ground between these two dogmatic and extreme views of nature.

These observations and opportunities are certainly a motivation for further applications of chaos theory in hydrology. Therefore, there is every reason to believe that chaos studies in hydrology will continue to grow in the future, both in width and depth. However, for such an endeavor to be particularly fruitful, it is also important to clearly identify the critical areas of application and the associated challenges. Such an identification, in turn, requires a sound knowledge of the current status. This chapter presents an overview of where we stand now in the applications of chaos theory in hydrology.

13.2 Reliable Identification

One of the major criticisms on chaos studies in hydrology, especially those conducted in earlier years, has been concerned with the identification of chaos in hydrologic time series. Questions on chaos studies in hydrology and the reported outcomes have been based either on the limitations of chaos identification methods or on issues associated with hydrologic data. The limitations in chaos identification methods are in regards to, for example, the assumptions involved in their development, the ability of the methods to provide reliable results when applied to hydrologic (or any other real) time series, and lack of clear-cut guidelines on the selection of parameters involved in the methods (e.g. delay time, embedding dimension, number of neighbors). The issues associated with hydrologic data include data size, data noise, and presence of zeros, among others. Oftentimes, however, it is the combination of the two that raises serious concerns.

As an example, the correlation dimension method was developed based on the assumption of infinite and noise-free time series (e.g. Grassberger and Procaccia 1983a, b). However, hydrologic time series are always finite (and often short) and contaminated by noise. It is indeed true that the concern has not been about the 'finite' and 'noisy' nature of the hydrologic time series, since an 'infinite' and 'noise-free' time series simply does not exist in hydrology (or in other real systems). Rather, the concern has been, for example, that hydrologic time series are not long enough to satisfy the guidelines on the minimum number of data (specifically linking data length to embedding dimension/correlation dimension) (e.g. Smith 1988; Nerenberg and Essex 1990) for a reliable estimation of the correlation dimension; see Schertzer et al. (2002) and Sivakumar et al. (2002a) for a discussion.

At the outset, concerns about the limitations of chaos identification methods and issues associated with hydrologic data may indeed have some merits. For instance, underestimation or overestimation of correlation dimension may occur when an improper selection of delay time is chosen for phase space reconstruction. The fact that there is no clear-cut guideline on the selection of delay time makes the phase space reconstruction problem complicated. If the time series under investigation is rather short, then this leads to further complications, as the number of reconstructed vectors is often small (especially when the delay time is large), which means that there may be an insufficient number of points in the phase space for a reliable estimation of the correlation exponent. These observations clearly indicate that criticisms about chaos studies and the reported outcomes cannot be dismissed altogether. What is basically needed is proper caution in the implementation of the methods and interpretation of the outcomes. Yet another way to gain more confidence in such studies is through applying different methods, examining data issues, and cross-verifying, and possibly confirming, the outcomes.

A large number of studies have employed more than one method in their investigation of chaos in hydrologic time series to have more confidence in their analysis and results regarding the presence/absence of chaos (e.g. Rodriguez-Iturbe et al. 1989; Jayawardena and Lai 1994; Porporato and Ridolfi 1996, 1997; Puente and Obregon 1996; Sivakumar et al. 1999a, 2000, 2002b; Elshorbagy et al. 2002a; Islam and Sivakumar 2002; Dhanya and Nagesh Kumar 2010; Kyoung et al. 2011; Khatibi et al. 2012). The methods employed include correlation dimension method, false nearest neighbor method, Lyapunov exponent method, Kolmogorov entropy method, Poincaré map, close returns plot, nonlinear prediction method (with an inverse approach for identification), and also surrogate data method and method of redundancy (for detection of nonlinearity). Some studies have used only the identification methods for cross-verification of results, while some others have used both identification and prediction methods. In almost all these studies, the results from the different methods have consistently indicated the presence (or absence) of chaos in the hydrologic time series under investigation. In yet other cases, the results reported by some studies have been subsequently verified by others, through application of additional methods or analysis of related data (e.g. Porporato and Ridolfi 1996, 1997; Sivakumar 2002a; Sivakumar and Wallender 2005; Hossain and Sivakumar 2006; Hill et al. 2008).

It is also important to note that several studies, even when employing only a single method (or more) for chaos identification in hydrologic time series, have addressed some important methodological and data issues, as appropriate, to obtain more reliable results. For instance, studies have investigated the effects of delay time, data size, data noise, and presence of zeros (e.g. Berndtsson et al. 1994; Jayawardena and Lai 1994; Koutsoyiannis and Pachakis 1996; Porporato and Ridolfi 1997; Kim et al. 1998, 1999; Sivakumar et al. 1998, 1999a, b, 2002c; Wang and Gan 1998; Pasternack 1999; Jayawardena and Gurung 2000; Sivakumar 2001a, 2005a, b; Jayawardena et al. 2002; Phoon et al. 2002; Koutsoyiannis 2006). In some of these studies, new approaches have also been developed to address these issues (e.g. Sivakumar et al. 1999b; Jayawardena et al. 2002; Phoon et al. 2002). In almost all of these cases, examination of the methodological limitations and data issues has helped further strengthen the basic results regarding the presence (or absence) of chaos in the time series under study, with additional insights into the dynamic behavior and predictability.

The above observations clearly indicate that many studies, if not all, on chaos identification and prediction in hydrologic time series have exercised sufficient caution in the implementation of the methods and interpretation of the outcomes. Indeed, the efforts made by many studies to overcome some of the potential limitations of chaos methods and data-related issues and to offer better interpretations, especially through proposal of new approaches, have certainly played an important role in advancing chaos studies in hydrology. There is no question that additional improvements can still be made, and are indeed necessary, to further enhance our understanding of the concepts and workings of chaos methods and to more accurately realize the potential of chaos theory in hydrology. Nevertheless, looking at the ability of the chaos tools at our disposal, quantity and quality of hydrologic data available, and the efforts we have already made, one can confidently say that we have the ability to reliably identify the presence (or absence) of chaos in hydrologic time series. This, in turn, suggests that, it should be possible to obtain, especially for short term, better predictions (than those possible using stochastic approaches, for example) of hydrologic time series, if the time series is identified to exhibit chaotic behavior.

13.3 Encouraging Predictions

A fundamental goal in identifying the presence of chaotic behavior in a time series is to obtain important insights into the evolution of the underlying system and the possibility of better predictions than that are already available through other approaches, especially in the short term. To this end, chaos studies in hydrology and the reported outcomes are certainly encouraging.

A large number of studies have employed chaos theory-based prediction of hydrologic time series, in particular the nonlinear local approximation prediction method (e.g. Farmer and Sidorowich 1987; Casdagli 1989, 1992). A majority of these studies have attempted chaos theory-based prediction only after the time series had been identified to exhibit chaotic behavior (e.g. Jayawardena and Lai 1994; Porporato and Ridolfi 1997; Sivakumar et al. 1999a, 2002b; Lambrakis et al. 2000; Islam and Sivakumar 2002; Sivakumar and Wallender 2005; Khatibi et al. 2012). This approach is certainly a more appropriate one to undertake for making predictions. Some other studies have employed chaos theory-based prediction method without initially checking for the presence of chaos in the time series, i.e. they have made no prior assumption about the presence (or absence) of chaos (e.g. Porporato and Ridolfi 2001; Sivakumar 2002a, 2003; Laio et al. 2003). It is important to note, however, that these studies have essentially aimed, in addition to chaos-based prediction, to identify the presence of chaos in the time series (through an inverse approach using the prediction results) or to compare the predictive ability of chaos-based methods with that of other approaches (e.g. stochastic methods, artificial neural networks, support vector machines).

Generally speaking, applications of chaos theory-based local approximation prediction methods have shown their suitability and effectiveness for predictions of hydrologic time series. The methods have yielded extremely good predictions for most of the streamflow, sediment, and water stage time series studied (e.g. Porporato and Ridolfi 1997; Lambrakis et al. 2000; Lisi and Villi 2001; Sivakumar et al. 2001b, 2002b; Jayawardena and Gurung 2000; Islam and Sivakumar 2002; Sivakumar 2002a; Laio et al. 2003; Sivakumar and Wallender 2005; Khatibi et al. 2012). However, this is not the case for some other hydrologic time series, such as rainfall and runoff coefficient (e.g. Jayawardena and Lai 1994; Sivakumar et al. 1999a, b, 2000), for which the prediction results are generally poor. As the presence of noise in the data often affects the performance of chaos-based prediction methods, some studies have employed noise-reduction techniques prior to predictions and reported further improvements for all the time series studied, including rainfall and streamflow (e.g. Porporato and Ridolfi 1997; Sivakumar et al. 1999b; Jayawardena and Gurung 2000; Elshorbagy et al. 2002b).

Since a basic aim in searching for chaos in hydrologic time series is to study if chaos theory-based methods can lead to better predictions, a number of studies have also compared the chaos-based predictions with those achieved using other approaches (e.g. Jayawardena and Lai 1994; Jayawardena and Gurung 2000; Lambrakis et al. 2000; Lisi and Villi 2001; Sivakumar et al. 2002b; Dhanya and Nagesh Kumar 2011; Tongal and Berndtsson 2014). Among the approaches used for comparison are stochastic methods, artificial neural networks, support vector machines, and wavelets. A majority of such comparison studies have reported that chaos theory-based methods outperform the other approaches in predictions of hydrologic time series, especially in short-term predictions.

The above observations are both encouraging and important in the context of hydrologic modeling and prediction, since they further emphasize the role of chaotic behavior in the evolution of hydrologic processes and the importance of identifying the presence of chaos in hydrologic time series, so as to develop more suitable, and perhaps alternative, modeling paradigms and methods. While there still remain many challenges in the prediction of hydrologic processes, whether using chaos theory or using other approaches, the utility of chaos theory-based prediction methods, especially for short-term predictions, is abundantly clear.

13.4 Successful Extensions

One of the main challenges in introducing and applying a new scientific theory, developed in some other field, in hydrology lies in identifying the areas/problems in which such a theory will have particular utility when compared to the already existing ones, and how. Since most of the ideas and methods of chaos theory are based on the concepts of 'system' and the time series representing it (well-known concepts in hydrology as well), their introduction in hydrology is fairly straightforward. In essence, as time series of data representing processes occurring in catchments (systems) are available, the task is basically to apply the methods to the time series to identify the behavior of the underlying system and to predict its evolution. This is why there has been a plethora of studies on identification of chaos in hydrologic time series and their prediction. The real challenge, however, is to identify important problems in hydrology where a straightforward application of chaos methods is not possible but their extensions or modifications could be found suitable. The following presents examples of such extensions.

As hydrologic data at very fine temporal (and spatial) scales are not widely available, disaggregation of data available at coarse resolution (e.g. daily) to fine resolution (e.g. hourly) is an important problem in hydrology. However, the methods available for chaos identification and prediction cannot be directly applied to the disaggregation problem. Sivakumar et al. (2001c) presented an extension of chaos theory for the purpose of disaggregation of rainfall (see Chap. 9, Sect. 9.3). They proposed a disaggregation approach that is somewhat similar to the approach adopted in local approximation prediction, but with modifications necessary for the disaggregation problem. Instead of using the actual rainfall time series at the two scales involved in disaggregation, they used the time series of distributions of weights of rainfall values between the scales. Application of this approach to rainfall data in the Leaf River basin, Mississippi, USA suggested its suitability and effectiveness for rainfall disaggregation; see also Sivakumar (2001b) for an argument in favor of a chaotic multi-fractal approach for rainfall. The approach was then also successfully applied for disaggregation of streamflow in the Mississippi River basin, at St. Louis, Missouri (Sivakumar et al. 2004) and for disaggregation of sediment load data in the same basin (Sivakumar and Wallender 2004); see Chaps. 9 (Sect. 9.3) and 10 (Sect. 10.3) for further details.

Classification of catchments has been an important problem in hydrology, as it is useful for selection of an appropriate model complexity and interpolation/extrapolation of data, among others. However, since chaos methods are generally used to identify/predict the behavior of a particular system (through single- or multi-variable time series of the system), their utility to compare different systems in the specific context of classification is rarely studied in most fields. This presented an opportunity to extend the ideas of chaos theory for the purpose of (catchment) classification in hydrology. Some early attempts have been made in this direction by studying streamflow time series from different catchments (e.g. Krasovskaia et al. 1999; Sivakumar et al. 2007; Sivakumar and Singh 2012; Vignesh et al. 2015), and, in some cases, a very large number of catchments. Employing phase space reconstruction, correlation dimension, and false nearest neighbor methods and using the associated qualitative/quantitative measures as indicators for comparison and classification, the studies have reported encouraging results on the utility and ability of the methods for classification purposes (see also Chaps. 9 (Sect. 9.4) and 10 (Sect. 10.4) for some details). Some other studies (e.g. Liu et al. 1998; Sivakumar 2003; Sivakumar et al. 2014), while not specifically addressing classification, have also used chaos methods to study spatial variability of hydrologic processes, which forms an important basis for classification. Indeed, these studies and the classification approach can supplement and complement the ideas underlying the dominant processes concept (DPC) in hydrology (e.g. Grayson and Blöschl 2000), and can also help in simplying the parameter estimation problem in hydrologic models (Sivakumar 2004b).

These observations clearly indicate that chaos studies in hydrology have not just been doing the 'same-old-thing' of identifying and predicting chaotic behavior in hydrologic time series, a perception that may still exist among a small section of the hydrologic community. Rather, such studies have been addressing some important and serious problems that are at the front and center of hydrology. Indeed, in addition to the above examples of disaggregation and catchment classification, extensions have also been attempted in the context of reconstruction of hydrologic system equations (e.g. Jinno et al. 1995; Zhou et al. 2002) and parameter estimation and uncertainty in hydrologic models (e.g. Hossain et al. 2004; Dhanya and Nagesh Kumar 2011; see also Sivakumar 2004b), among others. The success of these extensions certainly augers well for still other key problems in hydrology, including for studying the spatio-temporal dynamics of hydrologic processes, downscaling of global climate model outputs, and the interactions between hydrologic systems and other Earth systems.

13.5 Limitations and Concerns

Although, as the above sections indicate, significant progress has been achieved in understanding and applying the concepts of chaos in hydrology, potential limitations in the studies and the associated concerns deserve serious consideration. At the heart of this issue is the fact that we do not have complete prior knowledge of the exact structure and working of hydrologic systems, obtaining which is indeed the task at hand. This deficiency hampers, at least to a certain extent, our ability to verify if the chaos methods we employ are actually suitable and effective and if the results we obtain are accurate or reliable. This can be explained through an example in the context of estimation of correlation dimension and its reliability.

For synthetically-generated time series (chaotic or other), the dynamic properties of the underlying system, such as the correlation dimension, are known a priori. Since long and noise-free synthetic time series can be easily generated, determining the minimum length of data required for a reliable estimation of correlation dimension is fairly straightforward. Since the actual correlation dimension is known, one can simply estimate the correlation dimensions for different lengths of the time series and then identify the length at (or above/below) which the correlation dimension is equal to the actual dimension or based on other qualitative/ quantitative criterion (e.g. Sivakumar 2005a). Even when one may not have information on the optimum values of parameters (e.g. optimum delay time for phase space reconstruction), one can use a trial-and-error approach to identify the optimal parameters, since the task at hand is to essentially obtain the actual correlation dimension, which is already known.

For hydrologic (or any other real) time series, however, this procedure does not guarantee accurate identification of data size, since the actual correlation dimension in itself is not known. While one may indeed use a trial-and-error approach and use the criterion of 'no-change' in the correlation dimension to overcome this difficulty in the identification of minimum data size (e.g. Sivakumar et al. 1998, 1999a; Sivakumar 2005a), it is also important to recognize that such is reliable only in the absence of other influencing factors. As for hydrologic time series, since there usually are several other factors that influence the correlation dimension estimation, such as delay time selection for phase space reconstruction (there are no clear-cut guidelines) and noise in the data (whose level and effects are not accurately known), and the use of 'no-change' criterion, although useful, cannot be considered fool-proof; see also Sivakumar et al. (1999a, b) for details.

Indeed, all the existing chaos identification and prediction methods possess limitations, in their own ways, when applied to real hydrologic time series. The issues of minimum data size, noise in the data, and optimum delay time for phase space reconstruction are almost universal. All the methods are influenced, in one way or another, by these three issues, the only exception being the irrelevance of delay time in the close returns plot. However, the nature and extent of their effects may vary widely. For instance: (1) the correlation dimension method and the Lyapunov exponent method seem to require a longer time series when compared to the nonlinear prediction method and the close returns plot; (2) noise has more influence on the outcomes of the nonlinear prediction method than it does on the correlation dimension method; and (3) selection of an appropriate delay time seems to be more important in the correlation dimension method than in the nonlinear prediction method (where a very small delay time often seems desirable). Some other issues may be relevant only to specific methods. For instance, the issue of the presence of zeros in the data is relevant to the correlation dimension method (and a few others), since it may result in an underestimation of the attractor dimension in both methods (Tsonis et al. 1993, 1994; Sivakumar 2001a), but seems to have no relevance to the nonlinear prediction method.

On one hand, the above observations point out that there are indeed some genuine concerns in the applications of chaos methods in hydrology and emphasize the need to seriously consider such concerns for us to have more confidence in our studies and the reported outcomes. On the other hand, however, they also highlight that generalization of such concerns to all chaos studies in hydrology and rejection of the reported outcomes are unwarranted. In other words, it is important to offer/heed to 'warnings' on chaos studies in hydrology, but it is also equally important to make sure that such warnings are only 'proper' and not 'false alarms.'

13.6 Discussions and Debates

Since the introduction of chaos theory in hydrology in the late 1980s, there have been several discussions and debates on chaos studies in hydrology and the reported outcomes; see, for example, between Ghilardi and Rosso (1990) and Rodriguez-Iturbe et al. (1990), Sivakumar et al. (1999c) and Porporato and Ridolfi (1999), Schertzer et al. (2002) and Sivakumar et al. (2002a), and Sivakumar (2002b) and Elshorbagy et al. (2002c). Such discussions and debates have involved a whole range of issues, both methodology-related and data-related, and largely their combination. Overall, they have raised some important questions on chaos studies and, consequently, led to greater scrutiny of the methods employed, data used, and outcomes reported. This, no doubt, has also helped advance our understanding of the concepts and methods of chaos theory and their applications in hydrology, especially in identifying the limitations of the methods, in recognizing potential hydrologic problems for their applications, and in interpreting the outcomes (e.g. Sivakumar 2000, 2004a). Indeed, such discussions have also helped identify potential 'false alarms' about chaos studies in hydrology; see, for example, Schertzer et al. (2002) and Sivakumar et al. (2002a), regarding the study of Sivakumar et al. (2001a) on chaos in rainfall-runoff process.

An important consequence of the criticisms on chaos studies in hydrology and subsequent scrutiny is the finding that the limitations of chaos methods are not as serious as they are (at least were) generally perceived to be and that their applications to hydrologic time series can, and do, yield reliable outcomes. Of particular significance is the finding regarding the issue of data size, i.e. chaos methods can be reliable even when applied to a 'short' hydrologic time series, as long as the time series is long enough to adequately represent the system dynamics (e.g. Sivakumar et al. 2002a, c; Sivakumar 2005a). This finding has certainly played a crucial role in advancing chaos studies in hydrology in recent years, especially in identifying potential new areas and problems for applications, including the extensions highlighted earlier.

As of now, we have a reasonably good knowledge of the concepts and methods of chaos theory as well as their merits and limitations in hydrology. There indeed still remain some unresolved issues, and there could potentially be some more as we continue to expand the breadth and width of hydrologic applications. Therefore, there is no question that discussions and debates on chaos studies in hydrology will continue for at least some time to come. Nevertheless, it is fair to say that we have already established a strong base to build on. The new application areas provide great opportunities to further enhance the role of chaos theory in hydrology, and there may come some new challenges along the way as well. It is certainly helpful to keep an open mind and discuss and debate the role of chaos theory in hydrology.

13.7 Summary

During the past three decades or so, there has been an enormous progress in chaos theory in hydrology. Chaos theory methods have been applied to analyze different hydrologic data and to study different hydrologic problems. There is now clearly a great level of understanding of chaos concepts and methods. This chapter has presented an overview of the current status of chaos studies in hydrology. It has discussed the reliability of studies on chaos identification and prediction in hydrologic time series and the inroads we have made in extending the earlier studies to address many other hydrologic problems. Past and ongoing concerns about chaos studies in hydrology and the efforts undertaken to address such have also been highlighted. The status of chaos theory in hydrology is certainly encouraging, especially considering the fact that the theory is still in an exploratory stage when compared to the far more established deterministic and stochastic theories. The current status also augers well for the continued growth of chaos theory in hydrology. Some of the key areas for further advancement will be discussed in Chap. 14.

References

Berndtsson R, Jinno K, Kawamura A, Olsson J, Xu S (1994) Dynamical systems theory applied to long-term temperature and precipitation time series. Trends Hydrol 1:291–297

Casdagli M (1989) Nonlinear prediction of chaotic time series. Physica D 35:335–356

Casdagli M (1992) Chaos and deterministic versus stochastic nonlinear modeling. J Royal Stat Soc B 54(2):303–328

Dhanya CT, Nagesh Kumar D (2010) Nonlinear ensemble prediction of chaotic daily rainfall. Adv Water Resour 33:327–347

Dhanya CT, Nagesh Kumar D (2011) Predictive uncertainty of chaotic daily streamflow using ensemble wavelet networks approach. Water Resour Res 47:W06507. doi:10.1029/2010WR010173

Elshorbagy A, Simonovic SP, Panu US (2002a) Estimation of missing streamflow data using principles of chaos theory. J Hydrol 255:123–133

Elshorbagy A, Simonovic SP, Panu US (2002b) Noise reduction in chaotic hydrologic time series: facts and doubts. J Hydrol 256:147–165

Elshorbagy A, Panu US, Simonovic SP (2002c) Reply. Hydrol Sci J 47(3):528–531

Farmer DJ, Sidorowich JJ (1987) Predicting chaotic time series. Phys Rev Lett 59:845–848

Ghilardi P, Rosso R (1990) Comment on "Chaos in rainfall". Water Resour Res 26(8):1837–1839

Grassberger P, Procaccia I (1983a) Measuring the strangeness of strange attractors. Physica D 9:189–208

Grassberger P, Procaccia I (1983b) Characterisation of strange attractors. Phys Rev Lett 50(5):346–349

Grayson RB, Blöschl G (2000) Spatial patterns in catchment hydrology: observations and modeling. Cambridge University Press, Cambridge, UK

Hill J, Hossain F, Sivakumar B (2008) Is correlation dimension a reliable proxy for the number of dominant influencing variables for modeling risk of arsenic contamination in groundwater? Stoch Environ Res Risk Assess 22(1):47–55

Hossain F, Sivakumar B (2006) Spatial pattern of arsenic contamination in shallow wells of Bangladesh: regional geology and nonlinear dynamics. Stoch Environ Res Risk Assess 20(1–2):66–76
Hossain F, Anagnostou EN, Lee KH (2004) A non-linear and stochastic response surface method for Bayesian estimation of uncertainty in soil moisture simulation from a land surface model. Nonlinear Process Geophys 11:427–440
Islam MN, Sivakumar B (2002) Characterization and prediction of runoff dynamics: a nonlinear dynamical view. Adv Water Resour 25(2):179–190
Jayawardena AW, Gurung AB (2000) Noise reduction and prediction of hydrometeorological time series: dynamical systems approach vs. stochastic approach. J Hydrol 228:242–264
Jayawardena AW, Lai F (1994) Analysis and prediction of chaos in rainfall and stream flow time series. J Hydrol 153:23–52
Jayawardena AW, Li WK, Xu P (2002) Neighborhood selection for local modeling and prediction of hydrological time series. J Hydrol 258:40–57
Jinno K, Xu S, Berndtsson R, Kawamura A, Matsumoto M (1995) Prediction of sunspots using reconstructed chaotic system equations. J Geophys Res 100(A8):14773–14781
Khatibi R, Sivakumar B, Ghorbani MA, Kişi Ö, Kocak K, Zadeh DF (2012) Investigating chaos in river stage and discharge time series. J Hydrol 414–415:108–117
Kim HS, Eykholt R, Salas JD (1998) Delay time window and plateau onset of the correlation dimension for small data sets. Phys Rev E 58(5):5676–5682
Kim HS, Eykholt R, Salas JD (1999) Nonlinear dynamics, delay times, and embedding windows. Physica D 127(1–2):48–60
Koutsoyiannis D (2006) On the quest for chaotic attractors in hydrological processes. Hydrol Sci J 51(6):1065–1091
Koutsoyiannis D, Pachakis D (1996) Deterministic chaos versus stochasticity in analysis and modeling of point rainfall series. J Geophys Res 101(D21):26441–26451
Krasovskaia I, Gottschalk L, Kundzewicz ZW (1999) Dimensionality of Scandinavian river flow regimes. Hydrol Sci J 44(5):705–723
Kyoung MS, Kim HS, Sivakumar B, Singh VP, Ahn KS (2011) Dynamic characteristics of monthly rainfall in the Korean peninsula under climate change. Stoch Environ Res Risk Assess 25(4):613–625
Laio F, Porporato A, Revelli R, Ridolfi L (2003) A comparison of nonlinear flood forecasting methods. Water Resour Res 39(5). 10.1029/2002WR001551
Lambrakis N, Andreou AS, Polydoropoulos P, Georgopoulos E, Bountis T (2000) Nonlinear analysis and forecasting of a brackish karstic spring. Water Resour Res 36(4):875–884
Lisi F, Villi V (2001) Chaotic forecasting of discharge time series: a case study. J Am Water Resour Assoc 37(2):271–279
Liu Q, Islam S, Rodriguez-Iturbe I, Le Y (1998) Phase-space analysis of daily streamflow: characterization and prediction. Adv Water Resour 21:463–475
Nerenberg MAH, Essex C (1990) Correlation dimension and systematic geometric effects. Phys Rev A 42(12):7065–7074
Pasternack GB (1999) Does the river run wild? Assessing chaos in hydrological systems. Adv Water Resour 23(3):253–260
Phoon KK, Islam MN, Liaw CY, Liong SY (2002) A practical inverse approach for forecasting of nonlinear time series. J Hydrol Eng 7(2):116–128
Porporato A, Ridolfi L (1996) Clues to the existence of deterministic chaos in river flow. Int J Mod Phys B 10:1821–1862
Porporato A, Ridolfi R (1997) Nonlinear analysis of river flow time sequences. Water Resour Res 33(6):1353–1367
Porporato and Ridolfi (1999) Reply to "Comment on 'Nonlinear analysis of river flow time sequences' by Amilcare Porporato and Luca Ridolfi". Water Resour Res 35(3):899–901
Porporato A, Ridolfi R (2001) Multivariate nonlinear prediction of river flows. J Hydrol 248(1–4):109–122

Puente CE, Obregon N (1996) A deterministic geometric representation of temporal rainfall. Results for a storm in Boston. Water Resour Res 32(9):2825–2839
Rodriguez-Iturbe I, De Power FB, Sharifi MB, Georgakakos KP (1989) Chaos in rainfall. Water Resour Res 25(7):1667–1675
Rodriguez-Iturbe I, De Power FB, Sharifi MB, Georgakakos KP (1990) Reply. Water Resour Res 26(8):1841–1842
Schertzer D, Tchiguirinskaia I, Lovejoy S, Hubert P, Bendjoudi H (2002) Which chaos in the rainfall-runoff process? A discussion on 'Evidence of chaos in the rainfall-runoff process' by Sivakumar et al. Hydrol Sci J 47(1):139–147
Sivakumar B (2000) Chaos theory in hydrology: important issues and interpretations. J Hydrol 227(1–4):1–20
Sivakumar B (2001a) Rainfall dynamics at different temporal scales: a chaotic perspective. Hydrol Earth Syst Sci 5(4):645–651
Sivakumar B (2001b) Is a chaotic multi-fractal approach for rainfall possible? Hydrol Process 15(6):943–955
Sivakumar B (2002a) A phase-space reconstruction approach to prediction of suspended sediment concentration in rivers. J Hydrol 258:149–162
Sivakumar B (2002b) Discussion—analysis of cross-correlated chaotic streamflows. Hydrol Sci J 47(3):523–527
Sivakumar B (2003) Forecasting monthly streamflow dynamics in the western United States: a nonlinear dynamical approach. Environ Model Softw 18(8–9):721–728
Sivakumar B (2004a) Chaos theory in geophysics: past, present and future. Chaos Soliton Fract 19(2):441–462
Sivakumar B (2004b) Dominant processes concept in hydrology: moving forward. Hydrol Process 18(12):2349–2353
Sivakumar B (2005a) Correlation dimension estimation of hydrologic series and data size requirement: myth and reality. Hydrol Sci J 50(4):591–604
Sivakumar B (2005b) Chaos in rainfall: variability, temporal scale and zeros. J Hydroinform 7(3):175–184
Sivakumar B, Singh VP (2012) Hydrologic system complexity and nonlinear dynamic concepts for a catchment classification framework. Hydrol Earth Syst Sci 16:4119–4131
Sivakumar B, Wallender WW (2004) Deriving high-resolution sediment load data using a nonlinear deterministic approach. Water Resour Res 40:W05403. doi:10.1029/2004WR003152
Sivakumar B, Wallender WW (2005) Predictability of river flow and sediment transport in the Mississippi River basin: a nonlinear deterministic approach. Earth Surf Process Landf 30:665–677
Sivakumar B, Liong SY, Liaw CY (1998) Evidence of chaotic behavior in Singapore rainfall. J Am Water Resour Assoc 34(2):301–310
Sivakumar B, Liong SY, Liaw CY, Phoon KK (1999a) Singapore rainfall behavior: chaotic? ASCE J Hydrol Eng 4(1):38–48
Sivakumar B, Phoon KK, Liong SY, Liaw CY (1999b) A systematic approach to noise reduction in chaotic hydrological time series. J Hydrol 219(3–4):103–135
Sivakumar B, Phoon KK, Liong SY, Liaw CY (1999c) Comment on "Nonlinear analysis of river flow time sequences" by Amilcare Porporato and Luca Ridolfi. Water Resour Res 35(3): 895–897
Sivakumar B, Berndtsson R, Olsson J, Jinno K, Kawamura A (2000) Dynamics of monthly rainfall-runoff process at the Göta basin: a search for chaos. Hydrol Earth Syst Sci 4(3): 407–417
Sivakumar B, Berndttson R, Olsson J, Jinno K (2001a) Evidence of chaos in the rainfall-runoff process. Hydrol Sci J 46(1):131–145
Sivakumar B, Berndtsson R, Persson M (2001b) Monthly runoff prediction using phase-space reconstruction. Hydrol Sci J 46(3):377–388
Sivakumar B, Sorooshian S, Gupta HV, Gao X (2001c) A chaotic approach to rainfall disaggregation. Water Resour Res 37(1):61–72

Sivakumar B, Berndtsson R, Olsson J, Jinno K (2002a) Reply to 'which chaos in the rainfall-runoff process?' by Schertzer et al. Hydrol Sci J 47(1):149–158

Sivakumar B, Jayawardena AW, Fernando TMGH (2002b) River flow forecasting: use of phase-space reconstruction and artificial neural networks approaches. J Hydrol 265(1–4): 225–245

Sivakumar B, Persson M, Berndtsson R, Uvo CB (2002c) Is correlation dimension a reliable indicator of low-dimensional chaos in short hydrological time series? Water Resour Res 38(2). doi:10.1029/2001WR000333

Sivakumar B, Wallender WW, Puente CE, Islam MN (2004) Streamflow disaggregation: a nonlinear deterministic approach. Nonlinear Process Geophys 11:383–392

Sivakumar B, Jayawardena AW, Li WK (2007) Hydrologic complexity and classification: a simple data reconstruction approach. Hydrol Process 21(20):2713–2728

Sivakumar B, Woldemeskel FM, Puente CE (2014) Nonlinear analysis of rainfall variability in Australia. Stoch Environ Res Risk Assess 28(1):17–27

Smith LA (1988) Intrinsic limits on dimension calculations. Phys Lett A 133(6):283–288

Tongal H, Berndtsson R (2014) Phase-space reconstruction and self-exciting threshold modeling approach to forecast lake water levels. Stoch Environ Res Risk Assess 28(4):955–971

Tsonis AA, Elsner JB, Georgakakos KP (1993) Estimating the dimension of weather and climate attractors: important issues about the procedure and interpretation. J Atmos Sci 50:2549–2555

Tsonis AA, Triantafyllou GN, Elsner JB, Holdzkom JJ II, Kirwan AD Jr (1994) An investigation on the ability of nonlinear methods to infer dynamics from observables. Bull Amer Meteor Soc 75:1623–1633

Vignesh R, Jothiprakash V, Sivakumar B (2015) Streamflow variability and classification using false nearest neighbor method. J Hydrol 531:706–715

Wang Q, Gan TY (1998) Biases of correlation dimension estimates of streamflow data in the Canadian prairies. Water Resour Res 34(9):2329–2339

Zhou Y, Ma Z, Wang L (2002) Chaotic dynamics of the flood series in the Huaihe River Basin for the last 500 years. J Hydrol 258:100–110

Chapter 14
The Future

Abstract The tremendous progress that has been achieved, through three decades of research, in the applications of chaos theory in hydrology inevitably leads to questions regarding the future of chaos theory in hydrology. Of particular interest is to identify potential areas for further applications and advancement of the theory and possible ways to achieve fruitful outcomes. This chapter addresses these questions. In light of some of the research questions at the forefront of hydrology at the current time and will be in the future, and also looking at some studies that have already addressed these questions from the perspective of chaos theory (albeit rudimentary), several different areas are identified to further advance chaos theory in hydrology. These are: parameter estimation in hydrologic models, simplification in hydrologic model development, integration of different concepts in hydrology, development of catchment classification framework, extensions of chaos studies using multiple hydrologic variables, reconstruction of hydrologic system equations, and downscaling of global climate models. Finally, the need and the potential to establish reliable links between chaos theory, hydrologic data, and hydrologic system physics are also discussed.

14.1 Introduction

As highlighted in Chap. 13, the last three decades of research on the applications of chaos theory in hydrology has provided a strong base to build on further. In addition to the basic problem of chaos identification and prediction in hydrologic time series, we have made notable progress in the applications of chaos theory to several other areas and problems that are at the forefront of hydrologic research, including scaling and catchment classification framework. Nevertheless, some important areas in hydrology remain almost untouched.

Looking at recent and current studies in hydrology, it is fair to say that hydrologic research is struggling to find a balance between competing paradigms, motives, and interests. For instance, there is an increasing realization on the need to find a middle ground between (or at least interpretations to link) our deterministic

B. Sivakumar, *Chaos in Hydrology*, DOI 10.1007/978-90-481-2552-4_14

worldview on one hand and stochastic worldview on the other (e.g. Vogel 1999; Sivakumar 2004a, 2009). There is a growing tendency to develop more and more complex hydrologic models for individual catchments to represent as much of their details as possible (e.g. Beven 2002), but recognition of the difficulties in obtaining relevant data for these models and the associated issues (e.g. parameter estimation) is leading to general frameworks and simplified models (e.g. Grayson and Blöschl 2000; McDonnell and Woods 2004). There is enormous interest in applying specific scientific concepts and sophisticated mathematical methods, but also great interest in integrating different concepts and methods for a broader and more inclusive perspective (e.g. Sivakumar 2008a). A brief commentary on these issues can be found in Sivakumar (2008c).

In view of these, this chapter identifies some key areas where chaos theory can play a crucial role in the future in advancing hydrology further. These areas address model development, modeling issues, data analysis, and finally linking models, data, and physics. Indeed, not all of these areas are completely new to chaos theory, as some have been addressed in the past. However, considering the significance of these areas in current and future hydrologic research and also the relevance and ability of chaos tools to study them, one can only conclude that chaos theory-based studies are nowhere near where they can be and need to be.

14.2 Parameter Estimation

With significant technological and methodological advances over the last few decades, there has been a growing tendency to develop more and more complex hydrologic models. Many of the so-called 'physically-based' models are an excellent example for this. While these models are indeed useful for a more reliable representation of the hydrologic systems, they also require more details about processes and more parameters to be calibrated, which makes the parameter estimation problem extremely challenging. Constructive discussions and debates on this issue, especially on the identification of the best optimization technique and on the estimation of uncertainty in hydrologic models, have been on the rise in recent years (e.g. Beven and Young 2003; Gupta et al. 2003; Beven 2006; Sivakumar 2008b; Beven et al. 2012; Clark et al. 2012). Indeed, parameter estimation in hydrologic models is currently among the most important topics in hydrologic research. As there seems to be no end to our tendency to develop more and more complex models, the problem of parameter estimation will be even more important in the future.

A plethora of approaches for parameter estimation and uncertainty exists and is used in hydrologic models. Some of these approaches that have been widely used in the hydrologic literature in recent years include: the generalized likelihood uncertainty estimation (GLUE) framework (Beven and Binley 1992), Bayesian recursive estimation technique (BaRE) (Thiemann et al. 2001), Shuffled Complex Evolution Metropolis (SCEM) algorithm (Vrugt et al. 2003), dynamic identifiability analysis

(DYNIA) framework (Wagener et al. 2003), and data assimilation (DA) framework (Liu and Gupta 2007). These approaches and their many variants, among themselves, involve different assumptions, scientific concepts, and mathematical sophistication, and, consequently, possess different merits and limitations. There is certainly scope for alternative approaches, either as stand-alone approaches or as supplemental approaches, to further improve the efficiency and effectiveness of parameter estimation in hydrologic models. To this end, ideas from chaos theory can be one suitable alternative.

Research into the application of the ideas of chaos theory for parameter estimation in hydrologic models is almost non-existent. The very few attempts made thus far in this direction are certainly encouraging. For instance, Hossain et al. (2004), in their study of Bayesian estimation of uncertainty in soil moisture simulation by a land surface model, presented a simple and improved sampling scheme (within a Monte Carlo simulation framework) to GLUE by explicitly recognizing the nonlinear deterministic behavior between soil moisture and land surface parameters in the stochastic modeling of the parameters' response surface. They approximated the uncertainty in soil moisture simulation (i.e. model output) through a Hermite polynomial chaos expansion of normal random variables that represent the model's parameter (model input) uncertainty. They reported that their new scheme was able to reduce the computational burden of random Monte Carlo sampling for GLUE in the range of 10–70 % and about 10 % more efficient than the nearest neighborhood sampling method in predicting a sampled parameter set's degree of representativeness. A similar Hermite polynomial chaos expansion-based approach by Hossain and Anagnostou (2005) for uncertainty analysis of streamflow prediction by the TOPMODEL (Beven and Kirkby 1979) also yielded very encouraging outcomes, with about 15–25 % reduction in computational burden when compared to the uniform sampling for GLUE. The study by Sivakumar (2004b), addressing model simplification, proposed an approach that incorporates and integrates chaos theory (especially the correlation dimension method) with expert advice and parameter optimization to alleviate certain difficulties associated with conventional parameter estimation (see Sect. 14.3 for details). In view of these developments and encouraging outcomes, it is foreseeable that chaos theory will find an important place in parameter estimation in hydrologic models.

14.3 Model Simplification

In view of the issues and concerns associated with highly complex hydrologic models for individual catchments (including data constraints, parameter estimation, and extensions to other catchments and generalization), many studies during the past decade or so have emphasized the need for simplification in modeling as well as a common framework in hydrology (e.g. Young et al. 1996; Grayson and Blöschl 2000; Beven 2002; McDonnel and Woods 2004; Wainwright and Mulligan 2004). They advocate, in one way or another, moving beyond the notion of 'modeling everything' and adopting the notion of 'capturing the essential features.' Their argument, in essense, is: "... we should be developing methods to identify dominant processes that control hydrologic response (in different environments, landscapes and climates, and at different scales) and then developing models to focus on these dominant processes."

The above argument for model simplification can also be supported by, among others: (1) our knowledge through general observations that often only a few processes dominate hydrologic response in a given catchment, depending on the climate and other factors; and (2) our experience through modeling, parameter estimation, and prediction that simple models with only a few dominant parameters could capture the essential features of a given catchment's response to hydrologic events. For example, the case studies presented in Grayson and Blöschl (2000), representing an extraordinary range of environments, dominant processes, catchment sizes, data types, and modeling approaches, reveal that a single process dominates the hydrologic responses. Hydrologic literature is replete with further support to the role of one or a few dominant processes, regardless of the systems studied and the nature of the concepts and methods adopted (e.g. Michaud and Sorooshian 1994; Young and Beven 1994; Hsu et al. 1995; Coulibaly et al. 2001; Young and Parkinson 2002). It is indeed appropriate to note that this is also a fundamental idea of chaos theory, i.e. seemingly complex and random phenomena might also be the result of simple systems with only a few nonlinear interdependent variables with sensitivity to initial conditions.

While the realization that consideration of only a "few" dominant processes may be sufficient for modeling is certainly encouraging, determination of this "number" and the identification of the processes themselves are not straightforward. A logical way to deal with this is by evaluating the sensitivity of the system to each of the individual processes that are believed to have influence and for which data are also available (or can be measured). In essence, this procedure starts with the "most possible complex situation" (i.e., combination of all processes) and moves towards the "simplest reliable solution" (i.e., combination of dominant processes), through a trial-and-error elimination method. This procedure has been the cornerstone of conventional sensitivity analysis and parameter estimation studies in hydrology. Although generally reliable, this procedure is expensive from the perspectives of data, time, and computer requirements. In cases where a much larger number of relevant influencing processes (and hence data) are involved, implementation of the conventional procedure becomes tremendously difficult.

These observations clearly reflect the need for a better procedure for identification of the dominant processes and, hence, simplification of models. Obviously, such a procedure should not only be able to overcome some of the complexities and costs involved in the conventional procedure but also provide results that are comparable. One possible way to achieve this is through devising a procedure that starts with the "simplest reliable situation" and moves towards the "most complex potentially required solution." This does not mean that one must always start with just one process and include additional ones as needed on a trial-and-error basis, since this procedure may also become highly inefficient. Rather, it means that one must find a suitable method to first reliably determine the number of dominant processes from only the available (often limited amount of) data representing the system, so that this number can serve as a reliable starting point for data collection and sensitivity analysis; it must be noted, however, that this number could well end up being the optimum one too.

A number of methods already exist in the literature for determining the number of dominant processes from only a limited amount of available data. Indeed, many of the time series techniques (or data-based methods more broadly) belong to this category, including those that are based on the concept of 'dimensionality.' As highlighted in Chaps. 5 and 6, the dimension of a time series is, in a way, a representation of the number of dominant variables present in the evolution of the corresponding dynamic system. Since many of these methods can often be used to describe a multi- (and often large-) dimensional system using a single-variable series through data reconstruction concept, such as phase space reconstruction (Packard et al. 1980), they are particularly useful for model simplification. There is ample proof in the literature for the appropriateness and usefulness of data reconstruction concept for systems that are highly nonlinear and as complex as that of fluid turbulence and weather (e.g. Takens 1981; Fraedrich 1986; Tsonis and Elsner 1988), and hydrologic processes, as described in the preceding chapters.

In light of this, Sivakumar (2004b) used the ideas of chaos theory, especially the correlation dimension method, to address the issue of model simplification, proposing three steps: (1) determination of the number of dominant processes governing the system, using the correlation dimension method; (2) identification of the dominant processes through expert (especially field) knowledge; and (3) sensitivity analysis to arrange the dominant processes in the order of their extent of dominance on the system. These three steps have advantages of, respectively, requiring data of only a single variable representing the system, honoring practical (field) reality, and significantly reducing data collection, time, and computer costs. The fact that the correlation dimension method provides a reliable estimate of the number of dominant processes (e.g. Sivakumar 2000, 2005a; Sivakumar et al. 2002; Hill et al. 2008), this procedure can be very efficient, as it starts with the "simplest reliable situation" and moves towards the "most complex potentially required solution," if at all inclusion of additional information is needed. Although its effectiveness still remains to be tested and its superiority over others to be verified, the approach is clearly an example as to how the dimension concept can be useful, in tandem with other concepts, for model simplification in hydrology. The fact that chaos theory is fundamentally a simplified view of studying complex systems provides further support to the role of this theory for model simplification. With concerns about the development of highly complex hydrologic models growing and the need for simplification realized, there is every reason to believe that chaos theory will play a prominent role in future research in this direction.

14.4 Integration of Concepts

As of now, there exists no single scientific concept that can accurately describe everything (e.g. structure, function) about hydrologic systems. The different scientific concepts that exist now make different assumptions and, thus, possess different advantages and limitations in the study of hydrologic systems. In this situation, a sensible way to advance our understanding of hydrologic systems may be by combining or integrating different concepts, as this could help maximize the advantage and minimize the limitation. In other words, the "probability" of success achieved from integration of concepts would generally be greater than that can be achieved from one particular concept, provided the advantages and limitations of different concepts are well understood. This realization has, in recent years, led many studies to integrate two or more different concepts for modeling and prediction of hydrologic systems (e.g. See and Openshaw 2000; Chen and Adams 2006; Jain and Srinivasulu 2006; Nasr and Bruen 2008; Wu et al. 2009; Alvisi and Franchini 2011), some of which also address combining the so-called 'black-box' models with 'conceptual' or 'physically-based' models. Indeed, there have been arguments even in favor of our largely opposite worldviews and concepts (e.g. deterministic versus stochastic) for their supplementary and complementary roles (e.g. Vogel 1999; Sivakumar 2004a, 2009).

Despite the numerous applications of chaos theory in hydrology and recognition of its potential role in supplementing and complementing other theories (e.g. Sivakumar 2004a, 2009), there has not been much effort to combine chaos concepts with others to study hydrologic systems. Thus far, only a very small number of studies have attempted or proposed such an integration, including for model simplification, parameter estimation, and prediction uncertainty (e.g. Hossain et al. 2004; Sivakumar 2004b; Hossain and Anagnostou 2005; Dhanya and Nagesh Kumar 2011), but such attempts are still in the very early stages. Nevertheless, there is every reason to believe that research in this direction will soon start to grow fast, as there is a growing need for integrating different concepts to optimize our data collection as well as time and computational resources. To this end, both the parallels and the non-parallels many of the chaos-theory based concepts (e.g. nonlinearity, determinism, dimensionality, attractor, bifurcation, fractal, sensitive dependence, predictability) have with others (e.g. linearity, random, principal component, threshold, self-organized criticality, scaling, information content) should certainly help identify where and how effective integration is possible.

Having said that, there are some challenges in integrating different concepts. For instance, any attempt at such integration requires us to have an adequate knowledge of the different concepts/methods in the first place, so that we will be in a position to choose the appropriate ones for integration. This, however, is turning out to be very difficult because of the existence of numerous concepts and our tendency to focus on specific ones ("specialization"). With different concepts often adopting different terminologies (even to represent similar ideas and procedures), communications among researchers in hydrology has become increasingly difficult. Sivakumar

(2005b) explains this difficulty with an example of the role of “thresholds” in hydrologic systems and the various implicit forms it takes in hydrologic literature, depending up on the method/area of study, such as “critical states” in studies on self-organization and criticality (e.g. Rigon et al. 1994; Rodriguez-Iturbe and Rinaldo 1997), “characteristic patterns” in studies on self-organizing maps and artificial neural networks (e.g. Hsu et al. 2002), and “regimes” in studies on non-linear determinism and chaos (e.g. Sivakumar 2003), in addition to the explicit form it takes in some studies (e.g. Crozier 1986; Caine 1990; Reichenbach et al. 1998). The situation is not very different when it comes to, for example, the definitions and the modeling procedures adopted under different areas/methods of hydrologic research, or even within the same; see Refsgaard and Henriksen (2004) for details. Addressing these issues, and related ones, is important to achieve proper progress in the integration of concepts in hydrology.

14.5 Catchment Classification Framework

Catchment classification has been considered as an important means to achieve a common modeling framework in hydrology (e.g. McDonnell and Woods 2004). The basic idea in catchment classification is to streamline catchments into different groups and sub-groups based on their salient characteristics (e.g. system, process, data properties) and to develop suitable methods/models so that the outcomes can then be used for prediction, decision-making, and other catchment-related purposes. Catchment classification is particularly useful for identification of appropriate complexity of models for different types of catchments and for interpolation/extrapolation, including predictions in ungaged basins. Although catchment classification framework had been addressed as early as in the 1930s (Pardé 1933) and more so since the 1960s (e.g. Beckinsale 1969; Budyko 1974; Gottschalk et al. 1979; Haines et al. 1988; Nathan and McMahon 1990), there has been particular interest since the beginning of this century (e.g. Olden and Poff 2003; Snelder et al. 2005; Isik and Singh 2008; Moliere et al. 2009; Kennard et al. 2010; Ali et al. 2012; Sivakumar and Singh 2012). This interest has been driven by the need to address the concerns in our tendency to develop highly complex models for individual catchments and to offer better communication among researchers within and across different scientific disciplines (e.g. McDonnell and Woods 2004; Sivakumar 2008a; Young and Ratto 2009), among others.

Research into the development of a catchment classification framework has resulted in different approaches and methods for catchment classification. These include river/flow regimes, hydroclimatic factors, river morphology, hydrologic similarity indexes, hydrologic signatures, landscape and land use parameters, ecohydrologic and geomorphic factors, hydropedological factors, geostatistical properties, entropy, scale properties, data-based mechanistic strategies, data-driven methods, and many others; see Olden et al. (2012), Razavi and Coulibaly (2013), and Sivakumar et al. (2015) for some recent accounts of such approaches, challenges in their applications, and directions for further research on catchment classification. As for the specific role of chaos theory, some attempts have been made, in recent years, to apply the ideas of chaos theory for catchment classification purposes (e.g. Krasovskaia et al. 1999; Sivakumar et al. 2007; Sivakumar and Singh 2012; Vignesh et al. 2015) (see also Chap. 10, Sect. 10.4 for some details), or for spatial variability that can form an important basis for classification (e.g. Liu et al. 1998; Sivakumar 2003; Sivakumar et al. 2014). These studies have used phase space reconstruction, correlation dimension method, false nearest neighbor algorithm, and local approximation prediction method.

While the outcomes of the above studies are generally encouraging, answers to some key questions continue to elude: (1) what should be the basis for a catchment classification framework? (2) what components need to be included? (3) what is the appropriate methodology for formulation? and (4) how can a catchment classification framework be effectively formulated and verified? Sivakumar et al. (2015) have attempted to address these questions in more detail. In particular, as for the methodology, they have highlighted the usefulness of nonlinear dynamic and chaos theories and related concepts (e.g. complex network theory), especially based on past attempts (e.g. Sivakumar et al. 2007; Sivakumar and Singh 2012). Nevertheless, there remain important challenges in the implementation of the methodologies and interpretation of the results, as highlighted here.

A particular issue in chaos studies on catchment classification thus far is in regards to the use of single-variable time series (especially streamflow) for phase space reconstruction and subsequent analysis for classification, since what is essentially required is an analysis based on multiple variables (e.g. rainfall, streamflow, evaporation); see also Sect. 14.6. While a multi-variable chaos analysis in itself may not be a difficult task, as it has already been done in hydrology for chaos identification and prediction (e.g. Porporato and Ridolfi 2001; Laio et al. 2003; Jin et al. 2005; Sivakumar et al. 2005), its implementation and interpretation in the context of classification remains a question. Another relevant question to ask is whether catchment attributes (e.g. drainage area, elevation, slope) can be included in such an analysis. Finally, it is also important to verify if the classification achieved using one chaos method (e.g. correlation dimension method) is the same as that achieved from another chaos method (e.g. false nearest neighbor algorithm) or a method based on a different scientific concept. As the development of a generic catchment classification framework ranks among the most important and interesting topics in hydrology at the current time, there is no question that chaos theory will find a prominent role in research in this direction in the years to come.

14.6 Multi-variable Analysis

A particular advantage of chaos theory-based methods is that they are largely able to represent a multi- and (often large-) dimensional system using only a single-variable time series. This representation is normally done through a "pseudo" state space reconstruction, called "phase space reconstruction" (e.g. Packard et al. 1980); see Chap. 5 for additional details. The basic idea behind this reconstruction is that a nonlinear system is characterized by self-interaction, and that a series of a single variable can carry the information of the dynamics of the entire multi-variable system. There is ample proof in the literature for the appropriateness and usefulness of this data reconstruction concept for systems that are highly nonlinear and complex, including hydrologic systems, as presented in Chaps. 9 through 11. While this single-variable data reconstruction approach is commonly used (both in hydrology and in other fields), methods for multi-variable reconstruction do exist (e.g. Cao et al. 1998), and have been used in hydrology as well (Porporato and Ridolfi 2001; Laio et al. 2003; Jin et al. 2005; Sivakumar et al. 2005). It is appropriate to note, at this point, that use of a single-variable time series to study complex systems is not just limited to chaos-based methods but common to almost all "time series" or "data-based" methods.

The use of a single-variable time series to represent a multi- and large-dimensional hydrologic system can be defended based on the following: (1) Streamflow at the outlet (or any other point) of a catchment is essentially the outcome of whatever happens in the catchment (in the sense of hydrology). Therefore, streamflow time series alone should provide very useful information about the working of the catchment. The key factor to consider here is the selection of the variable, i.e. the variable that can adequately represent the system dynamics; and (2) Generally, it is not possible to observe all the variables relevant to the system. This could be either due to the lack of knowledge of all the influencing variables or due to resource constraints (e.g. measurement devices). Therefore, oftentimes, observations are made of only one (or a few variables), which are then used for studying the system dynamics.

Nevertheless, there are important concerns on the continued use of a single-variable time series to study complex hydrologic systems. Such concerns come at least from two angles: (1) There are no clear-cut guidelines on the selection of parameters (e.g. delay time, embedding dimension) involved in the phase space reconstruction method (see Chap. 7). Therefore, an adequate representation of a multi-dimensional system based on a single-variable time series through phase space reconstruction is often difficult to achieve; and (2) With developments in technology and measurement devices, for many hydrologic systems, observations of more than one variable are either available or at least can be estimated (e.g. rainfall, streamflow, evaporation). Therefore, use of only a single-variable time series, instead of the available multi-variable time series, for system representation does no longer make sense.

Most of the chaos studies in hydrology, thus far, have essentially used the single-variable phase space reconstruction approach. There is, therefore, certainly an enormous scope for use of a multi-variable phase space reconstruction approach for identification of chaotic behavior in hydrologic processes and for subsequent applications, including prediction, disaggregation, and classification. The outcomes of the very few studies that have employed such a multi-variable approach in hydrology are also encouraging (e.g. Porporato and Ridolfi 2001; Laio et al. 2003; Jin et al. 2005; Sivakumar et al. 2005). With the increasing availability of data representing multiple variables from hydrologic systems and better computational power on one hand and with the need to obtain far more accurate modeling and prediction outcomes for hydrologic systems on the other, multi-variable phase space reconstruction-based chaos studies will be an important part of future chaos studies in hydrology. Such studies would also help address, and hopefully alleviate, some of the concerns in the single-variable phase space reconstruction studies and the reported outcomes.

14.7 Reconstruction of System Equations

Arguably, the most fundamental challenge in hydrology (and any scientific field, for that matter) is an accurate derivation of the governing equation(s) for a given system. In general, there are two broad ways to addressing this problem: (1) Deduction—based on theory: and (2) Induction—based on observations. The deductive approach starts out with a theory or equation and then makes predictions based on this theory and finally uses the observations to verify, and confirm, if the theory was correct. The inductive approach is the opposite of the deductive approach. It starts out with making specific observations and then discerns patterns, and finally makes a generalization and infer a theory. With these, different modeling approaches make different assumptions about the system governing equations or adopt different data analysis methods, as the case may be.

It is fair to say that chaos studies in hydrology generally belong to the inductive approach. However, they are only partial in this regard, since they have mainly focused on the analysis of data, recognition of patterns, and chaos identification/prediction, without making any serious attempt to derive the governing equations. The only exceptions to this are the studies by Jinno et al. (1995) and Zhou et al. (2002). Jinno et al. (1995) attempted to reconstruct monthly sunspot numbers. They used a modified form of the Rössler equation as reference system equations, as the sunspot numbers' attractor, amplitude, and pseudoperiod were found to be similar to that of the Rössler attractor. Zhou et al. (2002) attempted reconstruction of the flood series in the Huaihe River Basin in China using the concepts of chaos theory and the inverted theorem of differential equations. Although these studies have highlighted how concepts of chaos theory can be used for reconstructing governing equations for hydrologic systems, there are also important questions about their assumptions and outcomes. For example, there are concerns about: (1) what assumptions to make for the reference system when hydrologic data do not exhibit attractors and other properties similar to that of any artificial chaotic system; and (2) the large number of coefficients in the reconstructed equations even in very low dimensions; see also Sivakumar (2004a).

Notwithstanding these concerns, the studies by Jinno et al. (1995) and Zhou et al. (2002) offer a good base and can lead to a more informed and realistic path to reconstruction of governing equations for hydrologic systems. For instance, a number of hydrologic time series from around the world, especially streamflow and sediment time series, are found to exhibit simple and well-defined attractors having a dimension less than three; see, for example, Sivakumar et al. (2007). In light of the methodological (and computational) developments during the past decade or so, especially in the field of complex systems science, it is possible to further refine the approaches employed in earlier studies on system reconstruction. To this end, our knowledge about yet other modeling concepts (both deductive-based and inductive-based), and especially the equations associated with them, should also help. While research in this direction will continue to be challenging, there is certainly scope and hope for advancement.

14.8 Downscaling of Global Climate Model Outputs

Global climate change is anticipated to have threatening consequences for our water resources, both at the global and at the local levels (IPCC 2014). Although the exact impacts of climate change are hard to predict, there is a broad consensus among scientists that the global hydrologic cycle will intensify and that extremes (e.g. floods, droughts) will occur more frequently and often with greater magnitudes. An important step in assessing the impacts of climate change on our water resources is the 'downscaling' of coarse-scale global climate model (GCM) outputs to fine-scale hydrologic data suitable for hydrologic predictions.

For downscaling GCM outputs, two broad approaches are employed: (1) Statistical downscaling—this approach uses an equation to represent the relationship between large-scale model behavior and small-scale phenomena, which may be obtained from change factors, regression models, weather typing schemes, and weather generators; and (2) Dynamical downscaling—in this approach, a high-resolution climate model is embedded within a GCM, in the form of a regional climate model (RCM) or a limited area model (LAM). Extensive details of these approaches are already available in the literature (e.g. Wilby and Wigley 1997; Fowler et al. 2007). Although either of these approaches can provide reasonable outcomes, the accuracy depends strongly on the quality of the GCM simulations and the nature of the transformation (i.e. downscaling) function. In general, however, the statistical approaches do not adequately take into account the nonlinear characteristics of connections between large-scale climate and small-scale catchment variables, while the dynamical approaches are computationally demanding. In addition, neither of these approaches gives sufficient consideration to the chaotic nature of relationship between the large-scale climate system and the small-scale hydrologic system, although evidence as to the presence of chaotic behavior in each of these systems independently is already well documented in the literature (e.g. Lorenz 1963; Elsner and Tsonis 1993; Sivakumar 2000, 2004a).

These observations clearly highlight the need for formulation of a downscaling approach that explicitly and sufficiently recognizes the chaotic behavior of climate-hydrology connections. However, this issue has not received any attention thus far. Indeed, there has been no study that has specifically examined the presence of chaotic behavior in the GCM outputs in the context of climate change, with the exception of the study by Kyoung et al. (2011). In their study, Kyoung et al. (2011) examined the dynamic characteristics of monthly rainfall in the Korean peninsula under conditions of climate change. Studying three rainfall series (present observed —1971 to 1999; present GCM-simulated—1951 to 1999; and future GCM-simulated—2000 to 2099) using chaos-theory based methods (phase space reconstruction, correlation dimension, and close returns plot), they reported that the nature of rainfall dynamics falls more on the nonlinear chaotic dynamic spectrum than on the linear stochastic spectrum. They also reported that the future GCM-simulated rainfall exhibits stronger nonlinearity and chaos compared to the present rainfall, with fewer variables dominantly interacting among themselves, although the overall rainfall will be greater in amount and intensity.

The study by Kyoung et al. (2011) certainly provides encouragement as to the role chaos theory can play in downscaling GCM outputs and also an opportunity to pursue research in this direction further. To this end, the study by Sivakumar et al. (2001), proposing a chaotic dynamic approach for rainfall downscaling in time, may also provide some useful clues. However, some significant modifications/ extensions to this methodology are required, since the primary interest in GCM outputs is spatial downscaling and the problem is also a much more complex spatio-temporal problem. On the other hand, advances in multi-variable analysis (discussed in Sect. 14.6) could also help in dealing with these problems.

14.9 Linking Theory, Data, and Physics

Hydrologic processes arise as a result of interactions between climate inputs and landscape characteristics that occur over a wide range of space and time scales. Due to the tremendous heterogeneities in climate inputs and landscape properties, hydrologic processes are also highly variable at different space and time scales. A proper understanding of catchment functions, therefore, requires observations of different catchment processes at many different spatial and temporal scales. However, due to various reasons (e.g. absence of knowledge and resource constraints), observations of different processes at different spatial and temporal scales are almost impossible to make. Consequently, oftentimes, observations of only one or a few selected processes at only one or a few selected scales are made. This situation inevitably gives rise to a 'mismatch' between the scale of the process and the scale of the observations and, hence, leads to difficulties in linking theory, data, and process. Indeed, linking data and physics has and continues to be a fundamental challenge in hydrology; see, for example, Beven (2002), Kirchner (2006).

As highlighted earlier, chaos-theory based methods are essentially data-based, and use data reconstruction and pattern recognition steps to identify and predict chaotic behavior. Therefore, at any given scale, the relevance and adequacy of chaos studies to represent the actual physical mechanisms and dynamics in catchments may be questioned. This is especially the case with studies that adopt phase space reconstruction using only a single-variable time series. The stakes become much higher when different scales are also considered.

Although the significance of linking the analysis and outcomes involved in chaos methods to catchment physics is abundantly clear, not much attention has been given to this issue thus far. Only a very few studies have addressed this issue, in slightly different ways. For example, there have been attempts to explain the delay time to seasonal cycle and attractor complexity to time of concentration (e.g. Sivakumar et al. 2007). There have been attempts to explain the methods and outputs through multi-variable analysis and reconstruction of system equations using attractor shape and dimension (e.g. Porporato and Ridolfi 2001; Jinno et al. 1995; Zhou et al. 2002). There have also been attempts to compare the results from chaos methods with results from other approaches towards explaining the relevance and reliability of chaos methods for representing catchment physics (e.g. Sivakumar et al. 2002; Hill et al. 2008). These studies, however, remain ‘bits and pieces,’ and there is a tremendous need and scope for advancing research in this direction. Indeed, serious efforts to establish links between chaos methods/outputs, data, and catchment physics are absolutely necessary, if chaos theory is to find a key role in hydrology.

Having said that, it is also crucial to recognize that our inability to bridge the gap between theory, data, and physics is not just specific to chaos theory-based methods but is common to literally all time series methods. Consequently, there have been attempts to link theory, data, and physics in the context of several other modeling concepts as well (e.g. Klemeš 1978; Salas and Smith 1981; Parlange et al. 1992; Wilby et al. 2003; Jain et al. 2004; Sudheer and Jain 2004). Nevertheless, there is still a long way to go. There is, therefore, a great opportunity to study the theory-data-physics link from a multi-concept perspective, with chaos theory as a key component. To this end, integration of concepts, discussed earlier (Sect. 14.4), can serve as one possible means.

14.10 Summary

A review of chaos theory studies in hydrology over the past three decades reveals that we have come a long way. Starting from basic identification of chaotic behavior in rainfall dynamics, we have explored many different data and problems associated with hydrologic systems from around the world. Nevertheless, several key areas, which are at the forefront of hydrology at the current time and will be in the foreseeable future, remain untouched. This chapter has identified some of these areas, highlighted the progress made thus far, and offered potential directions for further advancement. The need for establishing strong links between the concepts of chaos theory on one hand and the hydrologic system dynamics on the other, with hydrologic data serving as a medium, has also been emphasized. Chapter 15 will further discuss, both philosophically and pragmatically, the relevance and role of chaos theory in hydrology and how it can serve as a balanced middle-ground approach to our two dominant extreme views of determinism and stochasticity.

References

Ali G, Tetzlaff D, Soulsby C, McDonnell JJ, Capell R (2012) A comparison of similarity indices for catchment classification using a cross-regional dataset. Adv Water Resour 40:11–22
Alvisi S, Franchini M (2011) Fuzzy neural networks for water level and discharge forecasting with uncertainty. Environ Modell Softw 26(4):523–537
Beckinsale RP (1969) River regimes. In: Chorley RJ (ed) Water, earth, and man. Methuen, London, pp 455–471
Beven KJ (2002) Uncertainty and the detection of structural change in models of environmental systems. In: Beck MB (ed) Environmental foresight and models: a manifesto. Elsevier, The Netherland, pp 227–250
Beven KJ (2006) On undermining the science? Hydrol Process 20:3141–3146
Beven KJ, Binley AM (1992) The future of distributed models: model calibration and uncertainty prediction. Hydrol Process 6:279–298
Beven KJ, Kirkby MJ (1979) A physically based variable contributing area model of basin hydrology. Hydrol Sci Bull 24(1):43–69
Beven KJ, Young P (2003) Comment on "Bayesian recursive parameter estimation for hydrologic models" by M. Thiemann, M. Trosset, H. Gupta, and S. Sorooshian. Water Resour Res 39(5):1116
Beven KJ, Smith PJ, Westerberg IK, Freer J (2012) Comment on "Pursuing the method of multiple working hypotheses for hydrological modeling" by Clark et al. Water Resour Res 48:W11801
Budyko MI (1974) Climate and Life. Academic Press, New York
Caine N (1990) The rainfall intensity-duration control of shallow landslides and debris flows. Geogr Ann 62A:23–27
Cao L, Mees A, Judd K (1998) Dynamics from multivariate time series. Physica D 121:75–88
Chen J, Adams BJ (2006) Integration of artificial neural networks with conceptual models in rainfall-runoff modeling. J Hydrol 318(1–4):232–249
Clark MP, Kavetski D, Fenicia F (2012) Reply to comment by K. Beven et al. on "Pursuing the method of multiple working hypotheses for hydrological modeling." Water Resour Res 48:W11802
Coulibaly P, Bobee B, Anctil F (2001) Improving extreme hydrologic events forecasting using a new criterion for artificial neural network selection. Hydrol Process 15:1533–1536
Crozier M (1986) Landslides: causes, consequences and environment. Croom Helm, London
Dhanya CT, Nagesh Kumar D (2011) Predictive uncertainty of chaotic daily streamflow using ensemble wavelet networks approach. Water Resour Res 47:W06507. doi:10.1029/2010WR010173
Elsner JB, Tsonis AA (1993) Nonlinear dynamics established in the ENSO. Geophys Res Lett 20:213–216

Fowler HJ, Blenkinsop S, Tebaldi C (2007) Linking climate change modeling to impacts studies: recent advances in downscaling techniques for hydrological modeling. Int J Climatol 27(12):1547–1578

Fraedrich K (1986) Estimating the dimensions of weather and climate attractors. J Atmos Sci 43(5):419–432

Gottschalk L, Jensen JL, Lundquist D, Solantie R, Tollan A (1979) Hydrologic regions in the Nordic countries. Hydrol Res 10(5):273–286

Grayson RB, Blöschl G (2000) Spatial patterns in catchment hydrology: observations and modeling. Cambridge University Press, Cambridge, UK

Gupta H, Thiemann M, Trosset M, Sorooshian S (2003) Reply to comment by K. Beven and P. Young on “Bayesian recursive parameter estimation for hydrologic models”. Water Resour Res 39(5):1117

Haines AT, Finlayson BL, McMahon TA (1988) A global classification of river regimes. Appl Geogr 8(4):255–272

Hill J, Hossain F, Sivakumar B (2008) Is correlation dimension a reliable proxy for the number of dominant influencing variables for modeling risk of arsenic contamination in groundwater? Stoch Environ Res Risk Assess 22(1):47–55

Hossain F, Anagnostou EN (2005) Assessment of stochastic interpolation based parameter sampling scheme for efficient uncertainty analyses of hydrologic models. Comput Geosci 31(4):497–512

Hossain F, Anagnostou EN, Lee KH (2004) A non-linear and stochastic response surface method for Bayesian estimation ofuncertainty in soil moisture simulation from a land surface model. Nonlinear Process Geophys 11:427–440

Hsu KL, Gupta HV, Sorooshian S (1995) Artificial neural network modeling of the rainfall-runoff process. Water Resour Res 31(10):2517–2530

Hsu KL, Gupta HV, Gao X, Sorooshian S, Imam B (2002) Self-organizing linear output map (SOLO): an artificial neural network suitable for hydrologic modeling and analysis. Water Resour Res 38(12). doi:10.1029/2001WR000795

IPCC (2014) Climate change 2014—impacts, adaptation and vulnerability. In: Field CB, Barros VR, Dokken DJ, Mach KJ, Mastrandrea MD, Bilir TE, Chatterjee M, Ebi KL, Estrada YO, Genova RC, Girma B, Kissel ES, Levy AN, MacCracken S, Mastrandrea PR, White LL (eds) Contribution of working group II to the fifth assessment report of the intergovernmental panel on climate change. Cambridge University Press, Cambridge

Isik S, Singh VP (2008) Hydrologic regionalization of watersheds in Turkey. J Hydrol Eng 13:824–834

Jain A, Srinivasulu S (2006) Development of effective and efficient rainfall-runoff models using integration of deterministic, real-coded genetic algorithms, and artificial neural network techniques. Water Resour Res 40(4):W04302. doi:10.1029/2003WR002355

Jain A, Sudheer KP, Srinivasulu S (2004) Identification of physical processes inherent in artificial neural network rainfall runoff models. Hydrol Process 118(3):571–581

Jin YH, Kawamura A, Jinno K, Berndtsson R (2005) Nonlinear multivariate analysis of SOI and local precipitation and temperature. Nonlinear Process Geophys 12:67–74

Jinno K, Xu S, Berndtsson R, Kawamura A, Matsumoto M (1995) Prediction of sunspots using reconstructed chaotic system equations. J Geophys Res 100(A8):14773–14781

Kennard MJ, Pusey BJ, Olden JD, Mackay SJ, Stein JL, Marsh N (2010) Classification of natural flow regimes in Australia to support environmental flow management. Freshwater Biol 55(1):171–193

Kirchner JW (2006) Getting the right answers for the right reasons: linking measurements, analyses, and models to advance the science of hydrology. Water Resour Res 42:W03S04

Klemeš V (1978) Physically based stochastic hydrologic analysis. Adv Hydrosci 11:285–352

Krasovskaia I, Gottschalk L, Kundzewicz ZW (1999) Dimensionality of Scandinavian river flow regimes. Hydrol Sci J 44(5):705–723

Kyoung MS, Kim HS, Sivakumar B, Singh VP, Ahn KS (2011) Dynamic characteristics of monthly rainfall in the Korean peninsula under climate change. Stoch Environ Res Risk Assess 25(4):613–625

Laio F, Porporato A, Revelli R, Ridolfi L (2003) A comparison of nonlinear flood forecasting methods. Water Resour Res 39(5). 10.1029/2002WR001551

Liu Y, Gupta HV (2007) Uncertainty in hydrologic modeling: Toward an integrated data assimilation framework. Water Resour Res 43:W07401

Liu Q, Islam S, Rodriguez-Iturbe I, Le Y (1998) Phase-space analysis of daily streamflow: characterization and prediction. Adv Water Resour 21:463–475

Lorenz EN (1963) Deterministic nonperiodic flow. J Atmos Sci 20:130–141

McDonnell JJ, Woods RA (2004) On the need for catchment classification. J Hydrol 299:2–3

Michaud JD, Sorooshian S (1994) Effect of rainfall-sampling errors on simulations of desert flash floods. Water Resour Res 30(10):2765–2775

Moliere DR, Lowry JBC, Humphrey CL (2009) Classifying the flow regime of data-limited streams in the wet-dry tropical region of Australia. J Hydrol 367(1–2):1–13

Nasr A, Bruen M (2008) Development of neuro-fuzzy models to account for temporal and spatial variations in a lumped rainfall-runoff model. J Hydrol 349(3–4):277–290

Nathan RJ, McMahon TA (1990) Identification of homogeneous regions for the purpose of regionalization. J Hydrol 121(1–4):217–238

Olden JD, Poff NL (2003) Redundancy and the choice of hydrologic indices for characterizing streamflow regimes. River Res Appl 19(2):101–121

Olden JD, Kennard MJ, Pusey BJ (2012) A framework for hydrologic classification with a review of methodologies and applications in ecohydrology. Ecohydrology 5(4):503–518

Packard NH, Crutchfield JP, Farmer JD, Shaw RS (1980) Geometry from a time series. Phys Rev Lett 45(9):712–716

Pardé M (1933) Fleuves et Rivières. Collection Armond Colin, Paris, France

Parlange MB, Katul GG, Cuenca RH, Kavvas ML, Nielsen DR, Mata M (1992) Physical basis for a time series model of soil water content. Water Resour Res 28(9):2437–2446

Porporato A, Ridolfi R (2001) Multivariate nonlinear prediction of river flows. J Hydrol 248 (1–4):109–122

Razavi T, Coulibaly P (2013) Streamflow prediction in ungauged basins: review of regionalization methods. J Hydrol Eng 18:958–975

Refsgaard JC, Henriksen HJ (2004) Modeling guidelines—terminology and guiding principles. Adv Water Resour 27:71–82

Reichenbach P, Cardinali M, De Vita P, Guzzetti F (1998) Regional hydrological thresholds for landslides and floods in the Tiber River Basin (central Italy). Environ Geol 35(2–3):146–159

Rigon R, Rinaldo A, Rodriguez-Iturbe I (1994) On landscape selforganization. J Geophys Res 99 (B6):11971–11993

Rodriguez-Iturbe I, Rinaldo A (1997) Fractal river basins: chance and self-organization. Cambridge University Press, Cambridge

Salas JD, Smith RA (1981) Physical basis of stochastic models of annual flows. Water Resour Res 17(2):428–430

See and Openshaw (2000) A hybrid multi-model approach to river level forecasting. Hydrol Sci J 45:523–536

Sivakumar B (2000) Chaos theory in hydrology: important issues and interpretations. J Hydrol 227 (1–4):1–20

Sivakumar B (2003) Forecasting monthly streamflow dynamics in the western United States: a nonlinear dynamical approach. Environ Model Softw 18(8–9):721–728

Sivakumar B (2004a) Chaos theory in geophysics: past, present andfuture. Chaos Soliton Fract 19(2):441–462

Sivakumar B (2004b) Dominant processes concept in hydrology: moving forward. Hydrol Process 18(12):2349–2353

Sivakumar B (2005a) Correlation dimension estimation of hydrologic series and data size requirement: myth and reality. Hydrol Sci J 50(4):591–604

Sivakumar B (2005b) Hydrologic modeling and forecasting: role of thresholds. Environ Model Softw 20(5):515–519

Sivakumar B (2008a) Dominant processes concept, model simplification and classification framework in catchment hydrology. Stoch Env Res Risk Assess 22(6):737–748

Sivakumar B (2008b) Undermining the science or undermining. Nature? Hydrol Process 22(6):893–897

Sivakumar B (2008c) The more things change, the more they stay the same: the state of hydrologic modelling. Hydrol Process 22:4333–4337

Sivakumar B (2009) Nonlinear dynamics and chaos in hydrologic systems: latest developments and a look forward. Stoch Environ Res Risk Assess 23(7):1027–1036

Sivakumar B, Singh VP (2012) Hydrologic system complexity and nonlinear dynamic concepts for a catchment classification framework. Hydrol Earth Syst Sci 16:4119–4131

Sivakumar B, Sorooshian S, Gupta HV, Gao X (2001) A chaotic approach to rainfall disaggregation. Water Resour Res 37(1):61–72

Sivakumar B, Persson M, Berndtsson R, Uvo CB (2002) Is correlation dimension a reliable indicator of low-dimensional chaos in short hydrological time series? Water Resour Res 38(2). doi:10.1029/2001WR000333

Sivakumar B, Berndtsson R, Persson M, Uvo CB (2005) A multi-variable time series phase-space reconstruction approach to investigation of chaos in hydrological processes. Int J Civil Environ Eng 1(1):35–51

Sivakumar B, Jayawardena AW, Li WK (2007) Hydrologic complexity and classification: a simple data reconstruction approach. Hydrol Process 21(20):2713–2728

Sivakumar B, Woldemeskel FM, Puente CE (2014) Nonlinear analysis of rainfall variability in Australia. Stoch Environ Res Risk Assess 28(1):17–27

Sivakumar B, Singh V, Berndtsson R, Khan S (2015) Catchment classification framework in Hydrology: challenges and directions. J Hydrol Eng 20:A4014002

Snelder TH, Biggs BJF, Woods RA (2005) Improved eco-hydrological classification of rivers. River Res Applic 21:609–628

Sudheer KP, Jain A (2004) Explaining the internal behaviour of artificial neural network river flow models. Hydrol Process 18:833–844

Takens F (1981) Detecting strange attractors in turbulence. In: Rand DA, Young LS (eds) Dynamical systems and turbulence, vol 898. Lecture notes in mathematics. Springer, Berlin, pp 366–381

Thiemann T, Trosset M, Gupta H, Sorooshian S (2001) Bayesian recursive parameter estimation for hydrologic models. Water Resour Res 37(10):2521–2535

Tsonis AA, Elsner JB (1988) The weather attractor over short timescales. Nature 333:545–547

Vignesh R, Jothiprakash V, Sivakumar B (2015) Streamflow variability and classification using false nearest neighbor method. J Hydrol 531:706–715

Vogel RM (1999) Stochastic and deterministic world views. J Water Resour Plan Manage 125(6):311–313

Vrugt JA, Gupta HV, Bouten W, Sorooshian S (2003) A shuffled complex evolution metropolis algorithm for optimization and uncertainty assessment of hydrologic model parameters. Water Resour Res 39(8):1201

Wagener T, McIntyre N, Lees MJ, Wheater HS, Gupta HV (2003) Towards reduced uncertainty in conceptual rainfall-runoff modeling: dynamic identifiability analysis. Hydrol Process 17:455–476

Wainwright W, Mulligan M (2004) Environmental modeling: finding simplicity in complexity. Wiley, London

Wilby RL, Wigley TML (1997) Downscaling general circulation model output: a review of methods and limitations. Prog Phys Geogr 21:530–548

Wilby RL, Abrahart RJ, Dawson CW (2003) Detection of conceptual model rainfall-runoff processes inside an artificial neural network. Hydrol Sci J 48(2):163–181

Wu CL, Chau KW, Li YS (2009) Predicting monthly streamflow using data-driven models coupled with data-preprocessing techniques. Water Resour Res 45:W08432. doi:10.1029/2007WR006737

Young PC, Beven KJ (1994) Data-based mechanistic modeling and rainfall-flow non-linearity. Environmetrics 5(3):335–363

Young PC, Parkinson SD (2002) Simplicity out of complexity. In: Beck MB (ed) Environmental foresight and models: a manifesto. Elsevier Science, The Netherlands, pp 251–294

Young PC, Ratto M (2009) A unified approach to environmental systems modeling. Stoch Environ Res Risk Assess 23:1037–1057

Young PC, Parkinson SD, Lees M (1996) Simplicity out of complexity in environmental systems: Occam's Razor revisited. J Appl Statis 23:165–210

Zhou Y, Ma Z, Wang L (2002) Chaotic dynamics of the flood series in the Huaihe River Basin for the last 500 years. J Hydrol 258:100–110

Chapter 15
Final Thoughts: Philosophy and Pragmatism

Abstract Research on chaos theory in hydrology over the past three decades offers new opportunities as well as challenges. These opportunities and challenges, in turn, provide interesting ways to further explore the relevance and role of chaos theory in hydrology. An obvious question to ask is: if, and how, chaos theory fits within our two dominant, but extreme, views of hydrology of the twentieth century: deterministic and stochastic? This chapter attempts to answer this question, from both philosophical and pragmatic perspectives. It is pointed out that the underpinning concepts of nonlinear interdependence, hidden determinism and order, and sensitivity to initial conditions of chaos theory provide the necessary means to represent both the deterministic and the stochastic characteristics of hydrologic systems. This also leads to the argument that chaos theory offers a balanced middle ground to bridge the gap between the two extreme views of determinism and stochasticity and, therefore, serves as a coupled deterministic-stochastic paradigm to study hydrology in a holistic manner.

15.1 Introduction

As discussed in the preceding chapters, chaos theory in hydrology has witnessed a tremendous growth during the past three decades. Concepts and methods of chaos theory have been applied to identify and predict the chaotic behavior of many different hydrologic data, including rainfall, river flow, rainfall-runoff, lake volume, sediment transport, groundwater, soil moisture, and others. In addition to identification and prediction of chaotic behavior, a host of other hydrologic problems, such as scaling, catchment classification, missing data estimation, and reconstruction of system equations, have been studied, to a small extent. With concerns on potential data-related limitations, issues regarding temporal correlation, data size, data noise and noise reduction, presence of zeros, and some others have also been addressed.

On one hand, this progress offers enormous opportunities to further advance chaos theory in hydrology; for instance, there remain many largely-unexplored areas (including some mentioned above), such as parameter estimation, model

B. Sivakumar, *Chaos in Hydrology*, DOI 10.1007/978-90-481-2552-4_15

simplification and integration, catchment classification, multi-variable analysis, reconstruction of system equations, and downscaling global climate model outputs. On the other hand, however, it also highlights the challenges in chaos theory studies in hydrology; for instance, there remain difficulties in establishing clear links between theory, data, and physics and in overcoming some data-related issues (e.g. length, noise, zeros).

These opportunities and challenges indeed provide some interesting ways to further discuss the relevance and role of chaos theory in hydrology (and Nature, more broadly) towards a future path. Here, I present my thoughts from two different perspectives: philosophy (Sect. 15.2) and pragmatism (Sect. 15.3). In Sect. 15.2, I discuss the philosophy behind our attempts to understand, model, and predict hydrologic systems. In Sect. 15.3, I discuss the pragmatic approach that is needed (and, indeed, is largely adopted) in studying hydrologic systems.

15.2 Philosophy

The last century witnessed the domination of two vastly contrasting approaches for studying hydrologic systems: deterministic and stochastic. The basic philosophy behind the deterministic approach is that systems can be represented fairly accurately by deterministic mathematical equations based on well-known scientific laws, provided sufficient detail can be included to explain the underlying physical processes. The philosophy behind the stochastic approach, on the other hand, is that systems do not adhere to any deterministic principles and, therefore, probability distributions based on probability concepts are required for their description.

Either of these two approaches, with its solid foundations in scientific principles/philosophies, verifiable assumptions for specific situations, and the ability to provide reliable results, has merits for studying hydrologic systems. The deterministic approach has merits considering the 'permanent' nature of the Earth, ocean, and the atmosphere and the 'cyclical' nature of the associated processes (i.e. the *hydrologic cycle* or *water cycle*)—for example, seasonal cycle in rainfall, annual cycle in river flow, and diurnal cycle in temperature. Similarly, the stochastic approach has merits considering the facts that hydrologic systems and processes exhibit 'complex and irregular' structures and that we have only 'limited ability to observe' the detailed variations—for example, catchment properties and processes (not to mention climate inputs) vary tremendously with respect to scale (both in space and in time) due to different interacting components and cannot be measured accurately at all scales. However, for these very same reasons, both the deterministic approach and the stochastic approach possess important limitations in the study of hydrologic systems when applied independently and, consequently, neither approach is sufficient for all situations.

Considering these, it is often meaningless to ask if the deterministic approach is better or the stochastic approach is better. Indeed, such a question has no general answer (e.g. Gelhar 1993). Despite this, however, much of the hydrologic research during the last century was driven to essentially find a general, and definitive, answer to such a question. While the deterministic approach dominated far more during the first half of the century (e.g. Richards 1931; Horton 1933, 1945) due to the various physical principles and laws established earlier (e.g. Dalton 1802; Darcy 1856), the stochastic approach assumed more prominence in the second half (e.g. Thomas and Fiering 1962; Yevjevich 1963, 1972; Klemeš 1978; Freeze 1980; Salas and Smith 1981) especially with the arrival of stochastic time series analysis methods (Cramer 1940; Box and Jenkins 1970). Indeed, during the past half a century or so, the two approaches largely went in parallel ways, and there has been far more competition between deterministic and stochastic approaches than contributions to bring them together towards a generic framework in hydrologic modeling.

This extreme view philosophy (i.e. either determinism or stochasticity) can take us only a little distance, as such a philosophy often does not suit all situations encountered in hydrology. The appropriate approach is often different for different hydrologic situations, which may be defined in terms of system, process, scale, and purpose of interest (e.g. Sivakumar 2008a). For some situations, the deterministic approach may be more appropriate; for some other situations, the stochastic approach may be more appropriate; and for still others, both approaches may be equally appropriate. Indeed, there may be some rare situations where neither approach may be appropriate or satisfactory. It is also reasonable to contend that the two approaches are complementary to each other, since oftentimes both deterministic and stochastic properties are intrinsic to hydrologic systems. For example, there is significant determinism in river flow in the form of seasonality and annual cycle, whereas stochasticity is also brought by the interactions of various mechanisms involved and by their different degrees of nonlinearity; see Chap. 2 for additional details.

These observations suggest that a coupled deterministic–stochastic approach, incorporating both the deterministic and the stochastic components, would yield better outcomes compared to when either approach adopted independently and, thus, would be more appropriate for most, if not all, hydrologic systems. Although the need for this combinatorial approach was recognized almost 50 years ago (Yevjevich 1968) and also reiterated from time to time in the decades that followed (e.g. Yevjevich 1974, 1991; Szöllősi-Nagy and Mekis 1988; Becker and Serban 1990; Vogel 1999), there is not much evidence in the literature that points out to any serious effort to this end; see also Sivakumar (2008b) for some comments. One may indeed argue that physically-based hydrologic models (which are essentially deterministic) with hydrometeorologic inputs (real or stochastically-generated) belong to this coupled deterministic-stochastic approach. The flaw in this argument is the basic assumption that hydrometeorologic inputs are stochastic (also see below for details).

It is precisely in the context of a coupled deterministic-stochastic approach, chaos theory is particularly relevant and can play an important role. Indeed, with its underpinning concepts of nonlinear interdependence, hidden determinism and order, and sensitivity to initial conditions, chaos theory can bridge the gap between our extreme views of determinism and stochasticity and also offer a balanced and more realistic middle-ground perspective for modeling hydrologic systems. The appropriateness of these concepts to hydrologic systems and the potential role of chaos theory in their modeling can be realized, for example, from the following situations: (1) nonlinear interactions are dominant among the components and mechanisms in the hydrologic cycle; (2) determinism and order are prevalent in river flow, especially at coarser temporal scales; and (3) contaminant transport in surface and sub-surface waters is highly sensitive to the time (i.e. rainy or dry season) at which the contaminants were released. The first represents the 'general' nature of hydrologic systems, while the second and third represent their 'deterministic' and 'stochastic' natures, respectively. These observations clearly suggest how chaos theory can serve as a coupled deterministic-stochastic approach; see also Sivakumar (2004, 2009) for additional details.

Implicit in the above underpinning concepts of chaos theory is the notion that 'complex and seemingly random' behavior need not necessarily be the outcomes of systems governed by a large number of variables but may *also* be the result of simple nonlinear deterministic systems governed by a few degrees of freedom (e.g. Lorenz 1963). This has obvious relevance for hydrologic systems (e.g. runoff in a well-developed urban catchment, despite its highly irregular and random-looking behavior, depends essentially on rainfall) and, consequently, has far reaching implications, since most outputs from such systems (e.g. time series of rainfall, river flow, water quality) are typically 'complex and random-looking.' A crucial implication is the need, first of all, to identify the dynamic nature of the given system towards selection of an appropriate approach, as opposed to our traditional and common practice of simply resorting to a particular approach based on certain preconceived notion (determinism or stochasticity) that may or may not be valid. Such an identification, in fact, has been the main goal or an important part of chaos theory studies in hydrology, as discussed in the earlier chapters; see also Sivakumar (2000, 2004, 2009) for reviews.

15.3 Pragmatism

Although chaos theory offers a coupled deterministic-stochastic approach and, thus, a middle-ground perspective to the extreme views of either determinism or stochasticity, there are also some challenges in its applications in hydrology. To offer a pragmatic perspective on this, I highlight two major challenges here.

A major challenge in the application of chaos theory in hydrology is in linking the concepts of chaos theory with the actual dynamics of hydrologic systems. For instance, there are obvious questions regarding the appropriateness of the phase space reconstruction concept (e.g. Packard et al. 1980) for hydrologic systems. A fundamental question is: can a single-variable (or even multi-variable) time series really represent the complete spatio-temporal dynamics of a complex heterogeneous hydrologic system? It is difficult to answer this question accurately, since the dynamic properties of hydrologic systems are not known a priori. This does not, however, mean that the phase space reconstruction concept is not at all relevant or appropriate for hydrologic systems. What is really needed is a good understanding of the concept and an honest assessment as to whether it could offer reasonable interpretations and explanations in the context of hydrologic systems. This is further explained here, with an example of the use of streamflow series for phase space reconstruction to represent the dynamics of a catchment.

It is indeed true that we do not know whether a single-variable streamflow time series can accurately represent the spatio-temporal dynamics of a catchment. However, we can, and indeed do, recognize that streamflow measured at the outlet of a catchment is essentially the outcome of whatever happens within the catchment (i.e. any and all kinds of interactions between rainfall/other inputs and catchment properties) and, thus, is a strong representation of the functions of the catchment. This is indeed the basis of the phase space reconstruction concept. Therefore, study of streamflow is reasonably sufficient to understand the catchment dynamics and, therefore, a single-variable streamflow time series alone can be used to reconstruct the catchment dynamics, with the inclusion of a delay parameter (e.g. Takens 1981) that can represent the important changes in the dynamics (i.e. neither redundance nor irrelevance); see Fraser and Swinney (1986) and Holzfuss and Mayer-Kress (1986) for some guidelines on the selection of the delay parameter. This is indeed the basis of phase space reconstruction. Such a reconstruction is particularly useful when other relevant processes influencing the catchment dynamics are either not known or data for which are not available. Therefore, the key question to ask is whether the variable/time series chosen for phase space reconstruction is a good representation of the dynamics of the system of interest. Such a question has enormous significance, in terms of actual system dynamics as well as in terms of practical considerations.

Another major challenge in chaos studies in hydrology is in addressing a host of issues associated with hydrologic data; see Chap. 12 for details. A basic assumption in the development of chaos theory methods is that the time series is infinite and noise-free. However, real hydrologic data are often short and always contaminated with noise; in many cases, they also contain a large number of zeros. Since insufficient length, presence of noise, and a large number of zeros in the data may influence the estimation of many chaotic invariants and lead to inaccurate outcomes, there are obvious concerns on the applications of chaos methods in hydrology and on the reported outcomes, including regarding the presence/absence of chaos. Among the many questions, a particularly serious one is concerned with the data length: i.e. what is the minimum length of data required for chaos analysis? This is addressed here.

Considering that data length may influence the outcomes of chaos methods, the above question certainly has some merit and, thus, cannot be dismissed altogether. However, it is also important to recognize that such a question has no general answer, since: (1) every hydrologic system and, hence, every hydrologic time series is unique in its own ways; and (2) different hydrologic systems undergo changes at different scales and, thus, attain repetition in dynamics ("cycles") or settle into certain forms ("attractors") over different periods. Therefore, different hydrologic systems generally warrant different lengths of data. Indeed, even the same hydrologic system often warrants different lengths of data, depending upon the scale. For instance, while a streamflow time series with as many as 1,000,000 values collected at 1-sec interval (i.e. about 12 days) is not at all sufficient to represent the changes in the flow dynamics of a large-scale river basin, a streamflow time series with as few as just 1,000 values collected at monthly intervals (i.e. about 83 years) may be more than sufficient to represent the changes in the flow dynamics. Therefore, the minimum data length required for chaos analysis needs to be assessed only in terms of individual hydrologic situations.

The key is to assess whether the time period the data covers is long enough to sufficiently represent the changes the system undergoes, rather than the data length in terms of the sheer number of values. Such an assessment is a more balanced and appropriate approach to the data length issue, as it takes into account any limitations in the methodology on one hand and gives due consideration to the actual system dynamics on the other. Any general guideline, such as the one linking minimum data length to embedding dimension used in phase space reconstruction (e.g. Nerenberg and Essex 1990; see also Schertzer et al. 2002), is inappropriate; see Sivakumar et al. (2002a, b) and Sivakumar (2005) for additional details. Indeed, for most catchments, streamflow series with even less than 500 values collected at monthly intervals (i.e. about 40 years) or with a few thousand values collected at daily intervals (i.e. about 10 to 20 years) may be sufficient to represent the dynamics and to obtain reliable results using chaos theory methods.

15.4 Closing Remarks

There have been two dominant approaches in hydrology: deterministic and stochastic. Both these approaches have solid foundations and clear merits for hydrologic systems. At the same time, however, their contrasting and extreme views hamper our ability to present a holistic perspective on hydrologic systems. This is because, while both determinism and stochasticity are intrinsic to hydrologic systems, the deterministic approach focuses mainly on their deterministic nature and the stochastic approach focuses mainly on their stochastic nature. Although the need to bridge the gap between these two extreme and parallel approaches in hydrology had been realized a long time ago, finding an appropriate middle ground that can bring these two approaches together has been tremendously challenging.

Chaos theory provides an answer. With its underpinning concepts of nonlinear interdependence, hidden determinism and order, and sensitivity to initial conditions and, consequently, recognizing that even simple systems can give rise to complex and seemingly random outputs, chaos theory provides the necessary means to represent both the deterministic nature and the stochastic nature of hydrologic systems. Consequently, it offers the much-needed middle ground to bridge the gap between the two extremes and serves as a coupled deterministic-stochastic approach for hydrologic systems.

The outcomes of the numerous studies on chaos theory in hydrology, as discussed in the earlier chapters, have been largely encouraging. There are great opportunities to further advance chaos theory in hydrology, but, at the same time, there are some major challenges as well. Some of these opportunities and challenges may be new in the specific context of chaos theory. However, our enormous experience in the application of deterministic and stochastic approaches in hydrology should offer useful clues to address them. Indeed, many of the challenges associated with the application of chaos theory in hydrology, especially those related to data issues (e.g. data length, noise, presence of zeros), have and continue to be encountered in deterministic and stochastic approaches as well.

Population growth and its associated consequences have already created significant challenges in water planning and management around the world. Global climate change will likely complicate this even further, especially with the anticipated increase in the frequency and magnitude of extreme hydrologic events (e.g. floods, droughts). With water playing a central role in our environment, ecosystem, and socio-economic development, study of water is taking an increasingly prominent stage in the global affairs than ever before. There is indeed an urgent need to find better ways to study hydrologic systems, especially in a more balanced manner than the ones our traditional deterministic and stochastic approaches can provide. Chaos theory offers an important avenue to this end. It is indeed an exciting time for research on chaos theory in hydrology!

References

Becker A, Serban P (1990) Hydrological models for water-resources system design and operation. Oper Hydrol Rep 34, WMO No. 740, Geneva

Box GEP, Jenkins G (1970) Time series analysis, forecasting and control. Holden-Day, San Francisco

Cramer H (1940) On the theory of stationary random processes. Ann Math 41:215–230

Dalton J (1802) Experimental essays on the constitution of mixed gases; on the force of steam or vapor from waters and other liquids, both in a Torricellian vacuum and in air; on evaporation; and on the expansion of gases by heat. Mem Proc Manch Lit Phil Soc 5:535–602

Darcy H (1856) Les fontaines publiques de la ville de Dijon. V. Dalmont, Paris

Fraser AM, Swinney HL (1986) Independent coordinates for strange attractors from mutual information. Phys Rev A 33(2):1134–1140

Freeze RA (1980) A stochastic-conceptual analysis of rainfall-runoff process on a hillslope. Water Resour Res 16:391–408

Gelhar LW (1993) Stochastic subsurface hydrology. Prentice-Hall, Englewood Cliffs, New Jersey

Holzfuss J, Mayer-Kress G (1986) An approach to error-estimation in the application of dimension algorithms. In: Mayer-Kress G (ed) Dimensions and entropies in chaotic systems. Springer, New York, pp 114–122

Horton RE (1933) The role of infiltration in the hydrologic cycle. Trans Am Geophys Union 14:446–460

Horton RE (1945) Erosional development of streams and their drainage basins: Hydrophysical approach to quantitative morphology. Bull Geol Soc Am 56:275–370

Klemeš V (1978) Physically based stochastic hydrologic analysis. Adv Hydrosci 11:285–352

Lorenz EN (1963) Deterministic nonperiodic flow. J Atmos Sci 20:130–141

Nerenberg MAH, Essex C (1990) Correlation dimension and systematic geometric effects. Phys Rev A 42(12):7065−7074

Packard NH, Crutchfield JP, Farmer JD, Shaw RS (1980) Geometry from a time series. Phys Rev Lett 45(9):712–716

Richards LA (1931) Capillary conduction of liquids through porous mediums. Physics A 1:318–333

Salas JD, Smith RA (1981) Physical basis of stochastic models of annual flows. Water Resour Res 17(2):428–430

Schertzer D, Tchiguirinskaia I, Lovejoy S, Hubert P, Bendjoudi H et al (2002) Which chaos in the rainfall-runoff process? A discussion on 'Evidence of chaos in the rainfall-runoff process' by Sivakumar. Hydrol Sci J 47(1):139–147

Sivakumar B (2000) Chaos theory in hydrology: important issues and interpretations. J Hydrol 227 (1–4):1–20

Sivakumar B (2004) Chaos theory in geophysics: past, present and future. Chaos Soliton Fract 19(2):441–462

Sivakumar B (2008a) Dominant processes concept, model simplification and classification framework in catchment hydrology. Stoch Env Res Risk Assess 22:737–748

Sivakumar B (2008b) The more things change, the more they stay the same: the state of hydrologic modeling. Hydrol Process 22:4333–4337

Sivakumar B (2005) Correlation dimension estimation of hydrologic series and data size requirement: myth and reality. Hydrol Sci J 50(4):591–604

Sivakumar B (2009) Nonlinear dynamics and chaos in hydrologic systems: latest developments and a look forward. Stoch Environ Res Risk Assess 23:1027–1036

Sivakumar B, Berndtsson R, Olsson J, Jinno K (2002a) Reply to 'which chaos in the rainfall-runoff process?' by Schertzer et al. Hydrol Sci J v.47(1) p 149–158

Sivakumar B, Persson M, Berndtsson R, Uvo CB (2002b) Is correlation dimension a reliable indicator of low-dimensional chaos in short hydrological time series? Water Resour Res 38(2). doi:10.1029/2001WR000333

Szöllősi-Nagy A, Mekis E (1988) Comparative analysis of three recursive real-time river flow forecasting models: deterministic, stochastic, and coupled deterministic-stochastic. Stochastic Hydrol Hydraul 2:17–33

Takens F (1981) Detecting strange attractors in turbulence. In: Rand DA, Young LS (eds). Dynamical systems and turbulence, Lecture notes in mathematics 898, Springer-Verlag, Berlin, Germany, pp 366–381

Thomas HA, Fiering MB (1962) Mathematical synthesis of streamflow sequences for the analysis of river basins by simulation. In: Mass A et al (eds) Design of water resource systems. Harvard University Press, Cambridge, Massachusetts, pp 459–493

Vogel RM (1999) Stochastic and deterministic world views. J Water Resour Plan Manage 125(6):311–313

Yevjevich VM (1963) Fluctuations of wet and dry years. Part 1. Research data assembly and mathematical models. Hydrology Paper 1, Colorado State University, Fort Collins, Colorado, pp 1–55

Yevjevich VM (1968) Misconceptions in hydrology and their consequences. Water Resour Res 4(2):225−232

Yevjevich VM (1972) Stochastic processes in hydrology. Water Resour Publ, Fort Collins, Colorado

Yevjevich VM (1974) Determinism and stochasticity in hydrology. J Hydrol 22:225–258

Yevjevich VM (1991) Tendencies in hydrology research and its applications for 21st century. Water Resour Manage 5:1–23

Index

B. Sivakumar, *Chaos in Hydrology*, DOI 10.1007/978-90-481-2552-4

N